Preface

This book is the outcome of a face-to-face debate among a group of outstanding cognitive scientists who were encouraged to describe and defend their theoretical positions about meaning: meaning as symbol-based processes or meaning as grounded on embodied processes. We invited cognitive psychologists, philosophers, roboticists, neuroscientists, and educational psychologists with the aim of gaining understanding of a subject that is fundamental for all cognitive sciences. The scientists (and chapter authors) were encouraged to engage in a debate, but with the goal of providing solutions or suggestions for research to resolve multiple issues such as the nature of symbol grounding. The content of the debate includes empirical evidence, computational models, and conceptual arguments. Thus, the topics addressed in the book provide a coherent overview of some of the most important research and ideas in the field of cognitive science.

Several important books have been recently published that address advantages of the embodied cognition approach (e.g. Pecher and Zwaan 2005), or the statistical symbolist approach (e.g. Landauer *et al.* 2007) to cognition, language, and discourse. What makes the present book different is that it is the first attempt of cognitive scientists adopting symbolist and embodied viewpoints to join their efforts to clarify how the mind works and the nature of linguistic meaning. The book is also cross-disciplinary and the authors made an effort to converge on solutions to substantial issues beyond specific vocabularies and techniques. Finally, it addresses applied questions such as the application of theoretical approaches to education and the design of robots.

The authors were asked to describe their own position concerning several questions we formulated as guidelines for the debate. Among the major questions were:

- What are the central claims and assumptions of the symbolist and embodied approaches to the representation and processing of meaning?
- How is symbol grounding achieved in meaning representations according to the embodied and symbolist viewpoints?
- How do brain imaging and neurophysiological data contribute to solving the question of meaning?
- Are current computer and robotic models of comprehension grounded on embodied processes, or are they necessarily symbolic, or both?

Before writing their chapters, all the authors, as well as a group of young researchers, met for a workshop in Tenerife (Canary Islands). In the calm environment of the historic village of Garachico, we devoted three intensive days in November 2005 to describing and debating ideas. However, the current book is not simply a record of the proceedings because the chapters were written anew after the workshop. Nevertheless, the book benefits from the workshop in several respects. First, the authors wrote their chapters after serious consideration of the other contributors' arguments and data, refining their own proposals in consequence. Second, the open debates that followed each oral presentation in the workshop were transcribed, edited, and incorporated at the end of the corresponding chapters. In this way, the formal chapters with their academic style are complemented by the discussions with a fresher and more spontaneous style, typical of face-to-face communication. The young researchers attending the workshop also contributed to these discussions.

The final product of the debate was a success in several respects. First, the authors assumed the risk of getting involved in a debate with potential disagreement about fundamental questions. Second, despite differing in theoretical backgrounds, disciplines, and positions in the debate, the authors shared enough conceptual common ground to converge on a relatively small number of issues in their arguments and discussions. Third, the analyses and conclusions differed sufficiently to make the chapters and debates intrinsically interesting and an inspiration to pursue new research avenues. We hope that the community of cognitive scientists to whom the book is addressed share our assessment of the debate.

The book begins with an introductory chapter ('Framing the debate') written by the editors in advance to help the authors focus on the main points of the debate. The following chapters are not ordered and clustered according to any rigid conceptual criteria. Some chapters take a clear symbolist position, some others defend an embodied approach, but many others have quite a hybrid perspective. Therefore, we ordered the chapters to enhance the contrasting views shown by the authors during the debates. As a colophon to the book, the final chapter ('Reflecting on the debate') offers a comprehensive view of the main topics of debate, and includes some positive reflections and conclusions.

The workshop and part of the editing process were supported by several agencies in the US and Spain. We are grateful to the Spanish *Ministerio de Educación y Ciencia* (grant SEJ2004-2221-E/PSIC), the *Gobierno de Canarias* (Conference Organization Program, 2005), the National Science Foundation (grant BCS-0445627), and the Institute for Intelligent Systems at the University of Memphis (Memphis, TN, USA) for their generous support to cover travel and lodging of the contributors. We are grateful to the University of La Laguna (San Cristóbal de La Laguna, Tenerife, Spain), the Faculty of Psychology, the *Sociedad Española de Psicologia Experimental* (SEPEX), and the Garachico town hall for their help. Finally, this book would not be possible without the organizational effort devoted by several colleagues and graduate students at the University of La Laguna: Alberto Domínguez, Elena Gámez, José M. Díaz, Hipólito Marrero, Inmaculada León, Yurena Morera, Vicente Moreno, and Mabel Urrutia. We also are in debt to David Havas (University of Wisconsin, Madison, WI, USA) for his work on the web page for the workshop.

Manuel de Vega, Arthur Glenberg, and Arthur Graesser
December 2007

References

Landauer T, McNamara D, Dennis S, Kintsch W, Eds (2007). *Handbook of Latent Semantic Analysis*. Mahwah, NJ: Erlbaum.

Pecher D, Zwaan RA, Eds (2005). *Grounding Cognition: The Role of Perception and Action in Memory, Language, and Thinking*. New York, NY: Cambridge University Press.

Symbols and Embodiment

Symbols and Embodiment

Debates on meaning
and cognition

Edited by

Manuel de Vega

Arthur M Glenberg

Arthur C Graesser

OXFORD
UNIVERSITY PRESS

OXFORD
UNIVERSITY PRESS

Great Clarendon Street, Oxford OX2 6DP

Oxford University Press is a department of the University of Oxford.
It furthers the University's objective of excellence in research, scholarship,
and education by publishing worldwide in

Oxford New York

Auckland Cape Town Dar es Salaam Hong Kong Karachi
Kuala Lumpur Madrid Melbourne Mexico City Nairobi
New Delhi Shanghai Taipei Toronto

With offices in

Argentina Austria Brazil Chile Czech Republic France Greece
Guatemala Hungary Italy Japan Poland Portugal Singapore
South Korea Switzerland Thailand Turkey Ukraine Vietnam

Oxford is a registered trade mark of Oxford University Press
in the UK and in certain other countries

Published in the United States
by Oxford University Press Inc., New York

© Oxford University Press 2008

The moral rights of the authors have been asserted

Database right Oxford University Press (maker)

First published 2008

British Library Cataloguing in Publication Data

Data available

Library of Congress Cataloging-in-Publication Data

Data available

Typeset by Cepha Imaging Private Ltd., Bangalore, India
Printed in Great Britain
on acid-free paper by
CPI Antony Rowe, Chippenham, Wiltshire

ISBN 978-0-19-921727-4

10 9 8 7 6 5 4 3 2 1

Contents

List of Contributors

Lawrence W Barsalou
Emory University, United States

Manuel de Vega
University of La Laguna, Spain

Arthur M Glenberg
Arizona State University,
United States

Robert Goldstone
Indiana University, United States

Antoni Gomila
University of the Balearic Islands, Spain

Arthur C Graesser
University of Memphis, United States

G Tanner Jackson
University of Memphis, United States

Patrick Jeuniaux
University of Memphis, United States

Marcel Adam Just
Carnegie Mellon University, United States

Walter Kintsch
University of Colorado, United States

Andreas Knoblauch
Honda Research Institute Europe GmbH,
Germany

David Landy
Indiana University, United States

Max Louwerse
University of Memphis, United States

Sarita Mehta
University of Wisconsin – Madison,
United States

Mitchell J Nathan
University of Wisconsin, United States

Friedemann Pulvermüller
Medical Research Council, Cognition and
Brain Sciences Unit, Cambridge, United
Kingdom

Deb Roy
MIT Media Laboratory, United States

Anthony J Sanford
University of Glasgow, United Kingdom

Ava Santos
Emory University, United States

Lawrence Shapiro
University of Wisconsin – Madison,
United States

W Kyle Simmons
National Institute of Mental Health,
United States

Ji Y Son
Indiana University, United States

Luc Steels
Sony Computer Science Laboratory,
France,
VUB AI Lab, Vrije Universiteit Brussel,
Belgium

Christine D Wilson
Emory University, United States

Rolf A Zwaan
Florida State University, United States
Note: In addition to the contributors in
this list, Francis Quek, Julio Santiago, and
several unidentified speakers actively
participated in the debates.

Chapter 1

Framing the debate

Arthur M Glenberg, Manuel de Vega,
Arthur C Graesser

The purpose of this chapter is not to force a particular set of definitions or questions. Instead, we hope to lay out a set of definitions as a starting point. Other writers can either agree with these definitions or offer their own. Our goal is to be as precise as possible about what we mean when we use the following terms: *symbol*, *grounding*, and *embodiment*. Following the proposed definitions are examples of the theories that are consistent with the definitions, and then a reiteration of the debate questions.

1.1 Symbols, grounding, and embodiment

CS Peirce's theory of signs is a good starting point both for how the theory helps and how it doesn't (see Clark 1996 for a discussion of Peirce). According to Peirce, a sign is a member of a triad consisting of an object, a sign, and an interpretant. An interpretant is in the mind of an observer; it is the idea of the object brought about by the sign. On this view, signs are a mechanism for grounding ideas, that is, the sign connects the idea to the object. Peirce proposed three types of signs: *icons*, *indices*, and *symbols*.

Icons connect the interpretant to the object by virtue of perceptual similarity. Photographs are prototypical icons. An index connects object and interpretant by virtue of a causal, spatial, or temporal relation. Thus, a weathervane is an index for the direction of the wind (a causal, spatial, and temporal connection), a pointing gesture is a spatial index for the pointed-to object, and the utterance of 'I' is an index for the person speaking (a spatial and temporal connection). In contrast to signs that have a type of natural connection to their objects, symbols are signs only by virtue of a rule or convention. Thus, ordinary words are symbols. The word 'chair' does not literally point to chairs or illustrate them in any way. Furthermore, the conventional (or arbitrary) nature of a symbol is easily illustrated; the word that designates the object called a chair in English changes from language to language.

Presumably, at least some icons and indices are less strongly tied to a language community (e.g., the meaning of a pointing gesture). Peirce also proposed that symbols (unlike icons and indices) are types. That is, a symbol by itself does not designate a particular thing, only types. Finally, Peirce noted that signs can be mixed: part symbol, part index, part icon.

For Peirce, all signs (icons, indices, and symbols) are mechanisms for grounding. That is, signs connect an idea (the interpretant) and the object. This notion is quite different from the use of the term symbol in much of cognitive science, and the use that is of concern here. In that use, symbols are taken to be a theoretical account of the interpretant, that is, a theoretical account of ideas. Nonetheless, cognitive scientists have adopted other aspects of the Peircean definition. Namely, symbols are types: they may be *arbitrary* (that is, arbitrarily related to objects), they may be *abstract*, and they may be *amodal*, much like the word 'chair' is the same symbol whether read or heard.

We offer one proposal for the definition of a symbol that has frequently been used in cognitive science. This definition has often been adopted by theoretical positions that emphasize the arbitrary relation between symbol and referent, as in the case of physical symbol systems (Barsalou 1999; Harnad 1990; Newell 1980).

> A symbol is a theoretical element that is arbitrary, abstract, and amodal. Collections of symbols connected in the appropriate manner constitute ideas. Symbols and collections of symbols can be manipulated by using explicit rules to derive new ideas or to take action.

However, others in cognitive science have adopted a less constrained definition of symbol. Researchers who develop symbolic models often speak of symbolic representations and procedures that have symbolic units directly linked to sensory transducers and motoric actuators (Just and Carpenter 1992; Meyers 2000; Kintsch 1998; Kosslyn 1994; Marr 1982; Miller and Johnson-Laird 1976; Norman and Rumelhart 1975). Some symbols do have such links to modalities; others don't. Some symbols have merely conventional links to referents; others are more tightly grounded to perception and action. However, a symbolic position allows for the possibility of purely arbitrary links from symbolic code to sensorimotor referents, and this opens the door wide open to an interesting debate with those who adopt a radical embodied position. In the interest of promoting clarity, therefore, it is necessary to qualify the term symbol with the type of symbol and its functional properties. There are amodal symbols, motoric symbols, perceptual symbols, indexical symbols, arbitrary symbols, and so on. Confusions are likely to develop without a more precise specification of the type of symbol and its properties.

The term *representation* is also a construct that needs a precise definition, but there is the worry that proposing such a definition will open a Pandora's box of disagreements. Some view representations as a level of code that 'stands for' the referent, whereas others consider representations as having constraints that are closely aligned with the psychological content or neurophysiological activities. Some researchers may take the position that it is best to not use the term 'representation' in any explanatory sense because it creates confusion and may end up simply passing the buck. If a collection of symbols represents an idea, than what is the idea itself? We take it that one of the goals of theory is to explicate what an idea is, not to produce representations of ideas. Words and language seem to be perfectly good representations of (signs for) ideas, so our question is how those words work. We believe that nothing is gained theoretically, or methodologically for that matter, by speaking of a representation without specifying more precisely what the nature of the representation is. In particular, for the purposes of this book, we need to know precisely the extent to which the representations are embodied and grounded in sensorimotor experience, as opposed to being arbitrary, abstract, and amodal (AAA).

With this initial definition of a symbol, we can immediately start to frame questions for the debate. For example, can collections of amodal arbitrary symbols alone constitute ideas? If the answer is 'yes,' there is no reason to suspect that ideas are the sole province of biological systems. Why? Because AAA symbols are the mechanism by which simple computers and abstract logical systems represent human ideas. If the answer is 'no,' then what else is needed to constitute ideas? We might need advanced computer architectures that capture the constraints of perception and action at a fine-grained level, as researchers have advocated and already developed in the areas of machine vision and robotics. Or we might need to consider the constraints of perception, action, and sensorimotor experience to be the fundamental guide in shaping our scientific understanding of language comprehension.

Very often the 'what else' is framed in terms of grounding. A brief survey of sources that use the term 'grounding' in their titles (Glenberg and Kaschak 2002; Harnad 1990; Pecher and Zwaan 2005; Roy 2005; Steels 2003) does not reveal much in terms of definition. Here is what Roy (2005) says, and we will offer it as our definition of the process:

> The term grounding will be used to denote the processes by which an agent (human or machine) relates mental structures to external physical objects.[1]

Thus, we can ask a variety of questions about symbols. What sorts of things can be done with AAA symbols that do not require grounding (e.g., syntactic transformations)? Can ideas composed of symbols be meaningful (be about something in particular, say, your favourite chair) without grounding? Do all symbols need to be grounded, or can some collections of symbols constitute meaningful ideas because they include some proportion of grounded symbols? What proportion is enough? Finally, when using rules to manipulate symbols, do the grounding processes necessarily play a role, or does grounding only matter for the initial induction of a symbol and the interpretation of the results of symbol manipulation?

Another answer to the 'what else' question is embodiment. There have been several attempts to characterize the notion of embodiment, and a notable one is provided by Wilson (2002). Here we try for a different sort of explanation that might be useful for investigations primarily concerned with language and robotics. Before presenting that definition, however, a preamble is necessary. There are several ways of defining embodiment that do not seem to advance thinking. First, it is commonly accepted that knowledge about the world depends (at least in part) on perception and action systems that are parts of the body. Thus, claiming that a theoretical approach is embodied just because it proposes that knowledge derives from perception does not serve to discriminate between theories or provide an advance. Second, there has been a tremendous influence of neuroscience on theories of cognition, and it is commonly accepted that cognitive processes are reflected in neural processes. Thus, claiming that a theoretical approach is embodied

---

[1] Note that this admirably clear definition seems to break down in various ways. For example, could a mentalistic term such as 'idea' ever be grounded in the sense offered above? Probably a matter for the debate.

because it invokes neural mechanisms does not provide an advance. Given this preamble, our definition of embodiment in regard to language is:

> Linguistic symbols are embodied to the extent that: (a) the meaning of the symbol (the interpretant) to the agent depends on activity in systems also used for perception, action, and emotion, and (b) reasoning about meaning, including combinatorial processes of sentence understanding, requires use of those systems.

Landauer and Dumais's (1997) Latent Semantic Analysis (LSA) theory is one prototype of an amodal symbol theory: the meaning of a word (a symbol) consists of its relations to other words that have occurred within a particular semantic space. These meanings are constructed by (a) defining the semantic space as a large (e.g., 30,000) collection of texts;(b) forming a matrix of the frequencies of occurrence of the individual word types in each of the texts; and (c)using a mathematical technique (singular value decomposition) to reduce the dimensionality of the matrix. The result is a matrix in which each word is coded with values on about 300 dimensions (in contrast to the 30,000 texts). That is, the meaning of each word is a vector consisting of about 300 numbers containing information about frequencies and co-occurrence. Landauer and Dumais state, '...we suppose that word meanings are represented as points (or vectors; later we use angles rather than vectors) in k dimensional space...' (page 215). Later in the paper they write, 'Given the strong inductive possibilities inherent in the system of words itself, as the LSA results have shown, the vast majority of referential meaning may well be inferred from experience with words alone' (page 227). The sense of referential meaning here is the generic referential meaning of a word, such as the meaning of *bicycle* in different texts and contexts. This meaning is not a particular indexical or deictic reference during the instantiation of a particular text or experience.

Thus, the central claim of LSA, which can be empirically tested, is that much of the generic referential meaning of a word W can be bootstrapped from a model that simply considers statistical properties of the family of other words that accompany word W, without having to consider the physical and social world that may accompany text. In this sense, the vast majority of meaning of a word W may well be inferred from experience with other symbols. The obvious implication of this claim is that experience with the world is not important for the vast majority of (generic) referential meaning of a word.

LSA vectors appear to the prototypical abstract, amodal, and arbitrary symbol, namely sets of numbers. But that characterization is at the *implementation* level, to use David Marr's terminology, and in this case the implementation level is statistical. The question arises as to what these vectors denote, if anything, at Marr's level of *computational theory* and the level of *representation and algorithm*. An answer to this question is not straightforward. Certainly it is the case that the sets of numbers are derived from perception (of the words). It is also the case that words do not appear together for arbitrary reasons; they appear together because the objects named appear together in the world. Nonetheless, all real perceptual information is stripped away before coding in LSA in that (a) words are descriptions not the objects themselves, and (b) it is co-occurrence frequencies of words that matter. That is, frequency is the currency, not say, redness, or loudness, or spatial extent.

Landauer and Dumais (1997) note that at least some of the abstract symbols need to be grounded, and they suggest one scheme for doing so. Consider how a word such as *rabbit* might be grounded by 'judiciously add[ing] numerous pictures of scenes with and without rabbits to the context columns in the encyclopedia corpus matrix, and fill[ing] in a handful of appropriate cells in the *rabbit* and *hare* word rows'. This would seem to potentially ground (potentially, because there are enormous practical problems with the suggestion of adding pictures) the LSA symbols in visual information. We don't wish to prejudge whether this would actually lead to grounding; that should be a matter of debate. We do wish to note, however, that the suggestion for adding pictures falls far short of creating an embodied theory as defined above. First, there is no need for activity in anything like a neural system because the meaning depends on frequencies of co-occurrences, not anything intrinsic to, say, visual properties or auditory properties or action properties. Second, reasoning in LSA is based solely on the frequencies (or the corresponding dimensional values), not directly on anything to do with perceptual or action systems. For example, judging that a rabbit and a hare are similar does not require accessing visual information. Instead, the judgment is based on a mathematical comparison of the similarity of the vector representations. The values of these vector representations are loosely constrained by the perceptual world, so there may be an indirect link with perception and action.

A second example of a symbolic theory (or at least an implementation of a symbolic theory) is the connectionist netword[2] described by Roger et al. This example is interesting because whether one concludes that the theory is embodied or symbolic depends on whether one analyzes the verbal statement of the theory or the implementation as a connectionist network[2]. They state on page 206.

> '...we suggest that the representations and processes underlying semantic memory are best understood within a theory in which semantic knowledge emerges from the interactive activation of modality-specific perceptual representations of objects and statements about objects. In contrast to some contemporary approaches, we argue that semantic representations do not need to extract,

---

[2] Is a connectionist theory an appropriate example of a symbolic theory? First, connectionist models are usually considered—by connectionists and symbolists alike—as something different from the symbolic approach. Thus, in many PDP models there are no individual symbols for particular concepts. Second, any attempt to implement a theory—included an embodied theory—forces one to fit the constraints of current computer and programming technologies: for instance, even connectionist models are a simulation of parallel processing in a serial computer. Third, computations in the brain could be ultimately described as patterns of activations in neural networks, such as in Pulvermüller's neo-Hebbian approach, which is not far away from connectionist intuitions. Nonetheless, there may be several reasons for treating some types of connectionists accounts as symbolic. First, as in Rogers et al, some encodings may be localist in that one node corresponds to one idea, which is very much in conformation to our definition of a symbol. Second, as illustrated shortly, in some connectionist accounts, the manipulations are based only on frequency of occurrence or co-occurrence irrespective of the putative meaning or perceptual system.

sort, and retrieve attributes, facts or propositions about objects to fulfil this role; they need only to allow such information to be produced as overt responses in particular task contexts... . The content of semantic memory is represented in the same regions of cortex that directly encode modality-specific regularities in the environment during perception and action... These acquired [semantic] representations do not code explicit semantic content, but they are structured in ways that facilitate the system's ability to generate appropriate responses when given perceptual inputs.'

Later in the paper they write 'Semantic knowledge may be constructed as a process that mediates the interactions among content-bearing perceptual representations, rather than as a repository of propositional facts about objects'. One could hardly ask for a better match with our definition of embodiment.

The commitment to ideas of embodiment disappears, however, in the implementation of the model. There are three subnetworks encoding (a) *visual* features, (b) *verbal* features (object names, features such color names, and functional descriptions such as *can fly*, etc.), and (c) a hidden-node semantic network. After training, the input of visual features, combined with the information in the semantic units, produces activation of appropriate verbal features, and input of verbal features, combined with the information in the semantic units, produces appropriate activation of visual features.

The reason that this network is less than fully embodied is that different types of features are all made of the same stuff. Each feature is an on/off activation of a node. Although one node may be designated as a color, and another node designated as a body part or functional feature, that is irrelevant to the model's processing. The only thing that counts is frequency, or better yet, co-occurrence frequency (coherent covariation to use Rogers et al.'s terminology). For example, if there was another subnetwork coding auditory features, exactly the same processing rules would apply. The fact that one node would designate *loud* and another *high pitched* and another phoneme /k/ is irrelevant. The activations on those nodes would be treated identically, and identical to the activation on the *visual* nodes. Thus, in the Rogers et al. implementation, there is nothing visual or verbal about the subnetworks; the only thing that matters is co-occurrence. Similarly, in terms of reasoning about meaning, the only information that matters is co-occurrence frequency; it is irrelevant that one node is labeled *leg* and another *loud*. This conclusion is stated in the Rogers and McClelland's (2004) exposition of the theory. 'What is important to the model's behaviour is the propensity for various sets of properties to covary across different items and contexts, regardless of whether the properties are conveyed through spoken statements or through other perceptual information. It is not the identity of the properties themselves, but their patterns of covariation that is essential to the model's behaviour' (p. 117).

It could be argued, however, that the model by Rogers and colleagues is in fact embodied at the level of combining the primitive units in the execution of action, but not at the level of identification and activation of the primitive units themselves. All models need to precisely declare what level of behaviour or cognition is being explained versus what levels are assumed as primitive units. It is extremely important, as we debate the issues, that we specify precisely what phenomena we are attempting to explain as a direct object of inquiry, versus what phenomena and components are merely offered as a simplification in our working assumptions.

1.1.2 **Embodiment theories**

The first example of an embodied theory is Barsalou's (1999) perceptual symbol system. We won't describe the model in detail; instead we note that Barsalou describes a perceptual symbol for an object as based on the neural activity in multiple perceptual systems when the object was initially perceived. Furthermore, manipulation of knowledge requires a simulation using those same perceptual systems. Perhaps because this theory is not implemented, it can remain true to its embodied roots.

The second example of an embodied system is Roy's (2005) robot, Ripley, 'a robotic manipulator that is able to translate spoken commands such as "hand me the blue one on your right" into situated action'. According to Roy:

> Ripley's representations and algorithms led to an approach that grounds the meaning of verbs, adjectives, and nouns referring to physical referents using a unified representational framework … Verbs are grounded in sensorimotor control programs similar to x-schemas. Adjectives describing object properties are grounded in sensory expectations relative to specific actions … For example, the meaning of 'red' is not simply a colour category, but rather a colour category linked to the motor program for directing active gaze towards an object. 'Heavy' is grounded in haptic expectations associated with lifting actions. In this way, all perceptual properties are related to appropriate actions. Locations are encoded in terms of body-relative coordinates. Objects are represented as bundles of properties tied to a particular location along with encodings of motor affordances for affecting the future location of the bundle.

Does the implemented Ripley really encode the environment using its perception and effector systems, or does the commitment to embodiment break down as in Rogers *et al.* (2004)? Fortunately, we were able to ask Deb Roy himself at the debate (see Chapter 11).

1.2 **Questions for debate**

Given these proposed definitions and examples, many of the questions we formulated for the debate are still operative. Thus, in terms of the need for symbol grounding, is it the case that:

(a) Searle's *Chinese room argument* (see chapter 4) is wrong; symbol systems need not be grounded to represent or produce meaning;

(b) Language needs to be grounded, but the mechanism of grounding plays little role in most language processing;

(c) Searle's argument is no longer relevant to contemporary symbolic models that are complex, multilayered at coarse- and fine-grained levels, and grounded in perception and action. Searle's argument is either incorrect or indeterminate;

(d) Grounding language in embodied representations is important for understanding some aspects of language (e.g., about concrete situations), but not others (e.g., descriptions of some very abstract situations); *or,*

(e) All levels of language understanding are grounded in action and perception.

In regard to the interpretation of neuroscientific data, is it the case that:

(a) Word meaning activates brain regions that partially overlap those responsible for perception and action. Therefore, the embodiment of meaning has been empirically demonstrated;

(b) Neurological data do not solve the question because, even if the processing of word meaning overlaps sensorimotor areas in the brain, this fact does not preclude that the processes themselves are symbolic; *or,*

(c) In addition to the above claim, the classical perisylvian areas of language in the left-brain hemisphere perform computations that are more symbolic-like (morpho-syntactical) than embodied.

In regard to whether computers can model comprehension that is embodied and symbolically grounded, which of the following can be most persuasively defended?

(a) Because computers are not biological systems, they cannot be embodied or even simulate embodied cognition. Nonetheless, computers might be able to model embodied cognition in the same sense that a computer can model a thunderstorm by solving complex equations;

(b) A computer program can simulate embodied cognition in humans, but cannot be literally embodied in the same way that human meaning is embodied;

(c) A computer program can be embodied, but the nature of the embodiment is different from humans because the world experienced by the computer is different; *or,*

(d) A computer/robot can be fully embodied in the same way that meaning is embodied in humans.

When considering Ripley in the context of our definition of embodiment, it would seem that the answer must be (c) or (d). However, this may be challenged by some researchers who question the adequacy of Ripley in truly capturing the computational theory, representations, algorithms, and neuroscience implementation of an embodied theory.

Of course, all of this is just wasted words unless we figure out how to decide these issues empirically. Thus, we encourage you all to address in your chapters how these questions can be resolved, and to include your own data that contribute to such a resolution. In addition to the debate questions, we encourage you to tackle the following:

(a) Describe an ideal experiment that would convince a symbolist that embodied representations are routinely activated to understand language.

(b) Describe an ideal experiment that would convince an embodiment theorist that symbolic representations are necessary to understand at least some kinds of linguistic expressions.

References

Barsalou L (1999). Perceptual symbols systems. *Behavioural and Brain Sciences*, 22, 577–609.

Clark HH (1996). *Using Language*. New York, NY: Cambridge University Press.

Glenberg AM, Kaschak MP (2002). Grounding language in action. *Psychonomic Bulletin and Review*, 9, 558–65.

Harnad S (1990). The symbol grounding problem. *Physica D*, 42, 335–46.

Just MA, Carpenter PA (1992). A capacity theory of comprehension: individual differences in working memory. *Psychological Review*, 99, 122–49.

Kintsch W (1998). *Comprehension: A Paradigm for Cognition*. Cambridge: Cambridge University Press.

Kosslyn SM (1994). *Image and Brain: The Resolution of the Imagery Debate*. Cambridge, MA: MIT Press.

Landauer TK, Dumais ST (1997). A solution to Plato's problem: the latent semantic analysis theory of acquisition, induction and representation of knowledge. *Psychological Review*, 104, 211–40.

Marr D (1982). *Vision*. San Francisco, CA: Freeman.

Meyer DE, Kieras DE (1997). A computational theory of executive control processes and human multiple-task performance: Part 1. Basic mechanisms. *Psychological Review*, 104, 3–65.

Miller GA, Johnson-Laird PN (1976). *Language and Perception*. Cambridge, MA: Harvard University Press.

Newell A (1980). Physical symbol systems. *Cognitive Science*, 4, 135–83.

Normal DA, Rumelhart DE (1975). *Explorations in Cognition*. San Francisco, CA: Freeman.

Pecher D, Zwaan RA, Eds (2005). *Grounding Cognition: The Role of Perception and Action in Memory, Language, and Thinking*. New York, NY: Cambridge University Press.

Pulvermüller F (1999). Words in the brain language. *Behavioural and Brain Sciences*, 22, 253–336.

Rogers TT, Lambon Ralph MA, Garrod P *et al.* (2004). Structure and deterioration of semantic memory: a neuropsychological and computational investigation. *Psychological Review*, 111, 205–35.

Rogers TT, McClelland JL (2004). *Semantic cognition*. Cambridge, MA: MIT Press.

Roy D (2005). Grounding words in perception and action: computational insights. *Trends in Cognitive Sciences*, 9, 389–96.

Steels L (2003). Evolving grounded communication for robots. *Trends in Cognitive Sciences*, 7, 308–12.

Wilson M (2002). Six views of embodied cognition. *Psychonomic Bulletin and Review*, 9, 625–36.

Chapter 2

The limits of covariation

Arthur M Glenberg and Sarita Mehta

2.1 Introduction

Covariation is a powerful source of information that is incorporated into several types of cognitive models. For example, Landauer and Dumais (1997) suggest that covariation between words and the texts in which those words occur leads to the learning of new word meanings, and Rogers *et al.* (2004) suggest that covariation among perceptible features can serve as the basis for concepts and semantic memory. In this chapter, we provide the first empirical test of the claim that people recover meaning from covariation alone. We begin with a review of two theories that feature covariation as an important component of learning the meaning of concepts (Landauer and Dumais 1997; Rogers *et al.* 2004), followed by reasons to question a reliance on covariation alone. We then present the results of three experiments that demonstrate limits on how much meaning can be recovered from covariation alone. Given these limits, we end with a discussion of an issue central to the debate, namely what sort of data lend support to the notion that arbitrary, abstract, amodal (AAA) symbols play a role in cognition.

As a prelude, it is important to note that in this chapter we are using the definition of an AAA symbol offered in Chapter 1, namely that it is abstract, arbitrarily related to its referent, and amodal.[1] An implication of this definition is that modal characteristics of the referent that are not explicitly encoded in the symbol play no role in cognitive processing. For example, if mathematical reasoning about the number five is done using an AAA symbol, then the fact that the number was initially conveyed as an Arabic numeral or a Roman numeral or a spelled-out word should play no role in the cognitive processing. That is, when the symbols are used, they are ungrounded; processing proceeds through the application of rules to the symbol irrespective of the perceptual and actionable features of the referent.

2.2 Theories of covariation

Landauer and Dumais's (1997) *latent semantic analysis* (LSA) is meant to be a theory of acquisition, induction, and representation of knowledge. The basic theoretical mechanism is the tracking of covariation between words and the texts in which those words occur.

---

[1] Some theorists make use of grounded symbols. For example, Barsalou (1999) discusses perceptual symbols composed of neural activity in perception and action systems similar to the neural activity generated by the referent object. The term 'symbol' used in this chapter refers only to AAA symbols.

The LSA mechanism begins by tracking the frequency of occurrence of about 60,000 words across some 30,000 texts, forming a matrix with words as rows, texts as columns, and frequencies as the cell entries. After some preprocessing, the matrix is submitted to a singular value decomposition to reduce the dimensionality. The result is a matrix in which the words are coded with values on about 300 dimensions (in contrast to the 30,000 texts). Landauer and Dumais state, '... we suppose that word meanings are represented as points (or vectors; later we use angles rather than vectors) in k dimensional space...' (p. 215). Later in the paper they write, 'Given the strong inductive possibilities inherent in the system of words itself, as the LSA results have shown, the vast majority of referential meaning may well be inferred from experience with words alone' (p. 227). The obvious implication of this claim is that experience with the world is not important for the 'vast majority of referential meaning' of a word. Note that all perceptual information is stripped away before coding in LSA such that: (a) words are descriptions, not the objects themselves, and (b) it is frequencies of words that matter. That is, frequency (or covariation in frequency) is the currency of meaning, not say, redness, or loudness, or spatial extent. Thus, the LSA vector is very close to an AAA symbol; it is abstract in that it refers to a class of events (whenever the word occurs); it is amodal; and the relation between the symbol and the referent is just short of arbitrary. That is, there is probably a loose connection between the frequency of some words in text and the frequency of experience of their referents.

Landauer and Dumais (1997) do recognize the need for some sort of symbol grounding, 'But still, to be more than an abstract system like mathematics words must touch reality at least occasionally' (p. 227). To ground a word such as 'rabbit' they suggest, 'judiciously add[ing] numerous pictures of scenes with and without rabbits to the context columns in the encyclopedia corpus matrix, and fill[ing] in a handful of appropriate cells in the rabbit and hare word rows.'

Nonetheless, frequency remains the currency because all processing in the LSA theory is based solely on the frequencies-based dimensional values from the singular value decomposition. For example, judging that a rabbit and a hare are similar, even with the addition of picture contents to the initial matrix, would not require accessing visual information. The addition of a picture is simply the addition of a new context in which to track covariation. The similarity judgement would still be based on a mathematical comparison of the vectors representing 'rabbit' and 'hare'.

There are many demonstrations that covariation as implemented in LSA could serve as a basis of meaning. For example, Landauer and Dumais (1997) report that similarity between LSA vectors can be used to pick out synonyms with about as much success as non-native English speakers applying for admission to US colleges. Nonetheless, all of the demonstrations are correlational, and consequently, the causal relations cannot be determined. Consider, for example, the finding that the angle between two LSA vectors predicts the extent to which the corresponding words will prime one another in a lexical decision task. One possibility is that covariation between words and contexts determines meaning, and hence covariation correlates with priming, a putative measure of relational meaning.

Another possibility, however, is that words that share meaning (e.g., 'cow' and 'bovine') tend to occur in similar texts because of the overlap in meaning. Thus, the covariation does not determine meaning; it is the shared meaning that determines the covariation. The experiments reported in this chapter determine whether people can use covariation between symbols alone to induce meaning.

Rogers *et al.* (2004) describe another sort of theory in which covariation serves as the basic mechanism of meaning (at least in the implemented model). They write, 'Semantic knowledge may be construed as a process that mediates the interactions among content-bearing perceptual representations...' (p. 233). However, the implementation of the model reveals its complete dependence on covariation rather than perceptual representations. In the model, there are three subnetworks consisting of localist encoding of: 64 visual features, 152 verbal features (object names, 'perceptual' features such as colour names, functional features such as 'can fly', etc.), and a 64-unit, distributed, hidden-node semantic network. The model is trained so that the input of visual features, combined with the information in the semantic units, produces activation of appropriate verbal features. Similarly, the training ensures that input of verbal features, combined with the information in the semantic units, produces appropriate activation of visual features. Note, however, that the 'visual' features are nothing of the sort. Each feature is an on/off activation of a node. The fact that one node may be designated as the colour red, and another node designated as 'leg', and a third designated as 'big' is irrelevant to the processing of the model. The only thing that counts is frequency, or better yet, co-occurrence frequency ('coherent covariation' to use Rogers *et al.*'s terminology). In fact, if there were another subunit coding auditory features, exactly the same processing rules would apply. The fact that one node would designate 'loud' and another 'high pitched' and another the phoneme /k/ is irrelevant. What counts in the model are the covariations amongst the nodes, not what the nodes are labelled.

Thus, in the Rogers *et al.* implementation, there is nothing visual or verbal about the subnetworks; the only thing that matters is co-occurrence. Similarly, when reasoning about meaning, the only information that matters is co-occurrence frequency; it is irrelevant that one node is labeled 'leg' and another 'loud.' This conclusion is stated clearly in Rogers and McClelland's (2004) exposition of the theory, 'What is important to the model's behaviour is the propensity for various sets of properties to covary across different items and contexts, regardless of whether the properties are conveyed through spoken statements or through other perceptual information. *It is not the identity of the properties themselves, but their patterns of covariation that is essential to the model's behaviour*' (p. 117, emphasis added).

2.3 **The limits of covariation**

In contrast to models that depend on covariation for meaning, there are data (e.g., Glenberg and Robertson 2000) and logical arguments (e.g., Harnad 1990; Searle 1980) suggesting that word meaning requires more than covariation. Consider Harnad's (1990) *symbol merry-go-round* argument. Imagine someone travelling to another country who does not speak the language, but who has a dictionary written in that language. The person sees a sign and wishes to translate it. He uses the dictionary to look up the first word in the sign, but of course the definition is only in terms of other words in the unknown language. Then the person looks up the first word in the definition, but it too is defined solely in terms of other unknown words. No matter how many words the person looks up, that is, no matter how much covariation the person tracks, he will never be able to induce the meaning of the first word in the sign.

In line with the symbol merry-go-round argument, consider the idea that words derive their meaning from being grounded in bodily activity, not simply from covariation with other words. There is strong evidence for the grounding of word meaning in perception and action. For example, Kaschak *et al.* (2005) demonstrated that watching displays of visual motion interferes with the comprehension of sentences describing visual motion, as if the comprehension process called on perceptual processes. Glenberg and Kaschak (2002) demonstrated that understanding a sentence about directional action interferes with literal action in an incompatible direction, as if the comprehension processes called on action processes. Hauk *et al.* (2004) found that listening to verbs produced enhanced activation in areas of motor cortex corresponding to the effectors used in the actions the verbs named. Based on these results, it would appear that word meaning is based on how the words are associated with perceptions and actions, not simply covariation with each other or texts.

In summary, there are good reasons to suppose that covariation plays a critical role in determining the meaning of words, but there is equally good evidence to suggest that that role may be limited. Importantly, there do not seem to be any direct tests of the claim that the 'vast majority of referential meaning' can arise from covariation alone. That is, the evidence supporting covariation is mainly correlational, and the evidence noting that words are grounded does not rule out the possibility that some (or even the vast majority of) meaning may be based on covariation. The experiments reported next are designed to fill the evidential gap.

2.3.1 Experiment 1

Experiment 1 had three goals: (1) to produce clear evidence that people can learn about covariation of ungrounded symbols, (2) to determine whether meaning can be inferred solely from the learnt covariation structure, and (3) to determine whether grounding some of the symbols (by naming them) allows people to use the covariation structure to infer the meaning of the other symbols.

We began by selecting a semantic domain familiar to the student participants in our experiments, the domain of two-wheeled vehicles. Then, over the course of several weeks, we made quasi-random observations of various locations on campus resulting in 102 instances of the domain, including 20 road bicycles (e.g., those with downward-curving handlebars), 30 mountain bicycles, 7 recumbent bicycles, 7 leg-powered scooters, 11 motorcycles, and 26 mopeds. We coded each observation on 29 binary features that could be cross-classified by mode of observation (e.g., auditory or visual) and the type of relation between the vehicle and the feature (e.g., a part of the vehicle, or a way in

Table 2.1 Cross categorization of feature types and relations used in all of the experiments

		Relations			
		ISA (category)	**Parts**	**Properties**	**How changes**
Feature types	Abstract (categories)	Two-wheeled*, motorized, road bike*, mountain bike, recumbent bike, scooter, motorcycle*, moped			
	Abstract (features)			Inexpensive*	Short distance, medium distance*, long distance, gets flat tires, easily, disassembled*, carry another person
	Visible		Chain visible, mirror*, red light in back, keys needed	Not achromatic*, Used on sidewalk*	
	Auditory			Noisy*	
	Haptic (touch)		Smooth tires	Hot, not heavy*	
	Proprioceptive (body position)			Legs still*, sit upright, lean forward*, recumbent	

* features that were labelled during learning in experiments 2 and 3.

which the vehicle changes over time). These features and the cross-classifications are given in Table 2.1. The four relations we used in our experiment are the same as those used by Rumelhart (1990) and Rogers and McClelland (2004) in their simulations.

The user interface we constructed is illustrated in Figure 2.1. Note, however, that in the main part of the experiment, most of the specific verbal labels next to the radio buttons (e.g., 'road bike,' 'inexpensive,' 'noisy') were not present. However, the groups of radio buttons were labelled on the interface using the terms 'item class,' 'relations,' 'category features,' 'abstract features,' 'visible features,' 'hearable features,' 'touchable features,' and 'body position features' (as in Figure 2.1). In addition, the names of the relations ('ISA category,' 'parts,' 'properties,' and 'how changes') were displayed.

Participants were assigned to either a 'learning' condition or a 'control' condition. Those in the learning condition were specifically instructed to learn which features, that is, which radio buttons, tended to occur together. Learning proceeded by having the participant click on an item class button and a relation button. The computer would display (a) an example number (e.g., the number corresponding to a particular

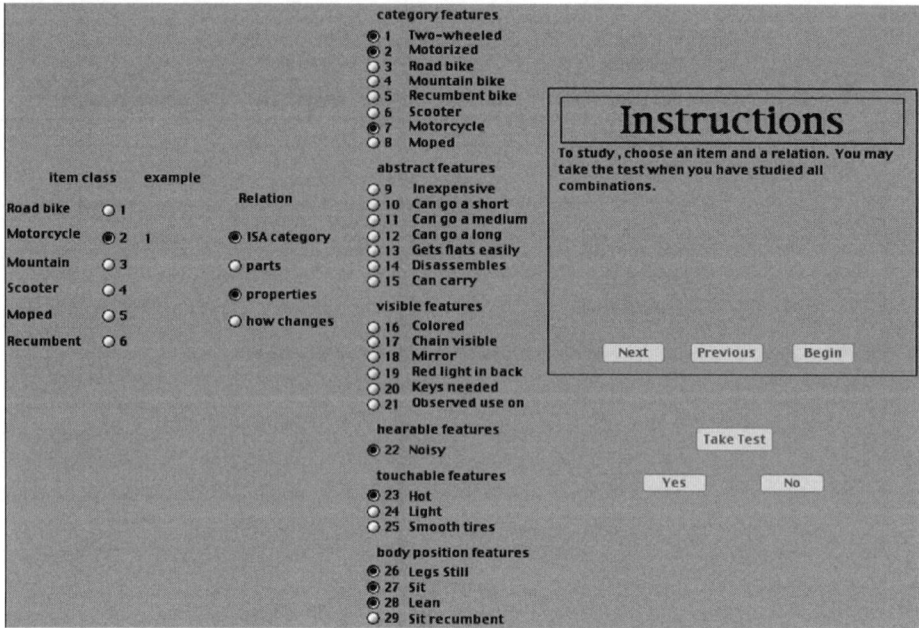

Fig. 2.1 User interface for participants in the learning condition.

road bike) and (b) highlight the buttons corresponding to the features coded for that example for the selected relation. For example, when the participant clicked on item class 5 (moped) and the ISA category, the computer would highlight category buttons 1 (two-wheeled), 2 (motorized), and 8 (moped). If the participant then choose the 'parts' relation, the computer might highlight buttons 17, 18, 19, 20, and so on, depending on whether those parts were coded for the actual observation of the example moped. Participants were free to change the example (by repeatedly clicking on a particular item class button), the item class (by clicking on a different item class button), and the relation. Participants had unlimited time to study.

The learning of at least part of the covariance structure was ensured by having participants study until two types of tests were passed. On each trial of the ISA test, both an item class button and the ISA category button were highlighted. The computer program also highlighted a selection of category radio buttons. Either all of the category buttons were correct (that is, the hidden labels described the item class and these buttons had been presented with each example from this item class) or one was changed. The participant made a decision as to whether all the selected categories were correct or not. There were 12 trials on the ISA test, two for each item class, with one trial containing correct categories and the other trial an incorrect category. To pass the ISA test, the participant had to make at least nine correct decisions. If the participant had fewer than nine correct, then he or she was asked to study again before retaking the test.

After meeting the ISA test criterion, the participant was given the 'feature test'. On each trial of this test, an item class was highlighted in addition to all of the correct category feature buttons. The participant would then cycle through the other relations (parts, properties, and changes). For each relation, either the correct feature buttons were highlighted or one was changed. The participant was to determine if all of the radio buttons for all of the relations were correct. Again, the criterion was nine correct out of twelve trials. If the criterion was not met, then the participant was asked to study again.

Meeting both the ISA test and feature test criteria ensures that the participant has learnt a good deal about the covariance structure. At this point, a series of five final tests ('domain', 'item 4', 'item 5', 'change feature', 'part feature') were given to determine whether or not the participant could infer meaning from the learnt covariance relations. For the domain test, the participant was told that he or she had been studying a coherent semantic domain, and that he or she was to indicate which of six choices was the domain studied and to give a confidence rating on a four-point scale ranging from guessing to positive. The domain choices consisted of two natural, non-biological categories (celestial objects and geological features), two natural, biological categories (mammals and single-celled organisms), and two artefactual categories (furniture and two-wheeled vehicles, the correct choice). The dependent variables were the choice made, the confidence with which the choice was made, and the time taken to make the choice. Although we cannot be sure, it is unlikely that the distractor categories (e.g., mammals) have the same covariance structure as two-wheeled vehicles. Thus, if covariance structure allows the induction of meaning, the choice of two-wheeled vehicles should be obvious.

Following the domain test, the participants were shown the names of four of the item class buttons (all but 'moped' and 'scooter'), as well as the names of all of the feature buttons except for the change feature 'easily gets flat tires' and the part feature 'needs keys.' That is, at this point in the experiment, the interface looked substantially like Figure 2.1. Thus, the question of interest is whether the learnt covariance structure, along with the given names, can be used to determine the meaning of the remaining, unnamed radio buttons.

For the item 4 final test, participants were asked to pick the name of the fourth item class radio button ('scooter') from six choices that included three human-powered vehicles ('scooter', 'skateboard', 'tandem bicycle') and three motorized vehicles ('moped', 'Segway', 'bicycle with motor'). The dependent variables were the probability of selecting from the correct category (human-powered), the confidence, and the time taken to make the selection.

For the item 5 final test, participants were asked to pick the name of the fifth item class radio button ('moped') from the same choices. Note that the choice of 'scooter' and 'moped' could have been trivially easy because these two names were presented as labels for the corresponding category features during these two final tests.

For the changes final test, the participant was given the names of all of the item class radio buttons so that only two feature buttons remained unnamed. The participant was asked to select the name of the missing change feature button ('easily gets flat tires') from among three features typical of human-powered vehicles ('easily gets flat tires,' 'easily falls on ice,' and 'can easily raise or lower seat') and three features typical of motorized vehicles ('new registration required yearly,' 'oil is added,' and 'can run out of gas'). The dependent variables were the probability of selecting a name from the correct category, confidence, and the time taken to make the decision.

On the part final test, participants were asked to select the name of the part feature button ('keys needed') from among three features typical of human-powered vehicles ('curved handlebars,' 'basket,' 'thin spoke') and three features typical of motorized vehicles ('keys needed,' 'speedometer,' 'exhaust pipe'). The dependent variables were the probability of selecting a name from the correct category, confidence, and the time taken to make the decision.

The performance on these final tests for participants in the learning condition was compared with the performance of participants in a control condition. Initially, control participants were treated identically to the participants in the learning condition (see procedure). However, instead of studying the covariance structure, these participants proceeded immediately to the final tests.

If learning the covariance structure of a set of stimuli can be used to (a) infer the semantic domain or (b) be used in conjunction with named features (e.g., the names of 27 of the 29 radio button features are provided after the domain test), then the participants in the learning condition should do substantially better than those in the control condition.

Method

The 74 participants were students enrolled in introductory psychology classes at the University of Wisconsin – Madison. Of these, 42 participants were assigned to the learning condition, and 36 of these completed the experiment. The others asked to be excused after repeated failures on the ISA or feature tests.[2] All 32 participants assigned to the control condition completed the experiment. Data were collected from a majority of the learning participants before we started collecting data from the control participants, although the two sets of data were collected in close proximity during the semester.

Procedure

All participants were treated identically at first. The computer-controlled instructions introduced the participants to the interface using three named item classes: 'pine tree', 'maple tree', and 'trout'. The 29 features were selected to fit these items (e.g., 'has leaves', 'grows', 'can be eaten'). The instructions explained how features could be cross-classified by 'how a person comes to know a feature,' such as through vision or touch, and 'how the feature relates to the item,' such as categories and parts. The participants were encouraged to observe how some features tend to occur together, such as 'living thing' (a category feature) and 'grows' (a change feature). They were also encouraged to attend to patterns of covariation rather than specific examples: 'It would be very difficult to learn all of the features for all of the examples in this experiment, so your goal should be to develop some general ideas, such as all living pine trees have needles, all living maple trees have leaves, but not needles.' At this point, the names of the example item

---

[2] Thus, there may be a selection bias in favour of the learning condition. That is, the data in the learning condition come from participants with sufficient motivation and learning ability to pass the ISA and Feature tests.

classes and features were erased from the computer screen, and the participants were instructed:

> Of course, you already know a lot about pine trees, maple trees, and fish. So, in this experiment you are not going to know the names of the items or the names of their features! (If you want to see the example features again, use the 'Previous' button.) Instead, your goal is to learn things such as: for item class 1, ISA categories are 1 and 2, and features 14 and 20 go together. For item class 2, ISA categories are 1 and 3, and features 17 and 18 go together.

The participants were encouraged to explore the interface by clicking on item classes and relations and observing the corresponding features. The participants were instructed further that the item classes they were studying were not fish or trees and that all of the features differed from those observed previously. Finally, they were instructed that before they could continue to the next stage of the experiment they had to study at least one example from each item class and all of the relations for that example. To begin the formal study, the participant clicked on a button labelled 'Begin.' At this point, the procedures for the learning and control conditions diverged.

Participants in the learning condition could begin studying the item classes in whatever order and to whatever extent they wished. After studying at least one example of each item class, participants were given the option of further study or of proceeding to the ISA test. There were 12 trials on the ISA test, two for each item class. One trial for the item class was correct (all of the categories were correct) and one trial for the item class was incorrect (one of the category features was incorrect). If the participant did not get at least 75% correct on the ISA test, he or she was directed to restudy the items. He or she had to study at least one example from each item class and all of the relations for that example before being able to re-take the ISA test. This procedure continued until the ISA test criterion was met, or the participant asked to be excused from the experiment.

Once the ISA test criterion was met, the participant was given the feature test. There were 12 trials on the feature test, two for each item class, and one trial was correct and the other incorrect. On incorrect trials, the computer randomly picked a relation (e.g., the 'part' relation) to modify and randomly chose a feature of that relation-type (e.g., another part) to substitute for the correct feature. If the participant did not get at least 75% correct on the feature test, he or she was directed to restudy the items. He or she had to study at least one example from each item class and all of the relations for that example before being able to re-take the test. This procedure continued until the feature test criterion was met, or the participant asked to be excused from the experiment.

After meeting the feature test criterion, the participant was given the series of five final tests as previously described. After the domain test, the domain was revealed along with the names of four of the six item class buttons and the names of 27 of the 29 feature buttons. The participant was then asked to select the names of the remaining two item class buttons (the item 4 and item 5 final tests) and the names of the remaining two feature buttons (the changes final test and the part final test).

For participants in the control condition, on clicking on the 'Begin' button, the following instructions were shown:

> You have been chosen to participate in a special version of this experiment. Other students study the item classes and features. They then take the same final test that you will take shortly. As a

special participant, your job is to help us to estimate the probability of educated guessing on the final test. Because you have not studied anything, your answers will have to be educated guesses. In what sense might your answers be 'educated?' There are several clues that you might use in making your guesses. For example: You are a college student. You are in an experiment. You know that the item classes are not trees or fish. From these hints and others, you may be able to deduce the correct answers, or at least eliminate some of the obviously wrong answers. Why should you bother? Why not just respond randomly? There are five questions on the final test. For each one you answer correctly, you will earn a bonus of $1. If you get all five correct, you will earn an additional bonus of $5. We hope that this money will motivate you to make educated guesses rather than random guesses.

After these instructions, the participants in the control condition proceeded immediately to the final tests.

Results

The extent of study, as well as performance on the ISA and feature tests, are an index of the degree to which participants in the learning condition mastered the covariance structure of the stimuli. On average, participants clicked 500 (standard deviation 446) times on item classes and 116 (117) times on different relations. Thus, participants seem to use a strategy of picking a relation and then cycling through the item classes. Participants needed an average of 1.75 (0.65) attempts to meet the ISA test criterion, and the average number correct on the last ISA test was 10.69 (1.12) out of 12. Participants needed an average of 4.56 (2.85) attempts to meet the feature test criterion, and the average number correct on the last test was 9.56 (0.70). These data testify to an impressive amount of study and accomplishment.

Clearly, people can learn something like the covariance structure of a set of ungrounded stimuli. Nonetheless, the question remains as to whether that learning can be a source of meaning. To answer that question, we turn to the data from the final tests. However, two caveats are required. First, the data from the learning condition can only be interpreted in relation to the data from the control condition. The reason is that it is unlikely that the six choices offered on the final tests are all equally attractive. Thus, the baseline or guessing rate can only be determined from the control condition. The second caveat is that it seems unreasonable to expect the participants in the learning condition to be able

Table 2.2 Data from experiment 1 final tests. Confidence was rated on a scale of 1 (guess) to 4 (positive). Time values represent the total time taken to select answer and enter confidence.

	Domain test		Item 4		Item 5		Change		Part	
	L	C	L	C	L	C	L	C	L	C
Correct (proportion)	0.11	0.43	0.75	0.57	0.72	0.53	0.56	0.50	0.64	0.57
Confidence	2.0	2.4	2.8	2.4	2.7	2.3	2.2	2.4	2.0	2.5
Time (s)	32.0	36.5	41.0	46.5	23.7	28.2	46.3	60.8	29.1	40.6

C = control condition; L = learning condition.

to infer specific item classes and features from the covariance structure. Consequently, except for the domain test, we scored whether or not the participant selected a choice within the appropriate class of motorized or non-motorized vehicles.

The results from the various final tests are given in Table 2.2. For the domain test, can the participants in the learning condition select the correct domain (two-wheeled vehicles) more accurately than participants in the control condition? In fact, just the opposite occurred. The participants in the learning condition selected the correct domain 11% of the time, whereas participants in the control condition selected the correct domain 43% of the time (χ^2 [1] = 8.89). We are uncertain as to how the control condition could have done so well on this test except if some of the domain choices seemed *a priori* unlikely. At the very least, this finding demonstrates that the participants in the learning condition cannot make much sense out of the covariance structure they learnt. There was no significant difference in the confidence with which the choices in either group were made (t [64] = −1.50).

Because of the positive skew in the selection times, the data were transformed into logarithms before the statistical analyses. The data reported in Table 2.2 are the means of the logarithms transformed back into seconds. Although the difference between the mean selection times was not significant, they are interesting for several reasons. First, in both conditions, the participants took a considerable amount of time to make their choices; apparently, they treated the task seriously. Second, the times provide little or no evidence that the participants in the learning condition were able to solve the task by simply referring to semantic memory. Instead, judging from the length of time required, it would appear that participants in both conditions were applying some sort of reasoning strategies.

For statistical analysis, we combined the data from the final tests on items 4 and 5. For each participant, we determined whether he or she chose an item from the correct category (motorized or not) for item 4 and for item 5, and then we summed the scores so that each participant received a score of 0, 1, or 2. There was a significant difference in favour of the learning condition (t[64] = 2.02). A similar difference was found for combined confidence (t[64] = 2.08), but not for combined time (t[64] = −1.01), with the means for the learning and control conditions being 48.53 and 65.61 seconds, respectively. Once again, the times are quite informative. The long durations are much more consistent with some sort of reasoning strategy rather than direct consultation of semantic memory. There is at least one good candidate for this reasoning process. At the time of the item tests, the names of all of the item classes were available as labels of the category features (Figure 2.1). Thus, a participant in the learning condition could map knowledge of the category buttons associated with a particular item class button to the newly labeled category buttons.

The last two tests involved identifying the change feature button 'easily gets a flat tire' and the parts feature button 'keys needed.' For each participant, we determined whether he or she chose an item from the correct category (motorized or not) for the two tests, and then we summed the scores so that each participant received a score of 0, 1, or 2. There was no significant difference in the average number of correct items (1.09 and 1.07

for the learning and control conditions, respectively; t[64] < 1). However, in regard to the average time to make a choice, participants in the control condition took longer than did the participants in the learning condition (mean 49.66 vs. 36.73 seconds; t[64] = −2.74), and the control condition was more confident than the learning condition (mean 2.47 vs. 2.11 seconds; t[64] = −2.03).

Discussion

The data are clear in two respects. First, people can learn at least part of the covariance structure of ungrounded symbols. This fact is demonstrated by performance on the ISA and features tests. Second, covariance structure cannot be used to determine the domain of study. In fact, the participants in the leaning condition did significantly worse than those in the control condition on the domain test.

The data are less clear in regard to the ability to use the covariance structure when some of the symbols are named. Participants in the learning condition were able to identify two of the item classes more successfully than those in the control condition. However, the very long times needed to make the identifications (over 30 seconds) suggests that people were applying some sort of reasoning strategy rather than relying on semantic memory. Participants in the learning condition were not able to identify features better than those in the control condition. Also, the control condition participants were more confident in their selection of features. Again, the long selection times argue for special strategies. Thus, the ability to use a covariance structure to infer meaning appears to be limited at best. That is, the covariance structure may be useful in a conscious, effortful reasoning process.

2.3.2 Experiment 2

In experiment 1, the symbols were named only after the covariance structure was learnt. Perhaps the covariance structure is more useful when some of the items in the structure are named (and presumably grounded) while learning takes place. Thus, in experiment 2, three of the item class buttons ('road bike,' 'mountain bike,' and 'motorcycle') were named at the beginning of the study period, as were 13 of the 29 features (see Table 2.1). Thus, if learning the covariance structure is useful when some of the features are named, participants in the learning condition should demonstrate a clear superiority over participants in the control condition. In addition, we might expect the selection times to be much shorter for the learning condition.

Several other changes were also made in the experiment. First, the numerical labels next to each feature were eliminated. (A commenter on a presentation of experiment 1 suggested that the numerals were grounded and that such grounding might interfere with the application of the covariance structure.) Second, the domain test was eliminated. Given that we provided the names of three of the item class buttons ('road bike,' 'mountain bike,' and 'motorcycle') the correct domain (*two-wheeled vehicles*) was obvious. Third, during the item 4 and item 5 final tests, the radio buttons corresponding to categories 'moped' and 'scooter' were not named, although the radio button corresponding to *motorized* was labelled.

Table 2.3 Data from experiment 2 final tests. Confidence was rated on a scale of 1 (guess) to 4 (positive). Time values represent the total time taken to select answer and enter confidence.

	Item 4		Item 5		Change		Part	
	L	**C**	**L**	**C**	**L**	**C**	**L**	**C**
Correct (proportion)	0.64	0.33	0.79	0.61	0.48	0.45	0.80	0.79
Confidence	2.4	2.1	2.6	2.0	2.2	2.3	2.3	2.4
Time (s)	55.6	81.1	23.5	45.7	45.5	73.3	26.2	42.8

C = control condition; L = learning condition.

Method

A total of 123 participants were quasi-randomly assigned to the learning (n=90) and control (n=33) conditions. We assigned more participants to the learning condition in anticipation of a greater drop-out rate. In all, 28 participants in the learning condition failed to finish the experiment. Thus, the data are based on 62 participants in the learning condition and 33 participants in the control condition. Except as noted above, the method was identical to experiment 1.

Results

On average, learning condition participants clicked 706 (standard deviation [SD] = 708) times on item classes and 246 (SD = 233) times on different relations. Participants needed an average of 1.66 (SD = 1.18) attempts to meet the ISA test criterion, and the average number correct on the last test was 10.44 (SD = 1.10) out of 12. Participants needed an average of 3.98 (2.83) attempts to meet the feature test criterion, and the average number correct on the last test was 9.43 (SD = 0.69). As in experiment 1, there is clear evidence of having learnt at least some of the covariance structure.

The data from the final tests are presented in Table 2.3, and in many respects, they mirror the data from experiment 1. Consider first the final tests on items 4 and 5. The participants in the learning condition were more accurate (t[92] = 3.23), more confident (t[92] = 2.58), and took less time (t[92] = −3.28) than did participants in the control condition.[3] In contrast, after combining performance on the two feature tests, there were no significant differences for accuracy or confidence (t[92] < 1); however, the learning participants took less time than did the control participants (t[92] = −3.61).

---

[3] Differences between the conditions in time to make a selection need to be interpreted with caution in this experiment. The participants in the learning condition had been exposed to about half of the names throughout the learning process (of some 950 clicks on radio buttons), whereas the participants in the control condition had only a brief exposure to the names during the latter half of the instructions. Hence, it is likely that the control participants are taking more time to read the names during the final tests.

Discussion

Once again, the data tell a mixed story, although the conclusion must be that there are limits on what can be gained from learning about covariance structure. The learning participants clearly out-performed the control participants on the tests of items 4 and 5. However, they took a fair amount of time in making their selections: 55.6 seconds for item 4 and 23.5 seconds for item 5. These times imply that the participants were doing a fair amount of thinking, rather than obtaining the information directly from semantic memory. What might that thinking have been? One hypothesis hinges on the fact that when the item 4 and 5 tests were given, the category radio button 'motorized' was named. Participants may have been aligning their memories for when that particular feature had been paired with item 5 (the mopeds) and not item 4 (the scooters). Thus, the accurate performance on the item tests may reflect a type of conscious hypothesis testing rather than an automatic access of semantic memory.

Regardless of the interpretation of the item tests, the final feature tests demonstrate constraints on learning from covariance structure even when that structure is partially grounded by providing names during study. In regard to accuracy, the two conditions were virtually identical (0.64 for learning and 0.62 for control) even though the participants in the learning condition had made approximately 950 clicks on item classes and relations and had attended to the results of those clicks as demonstrated by passing both the ISA and feature test criteria.

Note that the two unnamed radio buttons in the final feature test covaried with 13 named features over an average of 706 contexts (the average number of examples presented for study). Thus, the experiment is equivalent to presenting the LSA program with 706 contexts with each containing about 15 symbols (13 named radio buttons and two unnamed). Whereas the program could easily induce a similarity structure from this input, apparently people cannot.

2.3.3 Experiment 3

There may be a relatively simple reason why participants in the learning condition did not do better on the final feature tests. Namely, the covariances may not have closely matched real-world experiences. Although the number of examples of each item class was based on real-world observation, in experiments 1 and 2 the participants choose the frequency with which they studied the item classes. Thus, for example, although there were four times as many examples of mopeds as scooters, the participants may have studied the two classes approximately equally often by clicking on the respective radio buttons equally often. In this case, the relative frequencies of the item classes, and hence the covariances, obtained from the experiment would not match those obtained from the world.

In experiment 3, the interface was changed so that participants clicked a 'Next item' button, and the computer randomly selected an example from among the 102 examples in the database (the participant still selected the relations to study). Thus, the relative frequency of exposure in the experiment should approximate the relative frequency of exposure in real-world experience. A second change arose because of this change in interface. In experiments 1 and 2, before a participant could take or re-take the ISA test or the feature test, the participant had to study at least one example of each item class and all of the relations for each item class (either for that one example or distributed across multiple examples). Because the computer randomly selected items to study in experiment 3, using this study criterion might have produced very long (and frustrating) delays between successive attempts on the tests because the participant would be forced to wait until the computer selected an example from a low-frequency item class. Consequently, a different criterion was used: Participants had to study at least six examples before taking or retaking the ISA test and the feature test. A third change from the previous experiments was that the category feature *motorized* was not labelled until after the final tests on items 4 and 5.

Method

A total of 74 participants were quasi-randomly assigned to the learning (n=42) and control (n=32) conditions. We assigned more participants to the learning condition in anticipation of a greater drop-out rate. In total, 11 participants in the learning condition failed to finish the experiment. Thus, the data are based on 31 participants in the learning condition and 32 participants in the control condition. Except as noted above, the method was identical to experiment 2.

Results

On average, learning condition participants clicked 79.7 (SD = 86.3) times on item classes and 87.4 (50.24) times on different relations. These numbers are far smaller than the corresponding numbers from experiments 1 and 2. The small numbers may reflect faster learning, a different study strategy, or the more lenient criterion imposed before successive attempts at ISA and feature tests. Participants needed an average of 1.90 (1.14) attempts to meet the ISA test criterion, and the average number correct on the last test was 10.00 (1.00) out of 12. Participants needed an average of 6.84 (8.39) attempts to meet the feature test criterion, and the average number correct on the last test was 9.51 (0.76). The large mean and standard deviation for the number of attempts to pass the feature test are due to a few participants who persevered for many attempts. The median number of attempts needed to pass the feature test was 3.0. As in experiment 1, there is clear evidence of having learnt at least some of the covariance structure.

Table 2.4 Data from experiment 3 final tests. Confidence was rated on a scale of 1 (guess) to 4 (positive). Time values represent the total time taken to select answer and enter confidence.

	Item 4		Item 5		Change		Part	
	L	C	L	C	L	C	L	C
Correct (proportion)	0.61	0.56	0.68	0.50	0.55	0.47	0.65	0.63
Confidence	1.9	2.2	1.9	2.2	2.0	2.3	2.1	2.3
Time (s)	47.0	60.2	22.1	29.7	38.9	54.8	20.4	38.9

C = control condition; L = learning condition.

The data from the final tests are presented in Table 2.4. Combining performance on Items 4 and 5, there was little evidence for a difference between the learning and control conditions in regard to accuracy ($F[1,61] = 1.30$, MSE = 0.63), confidence ($F[1,61] = 2.33$, MSE = 2.4), or time ($F[1,61] = 2.75$, MSE = 0.32). Similarly, after combining performance on the two feature tests, there were no significant differences for accuracy ($F[1,61] = 0.37$, MSE = 0.42) or confidence ($F[1, 61] = 1.82$, MSE = 2.12); however, the learning participants took less time than the control participants ($F[1,61] = 9.15$, MSE = 0.32).

Arguably, this experiment provided the cleanest opportunity for the emergence of the ability to use covariance structure to determine meaning. That is, because of the random sampling of the 102 examples of two-wheeled vehicles, there should have been a close match between experimental and real-world relative frequencies. Nonetheless, the data show the least evidence for an ability to use covariance to determine meaning.[4]

2.4 Discussion

We began with three questions:

- Can people learn the covariance structure of ungrounded symbols?
- Can people derive meaning from that covariance structure alone?
- When part of the covariance structure is grounded by naming, can people use the covariance structure to infer meaning for the ungrounded/unnamed part?

We will consider the answer to these questions, and then discuss three more:

- If covariance plays a limited role in acquiring new meaning, how are those meanings acquired?
- What accounts for the impressive relations between covariance and meaning that have been documented?
- Given limits on the use of covariation in determining meaning, what sort of data could support the use of ungrounded symbols in cognition?

If we can trust the ISA and feature tests, then the data from the three experiments are clear: People can learn at least a portion of the covariance structure of ungrounded symbols. Should the ISA and feature tests be trusted? Not if the tests could have been passed by chance. Each test had 12 two-choice questions. With the probability of answering correctly by chance equalling 0.5, the probability of getting at least nine correct is 0.073. On average, participants took about two tries to pass the ISA test. The probability

----

[4] Two additional experiments using a methodology similar to that of experiment 3 were conducted after the debate and are reported in Glenberg AM and Mehta S. The limits of covariation (submitted). The first was a replication of experiment 3 except that the order of the final tests on items 4 and 5 was randomly determined for each participant. As in experiment 3, there were no significant differences between the learning and control groups on either the final item tests or the final feature tests. The other experiment was a replication of experiment 1 except that participants clicked on a 'Next item' button as in experiment 3. In this experiment, there were no differences between the learning and control groups on the final domain test, the final item tests, or the final feature tests.

of getting at least nine correct (by guessing) on either the first or second test is only 0.139. Participants needed about five attempts to pass the feature test. If they were guessing, the probability of passing the test on any of five tries is only 0.312. Thus, given that people succeeded in passing these tests, it seems clear that they have learnt something about the relations among the ungrounded symbols.

The answer to the second question also seems clear. People were unable to map the covariance structure of the ungrounded symbols to the correct general domain (*two-wheeled vehicles*). In fact, on the final domain test (experiment 1, and see footnote 4), the learning group was significantly less successful than the control group. Nonetheless, there are several constraints on this conclusion. First, we used only one domain. Perhaps it has a covariance structure that is difficult to map onto semantic memory structures, or we picked features that are not features normally attended. Second, it is not clear if the experiment meets the conditions needed to test the theory of Rogers *et al.* (2004). That theory proposes that information is processed by different perceptual systems and representations are resident in those systems. In the experiment, we labelled different features by perceptual system (e.g., visual, auditory, etc.), but in fact, all of the information was presented visually. The experiments do seem to meet the conditions needed to test the LSA theory of Landauer and Dumais (1997). Namely, words in texts are (until grounded) abstract symbols that are presented visually, and the operation of the theory is to compute the covariances among those symbols and their contexts. The theory would work exactly the same if the order of the words in each text used as input to LSA were randomized, or if each word was replaced by a random number (as long as the same random number was used each time that word was presented in any text). In fact, the theory would work exactly the same if it were given a coding representing the spatial location of each radio button in the user interface of the three experiments. That is, the LSA model would note the frequency that each radio button occurs with each example (treating an example as equivalent to a text), and the singular value decomposition could be applied to the resulting matrix. Thus, there is no obvious reason to suspect that the arrangement of stimuli in our experiment do not meet the boundary conditions of the LSA theory.

The answer to the third question – can covariance be used when some of the symbols are grounded/named? – is more complex. Participants in the learning condition were more successful than those in the control condition on the final item tests in experiments 1 and 2. This success might demonstrate some ability to use the covariance structure when part of that structure is grounded. However, several other findings argue against even this limited use of covariance structure. First, the participants may have used a single salient clue on the final item tests, namely the label 'motorized.' When that clue was eliminated in experiment 3, participants in the learning condition were no more accurate than those in the control condition. Second, the participants in the learning condition were never significantly more accurate than participants in the control condition on the final feature tests. Third, although the participants in the learning condition were sometimes significantly faster in selecting answers on the final tests than were participants in the control condition (but see footnote 3), it is still the case that the participants in the learning condition took much longer to make their choices than would be expected if they were simply consulting information in semantic memory. Finally, experiment 3 appears to most accurately simulate the conditions for learning a covariance structure that matches that of the world. Nonetheless, the evidence that people could use the covariance structure to induce meaning was the weakest in that experiment. Thus, in answer to the question of whether covariance structure can be used to induce meaning when some symbols are grounded/named, the data suggest that any such ability is severely limited. At best, that structure may be helpful in a slow, conscious, reasoning process that maps the most salient portions of the structure onto named stimuli such as the radio button labelled 'motorized.'

If covariance structure is not the source of concept acquisition and demonstrable fast mapping of linguistic structures to concepts (e.g., Casenhiser and Goldberg, 2005), what else might be needed? To expand on the rabbit/hare example from Chapter 1, if we already know a lot about rabbits, how is it that we can then learn a lot about hares from simply being told that hares are like large rabbits with elongated hind legs? Our proposal is that we use embodied knowledge about rabbits, such as their shape, to simulate or create a likely representation of hares; this is essentially Barsalou's (1999, Chapter 13, this volume) idea of creating a simulator. From this simulation, it is possible to derive affordances or inferences, such as the fact that hares run fast given the long hind legs. Consistent with this sketch, Smith (2005) has demonstrated that toddlers extend the name of an object to other objects that the toddler can manipulate in similar ways. Note that this sketch gives covariance an important role. Namely, covariance provides the opportunity for learning from embodied representations. The point of the experiments is to show that covariance alone is not useful for inducing meaning.

If covariance does not play a major role in determining meaning, what accounts for the impressive relations between aspects of covariance and meaning? As we suggested in the introduction, we believe that there is a simple answer, namely a confusion in the direction of the causal arrow. For example, Landauer and Dumais, propose that the meaning of word A is determined by its covariance with words B and C across different contexts. In contrast, we think that A happens to occur with B, C, and so on because A shares meaning with those other words. That is, (embodied) meaning causes covariance structure, not the other way around.

Our final questions are the ones of greatest theoretical importance: Is there any role for ungrounded, AAA symbols in the cognitive system? What sort of data would, in principle, support the claim that AAA symbols play a role in cognition? Certainly, a demonstration that meaning can be inferred from ungounded symbols, such as covariance structures, would help to secure the case. But, as the experiments reported here demonstrate, people do not appear to be able to use covariance structure alone to infer meaning in any significant way.

Other authors (e.g., Graesser and Jackson, Chapter 3; Louwerse and Jeuniaux, Chapter 15, this volume) have suggested several types of evidence for AAA symbols in cognition, namely, speed, abstraction, and output conforming to a rule. On closer examination, however, it is not clear why this sort of evidence supports the use of ungrounded symbols in contrast to fully grounded symbols.

Consider first speed of linguistic processing: are fluent readers necessarily eschewing embodied processes? That is, does fluency imply that readers are using symbolic processes that are not grounded in perception and action? A fluent reader may not consciously experience the sorts of simulations suggested by Barsalou (1999) or Glenberg and Kaschak (2002). However, simulation accounts do not posit that conscious experience of simulation is necessary. Furthermore, given what is known about the speed up of cognitive processes with practice (e.g., Logan, 1988), there is little reason to suspect that fully grounded processes cannot be fast. Finally, there is direct evidence to contradict the claim that grounded processes must be slow. For example, Pulvermüller (Chapter 6, this volume) reports that when listening to action verbs, motor areas of cortex are activated within tens of milliseconds of activation of auditory areas of cortex. Similarly, Wilson and Wilson (2005) suggest that turn-taking in conversation is coordinated down to tens of milliseconds not on the basis of a disembodied timer, but by entraining on speakers' rates of syllabification.

Graesser and Jackson (Chapter 3, this volume) note that some embodied processes, such as creating a detailed spatial situation model, or engaging in explicit mental imagery, is known to be time consuming. Nonetheless, (a) there is no evidence that all embodied processing must be slow (just as there is no evidence that all symbolic processes are fast), and (b) there is good evidence that at least some embodied processing is fast (e.g., Pulvermüller, Chapter 6, this volume). So unless a language comprehender were engaging in those embodied processes that are time-consuming, understanding could be both embodied and fast.

Because experiences are particular, it is easy to assume that cognition based on experiences must also be a cognition of particulars with no abstraction. However, there is no logical warrant for this assumption. For example, Hintzman (1986) has demonstrated how abstract prototypes can arise from the on-line combination of particular experiences, and Barsalou (1999) has discussed (a) how perceptual symbols and simulators can apply to a variety of experiences, and (b) how a cognitive system based on perceptual symbols can account for the acquisition of abstract concepts such as truth. Empirically, Glenberg and Kaschak (2002) have adduced evidence that the abstract notion of transfer (even transfer of information) is based on bodily actions such as reaching out with the arm. Thus, there is no need to consider abstraction as the exclusive domain of AAA symbol processing.

We know that rule systems that operate on ungrounded symbols (e.g., the rules of mathematics operating on symbols in a computer) produce invariant results, suggesting that any rule-like invariance is the result of ungrounded symbol manipulation. Again, however, the logic is faulty: there is no reason to suspect that systems of grounded symbols (e.g., perceptual symbols) cannot also lead to invariant results. Nonetheless, it would seem that output conforming to a rule in the face of significant changes in the embodiment of representations would suggest that those embodied representations are transduced into AAA symbols, or at least that the embodied nature of the representation plays no role in processing. In contrast, even in mathematics (the domain that many propose as most likely to be based on AAA symbol manipulation) the data demonstrate that the embodied nature of representations plays an important role in processing. Illustrations abound in the work of Campbell (e.g., 1994). For example, the speed and accuracy of verifying simple arithmetic problems (e.g., 8 x 7 = 56) depends on the format in which the problems are presented. Different formats (e.g., eight x 7 = ? and VIII x seven = ?) produce different errors. Thus, Campbell concludes that the perceptual format of the symbol affects both encoding and numerical computations (see also Campbell, 1999). That is, numerical computations are not performed with the same AAA symbol transduced from different perceptual input. Goldstone *et al.* (Chapter 16, this volume) use a different methodology to reach similar conclusions.

Thus, the answer to our final question (is there any role for ungrounded AAA symbols in cognition?) is a bit indirect. First, data briefly reviewed in the introduction support the claim that representations are grounded. Second, many intuitive appeals for AAA symbols (e.g., based on speed of processing or abstraction) are unwarranted. Third, two types of direct tests for ungrounded AAA symbols have failed. As demonstrated in the experiments reported here, people cannot use ungrounded symbols to readily induce meaning, and as demonstrated by Campbell (1994, 1999) and Goldstone *et al.* (Chapter 16, this volume) performance in well practiced, rule-like domains that could (logically) be based on ungrounded AAA symbols is in fact based on grounded symbols. Thus, there seem to be no data that uniquely point to any role for ungrounded symbols in human cognition.

Debate

Max Louwerse: Would the indexical hypothesis accept the weak meaning claim?

Arthur Glenberg: *(In the presentation, Glenberg distinguished between the strong meaning claim that covariation alone can determine meaning, and the weak meaning claim that covariation along with some grounding can determine meaning.)* As currently formulated, the indexical hypothesis does not accept either claim.

Robert Goldstone: In going from experiments 1 to 2, the numbering of the radio buttons was dropped. But this simply makes it more difficult to distinguish the radio buttons. Perhaps performance would be better if the radio buttons were more easily distinguishable.

Glenberg: We need to do experiments like that. On the other hand, in all of these experiments we demonstrate that people do learn at least part of the covariance structure (the ISA and feature tests), but they are generally unable to apply that learning to ground the symbols.

Goldstone: Yes, but the literature on associative learning shows that associations are formed more easily under some conditions than others.

Deb Roy: In describing experiment 2, you mention that half of the symbols are grounded, but in fact you only present words – other symbols – to do the grounding. Thus, what do you mean by grounding?

Glenberg: I think that the participants are using the words to access perceptual symbols (as in Barsalou, 1999), that they are creating perception- and action-based mental models, and that it is those models that are used to ground the word symbols.

Unidentified person: Why can't the grounding be accomplished by projecting the symbol into a 300-dimension LSA space?

Glenberg: If the participants were projecting into an LSA space, then the answers to the questions on the final item and final feature tests should have been obvious. Because performance on those tests was poor, I conclude that people were unable to project into an LSA-type space.

Lawrence Barsalou: Perhaps some words symbols are grounded in experience, but then other word symbols can be grounded based on their relations to the grounded words. Do you have any evidence for or against this idea?

Glenberg: The evidence is all indirect. Namely, many laboratories have demonstrated in various ways that at least some words are grounded in perception and action at least some of the time, but as far as I know, there is no convincing evidence that all words are grounded in this way all of the time.

Friedemann Pulvermüller: You seem to be saying that (a) much of the learning of word meaning is what I would call 'primary,' that is the direct pairing of words with objects or actions, and (b) that some word meaning could arise from 'secondary' or 'parasitic' grounding with the symbols already grounded in a primary fashion. But, isn't that close to what LSA is claiming? Can't we have a compromise?

Glenberg: Yes, I think that we can have a compromise, with a caveat. My data are showing that secondary grounding cannot be based on the mere fact that the to-be-grounded symbol co-occurs with a symbol that has received primary grounding. Instead, the covariation provides the opportunity for some process to occur, such as the creation of perception- and action-based mental models.

Walter Kintsch: Some 70–80% of the words we know we know only through reading. Where do they come from?

Glenberg: The point of the experiments is that it can't be just the covariation that allows us to induce the meaning of words like 'democracy.' It can't be that we know the meaning of 'democracy' because it occurs X times with 'vote' and Y times with 'Republican.' Whatever the process is, it needs to be investigated.

Author note

This work was supported by National Science Foundation grants 0233175, 0315434, and 0445627. Any opinions, findings, and conclusions or recommendations expressed in this material are those of the author(s) and do not necessarily reflect the views of the funding agencies.

References

Barsalou L (1999). Perceptual symbols systems. *Behavioural and Brain Sciences*, 22, 577-609.

Campbell JI (1994). Architectures for numerical cognition. *Cognition*, 53, 1–44.

Campbell JI (1999). The surface form x problem size interaction in cognitive arithmetic: evidence against an encoding locus. *Cognition*, 70, B25–33.

Casenhiser D, Goldberg AE (2005). Fast mapping between a phrasal form an meaning. *Developmental Science*, 8, 500–8.

Glenberg AM, Kaschak M (2002). Grounding language in action. *Psychonomic Bulletin and Review*, 9, 558–65.

Glenberg AM, Robertson DA (2000). Symbol grounding and meaning: a comparison of high-dimensional and embodied theories of meaning. *Journal of Memory and Language*, 43, 379–401.

Harnad S (1990). The symbol grounding problem. *Physica D*, 42, 335–46.

Hauk O, Johnsrude I, Pulvermüller F (2004). Somatotopic representation of action words in human motor and premotor cortex. *Neuron*, 41, 301–7.

Hintzman DC (1986). 'Schema abstraction' in a multiple-trace memory model. *Psychological Review*, 93, 411–28

Kaschak MP, Madden CJ, Therriault DJ *et al.* (2005). Perception of motion affects language processing. *Cognition*, 94, B79–89.

Landauer TK, Dumais ST (1997). A solution to Plato's problem: the latent semantic analysis theory of acquisition, induction and representation of knowledge. *Psychological Review*, 104, 211–40.

Logan GD (1988). Toward an instance theory of automatization. *Psychological Review*, 95, 492–527.

Rogers TT, Lambon Ralph MA, Garrod P *et al.* (2004). Structure and deterioration of semantic memory: a neuropsychological and computational investigation. *Psychological Review*, 111, 205–35.

Rogers TT, McClelland JL (2004). *Semantic Cognition*. Cambridge, MA: MIT Press.

Rumelhart DE (1990). Brain style computation: learning and generalization. In SF Zornetzer, JL Davis, C Lau, Eds. *An Introduction to Neural and Electronic Networks* (pp. 405–20). San Diego, CA: Academic Press.

Searle JR (1980). Minds, brains and programs. *Behavioural and Brain Sciences*, 3, 417–24.

Smith LB (2005). Action alters shape categories. *Cognitive Science*, 29, 665–79.

Wilson M, Wilson TP (2005). An oscillator model of the timing of turn-taking. *Psychonomic Bulletin and Review*, 12, 957–68.

Chapter 3

Body and symbol in AutoTutor: Conversations that are responsive to the learners' cognitive and emotional states

Arthur C Graesser and G Tanner Jackson

3.1 Introduction

Our discussions about embodiment and symbolic representations will centre around a computer system that holds conversations with people in natural language. The system is called AutoTutor, an intelligent tutoring system that has been built and tested in the interdisciplinary Institute for Intelligent Systems at the University of Memphis (Memphis, TN). AutoTutor has an animated conversational agent with synthesized speech, facial expressions, and gestures. AutoTutor attempts to interpret the language and emotions of the learner in an effort to be dynamically adaptive. AutoTutor holds a mixed initiative dialogue by asking and answering questions, giving hints, filling in missing pieces of information, and correcting misconceptions. Some versions of AutoTutor have speech recognition. Other versions have interactive three-dimensional simulations in order to enhance embodiment, the learner's engagement, and hopefully depth of the learning. At this point, AutoTutor has been used to tutor college students on such abstract topics as computer literacy, Newtonian physics, and critical thinking.

Our AutoTutor project provides a vivid backdrop for framing some of the issues for debate in this book. Our goal is really to stir questions and uncertainty more than to answer questions and offer solutions. Our research team has spent a decade building what many in the field of computer science call an 'embodied' conversational agent, a computer system that attempts to comprehend, think, speak, and perform actions. We have struggled with alternative theories of representation, discourse, adaptation, and planning in our efforts to build a reasonable conversational agent that mimics human tutors. Some of the theoretical models we have used lean toward the embodied end of the spectrum and others to the symbolic end. Our reflections raise some questions that readers will hopefully find worthy of discussion.

This paper is divided into three sections: Section 3.2 gives a snapshot of what AutoTutor is and does, Section 3.3 addresses the question of whether it is possible to have a computer generate and interpret cognition, action, and emotions, and Section 3.3 identifies some components of AutoTutor that are most naturally construed as symbolic and that might present some challenges to those who are pursuing an embodied theoretical framework. However, when push comes to shove, we are really inviting the following question: in what sense is a component embodied versus symbolic? If successful, we will plant some uncertainties in the debates.

It is important to briefly clarify the sense in which we consider a representation as being embodied. We consider *strong embodiment* as existing to the extent that the representation incorporates the constraints of an organism's body, its location in the world, its perspective in perceiving the world, and its interactions with the world. So perceptions, actions, and point of view are richly represented and are systematically aligned with perceptual–motor interactions in the world. We consider *weak embodiment* as existing when there are vestiges of perceptions, actions, and perspectives in the representation, but the components are less detailed or underspecified, yet to some extent systematic or recoverable. A representation is *not embodied* when the symbols have an arbitrary relationship with the various components of potential perceptual–motor interactions with the world, be they real or simulated.

It is also important to clarify the sense in which a representation is symbolic. A symbolic representation is a structured set of symbols, each of which stands for some aspect of a referential domain. What it stands for may or may not be embodied. Whereas a *modal* symbolic representation has links to perceptual–motor experience, either indirectly through interpretive mechanisms or directly through sensory transduction and motoric actuators, an *amodal* symbolic representation is not grounded in any embodied representation. It follows from our definitions that embodied and symbolic representations are not necessarily mutually exclusive; amodal symbolic representations are clearly different from strong and weak embodied representations. However, there is no clear-cut distinction between modal symbolic representations and weak or strong embodied representations. Indeed, many researchers have developed symbolic models of perceptual and motor representations that are situated in real-world scenarios. It may take breakthroughs in neuroscience to differentiate modal symbolic representations from embodied representations; however, it is beyond the scope of this chapter to consider the constraints of the brain and neurons.

3.2 **What is AutoTutor?**

AutoTutor is a computer tutor on the internet that simulates human tutors and that helps individuals learn about abstract difficult topics by holding a conversation in natural language (Graesser *et al.* 1999, 2001, 2004; Graesser, Chipman *et al.* 2005). Learners express their contributions through a keyboard or through speech, whereas AutoTutor communicates through an animated conversational agent with speech, facial expressions, and some rudimentary gestures. AutoTutor presents a series of challenging questions that requires deep reasoning and approximately a paragraph of information for an ideal answer, assisting the learner in the evolution of a complete answer from their initial response, which is normally very short (typically from one word up to two sentences). It tries to draw out more of the learner's knowledge (through hints and prompts), helps fill in missing information, repairs misconceptions, and answers student questions. The dialogue between AutoTutor and the student takes approximately 100 dialogue turns to answer a single challenging question, approximately the length of a conversation with a human tutor.

Perhaps the best way to convey what AutoTutor can do is through an example. Consider the conceptual physics problem below:

When a car without headrests on the seats is struck from behind, the passengers often suffer neck injuries. Why do passengers get neck injuries in this situation? Explain why.

The tutorial dialogue of AutoTutor is guided by the interface shown in Figure 3.1. The main question is presented in the top window and remains at the top of the page until it is finished being answered during a multi-turn dialogue. The students use the bottom-right window to type in their contributions for each turn, although speech recognition is being tested in a current version of AutoTutor. The animated conversational agent resides in the upper-left area. The dialogue history between AutoTutor and the student is shown in the bottom-left window (this is absent in versions with speech recognition). The version of AutoTutor in Figure 3.1 has an embedded interactive three-dimensional simulation that depicts a physics micro-world with people and objects in a spatial scenario. The student can manipulate parameters of the situation (e.g., mass and speed of vehicles), observe a simulation of what will happen, repeat this process of manipulation and

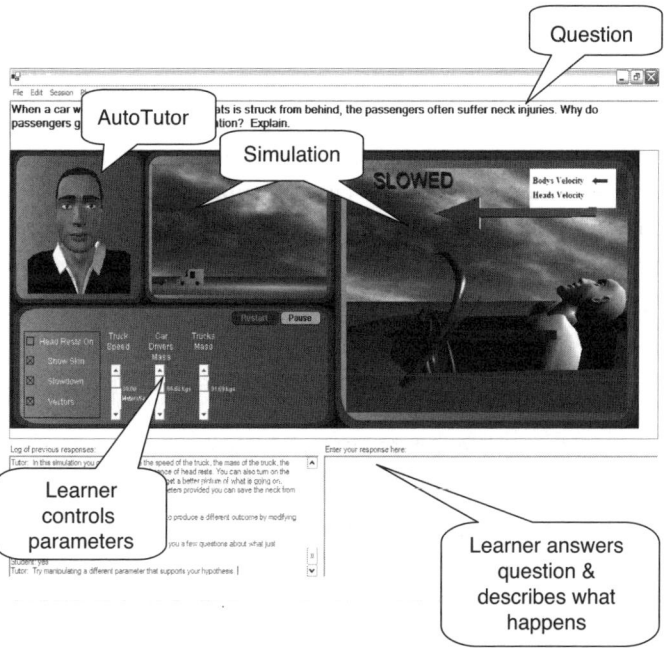

Fig. 3.1 A computer screen of AutoTutor on conceptual physics with interactive three-dimensional simulation.

observation several times, and explain what they see. The interactive simulation grounds the physics knowledge in perceptual and motor activities, one of the hallmarks of embodied theories (Barsalou 1999; Glenberg 1997; Glenberg and Kaschak 2002; de Vega *et al.* 2004; Zwaan 2004). Deep learning of physics is believed to emerge from the combination of interactivity, perceptual simulation, feedback on the simulation, descriptions of what happens, and explanations.

It is important to point out that flawed reasoning is often manifested during the answering of these conceptual physics questions. Adults are often seduced by the naïve physics that explain their perceptions of everyday events in the world and their everyday actions that affect the world (di Sessa 1993; Hunt and Minstrell 1996; Ploetzner and VanLehn 1997). For example, many college students believe that a rear-end collision, as in the Figure 3.1 example, will directly push the head of the victim forward – at times, unfortunately, through the windshield. Their mental model is possibly that the head goes forward much like a billiard ball goes forward when hit from behind, or perhaps they have a memory of a person's head going through the windshield from a movie or personal experience. However, this is flawed reasoning. The head initially goes backwards after the impact of the collisions by virtue of the forces underlying Newton's laws, which explains whiplash in accidents. The head subsequently goes forward after a recoil. Deep comprehenders identify the initial stage of the head going back when asked to draw pictures (Lu, personal communication), but shallow comprehenders miss this step.

It is not until the student can reason with abstract vectors, forces, and Newton's laws that a correct answer emerges. Table 3.1 lists some of the good answers (called *correct principles*) and the misconceptions that are affiliated with this rear-collision problem. It is a stretch to imagine how an embodied theory could go the distance in accounting for the full set of correct principles, whereas many of the misconceptions are grounded in embodied everyday experience. Perhaps the correct principles can be captured in a mental model that is embodied through educational activities or through a proxy representation that is enriched with a weak sense of embodiment. For example, the force dynamics of Talmy (1988) is a promising approach that may lead to a weak embodied account of the correct constraints of physics. However, no one that we know has come forward in specifying an embodied version of Newtonian physics that adequately

Table 3.1 Expectations and misconceptions associated with the rear-collision problem

Correct principles	The magnitudes of the forces exerted by A and B on each other are equal.
	If A exerts a force on B, then B exerts a force on A in the opposite direction.
	The same force will produce a larger acceleration in a less massive object than a more massive object.
Misconceptions	A lighter/smaller object exerts no force on a heavier/larger object.
	A lighter/smaller object exerts less force on other objects than a heavier/larger object.
	The force acting on a body is dependent on the mass of the body.
	Heavier objects accelerate faster for the same force than lighter objects.
	Action and reaction forces do not have the same magnitude.

captures a host of abstract constraints. This is very much an example of how abstract symbolic reasoning can transcend the limitations of embodied cognition and can empower humans with forms of intelligence that move us beyond what used to be called the infra-humans. Symbolic reasoning can often emancipate humans from being prisoners of naïve physics and of exclusive reliance on embodied thought.

Table 3.2 shows excerpts from an example tutorial session between a college student and AutoTutor on Newtonian physics. The conversation begins with a main question selected by AutoTutor from the repertoire of questions in its curriculum. The example illustrates the mixed-initiative dialogue of AutoTutor. AutoTutor attempts to accommodate virtually any student question, assertion, comment, or extraneous speech act. Each turn of AutoTutor requires the generation of one or more dialogue moves that adaptively responds to what the student just expressed and that advance the conversation in a constructive fashion. The dialogue moves of the tutor within a turn are connected by discourse markers (see Louwerse and Mitchell 2003), as illustrated in the example excerpts. Some dialogue moves are very responsive to the student's preceding turn, such as short feedback (positive, neutral, versus negative), answers to student questions, and corrections of student misconceptions. Other dialogue moves push the dialogue forward in an attempt to cover the correct principles in the answer to the main question. These forward-directed dialogue moves include hints, prompts for the student to fill in missing words, assertions of correct principles, and requests for the student to engage in an interactive simulation.

The pedagogical framework of AutoTutor was inspired by two bodies of theoretical and empirical research that go beyond the normal ITSs. These include explanation-based constructivist theories of learning (Aleven and Koedinger 2002; VanLehn et al. 1992) and empirical research that has documented the collaborative constructive activities that routinely occur during human tutoring (Fox 1993; Graesser et al. 1995; Chi et al. 2001). According to the explanation-based constructivist theories of learning, learning is more effective and deeper when the learner must actively generate explanations, justifications, and functional procedures than when being lectured or reading text. Regarding the empirical research on tutorial dialogue, the patterns of discourse uncovered in naturalistic tutoring were imported into the dialogue management facilities of AutoTutor. The discourse patterns were not confined to verbal content, but also included facial expressions and gestures with the hand, head, and body. It is well documented that there is substantial communicative and pedagogical value in facial expressions and gestures (Alibali and Nathan 2005; Graesser et al. 1995; Nathan, Chapter 18, this volume).

A few words should be devoted to the question of whether AutoTutor improves learning. One plausible prediction would be that learning is facilitated by embodied cognitive activities. We have found that the version of AutoTutor without interactive simulation has been shown to improve learning by nearly a letter-grade (effect size averages 0.8 sigma) in 12 experiments on physics and computer literacy (Graesser et al. 2004; VanLehn et al. 2007). In these assessments of learning gains, post-test scores from learning with AutoTutor were compared with reading a textbook, with pretests, and a variety of other comparison conditions (Graesser et al. 2003; VanLehn et al. 2007).

Table 3.2 Excerpts from an example dialogue with AutoTutor

AutoTutor: When a car without headrests on the seats is struck from behind, the passengers often suffer neck injuries. Why do passengers get neck injuries in this situation? Explain why. (MAIN QUESTION)

Student: *I don't know.* (This expression by the student is classified as a Metacognitive speech act because it conveys what the student knows about his/her own knowledge.)

AutoTutor: Could you at least try to give me an answer? That way, we'll have something as a starting place to work from. (PUMP. This dialogue move encourages the student to say more, but does not direct the student to express particular information.)
[At this point, the dialogue continues until AutoTutor believes it has detected that an important expected piece of information is missing.]

AutoTutor: Let's try this together. (DISCOURSE MARKER. This is a discourse marker that connects different dialogue moves within a turn.) When the collision provides a large force to accelerate the head, what could happen to the neck? (HINT. This dialogue move leads the student to articulate a particular sentence, clause, or proposition.)
[At this point in the dialogue the student asks a question so AutoTutor needs to provide an answer.]

Student: *What is a force?*

AutoTutor: I'll take a crack at answering your question. (DISCOURSE MARKER) Force is any influence that tends to accelerate an object; a push or pull; measured in newtons. Force is a vector quantity. (ANSWER. AutoTutor answers the student's direct question. Answers are either accessed from electronic glossaries or a paragraph is dynamically selected from a textbook.) Do you have another question? (QUESTION SOLICITATION. AutoTutor allows the student the opportunity to rephrase his question.)
[At this point, the dialogue continues until AutoTutor believes it has detected a misconception]

AutoTutor: See if you know the answer to this. (DISCOURSE MARKER. AutoTutor prepares the student for a direct question.) When the head and body are moving at the same acceleration, they are moving at the same _____? (PROMPT. This dialogue move encourages the student to fill in a particular word or phrase of information, in this case the word 'rate.' Prompts are needed for students who are low verbal and to elicit a particular noun, verb, or adjective.)

Student: *Distance*

AutoTutor: Moving at the same rate. (PROMPT COMPLETION. AutoTutor splices in the correct information if the student provides an incorrect answer to its prompt.)
[At this point, the dialogue continues until AutoTutor launches an interactive simulation]

AutoTutor: Let's take a look at a simulation that models real-world physics. (DISCOURSE MARKER) This simulation demonstrates what will happen when a car without headrests on the seats is struck from behind; the passengers often suffer neck injuries. In this 3D simulation environment, you can see the physics that occurs in reality. (SIMULATION INTRODUCTION. AutoTutor introduces the student to the simulation environment as the simulation is displayed in the interface. This move smoothly transitions the student to the new mode of interaction.)
[At this point, the dialogue continues until all relevant principles are covered in the conversation, followed by a summary by the tutor]

One intriguing finding was a comparison in learning gains among three conditions: (1) learning from AutoTutor, (2) reading a textbook for a period of time yoked to the time interacting with AutoTutor, and (3) no training. Learning gains showed the following pattern: AutoTutor > reading textbook = no training = 0. AutoTutor training was significant, as we had expected. Perhaps more interesting was the result that reading the textbook was no different than doing nothing at all. This suggests that normal reading of physics from a textbook either (a) does not afford embodied cognitive activities or (b) does afford embodied cognitive activities, but such activities do not produce deep learning of physics. This same pattern of data was found for the acquisition of computer literacy (Graesser *et al.* 2004).

Our recent experiments on conceptual physics have compared the interactive three-dimensional version of AutoTutor (AutoTutor-3D) with the normal conversational program (Jackson *et al.* 2006). The student has the freedom to interact with the three-dimensional world several times per scenario by manipulating values of the parameters (e.g., mass, speed) and seeing what happens. This condition was designed to maximize the extent of embodied cognitive activities. We found that the difference between AutoTutor-3D and the normal conversational program was subtle (sigma = 0.20) and nonsignificant. This result was somewhat disappointing because we had hoped that intensifying embodied activities through three-dimensional interactive simulation would improve learning over and above the normal conversational AutoTutor. Perhaps normal AutoTutor affords embodied cognitive activities, so interactive three-dimensional simulation has small incremental gains.

The subtle effect of the three-dimensional simulation might also be explained by the fact that most of the students did not use the simulation very much. Students who *did* make a minimum number of simulation manipulations showed positive correlations between their learning outcomes and two important simulation variables: the number of simulations they received ($r = 0.51, p < 0.01$) and the relevance of their manipulations to the problem ($r = 0.47, p < 0.05$). These positive correlations could be readily explained by the embodied framework. However, they could also be explained by a host of third variables, such as intelligence, persistence, and motivation.

The disappointing results of the three-dimensional simulation are compatible with some claims that are relevant to the embodiment–symbolic debate. It may be insufficient to simply ground the physics in a visual display that has affordances to manipulate parameters – and then 'hey presto!', learning occurs. Many users have no wisdom on what to do next and how to effectively manipulate the parameters in route to learning. They need modeling and coaching for effective self-regulated inquiry and for intelligent use of simulation environments (Azevedo and Cromley 2004; Lee 1999; Winne 2001). It is also conceivable that much of the coaching can be accomplished by periodically having AutoTutor make abstract suggestions, hints, or requests, e.g., 'consider force vectors' or 'you need to run some additional simulations to rule out alternative hypotheses'.

Embodiment theorists would presumably argue that AutoTutor needs to model the strategies by embodied demonstrations, whereas disembodied strategists would suggest that abstract hints provide a good alternative. These alternative explanations need to be explored further in future research. It suffices to say, however, that embodied experience does not routinely spawn deep learning, and may even be the source of persistent misconceptions. It may sometimes take abstract strategies to elevate us to deep learning (see also Goldstone and Sakamoto 2003).

Early versions of AutoTutor were designed to adapt to learners' cognitive states and indeed were shown to dynamically adapt to the knowledge states of individual learners –no two conversations with AutoTutor are ever the same. Our most recent work, in collaboration with Roz Picard at the Massachusetts Institute of Technology (MIT, Cambridge, MA), attempts to identify the students' affective states (emotions) during a typical learning experience and to respond adaptively to the students' affect states (D'Mello *et al.* 2007). When the learner gets frustrated, AutoTutor provides a hint or prompt that gets them unstuck; when the learner gets bored, AutoTutor needs some razzle dazzle; when the learner is excited and engaged in a flow experience (Csikszentmihalyi 1990), AutoTutor should stay out of the learner's way and let the learner take control; and when the learner is confused, there are a number of possible paths for AutoTutor. A chronically confused learner needs guided scaffolding, whereas a temporarily confused student may need time to self-regulate their understanding. Confusion occurs when the learner is in cognitive disequilibrium and deep thinking is to be encouraged (Craig *et al.* 2004; D'Mello *et al.* 2007).

AutoTutor attempts to track emotions through different channels of non-intrusive sensing devices. AutoTutor has been successful in accurately classifying learners' affect states by mining: (1) the tutorial dialog for patterns of verbal interaction that reflect emotions, (2) video capture of the face, (3) body posture, and (4) parameters of speech (D'Mello *et al.*. 2007). The facial sensing devices were developed in Roz Picard's laboratory at MIT (Kapoor and Picard 2002; Picard 1997, 2001), whereas the sensing algorithms for the other channels were developed by the AutoTutor research group at Memphis.

Facial expressions and body motions are obvious vestiges of embodied thought. Ekman's facial action coding system (Ekman 2003) has specified elementary facial actions (e.g., left eyebrow raise, corner of the mouth upward) that are correlated with emotions. We have identified the facial actions associated with confusion, delight, boredom, surprise, flow, and frustration. For example, confusion has a brow lowerer (action 4) simultaneously with an eyelid tightener (action 7), whereas delight has a lid tightener and a lip-corner puller (action 12). Regarding body movements, our body pressure measurement system records how the learner sits in the chair, via what we affectionately call the 'butt and dorsal-lumbar sensor'. Learners are on the edge of the chair with the head near the monitor during the affective states of curiosity, interest, and flow/engagement. In contrast, learners are either fidgety or are slumped back with the head far away from the monitor in states of boredom and disinterest. Now that we understand how the different channels are correlated with emotions associated with learning (e.g., frustration, confusion, boredom, excitement, flow/engagement, delight, and surprise), we are in the process of designing AutoTutor to respond with the appropriate conversational, emotional, and pedagogical dialogue that scaffolds the learners to new levels of learning and motivation.

The relations between emotions and the actions of the face and body are most naturally construed through an embodied framework. It is a stretch to imagine that a symbolic framework would provide any explanatory account of such relationships. In contrast, the verbal dialogue history also predicts emotions and is most naturally explained by symbolic models. The aspects of dialogue that predict emotions include feedback to the learners (positive, negative, neutral), the directness of AutoTutor's feedback (e.g., indirect hints versus direct assertions), and the quality of the student's contributions (D'Mello *et al.* 2006, in press). As one might expect, emotions are best predicted by a combination of features from embodied and symbolic foundations (D'Mello *et al.* 2007).

It should be quite apparent that the AutoTutor project covers a broad landscape of embodied and symbolic representations. AutoTutor's mechanisms are neither fully embodied nor fully symbolic, but rather a hybrid. AutoTutor's impact on learning processes and learning gains is partially embodied and partially symbolic. We next turn to the question of whether computer models can capture the embodied representations and activities.

3.3 Can a computer simulate embodied representations and activities?

One of the initial questions that motivated the debate is whether a computer is capable of simulating embodied mechanisms. Psychologists have typically equated computer models with amodal symbolic models because the computers being used are symbolic at some point in the system's implementation. However, we prefer a broader definition of computer modelling that takes a step up from this implementation level. We are primarily interested, to use Marr's (1982) terminology, in the level of 'computational theory' and the level of representations and algorithms. Many computational theories, representations, and algorithms in computer systems are not amodal, but rather attempt to capture the constraints of embodied mechanisms. It is possible to have a symbolic model that attempts to capture vision, action, and many other modalities. For the present purposes, we will define computer models as including amodal symbolic models, modality-specific symbolic models, neural networks, statistical models, and dynamical systems. Some colleagues view all of these as symbolic models whereas others include only the amodal symbolic models as symbolic models. We will not attempt to defend either of these positions on the matter of the meaning of symbolic, but we do want to accept a broad definition of computer model.

One position is that it is possible to simulate any complex mechanism with a family of computer architectures if there is enough design time and sufficient computational resources to do so. A second position is that biological mechanisms are constrained in ways that make it impractical or impossible for computer simulation. There is a third position – one that we find very worthwhile to explore – namely whether the computer models have characteristics that are naturally aligned with the phenomenon being simulated. If the computer models are *ad hoc* in accounting for the mainstream empirical findings affiliated with the phenomenon, then such models are limited and hardly explanatory. However, if the computer models generate data, particularly counterintuitive data, that are naturally aligned with the phenomenon, then the computer models would be persuasively explanatory. On the flipside, if an embodied model manages only *ad-hoc* fits to the data or indecisive predictions, than the value of such a model is limited.

Consider the phenomenon of whether a computer can simulate the bodily actions (torso, arm, fingers, face) of a person that would be needed in any embodied theory of cognition. When we started our AutoTutor project, many of our colleagues were very skeptical of this proposition; now the skepticism has waned. We would argue that designers of animated conversational agents have made major strides in simulating the facial expressions, bodily actions, and gestures of agents (Cassell and Thorisson 1999; Cassell *et al.* 2001; Cole *et al.* 2003; Gratch *et al.* 2003; Johnson 2001; Johnson *et al.* 2000; Massaro and Cohen 1995; Quek 2004; Rickel and Johnson 1999). All of these systems use modality-specific symbolic and statistical models that are implemented on computer.

The question arises how well the mechanisms of the animated agents are aligned with the biological mechanisms. This is a very important question, from our point of view. Let's consider three alternative families of animated agents, which we will label *kinematics agents, aesthetic agents*, and *engineered facsimile agents*. Kinematics agents have facial and bodily motions guided and constrained by the physical layout of the skeletal and cartilage structure, and the dynamic activities of the muscles. This class includes agents governed by the facial action coding system of Ekman (2003) and by kinematics models of body movement (e.g., Gratch *et al.* 2002 and several others).

Aesthetic agents follow algorithms that are not inspired by kinematics but by stylistic constraints, such as the agents designed by Disney, Dreamworks, and movie production companies in the entertainment industry. Rather surprisingly, Disney characters are rendered from a separate tradition that is insulated from kinematics. Dreamworks' *Shrek* animated character, for example, has three fingers rather than four and a gait that is stylistically heart-warming but would play havoc on a human spinal cord. Simply put, kinematics and Disney are not at all married to one another. Instead, Disney has the mission of generating characters that are interesting, cool, endearing, and sometimes discrepant with normal kinematics principles.

And what about the engineered facsimile agents? These evolve by examining a corpus of body movements, inducing patterns of movements, and copying these patterns into animated conversational agents. The movements of these computer agents are similar to their human counterparts, but the theoretical principles underlying these mirror algorithms are uncertain: engineering solutions use *ad hoc* methods to emulate surface dynamics.

We would argue that the kinematics agents are the proper candidates for simulating embodied action if we want to make the claim that animated agents simulate biological mechanisms. The aesthetic agents are governed by different principles, namely art and aesthetics, whereas engineered facsimile agents follow *ad hoc* constraints uncovered inductively from data mining methods. If we are correct in our claims, there is no crisp answer to the question of whether animated conversational agents simulate embodied action, it depends on the mechanisms and constraints of the computer model. We would need to examine the correspondence between the model and biological constraints.

It is important to acknowledge that symbolic models have the capacity for a fine degree of granularity and adaptation to the environment, particularly when they incorporate the soft constraints of statistical and fuzzy systems (Anderson *et al.* 1995; Just and Carpenter 1992; Kieras and Meyer 1998; Kosko 1992; Laird *et al.* 1987; Steels 2003; Roy, in press). Standard planners in robotic systems can invoke goal-driven procedures that decompose actions into small micro-steps that are so smooth and ballistic that the discrete operations are imperceptible to the human eye (Breazeal 2003; Gratch *et al.* 2002). Similarly, cybernetic modules that respond to the world can be very fine-grained. Our claim, therefore, is that psychological data that point out continuous gradients and ballistic actions can be handled by symbolic models, and are not the unique province of biological models. There are no principled limits to the grain-size and time-windows of symbolic computational models – this is not the place to look for ways to differentiate embodied and symbolic models.

The perception and interpretation of embodied activities can also be modelled by computer systems; there already exist a number of systems that attempt to automatically detect facial expressions, gestures, posture, and actions (e.g., Chu and Cohen 2005; Kapoor and Picard 2002; Kettebekov and Yeasin 2005). It would be an exaggeration to claim that these systems are robust and well-developed, but it is fair to say that consistent progress has been made in recent years. Moreover, some of these computerized classifiers are explicitly built with an architecture that simulates biological and neural systems.

Our AutoTutor team is currently exploring a biologically-inspired computational model that attempts to integrate input from five channels (dialogue, face, body posture, keyboard, and speech). The *K-set hierarchy* is a biologically inspired model of neural population dynamics developed by Freeman and colleagues (Kozma and Freeman 2001). Kozma and Freeman designed and parameterized a model that replicates experimentally recorded data of the dynamic behaviour in the olfactory system of rabbits. The olfactory system is interesting because there are no static features that can discriminate odours; instead, a complex dynamical system is needed. Models in the K-set network simulate the temporal and spatial properties of neural populations. One of the K-set models, the K-III network, operates as a classifier by creating low-dimensional local basins of attraction – each basin associated with each class – forming a global attractor landscape. These basins constitute dynamic memories and can be used to solve difficult classification problems in which the data are not linearly separable. The K-III network learns by Hebbian reinforcement, habituation, and normalization. A KA model developed by Harter (2004) is an engineering simplification of the original K-set hierarchy. A deterministic discrete-time model, its discrete nature is achieved by replacing the differential equations in the K-set model with difference equations, allowing the model to run three times faster.

There are several points to be made by introducing this K-set family of computational models. First, they are biologically-inspired computer models that attempt to handle the neural dynamics of complex classification. Second, the models can perform classification on datasets that are not linearly separable in real time. Third, the matter of whether the models are discrete or continuous is not a differentiating property of the success of the models; indeed, it is arguably an irrelevant characteristic. Perhaps our most general point is that no one has convincingly demonstrated that biological or neural systems are inherently unable to be simulated by computers, including those with symbolic architectures.

3.4 Components of AutoTutor and comprehension: embodied or symbolic?

AutoTutor is a complex system with several computational components, so it is beyond the scope of this paper to systematically scrutinize all of the components from the perspective of the embodied versus symbolic debate. Instead, we will focus on a few components that engender uncertainty.

3.4.1 AutoTutor's comprehension of the learner's verbal contributions

AutoTutor attempts to comprehend the learner's contributions by segmenting each turn into speech act units, classifying each unit into one of 22 categories, and semantically interpreting most of the speech acts. Of particular interest is the semantic interpretation of those speech acts that are classified as student assertions. AutoTutor does not use any embodied models of comprehension when interpreting student assertions. Instead, it uses *latent semantic analysis* (LSA) (Landauer and Dumais, 1997; Kintsch 1998; Landauer *et al.* 2007) and a metric that computes content word overlap in a manner that inversely weights words according to frequency of usage in the English language. In essence, statistical models in computational linguistics (Jurafsky and Martin, 2000) are used to evaluate the extent to which the student assertions cover the content of the correct principles and misconceptions (see Table 3.1). If a student assertion (or combination of assertions) has a high conceptual overlap with a particular correct principle, that principle is counted as covered. If there is a high overlap with a misconception, then AutoTutor immediately corrects the misconception. Therefore, pattern-matching mechanisms are executed by AutoTutor rather than a deep, embodied interpretation of student assertions.

It could be argued that AutoTutor's reliance on semantic pattern matching is a serious limitation. However, it is not obvious what alternatives would be feasible to test. Attempts to augment AutoTutor's interpretation mechanism with sophisticated syntactic parsers and structured semantic analysers have had modest incremental value in improving the accuracy of the pattern matches (Graesser *et al.* 2006). Unfortunately, most verbal expressions of the learner cannot be successfully interpreted by recent systems that have been developed in the fields of information extraction, computational linguistics, and artificial intelligence. There are no automated computational models of embodied comprehension, so there is no foundation to test an embodied theoretical framework. Indeed, all of the embodied theoretical frameworks of comprehension have been restricted to small families of sentences, generated by the experimenters, with a special composition to test particular theoretical claims (de Vega *et al.* 2004; Glenberg 1997; Glenberg and Kaschak 2002; Zwaan 2004). No one has assessed how well these embodied frameworks would scale up to handle a large corpus of naturalistic printed texts or a large corpus of naturalistic utterances (see de Vega *et al.* 2002 for an example of how this might be conducted). Our impression, based on informal attempts to analyse naturalistic corpora, is that embodied frameworks would not fare well in scaling up because the scope of the phenomena they explain is quite limited. *A priori* predictions are too sparse at this point in the science to accommodate a naturalistic corpus. However, we may be incorrect on this matter, so empirical corpus analyses are invited.

There are two fundamental challenges in applying embodied frameworks to interpreting learner assertions. The first challenge is that a very large proportion of the learner's turns are beset with poor grammar, speech disfluencies, vagueness, ambiguity, underspecification, and other characteristics that are not semantically well formed. Similarly, the world knowledge underlying the turns is open-ended, fragmented, ambiguous, and often contradictory. For example, consider the following naturalistic excerpt from an actual tutoring protocol on the topic of research methods:

Tutor: Do you know how to get the main effect?

Student: Yeah, and I understand that like its what the independent variable itself like a measurement of it itself. And um I know like that looking at these I know how to get it.

Tutor: Okay

It's not obvious how an embodied theoretical framework would offer advantages over the semantic pattern-matching algorithms of AutoTutor when interpreting the student's contribution, yet spoken utterances like these are ubiquitous in spoken tutoring sessions.

The second challenge is that abstraction, simplification, and disembodied representations are desired when interpreting student assertions because the tutor typically has only an approximate understanding of what the learner knows (Graesser *et al.* 1995). Human tutors and tutees share such little common ground that the two parties need to struggle to have hooks in understanding each others' worlds. Indeed, they frequently live in very different conceptual spaces, so there are very few alignments in conversational levels (Pickering and Garrod, 2004). This fact limits how stringent the criterion can be when determining whether each party of the conversation understands the other. We are convinced that adults comprehend only as deeply and thoroughly as the information becomes available and as the situation affords (Sanford and Graesser, 2006; Sturt *et al.* 2004). If AutoTutor insisted on constructing embodied representations, assuming that it could, then the system would be spinning its wheels most of the time to no avail. A good tutor knows how to manage the vagueness and underspecification of student contributions. This may be accomplished in a computer by fuzzy rules or by a more complex management of the dialogue in a fashion that is strategic, symbolic, and disembodied (Larsson and Traum 2000; Lochbaum 1998).

In conclusion, our core claim in this subsection is that AutoTutor's disembodied mechanism of interpreting the learner's contributions is more aligned with human tutoring than would be offered by an embodied theoretical framework. We would speculate, however, that an embodied framework would become progressively more valid to the extent that: (a) the common ground between the tutor and tutee increases, and (b) the expertise of the tutor and tutee increases. Under these conditions, the stakes and standards of deep, precise comprehension are higher. Embodiment would be one mechanism for enhancing the depth and precision.

3.4.2 **The learner's comprehension of AutoTutor's contributions**

In our view of the tutoring process, learners do not routinely comprehend the conceptual physics problems at a deep or embodied level. Nor do they do so when they comprehend AutoTutor's contributions, turn after turn. Tutors and teachers may want them to, but they don't. Instead, we are convinced that learners typically settle for shallow standards of comprehension (Baker 1985) and sustain an illusion of comprehension (Glenberg and Epstein 1985). They are satisfied if they can recognize the meaning of content words (nouns, adjectives, verbs) and can interpret sentence fragments. Moreover, the deep meaning of a conceptual physics problem does not come linearly and immediately in a series of well orchestrated flashes of insights, but rather unravels in hermeneutical cycles during struggles of comprehension (Winograd and Flores 1983). The referents of referring expressions and the meanings of clauses at time t_1 often end up being undone and reinterpreted at time t_2. The flow of comprehension moments is more akin to anarchy and opportunistic planning than it is to a well-timed, coordinated, embodied Swiss orchestra.

It is an understatement to say that the level in which most college students understand the conceptual physics problems is disappointing. Deep comprehension of this difficult material is not routinely achieved while reading text, as discussed earlier. Reading facilitates shallow comprehension, such as the meaning of key terms, properties of concepts, isolated facts, and other types of declarative knowledge; however, deep comprehension rarely occurs from reading. Deep comprehension requires problem solving in the face of obstacles to goals, contradictions, anomalies, and other events that place the learner in cognitive disequilibrium (Craig *et al.* 2004; Graesser, Lu *et al.* 2005). Multiple cycles of impasse, struggle, reinterpretation, and other nonlinear and non-monotonic mechanisms occur when most college students attempt to comprehend the conceptual physics problems.

If our analysis of the comprehension processes is correct, we believe it is unlikely that our college students construct a fully embodied representation of the physics problems and AutoTutor's turns during the tutoring process. Instead, they have partial, barely-embodied specifications of most referring expressions, clauses, and propositions. If the representations are embodied, the representations are anaemic rather than rich, weak rather than strong. The representations grow over multiple cycles but it is unclear how embodied the representations are. There may be periodic snapshots of perceptual images and motor activity, but it is doubtful that there is a coherent flow of mental choreography.

Of course, we may be incorrect. An alternative viewpoint is that the students immediately begin constructing a strong embodied representation of the conceptual physics problems. The embodied representations are often flawed with knowledge from everyday life. These embodied flaws are sometimes detected by the learner, which stimulates the achievement of a deeper understanding after multiple cycles and re-piecing of the correct knowledge, be it embodied or symbolic. So there is a question of whether the embodiment cup is half empty or half full: no one knows.

More generally, there are three fundamental challenges for the embodiment framework in explaining discourse comprehension. The first challenge is that it has difficulties scaling up in explaining how embodied representations can be constructed at a normal reading rate of 250–400 words per minute. For those in doubt, it is worthwhile to perform an exercise on a text that computes how long it would take to read it, based on available estimates of the time it takes to perform basic operations of embodied cognition. It takes approximately 300–1000 milliseconds to construct a new referent in a discourse space, several hundred milliseconds to move an entity from one location to another, a few hundred more milliseconds to have the mental camera zoom in on an entity within a crowded mental space, and so on (Millis *et al.* 2001). When there are multiple mental spaces, the representations potentially interfere with one another, which adds additional processing time. Research on mental models also suggests that it is difficult to envisage multiple mental models simultaneously. These considerations on timing and complexity make it doubtful that all referring expressions and clauses in the text have fully embodied representations during reading. It would take an order of magnitude increase in reading time to achieve this. A satisfactory resolution of this theoretical challenge to embodied frameworks would require researchers to perform systematic analyses analogous to a Goals, Operators, Methods, and Selection analysis in the human–computer interaction field (Card *et al.* 1983), a methods-time measurement analysis in the human factors field, or a task analysis with chronometric estimates of embodied operations during reading. Our intuition is that the embodied theoretical framework is unlikely to scale up to account for normal reading speed, but perhaps we are incorrect. That is an empirical question.

Assuming that our above concern is on the mark, the backup claim for an embodiment framework is that only some referring expressions and clauses (or propositions) end up being embodied during comprehension. This is precisely the position that we advocate. We believe that the embodied framework is essentially correct under the following conditions: (1) when tasks and tests encourage embodied activities, (2) when the stimuli are simple, e.g., few actors and objects in the mind's eye, (3) when there is visuospatial grounding, e.g., an established spatial layout, and (4) when there is sufficient time and cognitive resources to carry out these processing operations. These are the conditions in which the embodied theories reign supreme. Moreover, we believe that the content in the discourse focus (Grosz and Gordon 1999; Grosz and Sidner 1986) is an excellent candidate for being a recipient of such cognitive activities. In contrast, we believe that disembodied representations are more explanatory when the relevant task goals do not encourage embodied processing, when the stimulus is complex, when there is minimal visuospatial grounding, and when the reading rate is at the fast end of 250–400 words per minute. Much of the content that is presupposed and highly embedded will not be a good candidate for becoming a fully fleshed out embodied representation.

Our above analysis addresses the relatively time-consuming integration phase of Kintsch's construction integration model (Kintsch 1998) rather than the initial activation of representations. It is possible to have quick activations of many types of representations, both embodied and symbolic, during the initial activation of information associated with content words. Much of this automatic activation of representations end up dying away and never makes it to the integration phase that establishes a more coherent representation of the meaning of the text. Experiments that demonstrate a quick activation of embodied representations are therefore not pertinent to our current argument.

A third challenge for the embodiment framework is that it does not acknowledge the enormous value of symbols and abstractions. There are benefits to abstract representations (Goldstone and Sakamoto 2003), symbols (Koedinger *et al.* 2005), and minimally specified representations (Sanford and Graesser 2006; Sturt *et al.* 2004) in both text comprehension and learning. Symbolic representations and operations are sometimes faster, more accurate, and more useful than embodied representations. It is important to acknowledge the value of abstract representations, but to be fair to the embodiment theorists, the existence of abstract concepts does not *prima facie* challenge the generality of embodiment theories. Barsalou's work on perceptual symbol systems (Barsalou 2003; Barsalou and Wiemer-Hastings 2005) offers some evidence that a weak form of embodiment underlies many of the abstract concepts.

3.4.3 Embodiment in emotions and deep thinking

As discussed earlier, our current version of AutoTutor attempts to track the affective states of the learner on the basis of the dialogue history, facial expressions, body position, keyboard activities, and speech. We have performed three types of experiments that have tracked the learners' emotions while interacting with AutoTutor. One set of experiments simply had a trained judge record the emotions of the learner at random points during learning (Craig *et al.* 2004). A second set had college students 'emote aloud' while interacting with AutoTutor; we recorded their expressed emotions as well as their facial actions and AutoTutor's dialogue moves immediately before, during, and after a verbalized affect state (D'Mello *et al.*, in press). A third set of experiments first had college students learn with AutoTutor (uninterrupted) and subsequently replayed tapes of the interaction while the learner, a peer, and an expert made judgements about emotions at sampled points during the tape (D'Mello *et al.* 2007). Pre- and post-tests were administered on the tutored material so that we could assess learning gains and relate such gains to the incidence of affect states.

Our research uncovered some interesting trends in relating affect, embodiment, and learning. For example, we found that confusion was the most prominent emotion that correlated positively with learning gains. Deep thought is affiliated with confusion and is symptomatic of cognitive disequilibrium (Graesser, Lu *et al.* 2005). The face normally shows systematic movements during confusion, with changes in eyebrows, eyelids, and corners of the lip, as discussed earlier. Confusion normally occurs right after AutoTutor's short immediate feedback (negative or neutral) and indirect speech acts (such as hints). There is presumably a social significance to the signals of confusion on the face; it will presumably enhance the likelihood of receiving help from peers. It is interesting to note that there is a positive correlation between engagement/flow and learning, but there are no facial features that robustly discriminate this affect state from neutral. Without such signals, peers are less likely to interrupt the learner, which is precisely what one would want from a social functional perspective. Boredom is visibly noticeable, but frustration tends to be hidden unless it reaches extreme levels. The latter trends call for replication, but our findings on confusion and flow/engagement are consistent across experiments.

So far, it would appear that embodied facial actions and body actions are extremely important manifestations of learning and thinking. This is precisely what those who advocate an embodied framework would enjoy reading and reporting. However, there is one other finding that gives us pause. In particular, the vast majority of the time, learners do not exhibit much facial expression and bodily action during learning. Most of the time they appear neutral or there is no noticeable emotion at all that is manifested in their face and body. Perhaps most of the embodied activities take place in the mind or brain, rather than actions that can be detected by others. This possibility we leave to the neuroscientists to explore (Pulvermüller 2005).

3.5 **Discussion**

So where are we left in the embodiment versus symbolic debate? Some may view us as wishy-washy compromisers and advocates of ho-hum hybrid models. In truth, our long-term goal is to encourage researchers to identify the conditions under which representations are strong- or weak-embodied versus disembodied, and also whether symbolic models are modality-specific versus amodal. It is, of course, essential to clarify the explanatory foundations of whatever representation is applicable. However, our immediate goal is to place researchers in sufficient cognitive disequilibrium that confusion and more deep thinking might just prevail over the enthusiasm of carnal imperialism.

Debate

Two sets of issues were brought up during the presentation by Graesser and Jacskon at the debate. The first raised the question of how much embodied comprehension occurred during the reading of text and the interactions with AutoTutor. The fundamental data to be explained was the following pattern of learning gain scores: interactive three-dimensional simulation ≥ normal conversational AutoTutor > reading the textbook = no training = 0. The second set of issues raised the question of how embodied AuoTutor is, and whether it is similar to human tutors.

How embodied are the learners' cognitive representations while reading and while interacting with AutoTutor?

Robert Goldstone: When you are describing how AutoTutor does a better job than just reading the text, I was surprised to hear you say that [the data] was arguing against an embodied perspective. I think I see what you're saying. You're saying that when they're reading the text they're not creating these real full-blown mental simulations of what's going on. But from another interpretation of embodiment, that's exactly what you would predict. If you think of embodiment as using the world, that's part of your mind.

So the mind's not just this hermetically-sealed symbol-to-symbol system, but it's actually involving the world. And AutoTutor is part of the world, so it's helping your mind. So can you say more about why [the data] is evidence against an embodied perspective; that is, that the corporality helps.

Arthur Graesser: The main claim was that when people normally read text on these abstract topics, if they were trying to build an embodied sort of construction in their mind then presumably they would be doing some comprehending. However, they aren't. Now, what is going on in AutoTutor may be a combination of embodied experiences and symbolic, whatever that is. But the main claim was about the lack of difference between doing nothing and reading. If it were the case that when we are reading the abstract material we construct these mental spaces with visual and enactive representations, then you might expect some comprehension.

Arthur Glenberg: I think the reason why students reading the text aren't getting much out of it is because much of the text is not grounded for them. They have no idea what a force vector is. And so, as they're reading the text, they're unable to construct a good representation. And I think Rob is absolutely correct, what AutoTutor is doing is grounding these concepts for people. AutoTutor is making it possible for people to create an embodied representation of, for example, the physics domain.

Graesser: Okay, now what would you predict if you had a version of AutoTutor that was strictly conversational, that is, learners didn't have the pop-up simulation environments? How would you compare AutoTutor with those pop-up simulation environments versus without them? Which condition do you think would be best?

Glenberg: It's hard to answer in the abstract because it will depend on how well AutoTutor can use language, and it will depend on the domain. If it's a domain in which people can successfully ground concepts by applying their own experiences, there shouldn't be much difference between AutoTutor alone and AutoTutor with the pop-ups. But I would imagine that in something like physics, demonstrations with the pop-ups are going to be valuable.

Graesser: Well that's what we were hoping. We were hoping it would help for you to put that interactive simulation in there. It's visual. They're manipulating things. They get to see what happens. It's a very embodied experience. When you see the data, the difference [between AutoTutor-3D and conversational AutoTutor] is anaemic. Tanner Jackson and I really were wishing for a more dramatic effect, but we didn't find it, and that was a disappointment. Maybe we just didn't ground it in the right way.

How embodied is AutoTutor and how similar is it to human tutors?

Unidentified person: I think the right question to ask here is whether you think that AutoTutor understands the tutee. In other words, do we work the way that AutoTutor works? We know we can simulate many things with the computer, but the right question to ask is whether we work the way this machine works.

Graesser: Yeah, so one interesting follow-up to that question is that we are now running AutoTutor with speech recognition. And as you know, speech recognition systems are not too good. In fact, we think AutoTutor is getting only about 30–40% of the content words. However, there might just be enough hooks in AutoTutor where it seems like it's responding and occasionally connecting. So people might have the illusion that communication is taking place. Well you might say, 'Oh silly poor computer, it's not doing very well,' but I wonder if that's not too much different than a real human tutor. Under conditions when there's not a lot of successful communication, you still think you're in this communication process that appears like it's succeeding. Maybe that's normal in a tutoring environment. So if you compare human tutors with these machine tutors, they may not be that much different.

Unidentified person: Art, I want to push on this idea of what you said regarding 'this AutoTutor is *very* embodied'. When I think of learning physics and I think of watching simulations and engaging in dialogue, it's not completely unembodied. But there's a wide spectrum there, and I wouldn't classify it as *very* embodied. So imagine putting these students in this car and having them experience these jolts. I think what's happening [in the current AutoTutor] is that the conversation is across a very narrow range of what I think is a much wider range of embodied experience by which we learn a lot of things. When I think about the misconceptions that are so prevalent in physics, a lot of them can be linked to very embodied experiences. I think, for instance, of these issues with misunderstanding seasonal variation and the sun, and people say, 'well the earth is closer to the sun in the summer'. And you find out that they're often reasoning by analogy with things like, 'ooh, when I stand next to a fireplace, I'm much warmer than when I move away from it.' They start to draw these personal experiences and even visceral experiences into play when they make these kinds of judgements. I think when we just cast [embodiment] as observing simulations and getting involved in dialogue, we're sometimes using interactivity as a proxy for embodiment and that's not necessarily the same thing. When you talk about things like physics or mathematics, there's a lot more embodiment that could happen.

Graesser: You can consider that there may be a continuum. If you are just reading printed text, that's not much embodiment, but maybe as you add more of the conversation and the embedded simulation, that's more embodied. And then if you were in a virtual environment, that would be extremely embodied. And [consider when] you actually can have haptic sensors to get the sensations. So maybe it's a continuum.

Unidentified person: If it's a continuum, and let's say it's a logarithmic continuum, then we really have an issue here. Because every one of those leaps is enormous in terms of the amount of input that people can be using to make these conceptual gains.

Graesser: Most learners, when they receive the interactive simulations, don't really use it much. The ones who use it a lot, learn a lot. But many of them just kind of give up. They try a couple of times and then they give up.

Francis Quek: I have a comment on what you said about simulating conversation and understanding. I'm drawn to memory of conversations with foreign students who do not quite understand what I am talking about, but they give me all the correct signals. They nod and they do everything just right. Two weeks later I go back to them say, 'Well, did you do that?' [and the student says] 'No'. They didn't understand a single thing, but they were too polite to tell me that they didn't understand a single thing. So simulating the social interaction is by no means an effective way to communicate. In fact, it leads to more misunderstanding than anything else.

Author note

The Tutoring Research Group (TRG) is an interdisciplinary research team comprised of approximately 35 researchers from psychology, computer science, physics, and education (visit www.autotutor.org). The research on AutoTutor was supported by the National Science Foundation (SBR 9720314, REC 0106965, REC 0126265, ITR 0325428, REESE 0633918), the Institute of Education Sciences (R305H050169), and the Department of Defense Multidisciplinary University Research Initiative (MURI) administered by the Office of Naval Research under grant N00014-00-1-0600. Any opinions, findings, and conclusions or recommendations expressed in this material are those of the authors and do not necessarily reflect the views of the funding organizations.

References

Aleven V, Koedinger KR (2002). An effective metacognitive strategy: learning by doing and explaining with a computer-based cognitive tutor. *Cognitive Science,* 26, 147–79.

Alibali MW, Nathan MJ (in press). Teachers' gestures as a means of scaffolding students' understanding: Evidence from an early algebra lesson. In R Goldman, R Pea, BJ Barron, S Derry, Eds. *Video Research in the Learning Sciences.*

Anderson JR, Corbett AT, Koedinger KR, Pelletier R (1995). Cognitive tutors: lessons learnt. *Journal of the Learning Sciences,* 4, 167–207.

Azevedo R, Cromley JG (2004). Does training on self-regulated learning facilitate students' learning with hypermedia. *Journal of Educational Psychology,* 96, 523–35.

Baker L (1985). Differences in standards used by college students to evaluate their comprehension of expository prose. *Reading Research Quarterly,* 20, 298–313.

Barsalou LW (1999). Perceptual symbol systems. *Behavioural and Brain Sciences,* 22, 577–660.

Barsalou LW (2003). Abstraction in perceptual symbol systems. *Philosophical Transactions of the Royal Society of London: Biological Sciences,* 358, 1177–87.

Barsalou LW, Wiemer-Hastings K (2005). Situating abstract concepts. In D Pecher, R Zwaan, Eds. *Grounding Cognition: The Role of Perception and Action in Memory, Language, and Thingking* (pp. 129–63). New York, NY: Cambridge University Press.

Breazeal C (2003). *Designing Sociable Robots.* Cambridge, MA: MIT Press.

Cassell J, Bickmore T, Campbell L, Vilhjalmsson H, Yan H (2001). More than just a pretty face: conversational protocols and the affordances of embodiment. *Knowledge Based Systems,* 14, 55–64.

Cassell J, Thorisson K (1999). The power of a nod and a glance: envelope vs. emotional feedback in animated conversational agents. *Applied Artificial Intelligence,* 13, 519–38.

Chi MTH, Siler SA, Jeong H, Yamauchi T, Hausmann RG (2001). Learning from human tutoring. *Cognitive Science*, 25, 471–533.

Chu CW, Cohen I (2005). Posture and gesture recognition using body shapes decomposition. Presented at the IEEE Workshop on Vision for Human–Computer Interaction, June 21, 2005. Available at: http://iris.usc.edu/~icohen/pdf/Wayne-v4hci05.pdf

Cole R, van Vuuren S, Pellom B *et al.* (2003). Perceptive animated interfaces: First steps toward a new paradigm for human computer interaction. *Proceedings of the IEEE*, 91, 1391–405.

Craig SD, Graesser AC, Sullins J, Gholson B (2004). Affect and learning: an exploratory look into the role of affect in learning. *Journal of Educational Media,* 29, 241–50.

Csikszentmihalyi M (1990). *Flow: the Psychology of Optimal Experience*. New York, NY: Harper-Row.

di Sessa AA (1993). Towards an epistemology of physics. *Cognition and Instruction*, 10, 105–225.

de Vega M, Robertson DA, Glenberg AM, Kaschak MP, Rinck M (2004). On doing two things at once: temporal constraints on actions in language comprehension. *Memory and Cognition*, 32, 1033–43.

de Vega M, Rodrigo MJ, Ato M, Dehn D, Barquero B (2002). How nouns and prepositions fit together. An exploration of the semantics of locative sentences. *Discourse Processes*, 34, 117–43.

D'Mello SK, Craig SD, Sullins J, Graesser AC (2006). Predicting affective states through an emote-aloud procedure from AutoTutor's mixed-initiative dialogue. *International Journal of Artificial Intelligence in Education*,16, 3–28.

D'Mello SK, Craig SD, Witherspoon A, McDaniel B, Graesser AC (in press). Automatic detection of learner's affect from conversational cues. *User Modeling and User-Adapted Interaction*.

D'Mello SK, Picard R, Graesser AC (2007). Toward an affect-sensitive AutoTutor. *IEEE Intelligent Systems*, 22, 53–61.

Ekman P (2003). *Emotions Revealed: Recognizing Faces and Feelings to Improve Communication and Emotional Life*. New York, NY: Henry Holt & Co.

Fox B (1993). *The Human Tutorial Dialogue Project*. Hillsdale, NJ: Erlbaum.

Glenberg AM (1997). What memory is for. *Behavioural and Brain Sciences*, 20, 1–19.

Glenberg AM, Epstein W (1985). Calibration of comprehension. *Journal of Experimental Psychology: Learning, Memory, and Cognition*, 11, 702–18.

Glenberg AM, Kaschak MP (2002). Grounding language in action. *Psychonomic Bulletin and Review*, 9, 558–65.

Glenberg AM, Robertson DA (1999). Indexical understanding of instructions. *Discourse Processes*, 28, 1–26.

Goldstone RL, Sakamoto Y (2003). The transfer of abstract principles governing complex adaptive systems. *Cognitive Psychology,* 46, 414–66.

Graesser AC, Chipman P, Haynes BC, Olney A (2005). AutoTutor: an intelligent tutoring system with mixed-initiative dialogue. *IEEE Transactions in Education,* 48, 612–18.

Graesser AC, Jackson GT, Mathews EC et al.; The Tutoring Research Group (2003). Why/AutoTutor: a test of learning gains from a physics tutor with natural language dialog. *Proceedings of the 25th Annual Conference of the Cognitive Science Society*. Mahwah, NJ: Erlbaum.

Graesser AC, Lu S, Jackson GT *et al.* (2004). AutoTutor: a tutor with dialogue in natural language. *Behavioural Research Methods, Instruments, and Computers,* 36, 180–93.

Graesser AC, Lu S, Olde BA, Cooper-Pye E, Whitten S (2005). Question asking and eye tracking during cognitive disequilibrium: comprehending illustrated texts on devices when the devices break down. *Memory and Cognition*, 33, 1235–47.

Graesser AC, Penumatsa P, Ventura M, Cai Z, Hu X (in press). Using LSA in AutoTutor: learning through mixed-initiative dialogue in natural language. In T Landauer, DS McNamara, S Dennis, W Kintsch, Eds. *LSA: A Road to Meaning*. Mahwah, NJ: Erlbaum.

Graesser AC, Person NK, Magliano JP (1995). Collaborative dialogue patterns in naturalistic one-on-one tutoring. *Applied Cognitive Psychology*, 9, 359–87.

Graesser AC, VanLehn K, Rose C, Jordan P, Harter D. (2001). Intelligent tutoring systems with conversational dialogue. *AI Magazine*, 22, 39–51.

Graesser AC, Wiemer-Hastings K, Wiemer-Hastings P, Kreuz R; The Tutoring Research Group (1999). Auto Tutor: a simulation of a human tutor. *Journal of Cognitive Systems Research*, 1, 35–51.

Gratch J, Rickel J, Andre E, Cassell J, Petajan E, Badler N (2002). Creating interactive virtual humans: some assembly required. *IEEE Intelligent Systems*, 54–63.

Grosz BJ, Gordon PC (1999). Conceptions of limited attention and discourse focus. *Computational Linguistics*, 25, 617–24.

Grosz BJ, Sidner CL (1986). Attention, intentions, and the structure of discourse. *Computational Linguistics*, 12, 175–204.

Harter D (2004). Towards a model of basic intentional systems: nonlinear dynamics for perception, memory, and action in autonomous agents [doctoral thesis]. Memphis, TN: University of Memphis.

Hunt E, Minstrell J (1996). A collaborative classroom for teaching conceptual physics. In K McGilly, Ed. *Classroom Lessons: Integrating Cognitive Theory and the Classroom*. Cambridge, MA: MIT Press.

Jackson GT, Olney GT, Graesser AC, Kim HJ (2006) AutoTutor 3-D simulations: analyzing users' actions and learning trends. *Proceedings of the Cognitive Science Society*. Mahwah, NJ: Erlbaum.

Johnson WL (2001). Pedagogical agent research at CARTE. *AI Magazine*, 22, 85–94.

Johnson WL, Rickel JW, Lester JC (2000). Animated pedagogical agents: face-to-face interaction in interactive learning environments. *International Journal of Artificial Intelligence in Education*, 11, 47–78.

Jurafsky D, Martin JH (2000). *Speech and Language Processing: An Introduction to Natural Language Processing, Computational Linguistics, and Speech Recognition*. Upper Saddle River, NJ: Prentice-Hall.

Just MA, Carpenter PA (1992). A capacity theory of comprehension: individual differences in working memory. *Psychological Review*, 99, 122–49.

Kapoor A, Picard R (2002). Real-time, fully automated upper facial feature tracking. *Proceedings of the 5th International Conference on Automated Face and Gesture Recognition* (pp. 10–16).

Kettebekov SM, Yeasin M (2005). Prosody-based audio-visual co-analysis for coverbal gesture recognition. *IEEE Transactions in Multimedia: Multimedia Interfaces and Applications*, 7, 234–42.

Kintsch W (1998). *Comprehension: A Paradigm for Cognition*. Cambridge, UK: Cambridge University Press.

Koedinger KR, Alibali M, Nathan MJ (submitted). Trade-offs between grounded and abstract representations: evidence from algebra problem solving.

Kosko B (1992). *Neural Networks and Fuzzy Systems*. New York, NY: Prentice Hall.

Kozma R, Freeman WJ (2001). Chaotic resonance – methods and applications for robust classification of noisy and variable patterns. *International Journal of Bifurcation and Chaos*, 11, 1607–29.

Laird JE, Newell A, Rosenbloom P (1987). SOAR: an architecture for general intelligence. *Artificial Intelligence*, 33, 1–64.

Landauer TK, Dumais ST (1997). A solution to Plato's problem: the latent semantic analysis theory of the acquisition, induction, and representation of knowledge. *Psychological Review*, 104, 211–40.

Landauer T, McNamara D, Dennis S, Kintsch W (2007). *Handbook of Latent Semantic Analysis*. Mahwah, NJ: Erlbaum.

Larsson S, Traum D (2000). Information state and dialogue management in the TRINDI dialogue move engine toolkit. *Natural Language Engineering*, 6, 323–40.

Lee J (1999). Effectiveness of computer-based instructional simulation: A meta analysis. *International Journal of Instructional Media*, 26, 71–85.

Lochbaum KE (1998). A collaborative planning model of intentional structure. *Computational Linguistics*, 24, 525–72.

Louwerse MM, Mitchell HH (2003). Toward a taxonomy of a set of discourse markers in dialog: a theoretical and computational linguistic account. *Discourse Processes*, 35, 199–239.

Manning C, Schutze H (1999). *Foundations of Statistical Natural Language Processing*. Cambridge, MA: MIT Press.

Marr D (1982). *Vision*. San Francisco, CA: Freeman.

Massaro DW, Cohen MM (1995). Perceiving talking faces. *Current Directions in Psychological Science*, 4, 104–9.

Millis KK, King A, Kim J (2001). Updating situation models from descriptive texts: a test of the situational operator model. *Discourse Processes,* 30, 201–36.

Picard RW (1997). *Affective Computing*. Cambridge, MA: MIT Press.

Picard RW (2001). Toward computers that recognize and respond to user emotions. *IBM Systems Journal*, 39, 705.

Pickering MJ, Garrod S (2004). Toward a mechanistic psychology of dialogue. *Brain and Behavioural Sciences*, 27, 169–90.

Ploetzner R, VanLehn K (1997). The acquisition of informal physics knowledge during formal physics training. *Cognition and Instruction*, 15, 169–206.

Pulvermüller F (2005). Brain mechanisms linking language and action. *Nature Reviews Neuroscience*, 6, 576–82.

Quek F (2004). The catchment feature model: a device for multimodal fusion and a bridge between signal and sense. *EURASIP Journal on Applied Signal Processing*, 11, 619–36.

Rickel J, Johnson WL (1999). Animated agents for procedural training in virtual reality: perception, cognition, and motor control. *Applied Artificial Intelligence*, 13, 343–82.

Roy D (in press). Semiotic schemas: a framework for grounding language in action and perception. *Artificial Intelligence.*

Sanford AJ, Graesser AC (2006). Introduction: shallow processing and underspecification. *Discourse Processes*, 42, 99–108.

Steels L (2003). Evolving grounded communication in robots. *Trends in Cognitive Science*, 7, 308–12.

Sturt P, Sanford AJ, Stewart AJ, Dawydiak E (2004). Linguistic focus and good-enough representations: an application of the change detection paradigm. *Psychonomic Bulletin and Review*, 11, 882–8.

Talmy L (1988). Force dynamics in language and cognition. *Cognitive Science*, 12, 49–100.

VanLehn K, Graesser AC, Jackson GT, Jordan P, Olney A, Rosé CP (2007). When are tutorial dialogues more effective than reading? *Cognitive Science*, 31, 3–62.

VanLehn K, Jones RM, Chi MTC (1992). A model of the self-explanation effect. *Journal of the Learning Sciences,* 2, 1–59.

Winne PH (2001). Self-regulated learning viewed from models of information processing. In B Zimmerman, D Schunk, Eds. *Self-regulated Learning and Academic Achievement: Theoretical Perspectives* (pp. 153–89). Mahwah, NJ: Erlbaum.

Winograd T, Flores F (1986). *Understanding Computers and Cognition*. Norwood, NJ: Ablex.

Zwaan RA (2004). The immersed experiencer: toward an embodied theory of language comprehension. In BH Ross, Ed. *The Psychology of Learning and Motivation, Vol. 44* (pp. 35–62). New York, NY: Academic Press.

Zwaan RA, Stanfield RA, Yaxley RH (2002). Do language comprehenders routinely represent the shapes of objects? *Psychological Science*, 13, 160–71.

Chapter 4

Symbolism, embodied cognition, and the broader debate

Lawrence Shapiro

4.1 **Introduction**

This actually happened at the dinner table recently.

Me, to wife Athena: How's that new bar on King St.?

Athena: Too much attitude.

Sophia, our eight-year old: What does that mean?

Pause.

Me: Means that they're kind of snobby.

Sophia: What does that mean?

Me: Means that they have too much attitude.

Sophia: Oh, like that helps.

Sophia's disappointment in me, at least in this case, is well-deserved. Being told that the term X means the same as the term Y will not help anyone to understand what X means unless they already understand what Y means. I hope that this point is not controversial. Neither, I hope, is the following claim: Being told that the term X means the same as the term Y, and that the term Y means the same as the term 'Z', will not help anyone to understand what X means if one does not understand the meanings of either Y or Z.

Notice that the claims above need not be construed as *a priori*. Testing them would be easy enough. I could tell Sophia that being snobby means putting on airs. Though I did not actually say this to her, I hypothesize that she would then know nothing more about what it means to have too much attitude than she does presently. This makes the following induction very tempting: one cannot understand the meaning of a word or, more generally, a symbol, if one's only access to its meaning is via other symbols whose meanings are not understood.

Just to put a name on a target, I will call the view that a symbol can become meaningful to an interpreter solely in virtue of knowing its relations to other symbols whose meanings are not known, the *symbolist's folly*. Thus, if one were to assert that Sophia could eventually understand what 'having too much attitude' means simply by her knowing that it means the same as other expressions she does not understand, or in virtue of her knowing that people to whom the predicate 'has too much attitude' applies are often people to whom the predicate 'is condescending' (a term with which she is unfamiliar) also applies, but to whom the predicate 'is crunchy' (an expression with which she is unfamiliar) does not typically apply, and so on, then one would be engaged in the symbolist's folly.

But does anyone really believe in the symbolist's folly? Apparently some do, and at least one motivation among those preaching embodied cognition is to provide an account of meaning that avoids the symbolist's folly. For instance, Glenberg, De Vega, and Graesser, in their introduction to this volume (see Chapter 1), suggest that embodied approaches might provide the answer to the 'what else?' question regarding how amodal symbols might gain meaning. Glenberg especially (e.g., Glenberg and Kaschak 2002) has been pursuing embodied approaches to cognition in an effort to circumvent the so-called 'grounding problem', i.e. the problem about how symbols acquire their meaning. Barsalou, too, has looked to embodied cognition as a solution to the grounding problem, proposing perceptual symbols as a way out (1999, p. 614).

I think Searle's *Chinese room argument* (1980), under a fairly standard interpretation, can be taken as directed against the symbolist's folly. Although the Chinese room argument makes acute the symbol grounding problem, I shall argue that the symbolist's folly is largely a red herring. There are reasons to pursue research in embodied cognition, but these are independent of the need to respond to the symbolist's folly. As for Searle's Chinese room, there is not much beyond a trivial moral that we can draw from it.

Here's the plan. I will start with a discussion of Searle's Chinese room, showing how it is supposed to render obvious the folly inherent in the symbolist's folly. However, I'll argue, the Chinese room does *not* show that symbolic approaches (i.e. computational approaches) to cognition are deficient. That is, the Chinese room contains nothing that need worry a good old-fashioned symbolist. But, I then argue, the same resources a symbolist can marshal in response to Searle work equally well to deflect the charge that one must turn to embodied cognition to avoid the symbolist's folly. Here my focus will be on Barsalou's important work on perceptual symbol systems

In the final sections of the chapter, I suggest a positive motivation for embodied cognition. Embodied cognition proposes new ways to conceive of the role of representation in cognition, new ways that promise to provide better and simpler accounts of cognition than those that insist on traditional conceptions of representation. In this sense, I suggest that the debate between symbolism and embodied cognition is an instance of a more general debate within philosophy of science over the justification for theoretical posits.

4.2 Searle's Chinese room

4.2.1 Into the room

There sits Searle in the Chinese room. Slips of paper containing Chinese symbols enter through a slot in the door. Searle picks up the slips, consults a book of rules that specifies which Chinese symbols to write in response, and out goes his 'answer' to the initial 'question.' The quotes are necessary because Searle does not know he has provided an answer; he does not know that the original slip of paper contained a question. In fact, Searle may not even know that the symbols he is manipulating are Chinese. As far as Searle is concerned, he's playing a rather tedious game: here are some marks on a piece of paper; using your rule book, turn these marks into other marks. This, as far as Searle knows, is all that's going on.

To Tak Jun, who sits outside the room, something very different is happening. Tak Jun writes down questions, feeds them into the room, and a while later out come answers. Tak Jun writes, 'Why did the Yangtze overflow its banks this morning?' From the box appears an answer, 'Because of yesterday's heavy rains.' Tak Jun assumes that the box understands Chinese. There's no distinguishing it from a person who understands Chinese, so this is not an unreasonable thing to believe.

Searle supposes that we will all agree that he doesn't understand Chinese. He's in worse shape than Sophia. Sophia's stuck trying to understand one expression in terms of another that she doesn't understand, but at least she knows that what she doesn't understand are words. Words, of course, are symbols. Sophia is trying to learn the meaning of symbols. But Searle may not even know that the marks on the slips of paper are symbols. He may not know that the marks have any meaning at all.

I think we should agree that Searle does not understand Chinese. If unsure, we could ask him. He would tell us that he doesn't. We could ask Tak Jun to insult Searle, who's unlikely to be happy about this. If Searle remains calm, this is all the evidence we would need. Of course Searle doesn't understand Chinese.

4.2.2 Opening the door

It is not my intention to evaluate the significance of the Chinese room thought experiment for strong artificial intelligence (AI). My concern instead is with diagnosing why Searle does not understand Chinese. I think the reasons are fairly obvious, but I would like to place them in the context of a particular kind of theory of meaning. However, to do this, I am going to take Searle out of the Chinese room and put him in the Greek room. I do this because Chinese is Greek to me, but Greek is not.

The theory of meaning I have in mind is usually called a *causal theory of meaning*. Causal theories of meaning have been a mainstay in philosophy of mind since my colleague Dennis Stampe's seminal paper 'Toward a causal theory of linguistic representation" (1977). Others, notably Fred Dretske (1981) and Jerry Fodor (1987), have developed causal theories of meaning in their own ways, but they follow Stampe in analyzing meaning in terms of a representation's causes. The basic idea is that a representation is of X (that is, it means X) when the representation, under certain conditions – what Stampe calls 'fidelity conditions' – would be present because X is present. The fidelity conditions are those that, as Stampe puts it, govern the production of the representation. When appropriate fidelity conditions are present, it is 'reasonable to accept' that the presence of a representation of X implies the presence of X itself. Of course, this is only the barest sketch of the account; Stampe and others have much more to say about how to understand fidelity conditions so that they do not result in the representation of the wrong objects, or so that representation can sometimes be vacuous. For my purposes, I mention the causal theory of meaning because it has a compelling story to tell about how symbols acquire meaning.

According to this story, my thought 'Athens' means Athens because, in the context of certain conditions, it is reasonable to suppose that the presence of Athens would cause the thought 'Athens'. Obviously, the existence of an external environment is necessary for the meaningfulness of many of our thoughts. Thoughts acquire their meanings not through their associations with uninterpreted symbols, as the symbolist's folly would have it, but through their connections to the world or, often, their connections to other symbols that are connected to the world.

Consider now Searle's thought that Athens is in Greece and Thalia's thought that Athens is in Greece. Searle and Thalia have the same thought, but they express it differently. Searle uses the word 'Athens' to express his thought. Thalia uses the word 'Aθήνα'. Searle might also think '"Athens" has six letters.'; Thalia would think, '"Aθήνα" has five letters.' Searle and Thalia have different thoughts because their thoughts are no longer about the same thing. Searle's thought is about the English word 'Athens' and Thalia's thought is about the Greek word 'Aθήνα'.

Suppose now that Searle knows no Greek and we place him in the Greek room. Thalia writes down a question on a slip of paper which, translated into English, reads 'Is Athens the capital of Greece?' Searle receives the slip of paper, consults his English rules, writes down some Greek symbols that are as meaningless to him as the Chinese symbols are, and pokes the paper back through the hole in the door. Thalia reads the answer which, translated into English, is 'Yes. Athens has been the capital of Greece for a very long time.' Why does Thalia suppose herself to be having a conversation whereas Searle does not? The answer, I submit, is that the conditions that must be in place for 'Aθήνα' to mean Athens are in place for Thalia, but are not in place for Searle. For instance, Thalia belongs to a community in which utterances of 'Aθήνα' are caused by speech production mechanisms that in turn cause these utterances as a result of individuals' intentions to refer to Athens (Stampe 1977, pp. 52–3). Ignorant of these facts, Searle does not know that 'Aθήνα' means Athens. How could he? Consequently, Searle does not take himself to be having a conversation with the person slipping notes through the hole in the door.

The philosopher's distinction between use and mention provides another way to see how Searle and Thalia differ. Thalia can *use* the word 'Aθήνα', as she does when saying, in Greek, that Athens is in Greece. She can also *mention* the word, as she does when she says '"Aθήνα" has five letters.' Searle, on the other hand, can only mention Greek words. He can say and think '"Aθήνα" has five letters', but he cannot use the word 'Aθήνα' to say anything about Athens. In order to use Greek words, Searle needs to know their meanings, but he does not know their meanings because he knows nothing about the connections that Greek words bear to the world. For Searle, these words bear connections only to each other. Thalia can see *through* the Greek words, but Searle can never see *past* them.

But, if this diagnosis of why Searle does not understand Greek is correct, there is an obvious solution to his problem. One simply has to open the door to the Greek room and let Searle explore his Greek environment. In time, Searle will learn which words are appropriate to utter in which contexts. He'll learn to think 'σκύλος' when he sees a dog, to think 'σπίτι' when he sees a house, and so on. He'll understand Greek when he becomes attuned to the connections between Greek words, intentions, and objects.

4.3 **Symbolism and embodiment**

The lesson to draw from the Chinese room is, after all, not very interesting. Indeed, finding a non-trivial way to state the lesson is difficult. The lesson seems to be this: symbols do not acquire meanings merely through formal transformations. The symbolist's folly is folly; *of course* if one does not know the meanings of symbol *X* or symbol *Y* then knowing how *X* and *Y* are connected is not going to help. Sophia knew that. One needs to understand (tacitly or otherwise) how symbols are connected to the world in order to know what they mean.

Famously, Searle claims that thought cannot be just symbol manipulation, or symbol manipulation is, by itself, not sufficient for thought. This, too, seems right. If the operations of a computer involve only uninterpreted symbols, then a computer could never understand what it was 'thinking.' But this observation does not imply that the manipulation of *interpreted* symbols is not sufficient for thought. This point is significant for those who see embodied cognition as an answer to what ails symbolism. Consider two complaints that seem to motivate to some extent the embodied cognition turn:

(1) Symbol manipulation on its own cannot produce understanding.

(2) Symbols acquire their meaning only through embodiment.

My discussion of the Chinese room, I hope, makes clear that one can accept the first complaint without giving up a symbolist conception of cognition. That is, a symbolist can agree that the manipulation of mere, uninterpreted symbols is not sufficient for understanding. In addition to symbol manipulation, there must as well be a semantics that assigns meanings to the symbols.[1]

This easy answer to the first complaint might seem susceptible to the second complaint. Perhaps symbol manipulation does not create meaning because manipulation is in some sense insulated from the relations between mind, body, and world from which meaning emerges. Yet, from the perspective of the causal theory of meaning I described above, embodiment is irrelevant. Tree rings represent, or mean, a tree's age. Smoke means fire. Mercury being at a particular level means that it is 68 degrees. All of this is possible without it being true that trees, smoke, or mercury are parts of an embodied system. What matters is that symbols are connected in the right way to the things they represent. If embodiment provides such a way, all the better, but this would mean only that embodiment is one way to make the right connections.

It is worth exploring this idea a bit further in the context of Barsalou's theory of perceptual symbols, which, he believes, provides a solution to the grounding problem that is unavailable to the computational theorist. As we will see, the perceptual symbols that Barsalou introduces differ from traditional computational symbols less than he thinks.

---

[1] I do not wish to claim that the causal theory of meaning will, all by itself, suffice for understanding. Talk of understanding is vague. Understanding might involve a phenomenological component (e.g., the 'ahh!' feeling), and surely the causal theory of meaning has nothing to say about that. However, I would deny that understanding requires that one knows the meaning of the words one uses.

In essence, perceptual symbols are like traditional symbols except for having a 'built-in' grounding relation.

At the core of Barsalou's theory is a distinction he draws between amodal and modal symbols. Words are paradigms of amodal symbols; words are (typically) not similar to those things they represent. 'Car,' for instance, does not resemble a car: the word does not have four wheels, although the object does. Moreover, the connection between 'car' and cars is arbitrary in the sense that the vehicle might have been called by a different name. Indeed, names for a car differ across languages. In short, there is nothing like a law of nature that ties a car to a particular name.

In standard computational approaches to cognition, cognitive processes construct amodal symbols. Processing goes in stages, roughly as follows. First, stimulation hits the surfaces of sense organs. The perceptual system of which the sensory receptors are a part transduce this stimulation into a neural code. There is then a second stage of transduction in which a cognitive system converts perceptual neural symbols into non-perceptual, amodal symbols. Thus, for instance, perceiving a car requires first the stimulation of cells in the retina. The visual system then constructs a representation of the car. The properties of this visual representation correspond to information about the shape, size, and colour of the car. This representation is then delivered to cognitive systems that utilize frames, or feature lists, or other representational structures, in order to categorize the visual representation as being of a car. Thus, the cognitive system takes as input a visual representation and produces as output the judgement *car*. Importantly, this judgement consists of the construction of a word-like symbol that bears no more similarity to the car, or to the visual representation of the car, then the word 'car' bears to an actual car. Because there is no similarity between this final neural symbol and the visual properties that are encoded, the tie between the two is arbitrary. There is nothing that requires that the car be represented with the symbol *car* rather than some other symbol (Barsalou, 1999: 578-9).

Reinforcing this amodal conception of mental representation is Fodor's well-received defence of a language of thought. Fodor (1975, 1987) has argued that mental representations are, in many important respects, language-like. The beliefs, desires, and other attitudes we have are relations to language-like representations. Thus, my belief that Madison sits on an isthmus consists of a believing relation toward a language-like representation of the proposition that Madison sits on an isthmus. Because the representation is linguistic, it is compositional and thus capable of participating in the systematic and productive processes that are hallmarks of both thought and language. Crucially, adoption of the language of thought hypothesis seems to force a commitment to the same kind of amodality of mental representations that is true of linguistic representations (but see below).

Barsalou rejects this two-stage conception of cognition, proposing instead that there is no need for a separate processing stage in which perceptual representations are converted into amodal cognitive representations. 'Cognition is inherently perceptual,' he argues, 'sharing systems with perception at both the cognitive and neural levels' (1999, p. 577). In place of the amodal renderings of cognitive systems, Barsalou posits *perceptual symbols*. These symbols are reconstructions of the representations that were present in perceptual systems when these systems were initially transducing stimulation into a neural code. Barsalou calls these symbols modal, for they 'are represented in the same systems as the perceptual states that produced them' (1999, p. 578). For instance, the perception of a car involves the representation of visual, auditory, haptic, and proprioceptive information. Storage of these representations in the form of perceptual symbols underwrites future cognitive activities such as remembering a car, making inferences about cars, talking about cars, and so on. In the course of these activities, a representation of a car is produced anew in the very same perceptual systems that made possible the initial perception of cars.

Because my interest is in whether perceptual symbols offer a response to the grounding problem that is not available to, or is better than, the response I have already provided above to the computational theorist, I shall not discuss the evidence Barsalou collects in support of perceptual symbols. In short, my question is this: are perceptual symbols grounded in a way not available to traditional symbols? Barsalou clearly thinks that the answer to this question is 'yes.' However, I think Barsalou's insistence that perceptual symbols have an advantage over computational symbols rests on a confusion. The confusion resides in mistaking the modality of perceptual symbols with facts about their causal aetiology. Having made this mistake, Barsalou then attributes the groundedness of perceptual symbols to their modality, when in fact perceptual symbols owe their grounding to the same causal facts in virtue of which computational symbols may be grounded. Let me explain.

Consider first that perceptual symbols are modal only in the sense that they are present in a perceptual system that is dedicated to processing information of a particular mode (e.g. visual information, or auditory, or haptic). They are not modal in the sense that a perceptual symbol that represents red is in fact red, or one that represents vibration is in fact vibrating. Barsalou is rightly quite careful about this: perceptual symbols 'are modal because they are represented in the same systems as the perceptual states that produced them' (1999, p. 578). But Barsalou also tells us that perceptual symbols are analogical, because the 'structure of a perceptual symbol corresponds, at least somewhat, to the perceptual state that produced it' (1999, p. 578). However, he elaborates in a footnote (1999: p. 608), the structure of a perceptual symbol does not, typically, correspond to the structure of an object in the world. To claim that it did, presumably, would be to commit the error implicit in the idea that these representations *are* red or *do* vibrate. Instead, the correspondence exists only between features of the perceptual symbol and the perceptual *representation* of which it is a record.

Assuming that this is the extent to which perceptual symbols are modal and analogical, the question arises whether perceptual symbols differ from computational symbols in ways that are relevant to the symbol grounding problem. Like computational symbols, perceptual symbols are not similar to the things they represent. They may be structurally similar to the perceptual representations from which they are descended, but they do not *represent* these perceptual representations. Thoughts, inferences, and knowledge about cars are supposed to be about *cars*, not about representations of cars. And, as Barsalou cautions in the footnote cited above, perceptual symbols are not structurally similar to objects in the world.

But if perceptual symbols and computational symbols do not differ with respect to being analogical, perhaps their difference rests on the fact that computational symbols are arbitrary. However, amodal symbols of the sort that computational theories of cognition posit seem no more arbitrary than perceptual symbols. 'Let this salt shaker stand for the quarterback,' one might say when describing an interesting play observed on the grid iron earlier in the day. The salt shaker represents the quarterback, but this is just stipulative or accidental. The pepper shaker would have done as well. Barsalou is right that perceptual symbols are not arbitrary in this way. Given how the brain works, presumably there is a causal story to tell about why a given perceptual representation will leave the record that it does, will result in the particular perceptual symbol that it does. However, and this is the crucial point, the computational symbols in a language of thought are not arbitrary for the same reason. There is a causal process that leads from an object in the world to its neural representation. This is true whether the representations over which cognitive processes range are perceptual, as Barsalou thinks, or are derived from perceptual representations through further processes of transduction. Because there is a law-governed process leading from objects to their neural representations, it makes no sense to suggest that some other neural representation of, for example, a car, would have 'done as well.' Some other neural representation would not bear the correct causal relationship to cars, and would thus constitute a delusion of some sort ('you tell me that it's a car in front of me, but I'm seeing a giraffe!').

In light of this point, an objection that Barsalou raises to amodal symbols seems curious. Amodal symbols, but not perceptual symbols, confront the symbol grounding problem because we have no account of 'how amodal symbols become mapped back to perceptual states and entities in the world' (1999, p. 580). However, we do have an account, and it is one that Barsalou himself describes. Amodal symbols are caused by perceptual representations which are in turn caused by stimulation of the sensory organs. If Barsalou's account of how perceptual symbols are grounded is correct, then this account of grounding amodal symbols, which appears simply to involve an additional causal step, ought to be correct as well. If this account of how amodal symbols are grounded is *incorrect*, then it is hard to see why Barsalou's account of grounding perceptual symbols should not also be incorrect.

I have argued that perceptual symbols and computational symbols are not distinct with respect to being analogical or arbitrary. Neither resemble the objects that they represent. Because both are determined by causal processes that begin with stimulation, neither are accidentally connected to their objects. Despite these similarities, however, perceptual symbols may differ from traditional symbols with regard to the amount of processing they require. In particular, the processing of amodal symbols reveals a kind of indifference to aetiology that the processing of modal symbols does not. There is nothing in the syntax of an amodal symbol to mark its content as visual, auditory, or haptic, and so on. Consequently, theories of cognition that restrict themselves to amodal symbol manipulation might require more sophisticated and complex layers of processing.

In contrast, consider that because modal perceptual symbols declare their content as being of a particular perceptual kind, their processing is constrained insofar as visual content will be processed by the visual system, auditory content by the auditory system, and so on. This feature of perceptual symbols presumably brings with it gains in processing efficiency. Processing of amodal symbols would seem to require an extra step, *viz.* the identification of the symbol as having the mode of perceptual content that it does. In contrast, the fact that perceptual symbols are 'built' for particular kinds of perceptual processing eliminates the need for this additional step. We will see below that this tendency to conceive of representational features of the mind as built to reduce processing demands is a hallmark of embodied cognition.

If my comments so far are on track, then there is clearly a *wrong* reason to reject computational theories of cognition, and so to motivate the development of alternative theories of cognition. The following argument exemplifies this wrong way:

(1) Computational theories of cognition assume that symbols acquire their meanings by their associations only with other symbols;

(2) Symbols cannot acquire meaning through association only with other symbols;

(3) Therefore, computational theories of cognition are inadequate.

This argument is not sound because the first premise is false. The question whether cognition is computational is orthogonal to the question about how symbols acquire their meaning. For this reason, one can believe that cognition involves symbol manipulation but also believe that the second premise is true, i.e. that symbol manipulation all by itself does not generate meaning. One can be a computationalist without committing the symbolist's folly.

2.4 What's embodiment about?

So, it is possible to be a computational symbolist without falling victim to Searle's Chinese room. The Chinese room is little more than a dressed-up tautology: meaningless symbols don't mean anything. I conclude that if one pursues embodied cognition because one thinks that computationalism entails the symbolist's folly, then one has made a mistake. But if it is not because of the symbolist's folly that one should seek a motivation for embodied cognition, then what? The motivation, I submit, becomes clearer on recognition of the distinction between two claims. The first is that an association between symbols is by itself sufficient to provide them with meaning. The second is that cognition is symbolic in the traditional computational sense. As I argued in my discussion of Searle, one can accept the second of these claims without accepting the first. There are various theories of meaning, and a commitment to a symbolic view of cognition does not entail a commitment to the symbolist's folly.

As I see the significance of embodied cognition, among its important aims is to provide an alternative to the second claim – that cognition is symbolic in the traditional computational sense. We saw this aim in Barsalou's work, which attempts to build a case in favour of perceptual symbols in contrast to traditional computational symbols. Perceptual symbols differ from traditional symbols in announcing their aetiology, i.e. in declaring themselves to have a particular sensory origin. This feature of perceptual symbols short cuts the extra steps involved in processing amodal symbols. Moreover, perceptual symbols incorporate bodily facts into their syntax; their syntax marks them as visual, auditory, or haptic. But researchers in embodied cognition have offered approaches to cognition more radical than Barsalou's, approaches that attempt to eschew altogether the need for symbolic representation. These approaches, I shall now argue, do not succeed in showing that cognition can do without symbolic representation. However, like Barsalou's more tempered view, they do suggest an account of representation that, in virtue of embodiment, requires less or different kinds of processing than traditional symbolists have supposed.

The prominence of work by JJ Gibson, Rodney Brooks, and Esther Thelen in introductory discussions of embodied cognition suggest that embodied cognition has invested considerable interest in seeking to account for cognition in a way that does away with symbols altogether. Gibson, Brooks, and Thelen especially encourage a view I shall call *symbol antagonism*. Let's first see how symbol antagonism receives expression in Gibson.

A traditional approach to the problem of vision, such as David Marr's (1982), describes vision as a kind of 'inverse optics.' The three-dimensional world causes on the retina a two-dimensional representation. Vision is then a matter of converting this two-dimensional representation into a perception of a three-dimensional world. On this conception, vision is a process that starts and ends with representation. It is a process acting on symbols for the purpose of producing new symbols. This is the conception of vision that Gibson rejects.

In its place, Gibson (1979) contends that representations of the world are not required for vision because the ambient light bouncing back and forth between surfaces in our environment has an information-laden structure. This structure becomes apparent from the point of view of a perceiver, at which time the ambient light becomes an ambient optic array for this perceiver. Merely by moving through the environment, the perceiver is able to sample information contained in the ambient optic array in the form of invariants. Thus, for instance, surfaces in the environment will create discontinuities in the optic array that specify edges. These discontinuities remain invariant as the observer moves around the surface, providing information about the shape and composition of objects in the environment.

The novelty in Gibson's theory of perception lies in this shift from inverse optics. No longer is the task of the perceiver one of extracting a three-dimensional representation of the world from a two-dimensional one. Rather, the perceiver's task is to explore his environment, uncovering invariants in the ambient optic array that specify features of the world. The information is in the world and ready to be 'picked up.' There is, therefore, no need to construct a representation to play the role that the information in the optic ambient array already plays. For Gibson, perception is direct. It involves no symbolic middlemen.

One way to approach Rodney Brooks' (1991) work in robotics is to take it as implementing the Gibsonian theory of perception. Brooks' 'creatures' rely on what Brooks calls a subsumption architecture. This architecture comprises 'layers' that perform very simple tasks. Many of these tasks involve little more than immediate reactions to the detection of simple features of the environment. Thus, there might be a layer whose job is to stop the creature when it senses an object in front of it. Another layer might have the job of maintaining sensory contact with an object to the creature's left or right side, and another layer might simply direct the creature to move straight ahead. Making Brooks' work exciting is the behaviour his creatures exhibit simply through the summation of the multiple layers' behaviours. Creatures navigate through a busy environment, avoiding objects, changing courses, as if they were consulting a map. In fact, however, creatures do not construct map-like representations of their environments – of landmarks, of hazards, of hallways. There is no need for such a map, Brooks asserts, because his creatures are built to exploit the structure in their environment. A model of the world is not necessary, Brooks explains, because the world is its own best model.

Clearly, Brooks' work embodies Gibsonian principles. Just as Gibson urges the rejection of representations in favour of the detection or 'picking up' of information in the environment, so Brooks seeks to build robots that forsake rich representational content in exchange for the detection of informationally loaded variables. Instead of building a representation of an object in order to compute a course around it, a creature might just keep the object to its left as it moves forward. When it no longer senses the object, it can turn left again to continue on its course. In this way, the creature *uses* the object in 'figuring out' how to avoid it. It has no need for a representation of the object.

The third example of symbol antagonism I wish to discuss is Esther Thelen's (1995) study of motor development in infants. Thelen has shown that coordinated kicking or stepping behaviour in infants is modelled better in terms of a dynamical system than it is in terms of painstakingly calculated neural signals. The large muscles in the legs act as springs under the control of gravity. Stepping and kicking movements, rather than under the direction of a cognitive plan, are nothing more than the oscillatory movements of a spring. Similarly, the coordination involved in stepping, Thelen argues, does not involve any cognitive control, but results from the coupling between single-leg oscillators. Similarly, two pendula mounted on a single wall will, over time, synchronize their swing. The coordination in both cases is simply a product of unguided physical forces.

As with Gibson and Brooks, Thelen favours explanations that are antagonistic to the postulation of symbols. One can imagine her response to a symbolist who insists that an explanation of motor development must employ symbolic representations of trajectories, masses, and other variables that control motion. 'That's all very nice,' she might say, 'but what's really going on is simply the synchronization of oscillators.' To insist on a role for symbols would seem, from her perspective, simply gratuitous – an effort to fit the facts to the theory rather than the other way around.

The conclusion I draw from these brief descriptions of paradigm work is that embodied cognition distinguishes itself from symbolism through its promotion of a theory of mind that seeks to overturn traditional views of cognition as symbol manipulation. More specifically, where embodied cognition seems to depart from symbolism is in its preference for explanations of cognition that tie symbols more directly to idiosyncratic facts about organisms or interactions between organisms and their environments. This is evident in Barsalou's theory of perceptual symbols, which holds that symbols are copies of perceptual states and undergo processing in the same perceptual systems in which the initial perceptual states were produced, thereby tying the notion of representation to the particularities of organisms' sensory equipment. This is also evident in the work of Gibson, Brooks, and Thelen, all of whom try to exchange representation-heavy accounts of cognition for ones that let interactions between organisms and their environments do the work traditionally assigned to amodal symbolic representations.

4.5 **The debate in the context of philosophy of science**

I have so far defended a negative and a positive thesis. The negative thesis is that symbolists are not guilty of the symbolist's folly. There are theories of meaning, such as the causal theory of meaning, that symbolists might pursue to explain why Chinese symbols have no meaning for the man in the Chinese room despite having meaning for ordinary Chinese speakers. Accordingly, symbolists have a response to those embodied cognition researchers who attack symbolism for committing the symbolist's folly. The positive thesis is that embodied cognition distinguishes itself from traditional symbolic accounts of the mind through, in the case of researchers like Barsalou, an effort to reconceive the traditional notion of a symbol, or, more radically, in an effort to show that symbols play a much more anaemic role in cognition than symbolists have claimed, and that instead cognition emerges from the actions of an organism in an informationally laden environment.

I want now to consider how the symbolist might respond to the more radical views of Gibson, Brooks, and Thelen. Doing so will reveal that the debate between embodied cognition theorists and traditional symbolists is an instance of a broader debate in philosophy of science concerning realism. As such, future work in embodied cognition is likely to be of benefit from a philosopher of science's perspective.

The traditional symbolist is likely to look on the positive thesis of embodied cognition as a simple confusion. Perhaps, the symbolist might concede, there is more information in the environment than has been typically appreciated. Gibson's uncovering of invariances in the ambient optic array, Brooks' success building robots that spend less time planning and representing than traditional robots, Thelen's discovery that motor development does not require sophisticated neural representations, and so on, are all very interesting, the symbolist might agree. Nevertheless, the symbolist might insist, none of these discoveries impugns the view that cognition is symbolic.

The symbolist's apology would go like this. Whether there is little information in the environment or lots, the basic fact remains that information cannot move from the environment into the organism without an intervening representational step. It is all well and good to notice that the environment is chock full of information, but this information can only be of service to an organism if the organism is able to respond to it; and responding to information is possible only by way of representing that information. What's out there is of no use if it remains out there; but in order to bring it in here – into the mind – it is necessary to represent it.

The symbolist is free as well to be suspicious of claims like Gibson's that organisms simply 'pick up' information, or, as he will sometimes say, 'resonate' to information. These are just metaphors, the symbolist might charge, that when cashed out suggest a commitment to representation. One can grant Gibson the idea of invariants in the ambient optic array, but there remains the question how organisms come to use these invariants. The invariants are of no value unless an organism has a means by which to detect them, but what is detection if not representation? Likewise for Brooks' creatures. Perhaps they don't construct a map of their world, but they do represent objects in front of them, objects to the left or right, the command to go straight ahead, and other such simple things. Furthermore, perhaps rich motor plans are not necessary for motor development, as Thelen claims. Still, an infant's legs can move in synchrony only if able to respond to each other, and they can respond to each other only if there is some sort of representation of each leg's activity.

This symbolist response to the positive idea in embodied cognition raises many issues, and I cannot hope to resolve any of them in the present context. However, let me try to sort out some of them. First, the symbolist's claim that representation is always necessary to get information from 'out there' to a place where an organism can use it is, in some cases, simply false; but in other cases it misses the more important lesson that embodied cognition teaches.

The claim is false when the explanandum is the coordination of infant stepping behaviour. Assuming that Thelen's model of stepping behaviour is correct, there is no need to attribute representations for the purpose of understanding how the infant's legs synchronize. The principles responsible for the legs' synchrony are the same as those that explain the synchrony of two pendula on a wall. The pendula clearly affect each other, but this is a case of response without representation. It's closer in spirit to the effect a strong wind has on a palm tree than it is to a computation. I think the symbolist should be prepared to admit that some behaviours that might once have seemed to fall within their province are in fact behaviours that are better explained non-symbolically.[2]

The sense in which the symbolist response misses the point of embodied cognition becomes apparent when considering Gibson and Brooks' research. One might believe that Gibson and Brooks overstate their cases against representation, yet still believe that their work marks a significant departure from symbolic explanations of cognition. On a charitable reading of their work, they are less concerned with whether some aspects of cognition are representational than they are with investigating the extent to which the active exploration of an environment can play the role that has traditionally been assigned to symbolic computation. Much of Gibson's work is concerned with discovering what invariants are present in light such that an organism's computational burdens might be reduced. Similarly, Brooks' creatures are attempts to maximize behavioural versatility while minimizing computational load.

We can now see the sense in which the symbolist's response is off target. Researchers who pursue embodied cognition are calling into question the need for traditional

---

2 I have significant reservations about whether Thelen's work should be relevant in this dispute. Perhaps we should not consider the kinds of behaviour Thelen examines to be cognitive at all. What exactly is definitive of cognitive behaviour?

symbolic representation in cognitive explanation. Seen this way, embodied cognition is just the latest chapter in a debate that is very familiar in the broader context of philosophy of science. The debate concerns the justification of explanatory posits. Symbols, like atoms, forces, and other unobservables, will earn their keep insofar as they do the kind of work that these other unobservables have done. If the attribution of symbols is necessary for modelling cognitive processes, for predicting cognitive behaviour, for building cognitive systems, and so on, then the symbolists will have done as much as any scientist can do to establish the existence of their theoretical entities. On the other hand, if embodied cognition researchers are able to explain cognitive behaviour without assuming the existence of amodal, computational symbols, and if traditional symbolic explanations prove themselves to be as ill-founded as, for example, explanations of life that assume the presence of an *élan vital*, or of heat transfer that assume the existence of phlogiston, then so much the worse for symbolism.

Interestingly, Barsalou (1999) picks up on some of these broader issues. In commentary on Barsalou's theory of perceptual symbols, Murat Aydede (1999) raises an objection similar to one I raised earlier. Amodal symbols of the kind that the person in the Chinese room is manipulating *can* be grounded, Aydede assumes, given appropriate causal connections to the world. Thus, Aydede would presumably agree with me that Barsalou and others attack a straw man when arguing that amodal symbols face a special kind of grounding problem. In response to this criticism, Barsalou seems to soften his stance. He claims that 'Aydede notes that perceptual representations could indeed be a part of this [causal] sequence, yet he fails to consider the implication that potentially follows' (1999, p. 638). What follows, Barsalou argues, is that perceptual symbols might by themselves suffice for cognition – there is no need to go the extra step in positing the creation of amodal symbols from perceptual symbols. As Barsalou says, 'why include an additional layer of amodal symbols in the causal sequence, especially given all the problems they face? Nothing in Aydede's commentary makes a case for their existence or necessity' (1999, p. 638).

Barsalou's response to Aydede makes clear that the success of embodied cognition will depend in part on whether embodied theories of symbols and symbolic processing can offer better and simpler accounts of cognition than those currently available. It is in this context that the conceptual tools of the philosopher of science grow in importance. What counts as better? What counts as simpler? Answers to these questions requires a critical study of the research that participants in the debate, as well as many others, eagerly pursue.

Debate

Arthur Graesser: In the first draft of the framing chapter, I don't know whether it was you, Manuel, or Art who wanted to keep representation out of there, that it was bringing up a bunch of problems. And I wanted it in there. But here's the question: is it really true that the embodied position is denying representation? It was my understanding that it allowed representation. It was modality-specific and all this sort of stuff, but it was representational.

Lawrence Shapiro: Well, this is the question that I had for you when you were talking about embodied representations. If you have an embodied representation, what does that mean? Does that mean it's a non-symbolic representation? See, I thought that the work of people like Gibson, Brooks, and Thelen was intended to be antagonistic toward symbols. 'You say you need a cognitive map,' Brooks might say, 'but here, my creatures get around the environment – no cognitive map.'

Graesser: I see them not as symbol antagonists but as representational antagonists. But I'm just … well, I don't know, maybe Manolo can speak on this.

Manuel de Vega: I was surprised by your apparent identification of embodiment theories with non-representational theories. I think it is quite natural to consider that representations are embodied. First of all, you can establish a neurological description in which you find out some sort of overlapping between meaning and perception and action. This is clearly compatible with off-line embodied representations stored in the brain. Secondly, you can have some sort of behavioural study in which you demonstrate, for instance, interference or facilitation between perceptual and motor events and sentence meaning. In all these cases you receive a symbolic code which is language, and something occurs in your brain that we could call a 'representational activity', and this activity is functionally important because it interferes or overlaps with perceptual–motor activity, and it is embodied.

Shapiro: Well, now I'm getting confused what the contrast is supposed to be between embodied approaches to cognition and symbolic approaches. If embodied approaches are consistent with symbolic approaches, then what's the debate about? I'm really confused.

Antoni Gomila: May I continue? A historical point, which makes me feel old; you're younger. In the late 1970s, early 1980s, the discussion took the form of the imagist debate, whether there are images in the mind and the discussion was images against propositions.

Shapiro: Yes, between Pylyshyn and Kosslyn.

Gomila: Well, it still goes on, but I would say that the result of that debate was that we have to acknowledge the existence of non-propositional forms of representation. But not like pictures in the head that have to be seen. But, as a way to understand this imagistic representation, we have this perceptual systems theory that tries to account in terms of sensory or motor activity. The difficult issue here, or the terminological issue, is whether for this approach embodied meaning is just grounded symbols. Because if it's the same then it's a symbolist position, but the theory of the grounding of the symbols is very different. It cannot be a causal theory.

Shapiro: Well, again I want to resist this idea that to be a symbolist means you have to ground the symbols in a way that Searle's Chinese room suggests they have to be grounded. Now, I'm perfectly happy to accept, speaking as a symbolist – and you know, I don't know what I am – but if I were a symbolist, I'd be happy to accept this idea that some of my thoughts have meaning in virtue of my bodily interactions with the environment. But people in embodied cognition have to say more than that to distinguish themselves from traditional conceptions of cognition. I mean, no one doubts – or no one should doubt – that my perceptual states acquire meaning because there's a world outside me. I have to open my eyes and interact with the world in order to have perceptions with content. But embodied cognition people have to say something more distinctive than that. They have to say something more distinctive than our perceptions, our thoughts, get meaning by interaction with the environment. This is something a symbolist would easily accept. They'd say 'well, of course if your brain were in a black box, then you wouldn't have any thoughts.' So what more does embodied cognition say? How do you distinguish or make novel what embodied cognition is about? And my efforts to answer that question led me to this view that, well, given that everyone agrees that we have to interact with the environment, and it's our body that's going to allow this interaction, the real difference has to be in the role that embodied people think symbols play in cognition. And what they want to try to do – as I think Thelen, Gibson, and Brooks try to do – is to develop accounts of cognition that minimize the role of representation.

Anthony Sanford: Yes, I was interested in your ideas about phlogiston and *élan vital*. I mean, do you see the problem as being that we just don't know enough about how the brain actually does things? That is, do you think the distinction between symbol activity and embodied activity would disappear under one umbrella if one could get to the appropriate level of neuroscience? Is that what you mean, or do you mean something different, or something completely unknown?

Shapiro: Here's what I mean … if we could watch peoples' brains as they behave we wouldn't see any symbols in their brains, okay. When we attribute symbols to a cognitive agent or to the process of cognition, these are theoretical posits. We're attributing symbols. This is a way of describing what's going on and it's in virtue of this description that we're able to predict and make explanations. And in this regard I think cognitive science is like any other science. No one sees forces. No one sees gravity. These are posits that allow us to predict and model things. Symbols are like those. The question, as I see it, is, are these posits really valuable? Do we need these things? Is there a better way of explaining things? So, we no longer need to talk about angry gods to explain why lightning sometimes hits the earth. That's turned out to be a very unuseful posit. And I think it's still an open question whether symbols are a valuable posit. And the kind of ecumenical conclusion I like is that, well, it sure seems that they're necessary for some cognitive processes like language use and perhaps dead reckoning, but perhaps they're not … perhaps we can look at Brooks' work to see that, no? If we thought we needed a symbolic representation in the environment to navigate through it, we were wrong about that, so that's one case where they're more like *élan vital* or phlogiston. That's what I was trying to develop there.

Luc Steels: The game is to grab the microphone here. So, I just wanted to clarify this position of Brooks and others in behaviour-based robotics. I mean it was really as you said, you take a behaviour like following a wall or avoiding an obstacle … the point of this is actually very simple, which is you don't need to categorize something as an obstacle or something as a wall because you build a direct link between, let's say an infrared sensor, which is reflecting from the wall. So there is no categorization. I think that's the better way to think about this. You have an attraction to the wall or a repulsion from the wall, as an obstacle. And then if you put two things together and let them go, then you see emergent wall-following without having the category 'wall', so you're no longer talking about symbols or no symbols ... well, representations or no representations. I think the easiest way to understand is to ask how conceptual these categories are. All of this happened ten years ago, or fifteen years ago. So today, in robotics at least, I think we realize you can do many things this way. But as soon as you, let's say, do navigation and you want to, say, remember what was behind you, you need some sort of representation of that. So I think it's generally recognized that for sensorimotor behaviour not having these categorizations would be fine, but as soon as you go to something more interesting then you need categorization. Which is in fact what you said.

Shapiro: Yeah, it's about that line – at what point you need them, and at what point you don't ... I think the line used to be, you know, 'oh, you need them for everything'. And now, I don't know whether the line is moving up or down, but it's changing position, and this is why I was talking to Deb [Roy] last night about Mitchel Resnik's work. You know he's got this book *Turtles, Termites, and Traffic Jams*? It's full of these neat examples of behaviour that looked like, gee, you needed a representation to explain it and it turns out you don't.

Friedemann Pulvermüller: I have to disagree with one thing you said. You said that if we looked at the brain we wouldn't see any symbols. I think we have ... when looking at brain activation as people process symbols, we see very clear brain correlates of different types of symbols and, therefore, in the relevant sense we see brain correlates of symbols.

Shapiro: Okay, that's not something I want to take issue with. I mean, if you think of a word as a symbol, and suppose that corresponding to every distinct word there is some brain state, I'd say 'yes, these are correlates of symbols.' My objection was something like this: suppose that your best description of how bees find the source of nectar involves a Randy Gallistel-type explanation that attributes to bees all these sorts of vectors and positions of the sun and things like that. Well, this is a level of description, right? If we look in the brain we're not going to see vectors. And what we do is we impose this symbolic model onto this hardware. But are there really symbols in the brain? No.

Pulvermüller: I think we are very clear that such a symbol exists or has a clear framework or correlation. The question is how to describe it, whether you want to use vector description of the symbol or whether you want to use an embodied description, a characterization. I think that's the question as I understand it.

Shapiro: I think we're agreeing more than you think we are. Let's talk about this more later.

Author note

For useful discussion I would like to thank the participants in the Garachico workshop, as well as Ken Aizawa, Juan Comesaña, Dan Hausman, Carolina Sartorio, and Elliott Sober. Special thanks to Art Glenberg and Toni Gomila for comments that led to an extensive reworking of some of the ideas in this chapter.

References

Aydede M (1999). What makes perceptual symbols perceptual? *Behavioural and Brain Sciences*, 22, 610–11.

Barsalou L (1999). Perceptual symbol systems. *Behavioural and Brain Sciences*, 22, 577–609.

Brooks R (1991). Intelligence without representation. *Artificial Intelligence*, 47, 139–59.

Dretske F (1981). *Knowledge and the Flow of Information*. Cambridge, MA: MIT Press.

Fodor J (1975). *The Language of Thought*. Cambridge, MA: Harvard University Press.

Fodor J (1987) *Psychosemantics: The Problem of Meaning in the Philosophy of Mind*. Cambridge, MA: MIT Press.

Gibson J (1979). *The Ecological Approach to Visual Perception*. New York, NY: Houghton Mifflin.

Glenberg AM, Kaschak MP (2002). Grounding language in action. *Psychonomic Bulletin and Review*, 9, 558–65.

Marr D (1984). *Vision: A Computational Investigation in the Human Representation and Processing of Visual Information*. San Francisco, CA: WH Freeman.

Searle J (1980). Minds, brains, and programs. *Behavioural and Brain Sciences*, 3, 417–24.

Stampe D (1977). Towards a causal theory of linguistic representation. In P French, T Uehling, H Wettstein, Eds. *Midwest Studies in Philosophy 2: Contemporary Perspectives in the Philosophy of Language* (pp. 42–63). Minneapolis, MN: University Press.

Thelen E (1995). Motor development: a new synthesis. *American Psychologist*, 50, 79–95.

Chapter 5

What brain imaging can tell us about embodied meaning

Marcel Adam Just

5.1 Introduction

Brain imaging studies of language processing, using functional magnetic resonance imaging (fMRI), can indicate under what circumstances the embodied aspects of language representations become activated. In particular, the processing of language is distributed across a number of cortical centres, including not only classic language areas in association cortex (which might be involved in symbolic processing), but also sensory and motor areas. A set of fMRI studies on visual imagery in sentence comprehension reveals both the perceptual–motor and symbolic aspects of brain function that underlie language processing. Moreover, they indicate some of the conditions under which perceptual or motor representations are most likely to be activated. Another set of studies on word comprehension indicates that the neural signature of certain concrete semantic categories (*tools* and *dwellings*) and individual category exemplars can be identified by machine learning algorithms operating on fMRI data, and that perceptual and motor representations constitute part of the signature.

5.2 Functional magnetic resonance imaging studies

5.2.1 Visual imagery in sentence comprehension

Many types of thinking, in particular language comprehension, entail the use of mental imagery. Understanding a text on architecture or automobile design seems impossible without mental imagery. Language often refers to perceptually-based information. For example, to evaluate a sentence like 'The number eight when rotated 90 degrees looks like a pair of spectacles', a reader must process the content of the sentence, retrieve a mental image of the shape of the digit 8, mentally apply a rotation transformation to it, and then evaluate the resulting image. In this case, there seems little doubt that a perceptually-based representation is involved in the comprehension. This perceptually-based representation has sometimes been called the *referential representation* or the *situation model*. Our fMRI studies attempted to determine the characteristics of such representations and the conditions under which they are likely to be evoked or activated.

Previous studies have indicated that mental imagery generated by verbal instructions and by visual encoding activate similar cortical regions (Mellet *et al.* 1996, 1998, 2002;

Mazoyer *et al.* 2002). Several studies examining mental imagery have observed activation of the parietal area (*Just et al.* 2001; *Mellet et al.* 1996, 2000; *Deiber et al.* 1998; *Ishai et al.* 2000, *Kosslyn et al.* 1993, 1996, 1999), particularly around the intraparietal sulcus. Our imaging studies attempted to determine the conditions under which such activation occurs during language comprehension.

There is also a possibility that the neural activity underlying the imagery in language processing is affected by the presentation modality of the language (i.e., written versus spoken). For example, the neural activity elicited in primary visual cortex during mental imagery following verbal versus visual encoding was different (Mellet *et al.* 2000); there was less primary visual activation during imagery after visual encoding compared with verbal encoding, suggesting that presentation modality may indeed affect later imagery processing. Eddy and Glass (1981) examined how the visual processes in reading might be related to the visual imagery processes that a sentence engenders, comparing visual and auditory sentence presentation modes. High-imagery sentences took longer to verify as true or false than low-imagery sentences when the sentences were presented visually, but not when they were presented auditorally. These findings again suggest that the presentation modality of a sentence may affect the processing of the subsequent imagery.

Our studies examined mental imagery processes in the context of a language comprehension task (Just *et al.* 2004). One of the main goals was to examine the interaction between two somewhat separable neural systems, the mental imagery and language processing systems. In the context of the embodiment debate, these studies ask not whether embodied (perceptual or motor) activation occurs, but the circumstances under which it occurs and how it is related to other more symbolic activation. To accomplish this goal, we used fMRI to measure not only the activation levels, but also the functional connectivities of the regions believed to be involved in mental imagery, to determine the relations between the two systems. A second goal was to examine the effect of input modality, comparing the effect on the imagery-related activation when the sentences were either heard or read.

The study examined brain activation while participants read or listened to high-imagery sentences like 'The number eight when rotated 90 degrees looks like a pair of spectacles' or low-imagery sentences, and judged them as true or false. They included sentences requiring various types of spatial transformation or spatial assessment such as mental rotation (like the spectacles example), evaluation of spatial relations (e.g., 'On a map, Nevada is to the right of California'), combination of shapes (e.g., 'The number nine can be constructed from a circle and a horizontal line', a false example), and comparison of visual aspects of common objects (e.g., 'In terms of diameter, a quarter is larger than a nickel, which is larger than a dime'). Although these sentences generally required that a spatial transformation be mentally performed, pilot studies indicated that understanding a complex spatial description without a transformation produced similar results. The low-imagery sentences could be verified by referring to general knowledge, without the use of imagery (e.g., 'Although they are now a sport, marathons started with Greek messengers bringing news').

The sentence imagery manipulation affected the activation in regions (particularly the left intraparietal sulcus) that activate in other mental imagery tasks, such as mental rotation. Both the auditory and visual presentation experiments indicated much more activation of the intraparietal sulcus area in the high-imagery condition, as shown in Figure 5.1, suggesting a common neural substrate for language-evoked imagery that is independent of the input modality. There was more activation in the intraparietal sulcus area in the reading than in the listening condition (probably owing to the attentional demands of directing spatial attention and possibly eye movements to particular sentence locations during reading), but the magnitude of the imagery effect was comparable.

5.2.2 Functional connectivity and imagery

The various anatomical regions of the cortex involved in processing a task must be able to effectively communicate and synchronize their processes for the system to function. In a language task, this means that the areas responsible for executing subcomponent processes must collaborate to synthesize the information necessary for comprehension.

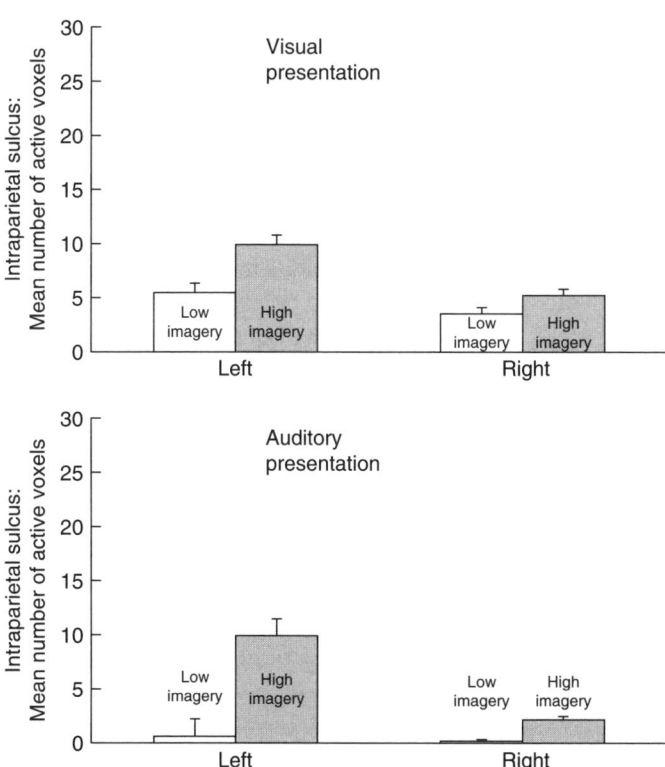

Fig. 5.1 The activation in the intraparietal sulcus area is higher for high imagery sentences (particularly in the left hemisphere), and the effect is similar regardless of whether the sentences are presented visually or auditorally.

Such collaboration can be measured in functional neuroimaging studies by computing the correlation of the activation time series in a given region with the activation time series of another region. The extent to which the activation levels of two regions rise and fall in tandem is taken as a reflection of the degree to which the two regions are functionally connected; the term that is widely used to refer to the activation time series correlation is *functional connectivity*. Previous research has provided some evidence that as task demands increase, functional connectivity also increases, for example, as a function of working memory load (e.g., Diwadkar *et al.* 2000), reflecting the need for tighter coordination in a more demanding condition. Functional connectivity has also been shown to be modulated by comprehension difficulty, and differentially so for people of different working memory capacity (Prat *et al.* 2007). A relation between functional and anatomical connectivity has been demonstrated in autism, where the functional connectivity between cortical regions is correlated with the size of the corpus callosum segment that anatomically connects them (Just *et al.* 2007).

In addition to exhibiting higher activation levels during the processing of high-imagery sentences, the left intraparietal sulcus also showed greater functional connectivity in this condition with other cortical regions – particularly with language processing regions – regardless of the input modality. The imagery manipulation affected the functional connectivity between the left intraparietal sulcus area and other brain areas that are activated in this task. It also affected the functional connectivity between the left superior temporal (Wernicke's) area and other brain areas. The left intraparietal sulcus and Wernicke's area are proposed to be involved in the imagery processing of high-imagery sentences and semantic (symbolic) processing, respectively. The five other activated brain areas were located in the left hemisphere, including areas centrally involved in language processing (pars opercularis, pars triangularis, dorsolateral prefrontal cortex, frontal eye fields, and the inferior parietal lobule). The left intraparietal sulcus had higher functional connectivities with the five activated brain areas when the sentences were high in imagery, whereas the left temporal area had higher functional connectivities for low-imagery sentences. This result applies to both the visual and auditory presentation conditions, as shown in Figure 5.2.

This result provides important converging evidence for implicating the intraparietal sulcus in imagery processing during sentence comprehension, and more directly indicates the higher degree of functional interaction between this embodied imagery representation and some of the other key activated regions in the high-imagery condition.

5.2.3 Sentence imagery in autism

In a recent study, we examined the processing of high- and low-imagery sentences in adults with high-functioning autism, in comparison with age- and IQ-matched controls. This study provides the opportunity to determine whether the use of imagery (or embodied representations) in sentence comparison might be disrupted in a special neurological population. Can meaning embodiment be disrupted or modulated by a neurological disorder?

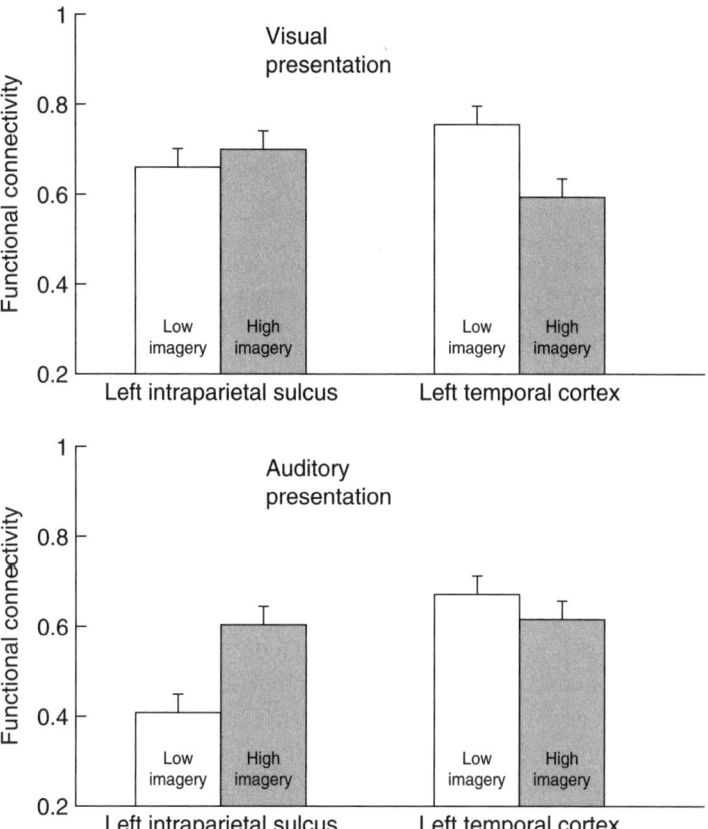

Fig. 5.2 The average functional connectivity of the left intraparietal sulcus region of interest (LIPS) and the left temporal region of interest (LT) and with five left-hemisphere regions of interest: pars opercularis, pars triangularis, dorsolateral prefrontal cortex, frontal eye fields, and the inferior parietal lobule. The functional connectivity between LIPS and these areas is greater for high-imagery sentences. The opposite is true for LT, which is not involved in imagery processing. The same pattern occurs for both visual and auditory presentation of sentences.

First, consider the effect of imagery on the control participants, which is very similar to the study described above. Plate 5.1 shows the imagery effect (high- minus low-imagery), displaying the prominent extra activation in the parietal area, as well as some in the prefrontal and inferior temporal areas.

But what of people with high-functioning autism? There have been frequent suggestions that spatial/perceptual processing is spared (or sometimes even enhanced) in autism. For example, Shah and Frith (1983,1993) found that participants with autism have an advantage at certain types of spatial tasks. When a task is amenable to either a visual or a verbal strategy, there is a suggestion that people with autism prefer a visual strategy. Indeed, there are many informal reports that individuals with autism are predominantly visual thinkers. For example, Temple Grandin, an accomplished professional with high-functioning autism, entitled her book *Thinking in Pictures* (Grandin, 1995). In an fMRI letter n-back study, Koshino *et al.* (2005) found more visual coding of letters in autism compared with a verbal coding strategy in the controls. Similar results were also found in a faces working memory task (Koshino *et al.* 2008). These studies indicate that there is a tendency in people with autism to use more visuospatial regions by recruiting posterior brain regions in accomplishing varied tasks, including language tasks.

The group with autism showed similar activation to the controls in the processing of high-imagery sentences (prominent activation in the parietal area, particularly around the intraparietal sulcus). It is reasonable to conclude that in the high-imagery condition, which can be said to make use of embodied (perceptual) representations, the processing of the autism group resembled the processing of the control group. However, the interesting new result was that unlike the control group, the autism group displayed a similar amount of parietal imagery-related activation even in the low-imagery condition. Plate 5.2 shows the minimal imagery effect (the subtraction of high minus low imagery). It appears that the autism group uses embodied representations even in contexts where control participants do not.

The tentative account for why this occurs is that the cortical connectivity of the neural system is compromised in autism (the main tenet of the underconnectivity theory of autism), particularly affecting the communication between frontal and more posterior areas, resulting in a greater reliance on the latter. Thus, in autism, there is often less frontal activation and more posterior activation in a number of different types of tasks. One might say that cognition is more embodied or concrete in autism, and less abstract.

5.2.4 Visual imagery in the comprehension of novel metaphors

Mason *et al.* (under review) compared the comprehension of frozen and novel metaphors and found an interesting difference that bears on embodiment. During the comprehension of a frozen metaphor passage, the same language processing areas are active as in normal reading (e.g., left middle and superior temporal lobe, left inferior frontal gyrus, dorsolateral prefrontal cortex bilaterally, right middle, superior temporal, and superior medial frontal areas). However, it is the contrast with novel metaphors that is important here. The novel metaphors evoked parietal activation around the intraparietal sulcus area bilaterally, suggesting that visual imagery processes were being used to instantiate and/or interpret the novel metaphors. The novel metaphors were generally visual in nature, such as a metaphor comparing a winding road to a ribbon. These results demonstrate the selective use of imagery in metaphor comprehension, suggesting that perceptual representations are used in the comprehension of novel, but not frozen, metaphors. It may be that the mappings between domains that are needed in novel metaphor comprehension (e.g. the mapping between the solar system and the atom, to take a clearly spatial example) are often mediated through spatial representations. A more general principle of neural function is that a given representation or process can be activated on an 'as-needed' basis. In this view, embodiment occurs only for those types of metaphors that require it for their appropriate comprehension.

5.2.5 Selectivity of the perceptual representations activated in a haptic imagery task

Imagery-related activation around the intraparietal sulcus can also be evoked in a haptic (touch) imagery task. Newman *et al.* (2005) questioned participants on either the geometric (visual) properties of two objects (e.g., 'Which is larger, a pumpkin or a cucumber?') or the haptic properties (e.g., 'Which is softer, a lemon or an egg?'). When the haptic properties (e.g. hardness) of the objects are interrogated via haptic imagery, there is extra activation in the inferior extrastriate region. Additionally, a region in the lateral occipital cortex is activated for either type of probe. Thus, thinking about the object in almost any way evokes *some* perceptual activation in a number of areas, but the amount of activation increases in areas whose information is probed. Thus, there is selectivity in the perceptually-related activation.

5.2.6 Summary of fMRI data

Our fMRI imagery studies have taught us many lessons about embodiment, particularly highlighting the circumstances under which embodied representations are likely to come into play. First, perceptual (embodied) representations are used when there is some mental action or assessment to be performed on the perceptual representation in order to comprehend the sentence or perform the task. Second, in the processing of sentence imagery, the parietal activation associated with the perceptual representation is synchronized with the symbolic language areas with which it is collaborating. Third, in autism, perceptual representations are used for comprehending even low-imagery sentences, which control participants process without the benefit of perceptual representations. Finally, perceptual representations are used in the comprehension of novel, but not frozen, metaphors. All of these lessons have an overarching theme: embodied representations are activated more often or to a higher level in situations in which perceptual information is particularly useful or salient.

5.3 The embodied neural signature of words referring to object concepts

In a multidisciplinary project done in collaboration with Tom Mitchell and our two research groups, particularly involving Robert Mason, Svetlana Shinkareva, Vincente Malave, and Francisco Peirera, we have been using machine learning techniques that can determine what activation pattern defines a cognitive state. Our investigation of the representation of words and sentences attempted to determine the neural signature of a semantic category using an engineering rather than a neuroscience approach. We started with concrete nouns as stimuli, taken from a small number of categories. In one study, we showed participants a total of 14 different words, including seven tools (e.g., 'hammer', 'hatchet', 'pliers', 'screwdriver') and seven dwellings (e.g., 'mansion', 'castle', 'palace', 'hut', 'apartment'). Although unconventional (as were the data analyses and goals), the experimental tasks attempted to isolate individual thoughts of brief duration, typically about 4 seconds. We first presented each word several times during a scanning session (six times in this experiment). We then trained the classifiers on all but one of the presentations, trying to classify the activation patterns in the omitted presentation, iterating through all the possible ways of leaving one presentation out. A classifier, of which logistic regression is a common example, is a decision system which takes as its input the values of some features, and produces as output a discrete label related to the input values. In this application, the input to the classifier is the activation level of a set of voxels measured during the reading of a word, and the output is the category that the word belongs to.

In our quest for the neural signature of concepts, we have discovered that the neural representations of these concepts at least contain perceptual and motor information that is pertinent to a category. The embodiment of a category, namely its representation of its perceptual and motor attributes, is part of the category's neural signature. For example, the representations of tools include voxels in the motor area and in the somatosensory area.

There are several ways in which our findings resemble those of two other fMRI projects reported in this volume (see Pulvermüller, Chapter 6). First, the neural signatures span a number of areas distributed over the cortex, including all four lobes in both hemispheres, as well as the cerebellum. Second, what differentiates the activation for different categories (what makes it a signature) is not which voxels are activated, but the intensity of their activation.

Our results extend beyond only these commonalities with the other studies. In addition, we have been able to classify with some accuracy the category of the word that is being read. Moreover, we have been able to identify the individual word that a person is reading from the neural signature. The machine learning techniques were able to identify which of two categories of words is being read by a participant, with a median rank accuracy over 12 participants of 79% (as of December 2005). (Rank accuracy is the percentile rank of the correct word within a list of predictions made by the classifier; chance rank accuracy would be 50%). For the participant with the best result, the classifier was extremely accurate in identifying the category of the word being read, with a rank accuracy of 93%. The classifier was also able to identify to some extent which of the 14 different words was being read, with a median rank accuracy over 12 participants of 69% (still far above chance). Rank accuracy for the best participant was 76%. The voxels used by the classifier to distinguish among categories and among words are distributed across many regions of the cortex, including the right hemisphere and motor areas. For example, Plate 5.3 shows a motor region in which several participants had diagnostic voxels, indicating that the motor representation of how a tool is used is part of the meaning of a word that names a hand tool.

The rank accuracy of the identification (particularly for distinguishing among the 14 words) is measured as follows: for each test input, the trained classifier outputs a rank-ordered list of the 14 words, ordered from most to least probable according to its learned model. A perfect identification produces a rank accuracy of 1 (the correct answer is ranked as the most likely) and random performance by the classifier produces a rank accuracy of 0.5. Classifier accuracy was measured by the percentile rank of the correct classification in the output sorted list.

The classifier that has been most expedient for our research is a Gaussian Naïve Bayes (GNB) classifier. Other classifiers that we sometimes use include artificial neural networks and logistic regression, with somewhat similar results across classifiers. The input to the classifier is represented as a feature vector, where each feature typically corresponds to the observed magnetic resonance intensity value at a specific voxel at a specific time, relative to a baseline. Although the classification can be done by providing input to the classifier about all of the voxels, the accuracy is generally higher if only a small subset of all the voxels (say 25–200 out of about 15,000) is used in the classification. We have used a number of different criteria to select the subset of diagnostic voxels. The most useful criterion has been the consistency of the voxels in how they respond to the different words across presentations.

The neural signature is based on many voxels, some of which are located in motor areas, some in perceptual areas, and some in frontal (conceptual) areas. Some informal sensitivity analyses were done to determine whether any subset of voxels is particularly important or unimportant for accurate classification. So far, no striking systematic differences have been found in how useful the voxels in various locations are. For example, voxels in the motor area are not very different from voxels elsewhere in terms of their contribution to classifying the signature.

5.4 Discussion

The neural evidence clearly indicates that, at least in some cases, perceptual and motor representations are activated during processing that is primarily conceptual. The scientific questions for the future might centre around the whens and hows of embodied cognition. Brain imaging provides a useful tool for addressing these questions, as well as providing substantiating evidence that embodied cognition is much more prevalent than was generally assumed a decade ago.

Debate

The debate corresponding to this chapter, as well as Chapters 6 and 7, is included at the end of Chapter 7.

Author note

This research was supported by the National Institute of Mental Health (grant MH029617) and the WM Keck Foundation.

References

Deiber M-P, Ibanez V, Honda MSN, Raman R, Hallett M (1998). Cerebral processes related to visuomotor imagery and generation of simple finger movements studied with positron emission tomography. *NeuroImage*, 7, 73–85.

Diwadkar VA, Carpenter PA, Just MA (2000). Collaborative activity between parietal and dorso-lateral prefrontal cortex in dynamic spatial working memory revealed by fMRI. *NeuroImage*, 12, 85-99.

Eddy JK, Glass AL (1981). Reading and listening to high and low imagery sentences. *Journal of Verbal Learning and Verbal Behaviour*, 20, 333–45.

Grandin T (1995). *Thinking in Pictures: And Other Reports from My Life with Autism*. New York, NY: Doubleday.

Ishai A, Ungerleider LG, Martin A, Haxby JV (2000). The representation of objects in the human occipital and temporal cortex. *Journal of Cognitive Neuroscience*, 12, 35–51.

Just MA, Carpenter PA, Maguire M, Diwadkar VA, McMains S (2001). Mental rotation of objects retrieved from memory: an fMRI study of spatial processing. *Journal of Experimental Psychology: General*, 130, 493–504.

Just MA, Cherkassky VL, Keller TA, Kana RK, Minshew NJ (2007). Functional and anatomical cortical underconnectivity in autism: evidence from an fMRI study of an executive function task and corpus callosum morphometry. *Cerebral Cortex*, 17, 951–61.

Just MA, Newman SD, Keller TA, McEleney A, Carpenter PA (2004). Imagery in sentence comprehension: an fMRI study. *NeuroImage*, 21, 112–24.

Koshino H, Carpenter PA, Minshew NJ, Cherkassky VL, Keller TA, Just MA (2005). Functional connectivity in an fMRI working memory task in high-functioning autism. *NeuroImage*, 24, 810–21.

Koshino H, Kana RK, Keller TA, Cherkassky VL, Minshew NJ, Just MA (2008). fMRI investigation of working memory for faces in autism: visual coding and underconnectivity with frontal areas. *Cerebral Cortex*, 18, 289–300.

Kosslyn SM, Alpert NM, Thompson WL *et al.* (1993). Visual mental imagery activates topographically organized visual cortex: PET investigations. *Journal of Cognitive Neuroscience*, 5, 263–87.

Kosslyn SM, Pascual-Leone A, Felician O *et al.* (1999). The role of area 17 in visual imagery: convergent evidence from PET and rTMS. *Science*, 284, 167–70.

Kosslyn SM, Shin LM, Thompson WL *et al.* (1996). Neural effects of visualizing and perceiving aversive stimuli: a PET investigation. *NeuroReport*, 7, 1569–76.

Mazoyer B, Tzourio-Mazoyer N, Mazard A, Denis M, Mellet E (2002). Neural bases of image and language interactions. *International Journal of Psychology*, 37, 204–8.

Mellet E, Bricogne S, Crivello F, Mazoyer B, Denis M, Tzourio-Mazoyer N (2002). Neural basis of mental scanning of a topographic representation built from a text. *Cerebral Cortex*, 12, 1322–30.

Mellet E, Bricogne S, Tzourio-Mazoyer N *et al.* (2000). Neural correlates of topographic mental exploration: the impact of route versus survey perspective learning. *NeuroImage*, 12, 588–600.

Mellet E, Petit L, Mazoyer B, Denis M, Tzourio N (1998). Reopening the mental imagery debate: lessons from functional anatomy. *NeuroImage*, 8, 129–39.

Mellet E, Tzourio N, Crivello F, Joliot M, Denis M, Mazoyer B (1996). Functional neuroanatomy of spatial mental imagery generated from verbal instructions. *Journal of Neuroscience*, 16, 6504–12.

Newman SD, Klatzky RL, Legerman SJ, Just MA (2005). Imagining material versus geometric properties of objects: an fMRI study. *Cognitive Brain Research*, 23, 235–46.

Prat C, Keller TA, Just MA (2007). Individual differences in sentence comprehension: an fMRI investigation of syntactic and lexical processing demands. *Journal of Cognitive Neuroscience*, 19, 1950–63.

Shah A, Frith U (1983). An islet of ability in autistic children: a research note. *Journal of Child Psychology and Psychiatry*, 24, 613–20.

Shah A, Frith U (1993). Why do autistic individuals show superior performance on the block design task? *Journal of Child Psychology and Psychiatry*, 34, 1351–64.

Chapter 6

Grounding language in the brain

Friedemann Pulvermüller

6.1 Introduction

Can a cognitive theory of embodiment be an abstract theory? Clearly, any theory is abstract in a relevant sense, as it is defined as a theoretical construct. But can a theory of embodiment remain at an entirely abstract level, or would it necessarily need to connect to the concrete mechanistic level of the brain, to nerve cells and circuits? Symbols and meaning are not grounded directly in experiences and actions — or only in a metaphorical sense. Rather, signs and symbols are mechanically based on, or grounded in, neuronal circuits in the brain. Importantly, as we have learned recently, some of these neuronal 'symbolic' circuits seem to play a relevant role in action and perception, too; in this sense, symbols are action-perception grounded. Therefore, the targets of a theory of embodiment include neuronal elements that realize thought, meaning, and language processes. They are not abstract entities but real objects following laws of nature and neuroscientific principles.

Here, a theory of the neuronal embodiment of language and conceptual processes is reviewed, which is built upon neuroscientific principles. *Action–perception networks* (APNs) in the brain are proposed to be the grounding machines realizing the binding of language, perceptual, and action-related information. There is evidence for this proposal from cognitive neuroscience research, and a review of this evidence will be given. It will also be mentioned that, in some cases, it was possible to address specific questions about embodied cognition using neuroscience experiments. The neuromechanics of embodiment and empirical neuroscience research targeting these mechanisms may therefore be welcome additions to cognitive theory.

6.2 Embodiment and the brain

A main idea immanent to cognitive theories of conceptual embodiment is that concepts and the meaning of symbols that function as vehicles of thought are embodied in perceptual and action information (Barsalou 1999; Clark 1996; Lakoff 1987; Lakoff and Johnson 1999; Varela *et al.* 1991). Meanings and concepts, be they abstract or concrete, can be explained, and therefore grounded (at least in part) by referring to concrete sensory information and to information about actions carried out by the individual. Although most cognitive theories of embodiment are at the – in one sense still abstract – level of cognitive representations and processes, it is clear that all cognitive entities are mechanistically based on brain circuits and their activation, as selective deficits after brain lesions demonstrate. From an embodiment perspective, it may be advantageous to look at brain correlates of conceptual and semantic processing, because the mechanistic processes and principles that emerge from the neuronal substrate could potentially contribute to experimental testing and explanation at the cognitive level.

The idea of embodiment goes back quite far in history, certainly to Aristotle, to Plato's cave allegory, and probably even further. A significant step was taken by Sigmund Freud who, when working on the neurology of organic language disturbances in his early career, drew the first diagrams of hypothetical networks of cortical neurons that might process symbols in the human brain (Freud 1891) (Figure 6.1). In one of these diagrams, he connected the neuronal representation of a spoken word, via reciprocal links, to an 'object association' network, which, as we might say today, would embody and cortically organize aspects of the meaning of the symbol. Freud's speculation about the embodiment of word meaning was rejected by neurologists of the 19th Century, but appears surprisingly modern today, as, on the basis of a multitude of neuroimaging results, distributed cortical networks are considered to be a likely basis of higher cognitive processes. A critical point might be that Freud was one of the first to propose an embodiment theory at the cognitive level with an anatomical and physiological basis in the brain. We may ask still today whether a cognitive theory of embodiment can be strengthened by supplying it with a brain basis, by spelling out its mechanisms in terms of neurons. Few would probably deny this, and important steps have been taken towards this goal (Gallese and Lakoff 2005). This chapter is one more contribution, focusing specifically on questions of embodiment in the context of a neuronal model of words, meanings and concepts (Pulvermüller 1992, 1999, 2001, 2005).

One might want to say that a theory of embodiment has to be embodied in the brain, but such a statement inevitably leads to confusion due to the different uses of the critical morpheme. From now on in this chapter, the usage of the word 'embody', and similarly of 'grounding', will be restricted to a relationship between concepts, meanings, signs, symbols, words, and their semantically related actions and perceptions. In contrast, these cognitive elements will be said to be realized, organized, implemented, wired, or laid down in the neural substrate in such and such a way.

Apart from supplying a cognitive theory with a brain basis, evidence from neuroscience may also help in answering burning cognitive questions. This stronger statement

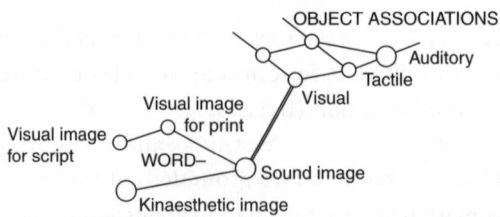

Fig. 6.1 Freud's brain-based model of the embodiment of word meaning in perceptual information. Multimodal 'object associations' were thought to be linked to the word form representation. This model was thought to depict connections and processing units at the level of the brain (Freud, 1891).

will also be addressed here: I will submit that cognitive neuroscience techniques have enabled us to provide strong evidence for an embodiment approach not previously available from behavioural experiments.

This chapter will first ask basic questions about how words are realized as neuron circuits that process actions and perceptions. Subsequently, the question of brain mechanisms of meaning grounded in action and perception will be addressed. Specific questions about phonological and semantic embodiment will be highlighted separately, in the light of recent findings from neuroimaging. Here, the case of category-specific semantic networks will be featured in detail as it turns out to be of particular theoretical importance. Critical questions about category-specific embodied semantic processing in the brain have been addressed in detail by research into the brain processes underlying action-related language. Multimodal neuroimaging work in this area will be reviewed, the emerging conclusion being that semantic processing in the brain may be realized, at least in part, by rapid, automatic, functionally relevant spreading of activation into sensory and motor areas, which reflects aspects of the reference of action and object words. From here, the scope of the chapter will widen, addressing fundamental issues in the brain embodiment of meaning and concepts, the role of correlations between words and world and among words, the discreteness of language representations, and the issue of abstract concepts. The proposal will be that a mere mapping of sensory and motor features is not enough for embodied meaning representation, but that logical operations are necessary as additional pre-wired ingredients.

6.3 Basic questions about language, thought, and the brain

How are words and their related meaningful concepts stored and processed in the brain? We may want to start with a simpler question, that of how objects and actions are processed in brain circuits, and especially in circuits in the brain structure most important for cognitive processes i.e., the cortex. It is well known that many neurons in cortex respond to elementary features of objects, for example form and colour features of visual images of the objects or acoustic length and frequency information of sounds characteristic of the object (Hubel 1995). In the motor domain, cortical neurons usually signal elementary features of muscle movements (Rizzolatti and Luppino, 2001), or, at higher levels, even elementary information about the goals connected with an action (Fogassi *et al.* 2005). As neurons appear to represent and process features of objects, the assumption that collections of neurons (*neuronal assemblies*), are the basis for the processing of sets of features characterizing objects and actions (Braitenberg 1978; Hebb 1949). Sets of neurons with strong links between their members would provide the ideal machinery for maintaining the memory of an object or action, and have therefore been proposed to underlie short-term or active memory (Fuster, 1995). In recent years, the cognitive neuronal equations 'feature = neuron' and 'object/action = neuronal assembly' have received support from a range of lines of neuroscientific research in both animals and humans (Braitenberg and Schüz 1998; Fuster 2003; Plenz and Thiagarajan 2007; Pulvermüller 2005; Singer and Gray 1995).

To bind individual neurons, which signal elementary and possibly more complex features of an object or action into a cell assembly, neural hardware and functional principles that drive the binding are necessary. From the hardware perspective, it must be noted that connections in cortex do not only run between adjacent neurons in the same area and local cortical patch, but also between distant sites. Plate 6.1 illustrates such long-distance connections, taking the example of links between different parts of the perisylvian language cortex (in green and blue) and the motor system (in red). Long-distance connections are indicated by arrows and, in addition, adjacent areas are also connected with each other. These links are, in part, evident from anatomical studies in humans (Brodmann 1909; Makris *et al.* 1999) or can at least be suggested on the basis of studies in monkeys (Pandya and Yeterian 1985; Young *et al.* 1994), taking into account the homology of cortical areas between species. As a multitude of long-distance links are available in cortex, information can be transmitted between distant cortical areas, for example, between superior temporal lobe (in blue), where acoustic information about speech sounds arrives, and inferior frontal cortex (in green), where the motor output is programmed and coordinated.

Apart from the anatomical principle that long-distance links connect different lobes and a range of distant cortical areas with each other, a functional principle is of utmost relevance: Nerve cells that are connected to each other and fire together frequently strengthen their connection (Artola and Singer 1993; Hebb 1949; Tsumoto 1992). In reverse, neurons that fire independently from each other, or in an antiphasic manner, tend to weaken their links, and even fine-grained sequential features of activation patterns can modify synaptic weights (Bi 2002). This means that the correlation of neuronal firing of cortical cells is translated into connection strengths. In essence, what fires together wires together, and the long-distance wiring in cortex guarantees that correlation-related links can even develop within sets of neurons distributed over distant areas and lobes. This has important implications for the way in which objects and actions are processed and stored in brain circuits; if an object is characterized by multimodal features, for example, shape, colour, and sound, which in many cases occur at the same time, it is plausible that correlated activity in different cortical lobes, in the visual and auditory systems, strengthens the connections between distant neurons processing these different information types. The thereby established distributed neuronal assembly would then bind information across sensory and motor modalities.

The neuroscientific principles of correlation learning and that of long-distance cortico-cortical connectivity have important implications for the brain basis of meaningful language units, words, and morphemes (Pulvermüller, 1999). When a word form is being articulated, this relates to neuronal activation in the motor cortex. Motor activation is, in turn, coordinated and controlled by premotor circuits, which are, in turn, linked to and influenced by activity in inferior prefrontal areas. In addition to activity in the inferior frontocentral cortex (violet areas), the speech produced leads to auditory input, which activates superior temporal auditory cortex, and, via short distance connections, the adjacent auditory belt and parabelt areas in superior temporal gyrus and sulcus (blue areas). As there are neuronal links between superior temporal lobe and inferior prefrontal areas, the co-activation of neurons in these areas, which is characteristic of the production of a spoken word, can lead to synaptic strengthening. A word-related cell assembly distributed over different parts of this perisylvian cortex (violet and blue areas) develops (Pulvermüller 1999; Pulvermüller and Preissl, 1991). As the inferior frontal and superior temporal neuron populations – which, at the start, had either been responsible for controlling the articulation movement or for specifically responding to the sounds characteristic of the word – the connected assembly can be considered an APN in which action-related and perceptual information is being bound together. This APN would represent and process a specific spoken word form, and therefore embody and ground it in its distinctive articulatory and acoustic features.

6.4 Action–perception networks

6.4.1 Cortical embodiment of words as action–perception networks

The speculation that spoken word forms such as 'crocodile' are grounded in action-perception networks, whereas meaningless but pronounceable and phonotactically legal pseudowords that are not being used in the language, such as 'crodobile', are not, might lead to fruitful research. One idea is that activation of a memory network leads to well coordinated reverberatory activity and synchronous oscillations at high frequencies in the so-called gamma band (>20 Hz) (von der Malsburg and Schneider 1986). Evidence for this comes from animal research (Singer and Gray 1995) and can also be found in noninvasive recordings, electroencephalography (EEG), and magnetoencephalography (MEG) (Lutzenberger *et al.* 1995; Tallon-Baudry and Bertrand 1999). These high frequencies may, in part, relate to the fact that most neurons in cortex conduct activity rather fast (5–10 metres per second; for discussion, see Pulvermüller 2000) so that reverberations in cortical neuron loops may take <50 milliseconds.

When investigating high-frequency cortical responses to words and meaningless word-like items, induced gamma-band responses were found to be enlarged for the lexical items and relatively small for the meaningless novel ones (Eulitz *et al.* 2000; Krause *et al.* 1998; Lutzenberger *et al.* 1994; Pulvermüller *et al.* 1995, 1996, 1997). This was true for different languages (e.g., English, German, Finnish), in both major language modalities (spoken and written language), and in a range of tasks and paradigms (lexical decision, reading, listening, and active and passive oddball tasks). The difference in gamma-band responses was usually most pronounced over, or in, the left language-dominant hemisphere. The frequency where differences were most pronounced ranged between 20 and 60 Hz. Similar effects to those reported earlier for words and pseudowords were also seen for phonemes of language versus non-language sounds and for familiar versus unfamiliar letters (Ihara and Kakigi 2006; Palva *et al.* 2002). Familiar objects and coherent visual patterns have been shown to elicit enhanced gamma-band activity in the human brain In the same way as familiar meaningful language elements (Gruber *et al.* 2006; Lutzenberger *et al.* 1995; Müller *et al.* 1996; Tallon-Baudry *et al.* 1996, 1998). These results indicate the existence of memory networks in the human brain generating coordinated high-frequency responses. Such circuits seem to develop for meaningful elements that have been learned, including words and objects (Pulvermüller *et al.*, 1997).

Different tests of the neuronal assembly model of word processing can be performed using other measures of cortical activity. An obvious prediction is that a memory network in cortex should act like an amplifier of cortical activity, so that input activating a memory network leads to a stronger brain response (input plus neuronal assembly activation) than an input that fails to activate such a network (response to sensory input only). A well known indicator of cognitive processes is the *mismatch negativity* (MMN) elicited by auditory stimuli. The MMN is larger to familiar sounds of one's own language than to phonemes of a foreign language (Näätänen *et al.* 1997). In the same way, familiar non-language sounds, such as clicks or whistles, elicit a larger MMN compared with physically matched unfamiliar sounds (Frangos *et al.* 2005; Hauk *et al.* 2006). Crucially, if a syllable or language sound is placed in a context where it is critical for understanding a meaningful word, its MMN is enhanced compared with a condition in which the same stimulus completes a meaningless but pronounceable pseudoword (Endrass *et al.* 2004; Korpilahti *et al.* 2001; Kujala *et al.* 2002; Pettigrew *et al.* 2004; Pulvermüller *et al.* 2001, 2004; Shtyrov and Pulvermüller 2002; Sittiprapaporn *et al.* 2003). This lexical enhancement of the MMN is best explained by the full activation (ignition; Braitenberg 1978) of a cell assembly triggered by a meaningful word, but not by an unfamiliar meaningless item. A similar explanation in terms of memory networks for phonemes and other familiar sounds has also been established (Näätänen 2001).

Although these results support cell assembly activation following word presentation and the lack thereof when pseudowords are being presented, no direct evidence has so far been discussed for APNs linking inferior frontal output control circuits and superior temporal comprehension processors by means of distributed neuronal systems. Imaging studies have directly addressed the question of whether the left hemispheric inferior frontal and superior temporal language areas are modules specialized in either speech perception or production, or rather represent two local areas that house neural elements participating in interactive distributed cortical processes that contribute to both production and comprehension. During listening to syllables and words, the left inferior frontal and premotor cortex is active, along with the superior temporal areas in the vicinity of the auditory cortex (Pulvermüller *et al.* 2003; Wilson *et al.* 2004; Zatorre *et al.* 1992) (Figure 6.2). During speaking, the superior temporal cortex was active, along with areas in inferior motor, premotor, and prefrontal cortex, although it was ensured that self-produced sounds could not be perceived through the auditory channel (Paus *et al.* 1996; Watkins and Paus 2004). In addition, it is well known that lesions in superior temporal or inferior frontal cortex that lead to aphasia usually impair both speech production and comprehension (Pulvermüller and Preissl 1991). This indicates that interactive neural systems distributed over the inferior frontal and superior temporal cortex contribute to both speech production and perception. During spoken word recognition and understanding, these systems become active near-simultaneously and largely in parallel, with a peak activation delay in the inferior frontal cortex of ~20 milliseconds after peak activation in superior temporal areas (Pulvermüller *et al.* 2003). These results suggest tight and rapid

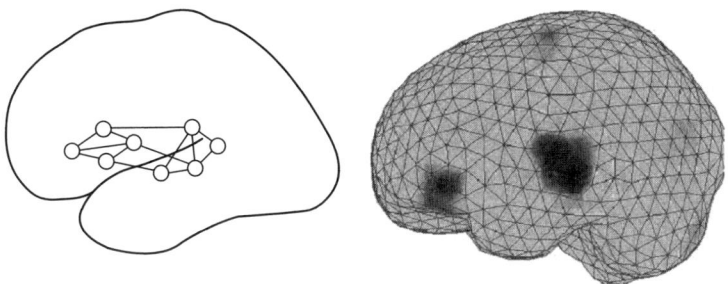

Fig. 6.2 Theory and data on perisylvian cell assemblies for words. The diagram on the left illustrates a cell assembly distributed over perisylvian areas, the kind of cortical network thought to represent and process spoken word forms at the cortical level (Pulvermüller and Preissl 1991). The perisylvian assembly can be said to embody the word as an action–perception network. The diagram on the right shows objective source estimates for the mismatch negativity brain response elicited by a spoken word, as they were recorded 130–150 milliseconds after the recognition point of the word. Two main sources in left perisylvian cortex could be distinguished: the posterior source lay in superior temporal areas close to auditory cortex and the anterior source in inferior frontal cortex anterior to the motor representation of the articulators. Activation peaks of these sources followed each other with a minimal delay of ~20 milliseconds (Pulvermüller *et al.* 2003).

functional links between speech perception and speech production processes, as postulated by neurobiological (Braitenberg and Schüz 1992; Fry 1966; Pulvermüller and Preissl, 1991) and psycholinguistic theories (Fowler 1986; Liberman *et al.* 1967).

That the links between superior temporal perceptual circuits and frontocentral speech production machinery are functionally effective has been demonstrated by experiments using transcranial magnetic stimulation (TMS). When spoken words and language sounds (phonemes) that strongly involve the tongue are being perceived, TMS applied to the inferior motor cortex elicits stronger muscle responses of the articulators compared with control conditions (Fadiga *et al.* 2002). Interestingly, this effect was most prominent when the critical phonemes were presented in meaningful word context, suggesting that cell assemblies for meaningful words play a role in linking articulatory gestures and auditory signals at the cortical level (see Pulvermüller and Preissl 1991). Converging evidence from functional connectivity studies on the basis of positron emission tomography (PET) and functional magnetic resonance imaging (fMRI) data indicates that the links between superior temporal and inferior frontal language areas depend on the amount of meaningful information being transmitted by words (Horwitz and Braun 2004).

The documented tight functional links between action and perception circuits of the left perisylvian language array (Pulvermüller 1999; Rizzolatti and Craighero 2004) cannot be explained in a straightforward manner within a modular approach according to which speech production and comprehension are situated in functionally separate encapsulated modules (e.g., Ellis and Young 1988). However, they meet the predictions of a distributed model postulating APNs and binding between specific acoustic speech patterns and the articulatory gestures that generate them (Plate 6.2A).

In summary, these neuroscience data provide support for the position that word forms are cortically based on APNs distributed over inferior frontal and superior temporal areas. At the purely cognitive level, this suggests that spoken words as entries of a 'mental lexicon' are embodied as complex articulatory gestures and as complex auditory spectrotemporal patterns, and, most importantly, as the specific functional connections between the two. Considering written language, additional links to visual pattern representations, written word forms, and writing gestures are evident. One may want to call the resulting APNs modality-unspecific representations of abstract word forms, but, as concrete and specific articulatory, gestural, acoustic, and visual features characterize word forms, it might also be accurate to speak of multi- or cross-modal representations.

6.4.2 Functional specificity of action–perception networks in phonological processing

The finding of strong functional links between superior temporal speech perception and comprehension circuits and inferior frontal action control circuits still leaves open the question of universality and functional specificity of APNs. When a speech sound is heard, its motor program might be automatically activated and the articulation action simulated mentally, in a similar way as has been proposed and observed for other action types (Buccino *et al.* 2001; Jeannerod 2001; Jeannerod *et al.* 1995). What has been shown so far, however, is merely an activation of frontal and motor/premotor cortex when speech information comes in. To demonstrate the specificity of APNs in speech perception, it would be necessary to demonstrate that activations in the frontal action system reflect sensory information. In the case of spoken language, it becomes important to ask whether articulatory features of speech sounds are reflected in the brain activation pattern in frontocentral cortex, and, if yes, whether these action-related activations would also occur during comprehension of specific speech sounds. Would the motor system activation reflect articulatory information about incoming speech?

To probe the possible involvement of specific motor circuits in the speech perception process, we used event-related fMRI and presented experimental subjects with spoken syllables including *[p]* and *[t]* sounds, which are respectively produced by movements of the lips or tongue. Physically similar nonlinguistic signal-correlated noise patterns were used as control stimuli. In localizer experiments, subjects had to silently articulate the same syllables, and, in a second task, move their lips or tongue. Speech perception most strongly activated superior temporal cortex. Crucially, distinct motor regions in the precentral gyrus sparked during articulatory movements of the lips and tongue, and also during nonlinguistic lip and tongue movements, were also differentially activated in a somatotopic manner when subjects listened to the lip- or tongue-related phonemes (Pulvermüller *et al.* 2006) (Plate 6.2).

These results indicate that, during speech perception, motor circuits are recruited that reflect phonetic distinctive features of the speech sounds encountered, thus providing direct neuroimaging support for specific links between the phonological mechanisms for speech perception and production. The conclusion might therefore be that APNs are specific to features of articulatory gestures and speech sounds. As could be made evident at the level of meaningful words, APNs would ground speech sounds in articulatory gestures and vice versa. The complex action–perception mapping appears crucial for the 'abstractness' – or cross-modality – of the representations.

6.4.3 Storing semantic information

Semantic category specificity

Where is word meaning represented and processed in the human brain? This question has been discussed controversially since 19th Century neurologists postulated a 'concept centre' in the brain that was thought to store the meanings of words (Lichtheim 1885). Today, the cortical loci proposed for a centre uniquely devoted to semantic binding between words and their meaning range from inferior frontal cortex (Bookheimer 2002; Posner and Pavese, 1998) to anterior, inferior, superior, or posterior left temporal cortex (Hickok and Poeppel 2004; Patterson and Hodges 2001; Price 2000; Scott and Johnsrude 2003). Others have proposed that the entire left frontotemporal cortex is a region equally devoted to semantics (Tyler and Moss 2001), or that the parahippocampal gyrus (Tyler et al. 2004) or the occipital cortex (Skrandies 1999) are particularly relevant. As there is hardly any area in the left language-dominant hemisphere for which there is no statement that it should house the semantic binding centre, these views are difficult to reconcile with each other (see Pulvermüller 1999). Is there a way to resolve this unfortunate diversity of opinions?

A way out might be pointed to by approaches to category-specific semantic processes (Daniele et al. 1994; Humphreys and Forde 2001; Warrington and McCarthy 1983; Warrington and Shallice 1984). The idea here is that different kinds of concepts, and different kinds of word meaning, draw upon different parts of the brain. Hearing the word 'crocodile' frequently together with certain visual perceptions may lead to strengthening of connections between the activated visual and language-related neurons. Specific form and colour detectors in primary cortex, as well as neurons responding to more complex features of the perceived gestalt higher up in the inferior temporal stream of visual object processing, will become active together with neurons in the perisylvian language areas that process the word form. These neurons would bind into distributed networks now implementing word forms together with aspects of their referential semantics. In contrast, learning of an action word, such as 'ambulate', critically involves linking an action type to a word form. In many cases, action words are learned in infancy when the child performs an action and the caretaker uses a sentence including an action word describing the action (Tomasello and Kruger 1992). As the brain circuits for controlling actions are in motor, premotor, and prefrontal cortex, it is clear that in this case correlated activation should bind perisylvian language networks to frontocentral circuits processing actions.

The cell assembly model and other theories of perception and action-related category specificity predict differential distribution of the neuron populations organizing action- and object-related words and similar differences can be postulated for other semantic categories (Pulvermüller 1996, 1999) (Figure 6.3). Many nouns refer to visually perceivable objects and are therefore characterized by strong semantic links to visual information, whereas most verbs are action verbs and link semantically to action knowledge. Like action verbs, nouns that refer to tools are usually also rated by subjects to be semantically linked to actions, and a large number of animal names are rated to be primarily related to visual information (Preissl *et al.* 1995; Pulvermüller, Lutzenberger *et al.* 1999; Pulvermüller, Mohr *et al.* 1999). Range of neuroimaging studies using EEG, MEG, PET, and fMRI techniques found evidence for category-specific activation in the human brain for the processing of action- and visually-related words and concepts (e.g., Cappa *et al.* 1998; Chao *et al.* 1999; Kiefer 2001; Preissl *et al.* 1995; Pulvermüller, Lutzenberger *et al.* 1999; Pulvermüller, Mohr *et al.* 1999). The results were largely consistent with the model of semantic category-specificity. Processing of action-related words, be they action verbs, tool names, or other action-related lexical items, tended to activate frontocentral cortex, including inferior frontal or premotor areas, more strongly than words without strong semantic action links. The same was found for temporo-occipital areas involved in motion perception. On the other hand, words with visual semantics tended to activate visual and inferior temporal cortex or temporal pole more strongly than action-related words. This differential activation was interpreted as evidence for semantic category specificity in the human brain (Martin and Chao 2001; Pulvermüller 1999).

Some problems with semantic category specificity

The results from metabolic and neurophysiological imaging demonstrate the activation of neuronal assemblies with different cortical distributions in the processing of action- and visually-related words and concepts. However, it has been asked whether the reason for the differential activation observed would necessarily be semantic or conceptual in nature. Could there be alternative explanations?

Although the broad majority of the imaging studies of category specificity support the idea that semantic factors are crucial, there is work that could not provide converging evidence (Devlin *et al.* 2002; Tyler *et al.* 2001). These studies used particularly

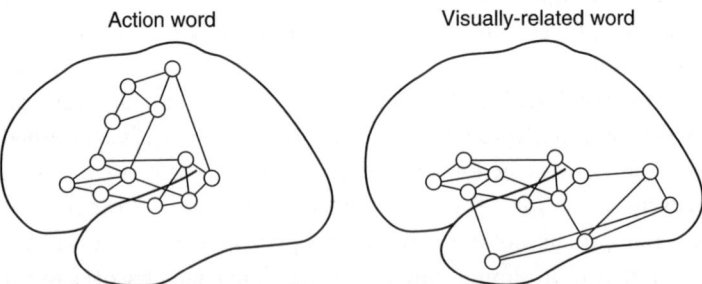

Fig. 6.3 A brain-based model of category-specific processing of words with different semantics. Words semantically related to actions may be cortically processed by distributed neuron ensembles linking together word forms and action programs. Words referring to objects that are perceived through the visual modality may be processing by neuron sets distributed over language areas and the visual system (Pulvermüller 1996).

well-matched stimuli, so that word length, frequency, and other psycholinguistic factors could not account for any possible differences in brain activation. Therefore, these authors argued that these factors might account for differences between 'semantic' categories reported previously. Although some earlier studies reporting semantic category differences performed meticulous stimulus matching for a range of psycholinguistic factors – word length and frequency included (Kiefer 2001; Preissl *et al.* 1995; Pulvermüller, Lutzenberger *et al.* 1999; Pulvermüller, Mohr *et al.* 1999) – a number of studies did not control for these factors. As pointed out previously (Bird *et al.* 2000), nouns tend to have more highly imageable meaning than verbs, whereas verbs tend to have higher word frequency. Any difference in brain activation, and also any differential vulnerability to cortical lesion, could thus be explained as an imageability–frequency dissociation, rather than in terms of semantic categories. Similarly, animals tend to be more alike than tools from both a visual and a conceptual point of view, and it has therefore been argued that perceptual and conceptual structure could contribute to the explanation of category dissociations (Humphreys and Riddoch 1987; Rogers *et al.* 2004; Tyler *et al.* 2000). On these grounds, at least some evidence for category specificity has therefore been criticized for not being fully convincing.

What makes things worse is that predictions on where category-specific activation should occur in the brain have not always been very precise. Whereas rough estimates, such as the prediction that action semantics should involve frontal areas and visual semantics temporo-occipital areas, could be provided and actually confirmed, more precise localization was sometimes surprising and not *a priori* predictable. For example, semantic information related to processing of colour and motion information semantically linked to words and pictures was reported to occur ~2 centimetres anterior to the areas known to respond maximally to colour or motion, respectively (Martin *et al.* 1995). It would be desirable to have evidence for category-specific semantic activation at precisely the locus a brain-based action–perception theory of semantic processing would predict. Such a perspective is opened by looking at subtypes of action words.

Action words

Action words are defined by abstract semantic links between language elements and information about actions. These words refer to actions and the neurons that process the word forms are likely interwoven, with neurons controlling actions. The motor cortex is organized in a somatotopic fashion with the mouth and articulators represented close to the sylvian fissure, the arms and hand at dorsolateral sites and the foot and leg projected to the vertex and interhemispheric sulcus (Penfield and Rasmussen 1950) (Figure 6.4). Additional somatotopic maps exist in the frontocentral cortex (He *et al.* 1993), among which one of the more prominent ones lies in the premotor cortex in the lateral precentral gyrus, and resembles the map in the primary motor cortex (Matelli *et al.* 1986; Rizzolatti and Luppino 2001). As many action words are preferably used to refer to movements of the face or articulators, arm or hand, or leg or foot, the distributed neuronal assemblies would therefore include semantic neurons in perisylvian (face words), lateral (arm words), or dorsal (leg words) motor and premotor cortex (Pulvermüller 1999).

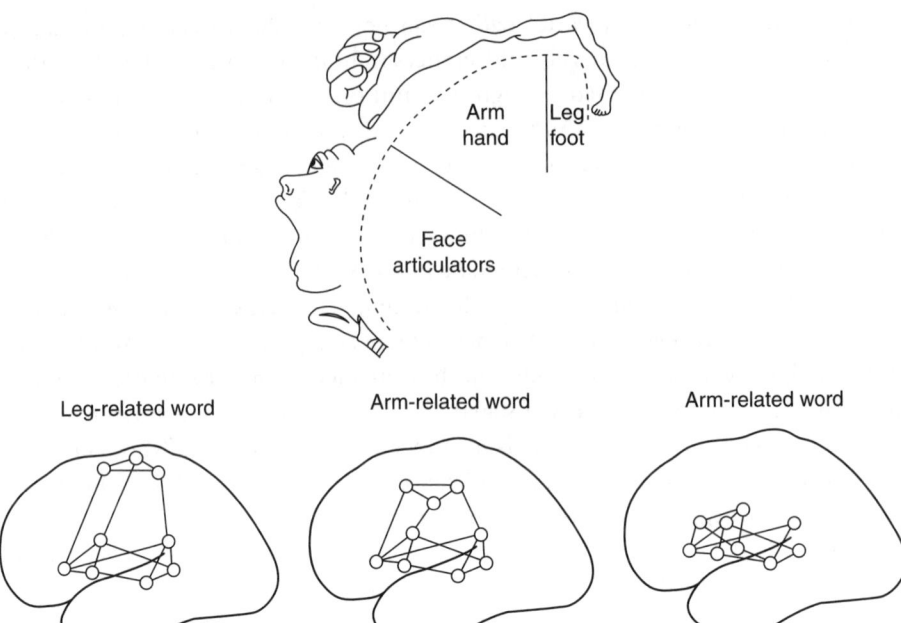

Fig. 6.4 Semantic somatotopy model of action word processing: distributed neuronal assemblies bind information about word forms and the actions they refer to semantically. Because action words can relate to different parts of the body (examples: 'lick', 'pick', 'kick'), the cortical distributions of their action–perception networks differ between each other (Pulvermüller 2001). The inset shows the somatotopy of the primary motor cortex as revealed by Penfield and Boldrey (1937).

This is the essence of the *somatotopy of action word* model, which implies differently distributed networks for the English words 'lick', 'pick', and 'kick' (Figure 6.4). The model allows for general predictions on action word-related cortical activity within the limits of the well known interindividual variation of cortical maps, most notably as a result of practice-related reorganization (Elbert *et al.* 1995), and is open to further elaboration by taking into account additional mapping rules, for example the topography of coordinated actions in a body-centred workspace suggested by recent work (Graziano *et al.* 2002).

Crucial predictions of the semantic somatotopy model is that perception of spoken or written action words should activate cortical areas involved in action control and execution in a category-specific somatotopic fashion, depending on the semantics of the action words. As the cortical areas of action control and execution can be defined experimentally, one could in principle use such action localizer experiments to predict exactly where semantic activation should occur for different aspects of action-related meaning.

In functional imaging experiments, elementary repetitive movements of single body parts activate motor and premotor cortex. For example, Hauk *et al.* reported fMRI data showing that tongue, finger, and foot movements lead to the somatotopic activation pattern illustrated in Plate 6.3 (diagram on the left; Hauk *et al.* 2004). When the same subjects were instructed to silently read action words related to the face, arm, and leg that were otherwise matched for important psycholinguistic variables (such as word frequency, length, and imageability), a similar pattern of activation emerged along the motor strip (Figure 6.3, right; Hauk *et al.* 2004). Consistent with earlier findings, all words equally activated areas in the temporal cortex and also in the inferior frontal cortex (Pulvermüller *et al.* 2003; Wilson *et al.* 2004; Zatorre *et al.* 1992). The additional category-specific somatotopic activation in response to face-, arm-, and leg-related words seen in the motor system was close to and overlapped with the motor and premotor representations for specific body part movements obtained in the motor localizer tasks. These results indicate that specific action representations are activated in action word understanding. The fact that the locus of semantic activation could be predicted by a theory of APNs provides strong evidence for this theory in particular and for the embodiment of aspects of semantics in action mechanisms in general.

A similar experiment was carried out with action words embedded into spoken sentences. In this case, subjects heard action descriptions such as 'The boy kicked the ball' or 'The man wrote the letter' while their brain metabolism was monitored (Tettamanti *et al.* 2005). Specific premotor areas reflecting the differential involvement of body part information in the semantic analysis of the language input were again found active. Taken together, these fMRI results indicate that somatotopic activation of motor circuits reflects aspects of word and sentence meaning, and that such activation can be elicited by spoken and by written language.

6.4.4 **Somatotopic activation: semantic or epiphenomenal?**

Although language-related somatotopic cortical activation could be demonstrated, the low temporal resolution of haemodynamic imaging makes it impossible to decide between two interpretations of this finding: One possibility is that the activation of specific action-related networks directly reflects action word recognition and comprehension, as the somatotopy of action word model would suggest. An alternative possibility has been pointed out by Glenberg and Kaschak (2002) in the context of behavioural work on embodiment. It is possible that thoughts about actions actually follow the comprehension process and behavioural, but also brain physiological, effects relate to such *post-comprehension inference*. Inferences would be triggered by the comprehension of a word or sentence, but would not necessarily reflect processes intrinsically linked to language comprehension. Importantly, earlier fMRI research has shown that observation of action-related pictures, but also mere voluntary mental imagery of actions, can activate motor and premotor cortex in a somatotopic fashion (Buccino *et al.* 2001; Jeannerod and Frak 1999). Therefore, it is important to clarify whether motor system activation to action-related language processing reflects the comprehension process *per se* or rather a later stage following language comprehension. Apart from mental imagery of actions, possible post-comprehension processes include planning of action execution, recalling an action performed earlier, and reprocessing the meaning of the language stimulus.

How is it possible to separate comprehension processes from subsequent inferences and other epiphenomenal mental activities? Let me propose that brain processes reflecting comprehension can be characterized as: immediate, automatic, and functionally relevant.

Early effects of lexical and semantic processing are known to occur around 100–200 milliseconds after critical stimulus information comes in (Pulvermüller 1996; Sereno *et al.* 1998). In contrast, late postlexical meaning-related processes are reflected by late components of the event-related potential (ERP) and field, which are maximal around 400 milliseconds after word onset (Holcomb and Neville 1990). If the activation of motor areas is related to semantic processes intrinsically tied to word form access, it should take place *immediately* (within the first 200 milliseconds) after stimulus information allows for the unique identification of an incoming word. *Automaticity* refers to the idea that when seeing or hearing a word it is hardly possible to avoid understanding its content; comprehension might even occur without intentionally attending to the stimuli. Therefore, brain processes reflecting comprehension might be expected to persist under distraction, when the subjects' attention is directed away from the critical language stimuli. In the case of *functional relevance*, if the presentation of action words leads to specific activation of motor systems relevant to word processing, one may expect that a change of the functional state of these motor systems leads to a measurable effect on the processing of words semantically related to actions. However, if somatotopic activation of motor systems did reflect a post-comprehension process, it can be late (substantially greater than 200 milliseconds) and absent under distraction, and functional changes in the motor system would be without effect on word processing. A series of experiments was conducted to investigate these three issues.

Immediacy

To reveal the time course of cortical activation in action word recognition and find out whether specific motor areas are sparked immediately or after some delay, neurophysiological experiments were conducted. Experiments using ERPs looking at silent reading of face, arm and leg words showed that category-specific differential activation was present ~200 milliseconds after word onset (Hauk and Pulvermüller, 2004). Consistent with the fMRI results, distributed source localization performed on ERPs revealed an inferior frontal source close to the motor representation of the face and mouth that was strongest for face-related words, and a superior central source close to the leg representation that was maximal for leg-related items (Hauk and Pulvermüller 2004). This dissociation in brain activity patterns supports the notion of stimulus-triggered early lexicosemantic processes. To investigate whether motor preparation processes co-determined this effect, experiments were performed in which the same response – a button press with the left index finger – was required for all words. The early activation difference between face- and leg-related words persisted, indicating that lexicosemantic processes rather than postlexical motor preparation were reflected (Pulvermüller *et al.* 2000).

In summary, the somatotopic activation that reflects word meaning aspects therefore appeared about one-quarter of a second after information about the words in the input was available. Earlier physiological studies of psycholinguistic processes had shown previously that the first brain responses reflecting comprehension and psycholinguistic information access at higher lexical and semantic levels appear at this point (Pulvermüller 1996; Sereno and Rayner 2003). Therefore, the early somatotopic mapping of meaning aspects reflects comprehension processes.

Automaticity

The earliness of word category-specific semantic activation along the sensorimotor cortex in passive reading tasks suggests that this feature might be automatic. To further investigate this possibility, subjects were actively distracted while action words were being presented and brain responses were measured (Pulvermüller, Shtyrov *et al.* 2005; Shtyrov *et al.* 2004). Subjects were instructed to watch a silent video film and ignore the language input while spoken face-/arm- and leg-related action words were presented. Care was taken to exactly control for physical and psycholinguistic features of the word material. For example, the Finnish words 'hotki' (eat) and 'potki' (kick) – which included the same recording of the syllable *[kI]* spliced to the end of each word's first syllable – were compared (Näätänen *et al.* 2001). In this way, any differential activation elicited by the critical final syllable *[kI]* in the context of *[hot]* or *[pot]* can be uniquely attributed to its lexicosemantic context. MEG results showed that a MMN that was maximal at 100–200 milliseconds after onset of the critical syllable was elicited by face/arm- and leg-word contexts (Plate 6.4). Relatively stronger activation was present in the left inferior frontal cortex for the face/arm-related word, but significantly stronger activation was seen in superior central areas, close to the cortical leg representation, for the leg-related word (Pulvermüller, Shtyrov *et al.* 2005).

These MEG results were confirmed with EEG using words from different languages, including, for example, the English word-pair 'pick' and 'kick' (Shtyrov *et al.* 2004). It is remarkable that peak activation of the superior central source followed that of the inferior frontal source with an average delay of only 30 milliseconds, consistent with the spread of activation being mediated by fast-conducting cortico–cortical fibres between the perisylvian and dorsal sensorimotor cortex. This speaks in favour of automatic activation of motor areas in action word recognition and therefore further strengthens the view that this activation reflects comprehension. It appears striking that differential activation of body-part representations in sensorimotor cortex to action word subcategories was seen across a range of cognitive paradigm, including lexical decision, attentive silent reading, and oddball paradigms under distraction. This further supports the idea that word-related, rather than task- or strategy-dependent, mechanisms are being tapped into.

Functional relevance

Even if action word processing sparks the motor system in a specific somatotopic fashion, and even if this activation is fast and automatic, it still does not necessarily imply that the motor and premotor cortex influence the processing of action words. Different parts of the motor system were therefore stimulated with weak magnetic pulses while subjects had to process action words in a lexical decision task (Pulvermüller, Hauk *et al.* 2005). To minimize interference between word-related activation of the motor system and response execution processes, lip movements were required while arm- and leg-related words were presented. Subthreshold TMS applied to the arm representation in the left hemisphere, where strong magnetic pulses elicited muscle contractions in the right hand, led to faster processing of arm words relative to leg words, whereas the opposite pattern of faster leg- than arm-word responses emerged when TMS was applied to the cortical leg area (Pulvermüller, Hauk *et al.* 2005). Processing speed did not differ between stimulus word groups in control conditions in which ineffective 'sham' stimulation or TMS to the right hemisphere was applied. This shows a specific influence of activity in the motor system on the processing of action-related words.

Further evidence for specific functional links between the cortical language and action systems comes from TMS-induced motor responses (Fadiga *et al.* 2002). Listening to Italian sentences describing actions performed with the arm or leg differentially modulates the motor responses brought about by magnetic stimulation of the hand and leg motor cortex (Buccino *et al.* 2005). It appears that effective specific connections of language and action systems can be documented for spoken or written language, at the word and sentence levels, and for a variety of languages (English, Italian, German, Finnish) using a variety of neuroscience methods (fMRI, MEG, EEG, TMS).

6.4.5 Interim summary

These experiments show that the activation of motor systems of the cortex occurs early in action word processing, is automatic to some degree, and has a semantically specific functional influence on the processing of action words. This provides brain-based support for the idea that motor area activation is related to comprehension of the referential semantic meaning of action words. In the wider context of a theory of embodiment of conceptual and semantic processing, the conclusion is that comprehension processes are related to, or embodied in, access to action information. It is noteworthy that neuroscience evidence was crucial in revealing this (Pulvermüller 2005). However, it is equally true that behavioural results are consistent with these conclusions and further strengthen the embodiment of language in APNs (Borghi *et al.* 2004; Boulenger *et al.* 2006; de Vega *et al.* 2004; Gentilucci *et al.* 2000; Glenberg and Kaschak 2002).

6.5 Fundamental issues in sensorimotor semantics

6.5.1 Can motor cortex map aspects of semantics?

One may question the idea that activity in the motor system might actually reflect semantic processes from a principled theoretical perspective. The idea that there should be a specific centre for semantics is still dominating (although there is little agreement between researchers about where the semantics area is situated, see section 6.4.3 on one or many semantics centres below). Areas that deal with the trivialities of motor movements, and equally those involved in elementary visual feature processing, are therefore thought by some to be incapable of contributing also to the higher processes one might be inclined to reserve for humans. In this context, it is important to point to the strong evidence that activation in motor systems directly reflects aspects of semantics.

Evidence that semantic features of words are reflected in the focal brain activation in different parts of sensorimotor cortex comes from MEG work on action words; there was a significant correlation between local source strengths in inferior face/arm-related and dorsal leg-related areas of sensorimotor cortex and the semantic ratings of individual words obtained from study participants (Pulvermüller, Shtyrov *et al*. 2005). This means that the subjects' semantic ratings were reflected by local activation strength and leaves little room for interpretations not of a semantic nature.

Even though the action-related and visually-related features discussed, and the associative learning mechanisms binding them to language materials, may not account for all semantic features of relevant word-related concepts, it seems clear that they reflect critical aspects of word meaning (Pulvermüller 1999): Crocodiles are defined by certain properties, including form and colour features, in the same way as the concepts of walking or ambulating are crucially linked to moving one's legs. Certainly, there is room for derived, including metaphorical, usage. As a big fish might be called the crocodile of its fish tank even if it is not green, one may speak of walking on one's hands or taking a stroll through the mind (thus ignoring the feature of body-part relatedness). One may even tell a story about a crocodile with artificial heart and kidneys, although it is generally agreed upon, following Frege, that these ingredients are part of the definition of an animal (or, more appropriately, a higher vertebrate) (Frege 1966). That writing is related to the hand may therefore be considered an analytical truth, in the same way as a crocodile is defined as having a heart, and in spite of the fact that it is possible to write in the sand with one's foot. This may simply be considered a modified type of writing, as the post-surgery crocodile is a modified crocodile. An instance of a heartless crocodile and leg-related writing is possible, but probably closer to metaphorical usage of these words than to their regular application. It seems safe to include perceptual properties such as green-ness and action aspects such as hand-relatedness in the set of possible semantic and conceptual features. Exceptions cannot prove a rule wrong.

6.5.2 Learning new meanings from language context: neuronal mechanisms of parasitic semantic feature extraction

The critical problem of learning new word meanings from context is frequently raised against embodied approaches to semantics, including the neuronal assembly model. Especially, the idea that word–world correlation provides a significant explanation of the acquisition of word meanings has been criticized, because it is well known that only a minority of words are actually being learned in the context of reference object perception and action execution (Kintsch 1974, 1998). However, after action–perception learning of aspects of word meaning has taken place for a sufficiently large set of words, it becomes feasible to learn semantic properties *parasitically* when words occur together in strings, sentences, or texts. A neuroscientific basis for this 'parasitic semantic learning' might lie in the overlap of word-related neuronal assemblies in the perisylvian language areas and the lack of semantic neurons related to action and perception information outside the perisylvian space of the networks processing new words with unknown semantic features (Pulvermüller 2002). In this case, a new word would activate its form-related perisylvian neuronal assembly, while neurons outside the perisylvian space are still actively process-ing aspects of the semantics of context words. The correlated activation of the semantic neurons of context words and the form-related perisylvian neurons of the new word lead to linkage of semantic features to the new word form. This provides a potential basis of second-order (parasitic) semantic learning and provides a putative neuroscience explanation for why correlation approaches to word meaning are successful in modelling semantic relationships between words (Kintsch 2002; Landauer and Dumais 1997).

However, it is important to note that this mechanism can only succeed if a sufficiently large set of semantic features and words is learned through correlation of perception, action, and language-form features in the first place. Otherwise, what Searle called the Chinese room argument, implying that semantic information cannot emerge from correlation patterns between symbols, cannot be overcome (Searle 1980). Action–perception correlation learning and the learning of correlations between language units are both indispensable for extracting semantic information for large vocabularies. Semantic knowledge is rooted in word-world and word-word correlation.

6.6 Early and late semantic activation

The time course of semantic activation in action word recognition was on a rather short scale. Relevant areas were seen to be active within 200 milliseconds after critical stimulus information came in (Pulvermüller, Shtyrov *et al.* 2005; Shtyrov *et al.* 2004). This suggests early semantic activation, as early as the earliest processes reflecting phonologi-cal or lexical information access (Hauk *et al.* 2006; Obleser *et al.* 2003; Shtyrov *et al.* 2005). However, the early neurophysiological reflection of semantic brain processes does not imply that meaning processing is restricted to the first 200 milliseconds after a word can be identified. There is ample evidence for neurophysiological correlates of semantic processes that take place later on (Coles and Rugg, 1995). These later processes may follow up on the early semantic access processes and may reflect reinterpretation or in-depth processing, which is especially important in circumstances where the context or other factors make comprehension difficult.

6.6.1 Abstract semantics from a brain perspective

These results summarized in section 6.4 demonstrate that action words activate the cortical system for action processing in a somatotopic fashion and that this somatotopy reflects word meaning. However, they do not imply that all aspects of the meaning of a word are reflected in the brain activation pattern it elicits. It is possible to separate brain correlates of semantic features specifying face-, arm-, and leg-relatedness, or, in the visual domain, of colour and form features (Moscoso Del Prado Martin *et al.* 2006; Pulvermüller and Hauk 2006; Simmons *et al.* 2007). It even became possible to provide brain support for the grounding of words referring to odours in olfactory sensation and evaluation mechanisms in brain areas processing olfactory and emotion-related information (de Araujo *et al.* 2005; González *et al.*, 2006). However, for other semantic features, the idea that their meaning can be extracted from sensory input, or deduced from output patterns, is more difficult to maintain. Although the question of how an embodiment perspective would explain abstraction processes has frequently been addressed (Barsalou 1999, 2003; Lakoff 1987), it is still not clear whether all semantic feature can – and have to – be extracted from input–output patterns.

A brain perspective might help to solve aspects of this issue. There are highly abstract concepts for which a deduction from sensory input is difficult to construe. Barsalou tried to ground the meaning of the word 'or' in the alteration of the visual simulations of objects (Barsalou 1999). However, if this view is correct, the disjunctive concept would be realized as the alteration mechanism allowing the brain to switch on and off specific representations alternatively. Looking at the brain theory literature, it is evident that every brain, even every primitive nervous system, is equipped with mechanisms for calculating disjunction, conjunction, negation, and other logical operation. This was the content of an early article by McCulloch and Pitts (1943) entitled 'A logical calculus of ideas immanent in nervous activity' that has since inspired much work in the theory of automata and language (e.g. Kleene 1956; Schnelle 1996).

The main points of this logical calculus theory of neuronal function still hold true, although neuronal models have significantly improved since the proposal was first made (e.g., Bussey et al., 2005). McCulloch and Pitts (1943) pointed out that a circuit including two neurons that project onto a third one will necessarily, given the activation threshold of neuron number three is adjusted in specific ways, give rise to the computation of a conjunction or disjunction function (Figure 6.5). Negation, identity, and either-or computations are equally straightforward. These examples demonstrate that our brain comes with built-in mechanisms relevant for abstract semantic processing. There is no need to construe the semantics of 'and', 'or', and other highly abstract words exclusively from sensorimotor information. The very fact that these mechanisms are built into nerve cell circuits may enable us to abstract away from the sensory input to more and more general concepts.

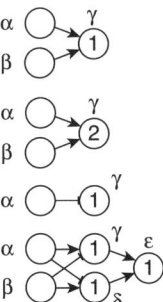

Fig. 6.5 Logical circuits immanent to a network of neurons as discussed by McCulloch and Pitts (1943). If two neurons project to a third neuron, the activation threshold of the third neuron will determine whether it acts like a logical element symbolizing 'or' (uppermost diagram) or 'and' (second from top). Circuits symbolizing 'not' and 'either-or' can also be implemented. Arrows stand for excitatory and T-shaped line endings for inhibitory connections. Numbers indicate activation thresholds (after Pulvermüller 2003).

One may argue that this proposal means watering down a 'radically embodied' perspective in the cognitive and brain sciences as the logical circuits are not derived from actions or perceptions but rather represent a neuronal a priori. However, the proposal here is to model language and concepts in mechanistic neuronal circuits. This implies to put to use those mechanisms that are evidently present (built-in) in nervous systems and to explore their implications for linguistic and semantic representations and processes. If the mechanisms of embodiment are brain mechanisms, we have little choice but to accept the functional principles immanent to neuronal function.

Abstraction by either-or computation may be the basis of action representation at different levels of the action description, corresponding to different levels of abstractness. Moving one's arm in such and such a way is a basic action; opening a door could imply exactly the same movement but with characteristic somatosensory and possibly auditory input; and freeing somebody could also be realized by performing the same basic action. To implement the aspects of the action semantics of 'open', it is possible to connect disjunction neurons with a range of action control neurons coordinating alternative action sequences that would allow one to open doors, boxes, and other objects. Similarly, in order to implement action semantics of the word 'free', higher order disjunction neurons would be needed that look at a range of different movement programs one could perform in the context of setting somebody or something free. Again, additional conditions would need to be met. The performer would need to assume that someone or something is captured, locked in, or contained in something else and would have the intention to get him/it out etc. Disjunction neurons receiving input from a range of concrete action representations may be located adjacent to motor and premotor sites and would be ideally placed in prefrontal cortex (see Pulvermüller 1999). This hypothesis might seem somewhat speculative on first appearance, and it is therefore important to point to the evidence that supports it. The prediction that words with more abstract action-related meaning are processed in areas adjacent and anterior to motor and premotor cortex, that is, in dorsolateral prefrontal areas, receives support from recent imaging work (Binder et al. 2005; Pulvermüller and Hauk 2006).

These hints towards abstraction mechanisms of different types might suffice here to point out some perspectives of a brain-based approach to embodied semantics. In a nutshell, I believe that a wide range of abstract concepts can be modelled, provided that the brain's built-in circuits are taken into account.

6.7 Language processes in the brain

6.7.1 Distributed processing

Similar to most current theories postulating distributed processing of language and concepts in the mind and brain (Braitenberg and Pulvermüller 1992; Rogers et al. 2004; Seidenberg et al. 1994), the proposal put forward here postulates that the cell assemblies processing language and concepts are widely distributed. This means that the neuronal ensembles are spread out over different areas of the brain, so that the areas would become more active in neuroimaging experiments when critical stimuli are under processing.

In the same way, the compartments of neuronal models simulating the relevant brain mechanism should show activation dynamics in a specific manner when the processing of language and concepts is being simulated (Garagnani *et al.* 2007; Pulvermüller 1999; Pulvermüller and Preissl 1991; Wennekers *et al.* 2006; Wermter *et al.* 2004). However, in contrast to most distributed processing accounts, the neuronal assemblies are conceptualized as functionally coherent networks that respond in a discrete fashion. This implies that the networks representing words, the 'word webs', are either active or inactive and that the full activation of one word's representation is in competition with that of other word-related networks. In this sense, neuronal assemblies are similar to the localist representations postulated by psycholinguistic theories (Dell 1986; Page 2000; Roelofs 1992). Still, as each of the distributed neuronal assemblies includes neurons processing features related to form or semantics, there can be overlap between neuronal assemblies representing similar words or concepts. This leads to an interplay of facilitatory and inhibitory mechanisms when word webs become fully active in a sequence. The full activation, or ignition, of a word-related neuronal assembly can be considered a possible cortical correlate of word recognition – or of the spontaneous pop-up of a word together with its meaning in the mind.

6.7.2 Discrete processing

If word webs can become active in a discrete fashion, this does not imply that each ignition is identical to all other full activations of the network. As word webs are linked to each other through the grammar network and also exhibit semantic and form overlap, the context of brain states and other cognitive network activations primes and therefore influences the way in which a given neuronal assembly ignites. This mechanism can be related to the observation that the meaning or 'sense' of a word in given contexts cannot be reduced to a 'core meaning', but should rather be conceptualized as a family of similar context-dependent semantic feature sets (Barsalou 1982; Wittgenstein 1953). Neuronal assemblies whose precise pattern of ignition depends on contextual priming may provide a mechanism for context-related selection of semantic features.

A further example illustrating the benefit of discrete representations is the contextual disambiguation of a semantically ambiguous word. The brain basis of an ambiguous word has been conceptualized as a set of two word webs overlapping in their form-related assembly part (Figure 6.6). Semantic context can, in this case, disambiguate by priming one of the semantic subassemblies of the two overlapping word representations. The two overlapping neuronal assemblies would be in both facilitatory (due to form overlap) and inhibitory (due to competition between neuronal assemblies) interaction. Most likely, the facilitatory effects would precede the inhibitory ones (Pulvermüller 2003). It is more difficult to envisage a mechanism for the separation of the meanings of ambiguous words in a fully distributed network without discrete representations.

6.7.3 Semantic conceptual binding sites

Although the results on the cortical correlates of semantic word groups cannot be explained if all semantic processes are restricted to one cortical area, they might still be compatible with the general idea of a central semantic binding site. This system would

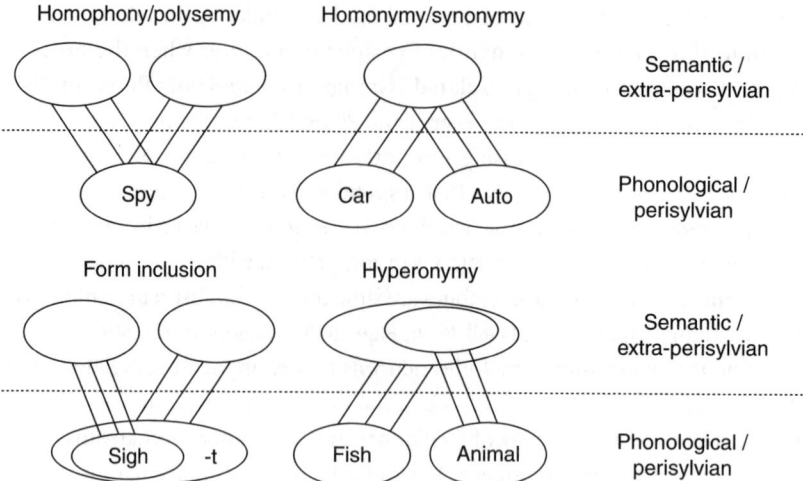

Fig. 6.6 Word forms and semantics are proposed to be processed in different parts of the distributed word-related cell assembly. Most semantic information is stored outside the perisylvian language areas, whereas most information related to the phonological word form is laid down in perisylvian space. Words related to each other phonologically or semantically would overlap in their perisylvian or extra-perisylvian sub-assemblies. Relationships between words with the same form but different meanings (homophones and polysemes), with the same meaning but different forms (homonyms and synonyms), with a form-inclusion and a hyperonomy/hyponymie relationship are illustrated (after Pulvermüller 2003).

be thought to manage dynamic functional links between multiple cortical areas processing word forms and conceptual semantic information. The idea of such a central 'concept area' or 'convergence zone' has a long tradition in the neuroscience of language and seems to be motivated by the believe that a central locus must exist, at which concepts are related to each other and abstract information is extracted from them. As I have tried to make clear, it may be possible to implement semantic binding by distributed neuronal assemblies which, as a whole, function as binding networks. In this case, the binding would not be attributable to one specific brain area but rather to a set of areas, those over which the assembly is distributed. Still, there are certainly more 'peripheral' areas, for example, primary motor and sensory cortices, where correlated activation patterns occur in the first place, and areas connecting the peripheral ones, with the major task of linking the correlation patterns together in the most effective manner. These higher or connection areas might naturally be more important for the binding of information from different modalities. Whether or not all these routes for multimodal information linkage necessarily go through the same convergence zone, or rather through a range of different areas of association cortex, as neuroanatomical studies might suggest (e.g., Braitenberg and Schüz 1998; Young *et al.* 1994), remains a matter of future research.

A possible route to answering the question of a centre for semantic and conceptual binding is offered by patient studies. Here it is remarkable that patients with semantic dementia usually have most severe neuronal degradation in the temporal pole. Therefore, this region was suggested as the area most important for semantic binding (see Patterson and Hodges 2001). However, the bilateral nature of neural degeneration usually seen in semantic dementia may suggest that one focal lesion is not enough to cause general semantic deficits (Patterson and Hodges 2001).

Multiple semantic binding sites are also supported by the specific semantic deficit in action-word processing seen in patients with motor neurone disease (Bak *et al.* 2001). More evidence for multiple semantic binding sites came from double dissociations between semantic word categories arising from lesions in right-hemispheric frontoparietal versus temporo-occipital areas (Neininger and Pulvermüller 2003), which complement similar observations made earlier for lesions to the left language-dominant hemisphere (Damasio and Tranel 1993; Daniele *et al.* 1994). Some of these lesions were so focal that they only affected motor and premotor cortex, but nevertheless specifically degraded the processing of action words in psychological experiments (Neininger and Pulvermüller 2001). Dissociations of these types are consistent with the existence of multiple semantic integration systems in both cerebral hemispheres (Pulvermüller and Mohr 1996).

6.8 Discussion

A model of embodiment of word forms and semantics at the level of neurons in the brain was outlined. APNs may bind together distinctive articulatory and acoustic information to yield transmodal representations of spoken word forms. The neuronal assemblies for word forms may be spread out over inferior frontal and superior temporal parts of left perisylvian language cortex (blue and green areas in Plate 6.1; see also Figure 6.2). Aspects of actions and perceptions relevant in the explanation of referential words may be linked to the word forms by way of neuronal correlation of the activations of the perisylvian word-form assembly and semantic neurons in action- and perception-related brain regions (Figure 6.3). Semantic category differences may be based on the differential involvement of neuronal sets in inferior temporal visual cortex, frontocentral action-related cortex, and other brain parts. Precise predictions on the cortical locus of specific semantic brain processes could be generated for subtypes of action words referring to face, arm, and leg movements, such as 'lick', 'pick', and 'kick' (Figure 6.4). Processing of these words lights up the motor system in a similar way as the respective actions would (Plate 6.3).

This semantic somatotopy is complemented by a phonological somatotopy; listening to speech sounds produced with the lips or tongue activates the areas specifically involved in lip or tongue movements, which also contribute to phoneme articulation (Plate 6.2). Specific activation of the motor systems takes place rapidly during speech and written language processing (Plate 6.4), is automatic and makes a functional contribution to word processing. This provides brain support that language is grounded in, and embodied by, action and perception mechanisms. In addition to elementary features of object and action reference, a brain model of abstract meaning can be built on the basis of known properties of neuronal circuits, for example their ability to implement logical operations (Figure 6.5). Abstract concepts may therefore develop in brain regions adjacent and anterior to areas involved in the processing of referent actions and objects (Plate 6.5).

The data summarized show that it is fruitful to model the brain basis of meaningful words as distributed neuronal assemblies binding phonological and semantic information about actions and perceptions at an abstract or cross-modal level. These distributed neuronal ensembles may function as discrete word-specific processors including neuron sets in different cortical areas. Different neuronal assemblies may overlap, thereby reflecting shared semantic or phonological features between words, and they may compete for full activation in the perception process. This view is consistent with context-dependent meaning processing and allows for modelling of word pairs that are semantically ambiguous or homophonous (Figure 6.5). In word recognition, activation of the distributed areas over which these neuronal assemblies are spread out is near-simultaneous, thereby binding information from different modalities (e.g., articulatory and acoustic) and linguistic functions (e.g., phonological and semantic). Apart from their role in language, these networks may play a role in conceptual processing.

These proposals and the reviewed neuroscience evidence backing them have important implications for constructing life-like APNs and robots with brain-like control systems (Knoblauch *et al.* 2005; Roy *et al.* 2005; Shastri *et al.* 2005; Wermter *et al.* 2004, 2005). A major conclusion here is that language models likening brain mechanisms have good reason to link up language processors to the body-related systems the language elements provide information about (e.g., which articulator is being moved, which colour is being referred to, which action is meant). Such embodied artificial models might succeed for the same reason why the biological originals they copy were successful in evolution. A main point here is the possibility to process cross-modal information in an extremely rapid manner.

Debate

The debate corresponding to this chapter, as well as Chapters 5 and 7, is included at the end of Chapter 7.

Author note

For comments and discussions, I am grateful to Larry Barsalou, Manuel de Vega, Arthur Glenberg, Arthur Graesser, Olaf Hauk, Markus Kiefer, Karalyn Patterson, and Yury Shtyrov. This work was supported by the UK Medical Research Council (grants U1055.04.003.00001.01, U1055.04.003.00003.01) and by the European Community under the Information Society Technologies and the New and Emerging Science and Technology programmes (EU IST-2001-35282, EU NEST-Nestcom).

References

Artola A, Singer W (1993). Long-term depression of excitatory synaptic transmission and its relationship to long-term potentiation. *Trends in Neurosciences*, 16, 4807.

Bak TH, O'Donovan DG, Xuereb JH, Boniface S, Hodges JR (2001). Selective impairment of verb processing associated with pathological changes in Brodmann areas 44 and 45 in the Motor Neurone Disease–Dementia–Aphasia syndrome. *Brain*, 124, 103–20.

Barsalou LW (1982). Context-independent and context-dependent information in concepts. *Memory and Cognition*, 10, 82–93.

Barsalou LW (1999). Perceptual symbol systems. *Behavioural and Brain Sciences*, 22, 577–609.

Barsalou LW (2003). Abstraction in perceptual symbol systems. *Philosophical Transactions of the Royal Society of London Series B – Biological Sciences*, 358, 1177–87.

Bi GQ (2002). Spatiotemporal specificity of synaptic plasticity: cellular rules and mechanisms. *Biological Cybernetics*, 87, 319–32.

Binder JR, Westbury CF, McKiernan KA, Possing ET, Medler DA (2005). Distinct brain systems for processing concrete and abstract concepts. *Journal of Cognitive Neuroscience*, 17, 905–17.

Bird H, Lambon-Ralph MA, Patterson K, Hodges JR (2000). The rise and fall of frequency and imageability: noun and verb production in semantic dementia. *Brain and Language*, 73, 17–49.

Bookheimer S (2002). Functional MRI of language: new approaches to understanding the cortical organization of semantic processing. *Annual Review of Neuroscience*, 25, 151–88.

Borghi AM, Glenberg AM, Kaschak MP (2004). Putting words in perspective. *Memory and Cognition*, 32, 863–73.

Boulenger V, Roy AC, Paulignan Y, Deprez V, Jeannerod M, Nazir TA (2006). Cross-talk between language processes and overt motor behaviour in the first 200 msec of processing. *Journal of Cognitive Neuroscience*, 18, 1607–15.

Braitenberg V (1978). Cell assemblies in the cerebral cortex. In R Heim, G Palm, Eds. *Theoretical Approaches to Complex Systems. Lecture Notes in Biomathematics Vol. 21* (pp. 171–88). Berlin: Springer.

Braitenberg V, Pulvermüller F (1992). Entwurf einer neurologischen Theorie der Sprache. *Naturwissenschaften*, 79, 103–17.

Braitenberg V, Schüz A (1992). Basic features of cortical connectivity and some considerations on language. In J Wind, B Chiarelli, BH Bichakjian, A Nocentini, A Jonker, Eds. *Language Origin: A Multidisciplinary Approach* (pp. 89–102). Dordrecht: Kluwer.

Braitenberg V, Schüz A (1998). *Cortex: Statistics and Geometry of Neuronal Connectivity* (Second edition). Berlin: Springer.

Brodmann K (1909). *Vergleichende Lokalisationslehre der Grobhirnrinde*. Leipzig: Barth.

Buccino G, Binkofski F, Fink GR *et al.* (2001). Action observation activates premotor and parietal areas in a somatotopic manner: an fMRI study. *European Journal of Neuroscience*, 13, 400–4.

Buccino G, Riggio L, Melli G, Binkofski F, Gallese V, Rizzolatti G. (2005). Listening to action-related sentences modulates the activity of the motor system: a combined TMS and behavioural study. *Brain Research: Cognitive Brain Research*, 24, 355–63.

Bussey TJ, Saksida LM, Murray EA (2005). The perceptual-mnemonic/feature conjunction model of perirhinal cortex function. *Quarterly Journal of Experimental Psychology B*, 58, 269–82.

Cappa SF, Perani D, Schnur T, Tettamanti M, Fazio F (1998). The effects of semantic category and knowledge type on lexical-semantic access: a PET study. *NeuroImage*, 8, 350–9.

Chao LL, Haxby JV, Martin A (1999). Attribute-based neural substrates in temporal cortex for perceiving and knowing about objects. *Nature Neuroscience*, 2, 913–19.

Clark A (1996). *Being There: Putting Brain, Body, and World Together Again*. Boston, MA: MIT Press.

Coles MGH, Rugg MD (1995). Event-related brain potentials. In MD Rugg, MGH Coles, Eds. *Electrophysiology of Mind: Event-related Brain Potentials and Cognition* (pp. 1–26). Oxford: Oxford University Press.

Damasio AR, Tranel D (1993). Nouns and verbs are retrieved with differently distributed neural systems. *Proceedings of the National Academy of Sciences, USA*, 90, 4957–60.

Daniele A, Giustolisi L, Silveri MC, Colosimo C, Gainotti G (1994). Evidence for a possible neuroanatomical basis for lexical processing of nouns and verbs. *Neuropsychologia*, 32, 1325–41.

de Araujo IE, Rolls ET, Velazco MI, Margot C, Cayeux I. (2005). Cognitive modulation of olfactory processing. *Neuron*, 46, 671–9.

de Vega M, Robertson DA, Glenberg AM, Kaschak MP, Rinck M (2004). On doing two things at once: temporal constraints on actions in language comprehension. *Memory and Cognition*, 32, 1033–43.

Dell GS (1986). A spreading-activation theory of retreival in sentence production. *Psychological Review*, 93, 283–321.

Devlin JT, Russell RP, Davis MH *et al.* (2002). Is there an anatomical basis for category-specificity? Semantic memory studies in PET and fMRI. *Neuropsychologia*, 40, 54–75.

Elbert T, Pantev C, Wienbruch C, Rockstroh B, Taub E (1995). Increased cortical representation of the fingers of the left hand in string players. *Science*, 270, 305–7.

Ellis AW, Young AW (1988). *Human cognitive neuropsychology*. Hove, UK: Erlbaum.

Endrass T, Mohr B, Pulvermüller F (2004). Enhanced mismatch negativity brain response after binaural word presentation. *European Journal of Neuroscience*, 19, 1653–60.

Eulitz C, Eulitz H, Maess B, Cohen R, Pantev C, & Elbert, T (2000). Magnetic brain activity evoked and induced by visually presented words and nonverbal stimuli. *Psychophysiology, 37*(4), 447-455.

Fadiga L, Craighero L, Buccino G, Rizzolatti G (2002). Speech listening specifically modulates the excitability of tongue muscles: a TMS study. *European Journal of Neuroscience*, 15, 399–402.

Fogassi L, Ferrari PF, Gesierich B, Rozzi S, Chersi F, Rizzolatti G (2005). Parietal lobe: from action organization to intention understanding. *Science*, 308, 662–7.

Fowler CA (1986). An event approach to the study of speech perception from a direct realist perspective. *Journal of Phonetics*, 14, 3–28.

Frangos J, Ritter W, Friedman D (2005). Brain potentials to sexually suggestive whistles show meaning modulates the mismatch negativity. *Neuroreport*, 16, 1313–17.

Frege G (1966). Der Gedanke (In German; first published 1918–1920). In G Patzig, Ed. *Logische Untersuchungen* (pp. 30–53). Göttingen: Huber.

Freud S (1891). *Zur Auffassung der Aphasien* [In German]. Leipzig: Franz Deuticke.

Fry DB (1966). The development of the phonological system in the normal and deaf child. In F Smith, GA Miller, Eds. *The Genesis of Language* (pp. 187–206). Cambridge, MA: MIT Press.

Fuster JM (1995). *Memory in the Cerebral Cortex: An Empirical Approach to Neural Networks in the Human and Nonhuman Primate*. Cambridge, MA: MIT Press.

Fuster JM (2003). *Cortex and Mind: Unifying Cognition*. Oxford: Oxford University Press.

Gallese V, Lakoff G (2005). The brain's concepts: The role of the sensory-motor system in conceptual knowledge. *Cognitive Neuropsychology*, 22, 455–79.

Garagnani M, Wennekers T, Pulvermüller F (2007). A neuronal model of the language cortex. *Neurocomputing*, 70, 1914–19.

Gentilucci M, Benuzzi F, Bertolani L, Daprati E, Gangitano M (2000). Language and motor control. *Experimental Brain Research*, 133, 468–90.

Glenberg AM, Kaschak MP (2002). Grounding language in action. *Psychonomic Bulletin and Review*, 9, 558–65.

González J, Barros-Loscertales A, Pulvermüller F *et al.* (2006). Reading cinnamon activates olfactory brain regions. *NeuroImage*, 32, 906–12.

Graziano MS, Taylor CS, Moore T (2002). Complex movements evoked by microstimulation of precentral cortex. *Neuron* 34, 841–51.

Gruber T, Trujillo-Barreto NJ, Giabbiconi CM, Valdes-Sosa PA, Muller MM (2006). Brain electrical tomography (BET) analysis of induced gamma band responses during a simple object recognition task. *NeuroImage*, 29, 888–900.

Hauk O, Davis MH, Ford M, Pulvermüller F, Marslen-Wilson WD (2006). The time course of visual word recognition as revealed by linear regression analysis of ERP data. *NeuroImage*, 30, 1383–400.

Hauk O, Johnsrude I, Pulvermüller F (2004). Somatotopic representation of action words in the motor and premotor cortex. *Neuron*, 41, 301–7.

Hauk O, Pulvermüller F (2004). Neurophysiological distinction of action words in the fronto-central cortex. *Human Brain Mapping*, 21, 191–201.

Hauk O, Shtyrov Y, Pulvermüller F (2006). The sound of actions as reflected by mismatch negativity: rapid activation of cortical sensory-motor networks by sounds associated with finger and tongue movements. *European Journal of Neuroscience*, 23, 811–21.

He SQ, Dum RP, Strick PL (1993). Topographic organization of corticospinal projections from the frontal lobe: motor areas on the lateral surface of the hemisphere. *Journal of Neuroscience*, 13, 952–80.

Hebb DO (1949). *The Organization of Behaviour. A Neuropsychological Theory*. New York, NY: John Wiley.

Hickok G, Poeppel D (2004). Dorsal and ventral streams: a framework for understanding aspects of the functional anatomy of language. *Cognition*, 92, 67–99.

Holcomb PJ, Neville HJ (1990). Auditory and visual semantic priming in lexical decision: a comparision using event-related brain potentials. *Language and Cognitive Processes*, 5, 281–312.

Horwitz B, Braun AR (2004). Brain network interactions in auditory, visual and linguistic processing. *Brain and Language*, 89, 377–84.

Hubel D (1995). *Eye, Brain, and Vision* (Second edition). New York, NY: Scientific American Library.

Humphreys GW, Forde EM (2001). Hierarchies, similarity, and interactivity in object recognition: 'category-specific' neuropsychological deficits. *Behavioural and Brain Sciences*, 24, 453–509.

Humphreys GW, Riddoch MJ (1987). On telling your fruit from your vegetables – a consideration of category-specific deficits after brain-damage. *Trends in Neurosciences*, 10, 145–8.

Ihara A, Kakigi R (2006). Oscillatory activity in the occipitotemporal area related to the visual perception of letters of a first/second language and pseudoletters. *NeuroImage*, 29, 789–96.

Jeannerod M (2001). Neural simulation of action: a unifying mechanism for motor cognition. *NeuroImage*, 14, S103–9.

Jeannerod M, Arbib MA, Rizzolatti G, Sakata H (1995). Grasping objects: the cortical mechanisms of visuomotor transformation. *Trends in Neuroscience*, 18, 314–20.

Jeannerod M, Frak V (1999). Mental imaging of motor activity in humans. *Current Opinion in Neurobiology*, 9, 735–9.

Kiefer M (2001). Perceptual and semantic sources of category-specific effects: event-related potentials during picture and word categorization. *Memory and Cognition*, 29, 100–16.

Kintsch W (1974). *The Representation of Meaning in Memory*. Hillsdale, NJ: Erlbaum.

Kintsch W (1998). *Comprehension: A Paradigm for Cognition*. New York, NY: Cambridge University Press.

Kintsch W (2002). The potential of latent semantic analysis for machine grading of clinical case summaries. *Journal of Biomedical Informatics*, 35, 3–7.

Kleene SC (1956). Representation of events in nerve nets and finite automata. In CE Shannon, J McCarthy, Eds. *Automata Studies* (pp. 3–41). Princeton, NJ: Princeton University Press.

Knoblauch A, Markert H, Palm G (2005). An associative cortical model of language understanding and action planning. In J Mira, JR Alvarez, Eds. *International Work-conference on the Interplay between Natural and Artificial Computation 2005* (vol. 3562, pp. 405–14). Berlin: Springer.

Korpilahti P, Krause CM, Holopainen I, Lang AH (2001). Early and late mismatch negativity elicited by words and speech-like stimuli in children. *Brain and Language*, 76, 332–9.

Krause CM, Korpilahti P, Porn B, Jantti J, Lang HA (1998). Automatic auditory word perception as measured by 40 Hz EEG responses. *Electroencephalography and Clinical Neurophysiology*, 107, 84–7.

Kujala A, Alho K, Valle S *et al.* (2002). Context modulates processing of speech sounds in the right auditory cortex of human subjects. *Neuroscience Letters*, 331, 91–4.

Lakoff G (1987). *Women, Fire, and Dangerous Things: What Categories Reveal About the Mind.* Chicago, IL: University of Chicago Press.

Lakoff G, Johnson M (1999). *Philosophy in the Flesh: The Embodied Mind and its Challenge to Western Thought.* New York, NY: Basic Books.

Landauer TK, Dumais ST (1997). A solution to Plato's problem: the latent semantic analysis theory of acquisition, induction, and representation of knowledge. *Psychological Review*, 104, 211–40.

Liberman AM, Cooper FS, Shankweiler DP, Studdert-Kennedy M (1967). Perception of the speech code. *Psychological Review*, 74, 431–61.

Lichtheim L (1885). On aphasia. *Brain*, 7, 433–84.

Lutzenberger W, Pulvermüller F, Birbaumer N (1994). Words and pseudowords elicit distinct patterns of 30-Hz activity in humans. *Neuroscience Letters*, 176, 115–8.

Lutzenberger W, Pulvermüller F, Elbert T, Birbaumer N (1995). Local 40-Hz activity in human cortex induced by visual stimulation. *Neuroscience Letters*, 183, 39–42.

Mahon, B.Z., and Caramazza, D. (2008). A critical look at the embodied cognition hypothesis and a new proposal for grounding conceptual content. J Physiol. Paris, 102 (1–3), 59–70.

Makris N, Meyer JW, Bates JF, Yeterian EH, Kennedy DN, Caviness VS (1999). MRI-based topographic parcellation of human cerebral white matter and nuclei II. Rationale and applications with |systematics of cerebral connectivity. *NeuroImage*, 9, 18–45.

Martin A, Chao LL (2001). Semantic memory and the brain: structure and processes. *Current Opinion in Neurobiology*, 11, 194–201.

Martin A, Haxby JV, Lalonde FM, Wiggs CL, Ungerleider LG (1995). Discrete cortical regions associated with knowledge of color and knowledge of action. *Science*, 270, 102–5.

Matelli M, Camarda R, Glickstein M, Rizzolatti G (1986). Afferent and efferent projections of the inferior area 6 in the macaque monkey. *Journal of Comparative Neurology*, 251, 281–98.

McCulloch WS, Pitts WH (1943). A logical calculus of ideas immanent in nervous activity. *Bulletin of Mathematical Biophysics*, 5, 115–33.

Moscoso Del Prado Martin F, Hauk O, Pulvermüller F (2006). Category specificity in the processing of color-related and form-related words: an ERP study. *NeuroImage*, 29, 29–37.

Müller MM, Bosch J, Elbert T *et al.* (1996). Visually induced gamma-band responses in human electroencephalographic activity: a link to animal studies. *Experimental Brain Research*, 112, 96–102.

Näätänen R (2001). The perception of speech sounds by the human brain as reflected by the mismatch negativity (MMN) and its magnetic equivalent (MMNm). *Psychophysiology*, 38, 1–21.

Näätänen R, Lehtokoski A, Lennes M *et al.* (1997). Language-specific phoneme representations revealed by electric and magnetic brain responses. *Nature*, 385, 432–4.

Näätänen R, Tervaniemi M, Sussman E, Paavilainen P, Winkler I (2001). 'Primitive intelligence' in the auditory cortex. *Trends in Neurosciences*, 24, 283–8.

Neininger B, Pulvermüller F (2001). The right hemisphere's role in action word processing: a double case study. *Neurocase*, 7, 303–17.

Neininger B, Pulvermüller F (2003). Word-category specific deficits after lesions in the right hemisphere. *Neuropsychologia*, 41, 53–70.

Obleser J, Lahiri A, Eulitz C (2003). Auditory-evoked magnetic field codes place of articulation in timing and topography around 100 milliseconds post syllable onset. *NeuroImage*, 20, 1839–47.

Oliveri, M., Finocchiarro, C., Shapiro, K., Gangitano, M., Canramazza, A., and Pascual-Leone, A. (2004). All talk and new action: a transcranial magnetic stimulation study of motor cortex activation during action word production. J Cogn. Neurosci., 16(3), 374–381.

Page M (2000). Connectionist modelling in psychology: a localist manifesto. *Behavioural and Brain Sciences*, 23, 443–67.

Palva S, Palva JM, Shtyrov Y *et al.* (2002). Distinct gamma-band evoked responses to speech and non-speech sounds in humans. *J Neurosci*, 22, RC211.

Pandya DN, Yeterian EH (1985). Architecture and connections of cortical association areas. In A Peters, EG Jones, Eds. *Cerebral Cortex, Volume 4: Association and Auditory Cortices* (pp. 3–61). London: Plenum Press.

Patterson K, Hodges JR (2001). Semantic dementia. In RF Thompson, JL McClelland, Eds. *International Encyclopaedia of the Social and Behavioural Sciences. Behavioural and Cognitive Neuroscience Section* (pp. 3401–5). New York, NY: Pergamon Press.

Paus T, Perry DW, Zatorre RJ, Worsley KJ, Evans AC (1996). Modulation of cerebral blood flow in the human auditory cortex during speech: role of motor-to-sensory discharges. *European Journal of Neuroscience*, 8, 2236–46.

Penfield W, Boldrey E (1937). Somatic sensory and motor representation in the cerebral cortex as studied by electrical stimulation. *Brain*, 60, 389–443.

Penfield W, Rasmussen T (1950). *The Cerebral Cortex of Man*. New York, NY: Macmillan.

Pettigrew CM, Murdoch BE, Ponton CW *et al.* (2004). Automatic auditory processing of english words as indexed by the mismatch negativity, using a multiple deviant paradigm. *Ear and Hearing*, 25, 284–301.

Plenz D, Thiagarajan TC (2007). The organizing principles of neuronal avalanches: cell assemblies in the cortex? *Trends in Neurosciences*, 30, 101–10.

Posner MI, Pavese A (1998). Anatomy of word and sentence meaning. *Proceedings of the National Academy of Sciences USA*, 95, 899–905.

Preissl H, Pulvermüller F, Lutzenberger W, Birbaumer N (1995). Evoked potentials distinguish nouns from verbs. *Neuroscience Letters*, 197, 81–3.

Price CJ (2000). The anatomy of language: contributions from functional neuroimaging. *Journal of Anatomy*, 197, 335–59.

Pulvermüller F (1992). Constituents of a neurological theory of language. *Concepts in Neuroscience*, 3, 157–200.

Pulvermüller F (1996). Hebb's concept of cell assemblies and the psychophysiology of word processing. *Psychophysiology*, 33, 317–33.

Pulvermüller F (1999). Words in the brain's language. *Behavioural and Brain Sciences*, 22, 253–336.

Pulvermüller F (2000). Cell assemblies, axonal conduction times, and the interpretation of high-frequency dynamics in the EEG and MEG. In R Miller, Ed. *Time and the Brain* (pp. 241–9). Chur: Harwood Academic Publishers.

Pulvermüller F (2001). Brain reflections of words and their meaning. *Trends in Cognitive Sciences*, 5, 517–24.

Pulvermüller F (2002). A brain perspective on language mechanisms: from discrete neuronal ensembles to serial order. *Progress in Neurobiology*, 67, 85–111.

Pulvermüller F (2003). *The neuroscience of language*. Cambridge: Cambridge University Press.

Pulvermüller F (2005). Brain mechanisms linking language and action. *Nature Reviews Neuroscience*, 6, 576–82.

Pulvermüller F, Birbaumer N, Lutzenberger W, Mohr B (1997). High-frequency brain activity: its possible role in attention, perception and language processing. *Progress in Neurobiology*, 52, 427–45.

Pulvermüller F, Eulitz C, Pantev C et al. (1996). High-frequency cortical responses reflect lexical processing: an MEG study. *Electroencephalography and Clinical Neurophysiology*, 98, 76–85.

Pulvermüller F, Härle M, Hummel F (2000). Neurophysiological distinction of verb categories. *NeuroReport*, 11, 2789–93.

Pulvermüller F, Hauk O (2006). Category-specific processing of color and form words in left fronto-temporal cortex. *Cerebral Cortex*, 16, 1193–201.

Pulvermüller F, Hauk O, Nikulin VV, Ilmoniemi RJ (2005). Functional links between motor and language systems. *European Journal of Neuroscience*, 21, 793–7.

Pulvermüller F, Huss M, Kherif F, Moscoso del Prado Martin F, Hauk O, Shtyrov Y (2006). Motor cortex maps articulatory features of speech sounds. *Proceedings of the National Academy of Sciences USA*, 103, 7865–70.

Pulvermüller F, Kujala T, Shtyrov Y et al. (2001). Memory traces for words as revealed by the mismatch negativity. *NeuroImage*, 14, 607–16.

Pulvermüller F, Lutzenberger W, Preissl H (1999). Nouns and verbs in the intact brain: evidence from event-related potentials and high-frequency cortical responses. *Cerebral Cortex*, 9, 498–508.

Pulvermüller F, Mohr B (1996). The concept of transcortical cell assemblies: a key to the understanding of cortical lateralization and interhemispheric interaction. *Neuroscience and Biobehavioural Reviews*, 20, 557–66.

Pulvermüller F, Mohr B, Schleichert H (1999). Semantic or lexico-syntactic factors: What determines word-class specific activity in the human brain? *Neuroscience Letters*, 275, 81–4.

Pulvermüller F, Preissl H (1991). A cell assembly model of language. *Network Computation in Neural Systems*, 2, 455–68.

Pulvermüller F, Preissl H, Lutzenberger W, Birbaumer N (1995). Spectral responses in the gamma-band: physiological signs of higher cognitive processes? *NeuroReport*, 6, 2057–64.

Pulvermüller F, Shtyrov Y, Ilmoniemi RJ (2003). Spatio-temporal patterns of neural language processing: an MEG study using Minimum-Norm Current Estimates. *NeuroImage*, 20, 1020–5.

Pulvermüller F, Shtyrov Y, Ilmoniemi RJ (2005). Brain signatures of meaning access in action word recognition. *Journal of Cognitive Neuroscience*, 17, 884–92.

Pulvermüller F, Shtyrov Y, Kujala T, Näätänen R (2004). Word-specific cortical activity as revealed by the mismatch negativity. *Psychophysiology*, 41, 106–12.

Rizzolatti G, Craighero L (2004). The mirror-neuron system. *Annual Review in Neuroscience*, 27, 169–92.

Rizzolatti G, Luppino G (2001). The cortical motor system. *Neuron*, 31, 889–901.

Roelofs A (1992). A spreading-activation theory of lemma retrieval in speaking. *Cognition*, 42, 107–42.

Rogers TT, Lambon-Ralph MA, Garrard P et al. (2004). Structure and deterioration of semantic memory: a neuropsychological and computational investigation. *Psychological Review*, 111, 205–35.

Roy D (2005). Grounding words in perception and action: computational insights. *Trends in Cognitive Science*, 9, 389–96.

Schnelle H (1996). Approaches to computational brain theories of language – a review of recent proposals. *Theoretical Linguistics*, 22, 49–104.

Scott SK, Johnsrude IS (2003). The neuroanatomical and functional organization of speech perception. *Trends in Neurosciences*, 26, 100–7.

Searle JR (1990). Minds, brains, and programs. *Behavioural and Brain Sciences*, 3, 417–57.

Seidenberg MS, Plaut DC, Petersen AS, McClelland JL, McRae K (1994). Nonword pronunciation and models of word recognition. *Journal of Experimental Psychology: Human Perception and Performance*, 20, 1177–96.

Sereno SC, Rayner K (2003). Measuring word recognition in reading: eye movements and event-related potentials. *Trends in Cognitive Sciences*, 7, 489–493.

Sereno SC, Rayner K, Posner MI (1998). Establishing a time line for word recognition: evidence from eye movements and event-related potentials. *NeuroReport*, 13, 2195–200.

Shastri L, Grannes D, Narayana S, Feldman J (2005). A connectionist encoding of parameterized schemas and reactive plans. In GK Kraetzschmar, G Palm, Eds. *Hybrid Information Processing in Adaptive Autonomous Vehicles*. Berlin: Springer.

Shtyrov Y, Hauk O, Pulvermüller F (2004). Distributed neuronal networks for encoding category-specific semantic information: the mismatch negativity to action words. *European Journal of Neuroscience*, 19, 1083–92.

Shtyrov Y, Pihko E, Pulvermüller F (2005). Determinants of dominance: is language laterality explained by physical or linguistic features of speech? *NeuroImage*, 27, 37–47.

Shtyrov Y, Pulvermüller F (2002). Neurophysiological evidence of memory traces for words in the human brain. *NeuroReport*, 13, 521–5.

Simmons WK, Ramjee V, Beauchamp MS, McRae K, Martin A, Barsalou LW (2007). A common neural substrate for perceiving and knowing about color. *Neuropsychologia*, 45, 2802–10.

Singer W, Gray CM (1995). Visual feature integration and the temporal correlation hypothesis. *Annual Review in Neuroscience*, 18, 555–86.

Sittiprapaporn W, Chindaduangratn C, Tervaniemi M, Khotchabhakdi N (2003). Preattentive processing of lexical tone perception by the human brain as indexed by the mismatch negativity paradigm. *Annals of the New York Academy of Sciences*, 999, 199–203.

Skrandies W (1999). Early effects of semantic meaning on electrical brain activity. *Behavioural and Brain Sciences*, 22, 301.

Tallon-Baudry C, Bertrand O (1999). Oscillatory gamma activity in humans and its role in object representation. *Trends in Cognitive Sciences*, 3, 151–61.

Tallon-Baudry C, Bertrand O, Delpuech C, Pernier J (1996). Stimulus specificity of phase-locked and non-phase-locked 40 Hz visual responses in humans. *Journal of Neuroscience*, 16, 4240–9.

Tallon-Baudry C, Bertrand O, Peronnet F, Pernier J (1998). Induced gamma-band activity during the delay of visual short-term memory tasks in humans. *Journal of Neuroscience*, 18, 4244–54.

Tettamanti M, Buccino G, Saccuman MC *et al.* (2005). Listening to action-related sentences activates fronto-parietal motor circuits. *Journal of Cognitive Neuroscience*, 17, 273–81.

Tomasello M, Kruger AC (1992). Joint attention on actions: acquiring verbs in ostensive and non-ostensive contexts. *Journal of Child Language*, 19, 311–33.

Tsumoto T (1992). Long-term potentiation and long-term depression in the neocortex. *Progress in Neurobiology*, 39, 209–28.

Tyler LK, Moss HE (2001). Towards a distributed account of conceptual knowledge. *Trends in Cognitive Sciences*, 5, 244–52.

Tyler LK, Moss HE, Durrant-Peatfield MR, Levy JP (2000). Conceptual structure and the structure of concepts: a distributed account of category-specific deficits. *Brain and Language*, 75, 195–231.

Tyler LK, Russell R, Fadili J, Moss HE (2001). The neural representation of nouns and verbs: PET studies. *Brain*, 124, 1619–34.

Tyler LK, Stamatakis EA, Bright P *et al.* (2004). Processing objects at different levels of specificity. *Journal of Cognitive Neuroscience*, 16, 351–62.

Varela FJ, Thompson E, Rosch E (1991). *The Embodied Mind: Cognitive Science and Human Experience*. Boston, MA: MIT Press.

von der Malsburg C, Schneider W (1986). A neural cocktail-party processor. *Biological Cybernetics*, 54, 29–40.

Warrington EK, McCarthy RA (1983). Category specific access dysphasia. *Brain*, 106, 859–78.

Warrington EK, Shallice T (1984). Category specific semantic impairments. *Brain*, 107, 829–54.

Watkins K, Paus T (2004). Modulation of motor excitability during speech perception: the role of Broca's area. *Journal of Cognitive Neuroscience*, 16, 978–87.

Wennekers T, Garagnani M, Pulvermüller F (2006). Language models based on Hebbian cell assemblies. *Journal of Physiology–Paris*, 100, 16–30.

Wermter S, Weber C, Elshaw M, Gallese V, Pulvermüller F (2005). Neural grounding of robot language in action. In S Wermter, G Palm, M Elshaw, Eds. *Biomimetic Neural Learning for Intelligent Robots* (pp. 162–81). Berlin: Springer.

Wermter S, Weber C, Elshaw M, Panchev C, Erwin H, Pulvermüller F (2004). Towards multimodal neural network robot learning. *Robotics and Autonomous Systems*, 47, 171–5.

Wilson SM, Saygin AP, Sereno MI, Iacoboni M (2004). Listening to speech activates motor areas involved in speech production. *Nature Neuroscience*, 7, 701–2.

Wittgenstein L (1953). *Philosophical Investigations*. Oxford: Blackwell.

Young MP, Scannell JW, Burns G, Blakemore C (1994). Analysis of connectivity: neural systems in the cerebral cortex. *Review in Neuroscience*, 5, 227–49.

Zatorre RJ, Evans AC, Meyer E, Gjedde A (1992). Lateralization of phonetic and pitch discremination in speech processing. *Science*, 256, 846–9.

Chapter 7

Symbols and embodiment from the perspective of a neural modeller

Andreas Knoblauch

7.1 Introduction and definitions

This paper contributes to the current debate about symbols and embodiment by pointing out the perspective of a neural modeller. I illustrate the default definitions of 'symbol', 'embodiment', 'meaning', and 'grounding' in the context of detailed neural network models, i.e., on a level more detailed than common connectionist approaches. My arguments are based on Hebbian *neuronal or cell assemblies* (Hebb 1949; Braitenberg 1978; Palm 1982, 1990) and detailed models of the cortical microcircuitry. These models have been employed to implement a large-scale cortical architecture to enable a robot to perform simple tasks such as understanding and reacting to simple spoken commands. More generally, I finally discuss the relations between embodiment, grounding, anchoring, binding, and the invariant recognition in distributed hierarchical systems.

7.1.1 Symbols

For a neural network modeller, one simple possible way to discern symbols from nonsymbols is to look at the inner structure of the representational units. Subsymbols have an inner structure which can be used to define a similarity metric relevant for the represented entity. In contrast, symbols have no relevant inner structure (i.e., symbols are abstract and arbitrary). For example, in simple object recognition systems, a nonsymbol or subsymbol may be a vector of sensory features, while a symbol may correspond to a single node representing an object category. These definitions are sufficient for a low-level (e.g., neural) description of a cognitive subsystem (e.g., for object recognition), but may not be adequate for the current debate which is about language and the representation of meaning. Here, the discussion includes higher-level symbols employed by a cognitive system that is able to think, to reason, and to manipulate these symbols in a flexible way.

According to Glenberg *et al.*'s default definition (Chapter 1) such a symbol is a 'theoretical element that is arbitrary, abstract, and amodal'. Before we proceed by discussing and adapting that definition, it may be useful to be aware of the different contexts in which we will use the word 'symbol'. The situation is illustrated in Figure 7.1. We live in a physical world W where systems or subjects S are part of that world and interact with the world. Some of the systems (namely we, the subjects) are somehow able to generate

Fig. 7.1 Different modelling
levels. We live in a physical
world W; systems (or subjects)
S are part of that world and
interact with the world. The
subjects are somehow able to
generate a (usually unique)
psychological or phenomenological space P, which we can
use, for example, to generate
theories T about all kind of
issues on all levels W, S, P,
and T.

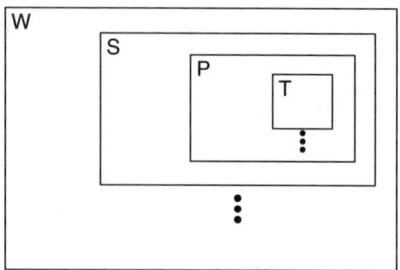

a usually unique psychological or phenomenological space P, which we can employ, for example, to generate ideas or theories T about all kinds of issues on all levels W, S, P, and T. In particular, we can make theories about S (predominantly done by biology, neuroscience, and artificial intelligence [AI]), P (psychology), or T (metamathematics or logics). Ideas or theories T essentially consist of a set of symbols (as defined above) and additional rules determining how the symbols can be 'manipulated'.

Since symbols are part of theories or ideas this implies two aspects of a symbol correspon-ding to levels S and T. As a neural modeller one is predominantly interested in the S-level implementation of P *in W*, i.e., in reducing the observable psychological and behavioural phenomena as much as possible to detailed neural and synaptic processes, and finally to the physical laws. (The underlying preliminary naturalistic working hypothesis assumes that this goal is actually possible.) But since models about S are ultimately also T theories, the neural modeller (and any other kind of S-modeller like many an AI researcher) has to discern between the two kinds of symbols within his theory: symbols to model the T-symbols of S, and symbols to model the implementation of the T-symbols of S. Thus, we will refer to these two kind of symbols as T-symbols and S-symbols, respectively. For exam-ple, for a cognitive system S capable of understanding language, a T-symbol is the represen-tation of a word, while S-symbols are finer-grained entities used for implementing the word representation. An S-symbol could be a node in a connectionist network (or, alternatively, a state or band variable in a Turing machine) while a T-symbol could be implemented by a set of distributed S-symbols (and possibly further dynamic processes).

The current debate is about the question whether cognitive systems (such as we) are or have to be either symbolic or 'embodied'. Of course, any kind of cognitive system must be symbolic in the trivial sense that we have symbolic language and theories T. Thus, any cognitive system must be T-symbolic. Correspondingly, the critical question is not about the reality of T-symbols, but about the way T-symbols are implemented in S-symbols. Note that, depending on how we define 'symbol' and 'embodiment', it might be thinkable that T-symbols could be implemented by non-symbolic processes at the S level (although this seems counterintuitive since we actually aim to develop a symbolic theory T about S).

7.1.2 **Embodiment**

Embodiment comes in several different flavours (cf. Wilson 2002). The strongest claim would be that embodiment could extend the qualitative, or at least quantitative, computational capabilities of a system S by exploiting the properties of W, e.g., described by the physical laws. If true, this would essentially negate the Church–Turing thesis that symbolic Turing machines can compute any 'naturally' (or physically) computable function. For example, 'embodied' analogue computers might be better in simulating physical systems, or in computing real numbers with infinite precision (see also Brooks 1990). Nevertheless, symbolic Turing computers can approximately simulate any physical system by numerically solving the differential equations of physics, although this can take considerable time and, for chaotic systems, infinite computing precision. Another example could be computers exploiting the quantum properties of the physical world. It has been shown that such quantum computers, if physically possible, could compute certain functions much faster than conventional computers or Turing machines (DiVincenzo 1995).

A less strong idea of embodiment is the dichotomy of embodied versus symbolic cognitive systems addressed by the current debate which attempts to classify cognitive systems according to the interface between the system S itself and the external world W. Obviously, any cognitive system S that deserves that name will have to interact with its environment (percept and act) and is therefore embodied in a trivial sense. Similarly, any cognitive system S must be symbolic in a trivial sense since it must explain our capabilities to use language and think in symbols (e.g., to develop theories T within our psychological space P). Thus, in this trivial sense any model of a cognitive system will be both embodied and symbolic.

Obviously, any cognitive system can be divided into sensors, actors, and internal machinery, such that the interaction with the environment is accomplished only via the sensors and actors. Strictly speaking such an interactive system cannot be adequately modelled by a Turing machine. The original 'autistic' Turing machine has been suggested as a model for computation only. That scenario assumes separate phases for: (1) providing the input from the environment to the Turing machine's band, (2) doing the computation independently of the environment, possibly for a very long time, and (3) returning the output of the computation from the band to the environment. In contrast, we rather have to think of an 'interactive' Turing machine that depends on the environment and can influence the environment at any time. Thus, it would be possible to define embodiment by the degree of interaction with the environment.

Our default definition of an 'embodied' system goes in a similar direction by demanding that the meaning of a symbol must depend on activity in systems also used for perception, action, and that emotion and reasoning must require the use of those systems (see Chapter 1). This form of weak embodiment is stronger than the trivial version of embodiment, but addresses only the high-level structure of the internal machinery, e.g., in Marr's (1982) terms, the algorithmic or computational levels but not the implementation level. As a consequence, any such 'embodied' system can be translated into a purely (S-)'symbolic' system (e.g., a computer program or Turing machine) with the same sensor/actor interface and vice versa. This is true according to the Church–Turing thesis, at least as long as our 'embodied' system does not exploit the physical world in a super-Turing manner as described before. We can also conclude that this form of embodiment will probably be neutral to such questions as whether 'ideas are the sole province of biological systems' (as discussed in Chapter 1). And, of course, the property of embodiment will be a gradual property. Nevertheless, the idea of embodiment might still prove useful, e.g., in building more efficient artificial cognitive systems, or in guiding the analysis of the brain.

7.1.3 Meaning

Our last definition of embodiment refers to the term 'meaning', which may require an explicit definition. A simple definition states that meaning is the 'content' of a sign or symbol. Here, 'content' refers to all the parts of an information processing systems theories T that have a relation to the symbol. For example, the meaning of the word symbol 'car' may include prototypical 'icons' of cars, knowledge about the corresponding *consists-of* and *is-a* ontologies, knowledge about actions that can be done with, to, or by a car, and episodic knowledge about particular experiences with cars. This very general meaning of a symbol must usually be strongly constrained by context.

An alternative, more behaviouristic, definition is this: the meaning of a symbol or sign to a cognitive system is the sum of the potential or actual behavioural changes after receiving the sign. For example, the meaning of the traffic sign 'stop' to a car driver is a propensity to apply the brakes to stop the car. Or the meaning of signs indicating bad politics to a voter is a propensity to no longer give a vote to the responsible politician in future elections.

The second definition has several advantages: first, it does not refer to the internals of the system which may be difficult to observe and interpret, e.g., in animal experiments. Instead, it solely refers to the system's or agent's observable behaviour or actions. Further, this direct reference to actions and agents is more relevant for our current debate about meaning and embodiment. Defining meaning in this way by particular actions then obviously implies (by definition) that establishing meaning is closely related to or even 'depends on activity in action systems' which is essentially our definition of embodiment (see Chapter 1).

Indeed, neurobiological experimenters investigating the brains of animals by correlating behaviour and neural activity actually have to define categories (which can then be symbolized) by actions. For example, experimenters can find out which one of two possible interpretations of an ambiguous figure a monkey is perceiving by training the monkey to move the right hand for interpretation 1 and the left hand for interpretation 2 (which already defines the meaning of the figures for the monkey). Similarly, it has been proposed that we humans, as well as other animals, can learn to discriminate two classes or categories only if the discrimination is behaviourally relevant. This is most obvious for lower animals with only a very limited behavioural repertoire (and therefore only a limited way of developing abstract classes such as prey, predator, or mate, corresponding to actions like feeding, fighting, or mating).

In summary, we may already conclude here that understanding the meaning of signs indeed requires embodiment defined as regular activation of action- and perception-related subsystems. However, this conclusion is not sufficient for understanding how the brain works, or how to create intelligent artificial systems. This requires a detailed theory on how perception, action, and learning of abstract categories is accomplished by the brain. The rest of this chapter focuses on delivering building blocks for this detailed theory by having a closer look at neural models of symbols and meaning.

7.1.4 The easy and the hard problems

With this essay I do not intend to tackle what has been called the 'hard problems' in the psychology of consciousness, for example, explaining how physical processes in the brain give rise to subjective experience (Chalmers 1996; Jackendoff 1987). Although I believe that, from a third-person perspective, symbols, embodiment, and meaning as defined above are easy problems (i.e., nothing 'mystic'), and that we will probably soon be able to realize artificial systems that can be said to be embodied and represent meaning in a similar way to humans, I admit that assuming full identity between the first-person subjective processes of humans (including feelings) and those of digital computers (Dennett, 1996; Metzinger, 2004) may have problematic consequences.

For example, if we were to ascribe subjective feelings to a robot then we would also have to ascribe feelings to the robot rid of its sensorimotor interface (just as we ascribe feelings to quadriplegia and locked-in syndrome patients). Then we ultimately would have to ascribe feelings to a digital computer, i.e., finite-state-machine (FSM). Or, to put it more exactly, we would ascribe feelings to particular states of the FSM. Adapting Searle's Chinese room argument (Searle, 1980) to feelings (instead of meaning), this seems strange because the FSM's states are arbitrary, i.e., it is unclear why one particular physical state should be associated with pain (and not with joy or something else).

One could answer that meaning and feelings are not associated with a single state but with particular *recurring* state sequences (for example, realizing a kind of monitoring structure). However, this will probably not help us since we can always construct an 'equivalent' FSM where such a sequence corresponds to a single state again (although this will require a large FSM with many states). We may accept this for the case of meaning defined in terms of 'potential behaviour' (see above) because the question 'Is that machine understanding this?' can be resolved, in principle, by looking at the FSM's past or future states. However, we are usually more reluctant in the case of feelings because the question 'Is that machine feeling pain?' must be answered in the present (and proposing that past or future states would make a difference contradicts the state concept of classical physics).

7.2 A neural modeller's perspective

When words referring to actions or visual scenes are presented to humans, distributed neural networks, including areas of the motor and visual systems of the cortex, become active (e.g., Pulvermüller 1999, 2003). The brain correlates of words and their referent actions and objects appear to be strongly coupled neuron ensembles in defined cortical areas. The theory of cell assemblies (Hebb 1949; Braitenberg 1978; Palm 1982, 1990) provides one of the most promising frameworks for modelling and understanding the brain in terms of distributed neuronal activity. It is suggested that entities of the outside world (and also internal states) are coded in groups of neurons rather than in single 'grandmother' cells, and that a cell assembly is generated by Hebbian coincidence or correlation learning where the synaptic connections are strengthened between co-activated neurons.

7.2.1 Local cell assemblies, associative memory, and neural S-symbols

The notion of neuronal assemblies as strongly coupled neurons leads to the concept of *neural (auto-)associative memory* (Willshaw *et al.* 1969; Palm 1980; Hopfield 1982). One simple model of neural associative memory has been proposed by Steinbuch and Willshaw (Willshaw *et al.* 1969; Steinbuch 1961; Palm 1980; Knoblauch 2005) consisting of McCulloch–Pitts-type threshold units and recurrent binary synapses (Figure 7.2). Here, the activity pattern of the cell population can be described by a binary vector, and we identify stored activity patterns with the cell assemblies. After learning a number of

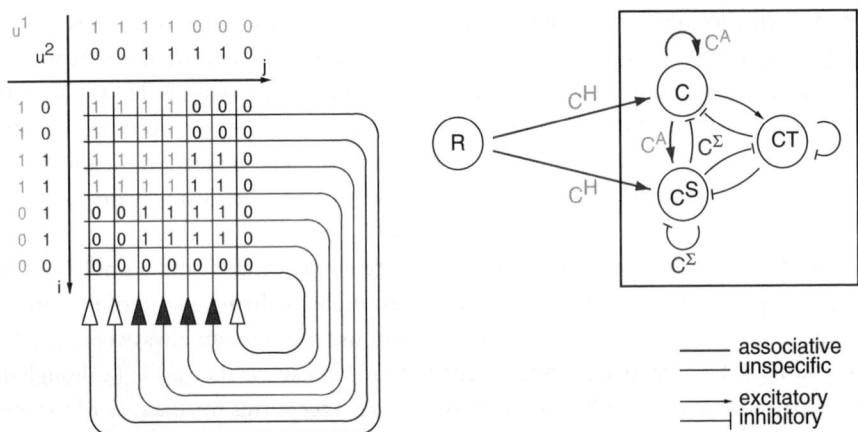

Fig. 7.2 Left: Neural (auto-)associative memory where two cell assemblies of size $k = 4$ have been stored (corresponding to the activation patterns u1 and u2) in the 'memory matrix' of synaptic connections. Filled units indicate the active neurons of pattern u2. Right: A more realistic implementation of associative memory, modelling a small patch (about 1 mm³) of cortical tissue (Knoblauch and Palm, 2001). The model comprises several populations of excitatory and inhibitory spiking neurons of the integrate-and-fire type. Here, C is the main excitatory population of pyramidal cells receiving input from another cortical patch, R. C^S and C^T are inhibitory interneuron populations controlling local excitation. Each neuron is modelled as a leaky integrator with excitatory and inhibitory conductances, where a spike is emitted as soon as the dendritic potential exceeds a threshold. The memory matrices are employed in several afferent and recurrent synaptic connections (cH, cA) to and from populations R, C, and C^S. The remaining inhibitory feedback connections are unspecific (i.e., independent of the learned activity patterns).

neuronal assemblies, the network can be described by a connection matrix A correspon-ding to a graph where the nodes correspond to the neurons and neuronal assemblies correspond to k-cliques of neurons (a k-clique is a subset of size k consisting of completely connected neurons).

Hetero-association works similar to auto-association except that the 'memory matrix' describes the synaptic connections between two different neuron populations. Hetero-associative connections can map neuronal assemblies of the first population (or sets or parts of them) to assemblies of the second population (or sets or parts of them).

The virtue of the binary model is that it is easy to understand, analyse, and implement, but the main results apply also to more realistic gradual and spiking models (Hopfield, 1984; Knoblauch and Palm, 2001). Neural associative memories have a couple of nice features. They achieve pattern completion, i.e., a neuronal assembly can be activated not only by the very same inputs that have been used for learning, but also by modified patterns that are 'sufficiently similar' to the original address pattern. For example, assem-bly u2 in Figure 7.2 will already be activated by addressing an arbitrary subset of size ≥ 3.

It can be shown that the number of storable patterns is almost proportional to the number of synapses if the patterns are sparse and have random character (i.e., a popula-tion of n neurons can store almost n^2 sparse cell assemblies with $k \ll n$). Access time is essentially independent of the number of stored patterns. The overlaps of different neuronal assemblies can be used to express the similarities of the represented entities. Neuronal assemblies thereby provide a very natural associative way of grounding new representations in the sensory inputs by means of bidirectional associative connections (cf. Barsalou 2003).

Associative memories have been used to model small volumes of cortical tissue (e.g., 1 mm^3) corresponding to a macrocolumn or the range where dense local recurrent connections between any cell pairs are possible (Braitenberg and Schüz, 1991). A step towards biological realism is to replace the single McCulloch–Pitts population by more realistic spiking neuron models and to incorporate known properties of cortical circuits (Figure 7.2). These models can extend the computational abilities of the standard model, for example, by making use of spike timings according to a latency code in that early spikes (relative to an external event or an underlying oscillation) are much more relevant than late spikes for activating an assembly (Knoblauch and Palm 2001; Knoblauch 2005).

Local cell assemblies can be seen as elementary neural (S-)symbols which can be 'allocated' or learnt to represent the inputs for further processing in downstream target populations. The symbolic character is most apparent if the assembly size is $k = 1$, corre-sponding to a localist code, or if the neurons that constitute a cell assembly are chosen at random, e.g., by noise. In the latter case the correlations (or overlaps) between two cells are minimal, which is required to store a maximal number of different activity patterns. Due to their singular or random character the neuronal assemblies could be said to be abstract and arbitrary, whereas the property of amodality depends on the loca-tion of the neuron population, e.g., a local population receiving visual inputs will develop visual perceptual symbols (cf. Barsalou 1999).

7.2.2 **Global cell assemblies, language, and T-symbols**

We have designed a large-scale brain model consisting of many interconnected cortical areas employing spiking associative memories. The model was implemented and tested on a robotic platform enabling the robot to understand and react to simple commands such as 'Bot show plum!' (Knoblauch *et al.* 2004). The language part of the model is illustrated by Figure 7.3 and the action part by Figure 7.4.

Each box in the figures corresponds to a spiking neural associative memory storing local cell assemblies, as described above. For illustration purposes, each area has been labelled according to the current activity pattern. (In general, a superposition of several stored local assemblies can be activated, e.g. to represent uncertainty or to represent new entities to be learned; here, the labels correspond to the neuronal assembly most similar to the current activation pattern). The resulting *global assembly*, for example, representing the *T*-symbol 'plum', stretches over many cortical areas (involving visual, auditory, action, and goal-related areas) and changes dynamically during the process of 'understanding' and reacting to the command. Thus, the global cell assembly as a whole works as a sign in Peirce's sense, i.e., as a mediator between the idea of a 'plum' and the real plum in the external world. The global assembly consists of parts, some of which can

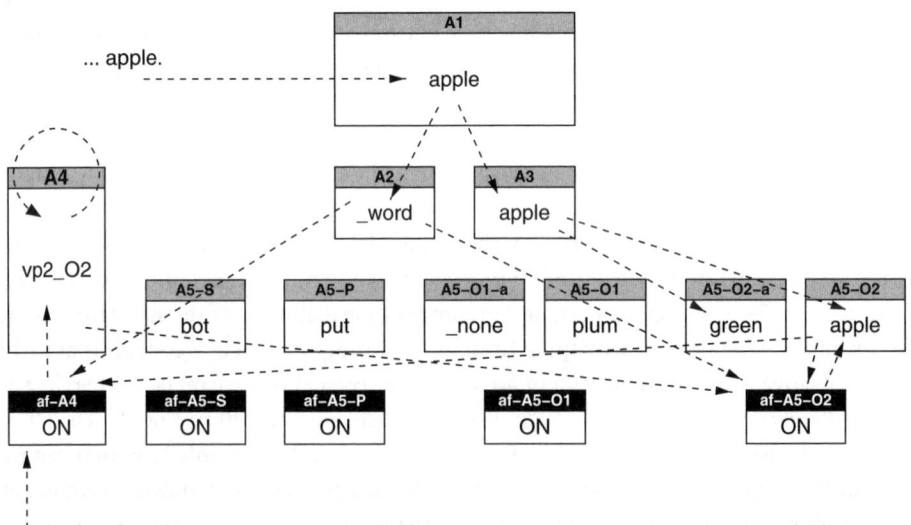

Fig. 7.3 The language part of an associative cortex model (see Knoblauch et al., 2004) at the end of processing the sentence 'Bot put plum to green apple'. Each box corresponds to a cortical (or subcortical) area modelled as a neural associative memory. The meaning of the sentence is represented by distributed cell assemblies comprising 'slot areas' for different grammatical roles to implement elementary productivity and systematicity. Auditory input enters the central areas via areas A1 and A3 and is distributed across the grammatical slots according to a logic controlled by a grammatical sequence memory (A4) (where basic sentence types are stored) and subcortical 'activation fields' (small boxes). Arrows indicate recently activated synaptic connections.

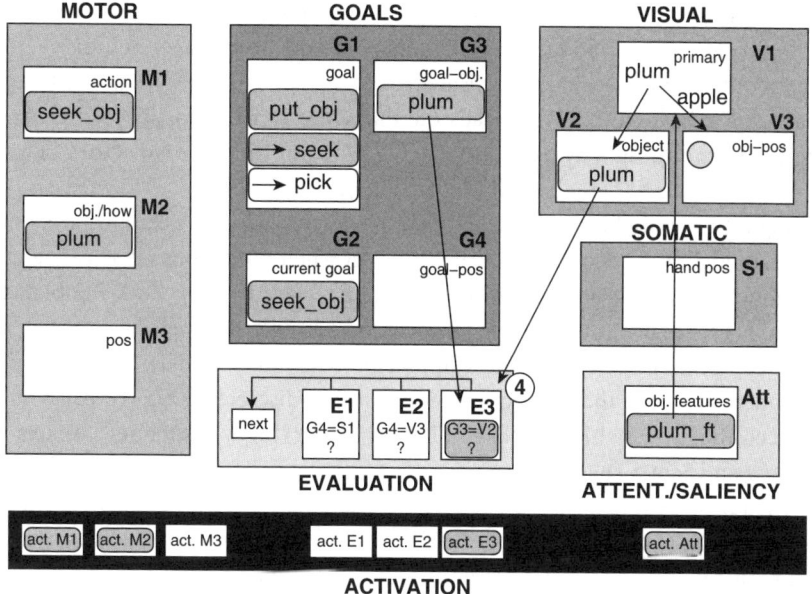

Fig. 7.4 Action part of an associative cortex model (see Knoblauch *et al.*, 2004) during performing the command 'Bot put plum to green apple'. The goal areas (G1–G4) received their inputs from the grammatical role areas (A5-S, A5-O1, etc., as illustrated in Figure 7.3) and divide the goals into a sequence of subgoals (i.e., seek plum, pick plum, move to the apple, drop plum). (High-level) motor areas receive inputs from the goal areas in order to perform the current subgoal. The completion of subgoals and switching to the next subgoal is controlled by 'evaluation fields' which check, for example, the consistency of visual perceptual activity patterns (e.g., in V4) with goal representations (e.g., in G3). At the shown system state the robot is about to finish the subgoal of seeking the plum.

be attributed as 'abstract' and 'amodal', e.g., the lexical representation of 'plum' in A3. But these symbolic parts are naturally grounded in the synaptically connected perceptual and action-related parts of the global assembly.

7.2.3 Cortical macrocolumns, prediction, and embodiment

Cognitive processes must be able to distinguish between different representational modi. For example, representational states may refer to present, future (or prediction), reality, wish, or signal (detailed and concrete), or symbol (abstract, amodal), perception, or action. Many cognitive architectures take a modular approach where these different representational modi are segregated into different cognitive subsystems or modules. (In general, an architecture can be said to be modular if it can be divided into subsystems such that there is much more communication between processes inside a subsystem than between processes of different subsystems.) For example, we could segregate a cognitive system into different modules for perceptions, actions, goals, memory, rule-based prediction systems, etc.

We have argued how global neuronal assemblies can implement and ground T-symbols (e.g., words of a language) by distributed activation stretching over many sensory, motoric, and associative cortical areas (Pulvermüller 1999). Thus, although the brain appears to have a modular character in that sense, hints to possible complementary strategies of grounding may be found when looking at the microstructure of a single cortical macrocolumn (Douglas and Martin 2004).

Although it has long been established that neocortical anatomy exhibits a six-layered structure, modellers have often neglected this fact when modelling a cortical patch by a single 'monolithical' neuron population (e.g., Palm 1982; Ritz *et al.* 1994; Knoblauch and Palm 2001). This may be attributable to the wish to focus on a single layer or the lack of adequate computational resources to simulate more detailed models, but also to doubting or underestimating the functional significance of discrete within- or between-layer synaptic connections which appear to have a rather 'fuzzy' character (Abeles 1991; Braitenberg and Schüz 1991).

In accordance with ideas developed by Körner *et al.* (1999) (see also Hawkins and Blakeslee 2004; Rao and Ballard 1999; Guillery 2003), we assume as a working hypothesis that the basic function of a cortical column is to represent and actively predict its sensory inputs. To achieve this in a self-organizing, autonomous way, it is necessary to have access to (at least some of) the different representational modi described above. We propose that different representational modi of the same entity are located in different layers within the same macrocolumn rather than monolithically in different columns or areas.

Figure 7.5 illustrates this functional model and our current implementation employing spiking associative networks similar to that discussed in Section 7.2.1. At each time the model must represent a state $v = (w, a)$ and use sensory input s to update the state v according to a function f. We found it meaningful to divide the state variable v into two independent entities: a variable w describing 'external' entities from the outside world and another variable a describing a local 'internal actor'. In addition to updating a state, the system should also be able to predict a future state w' without accessing sensory input. Note that the proposed circuitry provides the basic ingredients for simulating (or predicting over) larger time intervals.

By comparison with known anatomical facts we can match our functional model (Figure 7.5) with the layered organization of neocortex (Körner *et al.* 1999; Guillery 2003; Douglas and Martin 2004; Braitenberg and Schüz 1991; Felleman and Van Essen 1991). For example, it is well established that *feed-forward inputs* to a cortical column mainly target layer IV neurons, and that the feed-forward output to the next cortical stage leaves a cortical column via layer II/III neurons. In contrast to the feed-forward stream, *feed-back inputs* avoid layer IV and target mainly the upper and lower layers. Another remarkable feature of the cortical microcircuitry is that layer V pyramidal neurons, at any cortical site, project to subcortical regions closely related to action and behaviour (Guillery, 2003).

Based on these facts we believe that the forward recognition function f is located in the middle and upper layers, while the remaining functionality, related to behaviour and predictions, is located in the lower layers V and VI. Furthermore, we believe that the

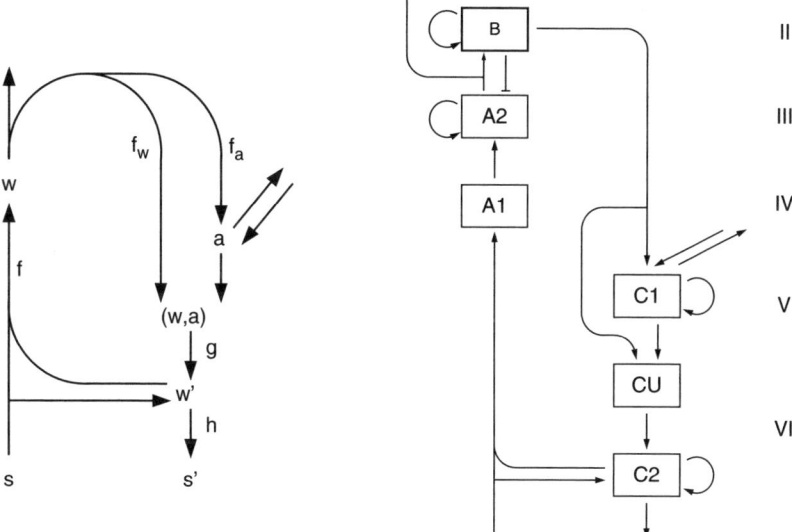

Fig. 7.5 Left: Basic functional circuit of a cortical column. Sensory input *s* is used to update the current world state *w*. This is used to choose an appropriate action *a*. World state and action can be used to predict the next world state *w'* and next sensory input *s'*. Right: Implementation of a cortical column by spiking associative memories (see Knoblauch *et al.* 2005, 2007; Körner *et al.* 1999). Sensory input activates signal-like representations (based on basis vectors) in the middle cortical layers (A1, A2), a symbol-like prototypical cell assembly in the upper layers (B), and finally action and prediction related representations in the lower layers (C1, C2).

recognition system of the middle and upper layers is split up into two subsystems, one for fast bottom-up recognition (the *A system*, layers IV and upper III) and another for refined recognition employing feedback (the *B system*, layers II/III). As an example we have implemented a model of several cortical and subcortical areas for learning saccadic object representations (including several visual cortical areas and the superior colliculus; see Knoblauch *et al.* 2005, 2007). In that particular case, the representational world states are object views (e.g., the retinal image when fixating on a particular key feature of a visual object) and the actions correspond to saccades.

What does this has to do with embodiment and grounding? Our model suggests that actions are grounded in perceptions already at the level of a single cortical macrocolumn (cf. Hawkins 2004; Körner *et al.* 1999; Guillery 2003; Young 1993). Thus, recognition of incoming signals will automatically induce an action-related process in the same macrocolumn and related subcortical structures. In our model, the perceptual representation *w* is on the one hand really symbolic in cortical layers II/III, is grounded in sensory input and a signal-like (basis vector-based) representation in layer IV, and is the result of a prediction $(w^{t-1}, a) \rightarrow w$ (layer VI) such that it is also grounded in action (layer V). Vice versa, for the same reason the produced action will be grounded in perception. Furthermore, the proposed circuitry provides the basic ingredients for simulating (or predicting) the represented state (*w*) over larger time intervals (at a microlevel), which has been suggested to be required for understanding meaning on the $(P,T\text{-})$macro-levels (cf. Barsalou 1999).

7.2.4 Hierarchies, invariances, binding, and grounding

The areas of the brain and particularly the cerebral cortex are organized in a distributed and hierarchically ordered manner. For example, the visual system of primates consists of about fifty areas and about fifteen processing stages, where each area is dedicated to a specific task (Felleman and Van Essen 1991). On the top of the hierarchy there are neurons in certain areas representing quite abstract entities coming close to $(T\text{-})$symbols. For example, in the mediotemporal lobe, cells have been found that respond in a specific and invariant way to a particular person, regardless of the stimulus being the person's face, a cartoon, or their written name (Quiroga *et al.* 2005).

There are many neural models (and AI systems) claiming to do object recognition in a similar way (e.g., Riesenhuber and Poggio 1999; Wersing and Körner 2003), although performance cannot yet be compared with that of the real brain. The basic principle of a hierarchical object recognition system is illustrated in Figure 7.6, left panel. Processing along the hierarchy usually changes gradually in two ways: on the one hand, the represented feature quality becomes more and more specific with increasing hierarchical level (e.g., from basic features to particular objects). On the other hand, the representations become more and more invariant against transformations in the sensory space (e.g., translation, scale, rotation, colour, lighting conditions, etc.). The first aspect is essentially an AND operation (e.g., an corner consists of a vertical edge *and* a horizontal edge), the second aspect an OR operation (e.g., a feature configuration may occur at one *or* another location, as indicated by the pooling in Figure 7.6, left panel).

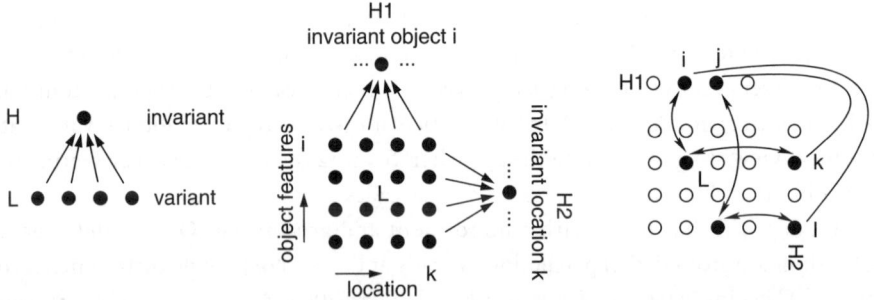

Fig. 7.6 Illustration of sign grounding, anchoring, or binding. Left: The development of invariant abstract (e.g., symbolic) representations implies a noninvertible mapping between a lower processing stage L and higher stage H (e.g., mapping feature configurations at different spatial locations to the same object label). This complicates adequate grounding, in particular if several objects are processed at the same time. Middle: To become invertible one needs further (possibly independent) mappings, e.g., an explicit representation of the object location. Right: In the presence of several objects, correct bindings between the higher-level symbols (e.g., object label and location) and grounding to the lower-level features may require additional mechanisms (e.g., temporary associations).

Thus, such an object recognition system must somehow map or bind, on each level of the hierarchy, a more abstract representation to particular configurations of lower-level features which may vary significantly. Realizing this in a bottom-up fashion is straightforward, for example, by means of a multi-layer perceptron. However, because of the invariance property (by the OR operation), this mapping between the two levels is necessarily noninvertible, which complicates binding between the two processing levels, in particular for more realistic scenarios requiring feedback processing (e.g., dynamical ambiguous scenes with multiple objects, occlusions, clutter, etc., as well as particular queries referring to low-level subsymbolic properties of the objects).

The invertibility of the mapping can be (almost) restored by further independent processing paths explicitly representing the varying factors (e.g., object type and object location in Figure 7.6, middle panel). However, this may lead to a further binding problem if several objects are processed at the same time; for example, in Figure 7.6 (right panel), it is problematic to ground the two objects i and j of layer H1 in the feature layer L because it may be unclear which object occurs at which of the two locations k and l represented in layer H2. Implicit binding over the lower layer L is possible but nontrivial; note that the segregated nodes of layer L may in fact correspond to distributed overlapping feature configurations. Neural experimenters and theoreticians have spent a lot of time investigating this form of the binding problem, and several explicit solutions have been suggested, for example by spike synchronization and oscillations, rapid reversible synaptic plasticity, dynamic routing, attention, coarse conjunctive coding (e.g., see von der Malsburg 1999; Shastri and Ajjanagadde 1993; Treisman 1998; Mel and Fiser 2000; Knoblauch and Palm 2002).

In summary, it appears that embodiment is very closely linked (if not identical) to the problems of symbol grounding (Harnad 1990), symbol anchoring (Coradeschi and Saffioti 2003), binding, and invariant recognition in distributed hierarchical systems. The underlying problem is always the coordination between the bottom-up and top-down streams of information in a hierarchical system, which still lacks a satisfactory solution in current artifical systems and theoretical models.

7.3 Conclusions

In general I am inclined to accept the major claims of the embodiment proponents, for example, that understanding the meaning of a sentence may require the ability to simulate or predict the situation described by the sentence, or that meaning is closely related to action and perception (which may cause interference effects as observed in experiments, see Glenberg and Kaschak 2002; Rizzolatti *et al.* 2001; Guillery 2003). But, in contrast to Searle's Chinese room argument (Searle 1980), I see no good reason why implementations on symbolic systems such as Turing machines or currently available digital computers should not do that job. Nevertheless, special hardware such as massively parallel neural networks – although in terms of price, speed, and flexibility still inferior to general purpose processors – could be of great advantage to this end (e.g., Hammerstrom *et al.* 2006).

Furthermore, I believe that many so-called symbolists would ultimately agree with these positions, and that probably much of the remaining disagreement results rather from imprecise definitions about what exactly is a symbol and what is embodiment. As discussed in Section 7.1.1, it is important here to distinguish between what I have called T-symbols and S-symbols. T-symbols are used by the cognitive system for thinking and communicating, whereas S-symbols may be used to implement the cognitive system (including T-symbols) on a symbolic substrate such as a digital computer, Turing machine, or in a neural network architecture (see Sections 7.2.1 and 7.2.2). For example, a T-symbol could be a word for an object, whereas an S-symbol could be a state of a Turing machine. Note that this distinction leads to the apparent contradiction that a system can be both (T-)embodied and (S-)symbolic. This is simply because embodiment in the sense of 'employing systems used both for action and perception' is actually a high-level property and therefore independent of the implementation substrate. Thus, any implementation of an embodied system on a digital computing device must be called (S-)symbolic.

We have also seen that the dichotomy between symbolic and embodied systems can be problematic. As argued in Section 7.1.2, any cognitive system must be necessarily both symbolic and embodied. Moreover, defining embodiment by activity in systems for action and perception implies that being T-embodied is a rather gradual property: a high-level T-symbol may be embodied more or less deeply in a hierarchy of perception- and action-related subsystems. In the brain at least, there certainly exists such a hierarchy consisting of many different cortical areas (Felleman and Van Essen, 1991). Understanding a particular sentence of T-symbols will require some, but probably not all, of these areas. Thus, one could define the degree of embodiment by the number and hierarchical levels of such areas necessary for understanding.

For example, for understanding 'the grass is yellow' a shallow embodiment is likely to be sufficient. Here, T-symbolic processing may be enough for understanding which is merely the association of two well-known T-symbols. Of course, cell assembly theory alone would predict that the T-symbol 'yellow' is so familiar that there will be strong synaptic links from the (T-)symbolic 'yellow' subassembly to low-level sensory neurons representing the colour yellow. Thus, hearing the word 'yellow' would inevitably activate those low-level neurons to a certain degree. Here, I think this priming activity would more likely be a side effect of neuronal assemblies and not actually necessary for understanding. One would predict that, in an experiment in which a number of low-level visual cortices could temporarily be deactivated or disturbed (e.g., by TMS) while hearing 'the grass is yellow', understanding is still possible.

In contrast, understanding or verifying the sentence 'the second house on the right has seven floors' requires relating several T-symbols to a currently perceived visual scene which may be represented in a (T-)subsymbolic format in early visual areas. This process includes identifying the 'second house on the right' among many other buildings and relating the number seven to the floors of the house. Here, one would predict that understanding is much more vulnerable to the deactivation of low-level cortices. Finally, on-the-fly understanding of sentences such as 'the woman crutched the goalie the ball' containing innovative denominal verbs will probably require even deeper embodiment including mental simulation of the described situation at many processing levels (see Glenberg and Kaschak 2002).

If we accept that embodiment is a gradual property within a processing hierarchy this idea becomes very closely related (if not identical) to long-discussed concepts and problems such as symbol grounding, symbol anchoring, and feature binding (see Section 7.2.4). We could give the following (maybe too) simple definition: a sign or symbol of a higher processing stage H is embodied in (or, synonymously, grounded in, anchored in, or bound with) the signs or symbols of a lower processing stage L iff it is possible to establish a one-to-one mapping between the H signs and L signs.

This definition of embodiment would have several advantages: first, it is a unified definition for seemingly different but closely related concepts. The definition remains independent of particular modalities (such as particular perceptions, actions, or emotions). Instead it is sufficient to distinguish between higher and lower processing stages. But we can still say that a symbol X is grounded in particular action representations Y, and we can still distinguish shallow embodied systems from deeply embodied systems where the one-to-one mapping is established across several processing stages.

However, the term 'mapping' may still be underspecified. Unfortunately, I cannot give a more specific answer, because in my opinion this is one of the unsolved core problems of current brain models and artificial cognitive systems as discussed in Section 7.2.4 (i.e., the problem of adequately integrating bottom-up information with top-down expectations is by no means satisfactorily understood in complex systems consisting of distributed processing hierarchies and capable of developing increasingly abstract invariant (and finally symbolic) representations). I believe that the circuitry of the cerebral cortex (see Section 7.2.3) can tell us a lot about how this integration happens in the brain, and how learning of abstract categories is linked to action and behaviour.

Debate

In addition to Andreas Knoblauch's contributions, this debate section encompasses the discussions surrounding the contributions of Marcel Just and Friedemann Pulvermüller (Chapters 5 and 6, respectively).

Arthur Glenberg: Alright, I will start. I have questions for Friedemann, and perhaps also for Andreas. I was very pleased with most of your talk showing that language is embodied from top to bottom and from bottom to top, from phonemes up to meaning, but then you said, right at the very end, that you think your cell assemblies might be implementing discrete symbols. So, a part of my question is to ask you to say a little bit more about why you believe it's the case that it's implementing discrete symbols. And in particular I'd like you to think about the following idea where you have people listening to a word like 'pick,' and so we know that we're going to get activation in perisylvian regions as well as regions related to hand movement. And the question is this: if it's a symbol that we're showing, then you would expect to see activation for 'pick' to be the same in all contexts, but I imagine you would agree that if we had people thinking about 'pick' in terms of a pen versus 'pick' in terms of a ball, that we would see different picks and then it doesn't seem to me like it's a symbol anymore.

Friedemann Pulvermüller: Well, I believe we have little choice when theorizing about language, to accept the discrete nature of language. There are discrete phenomena in language and other cognitive domains: a word is either in the dative case or the accusative case. There is nothing in between. It either refers to an object or not. It's either a meaningful item or not. Think for example of Andreas' simulations of the geometrical figures. In scene segmentation there is a big problem: how can we perceive a square partly hidden behind a triangle? How can we solve the problem of scene segmentation with non-discrete representations? Without them, there is activation all over the place and the overlapping figures are being meshed into each other. However, if perception is based on discrete representations, which have been learned separately, a particular complex scene can separately activate the pre-established representations, and activation can be maintained in two or more discrete object representations at a time. This seems to work, as neural network simulations demonstrate (see this chapter), even if the discrete representations overlap a little bit.

We have very similar problems in language, too. We need to know whether the word 'pick' or 'kick' has been pronounced, and of course there can be something in between a /p/ and a /k/. Still, there is categorical perception at the phonological level; and at the lexical and semantic levels as well. At the semantic level it could be two meanings of an ambiguous word, for example, take the word 'star', which could be an object in the sky or a person, and there seems to be no alternative in between, right? So I think that there are lots of problems – in language and cognition in general – which we can only model by using discrete symbols. At the mechanistic level of the brain, neuronal devices such as cell assemblies that behave in a discrete manner may 'ground' symbols. Cell assemblies, or as we sometimes call them nowadays, action–perception networks) are discrete because they are linked together so strongly, because of their strong internal feed-back and feed-forward connections. If they are activated to a certain degree, there is an explosion of activation within the cell assembly and the whole circuit gets active at once, which is very handy, especially for modelling discrete psychological processes.

Donald Hebb already pointed out that this is potentially an important process in Gestalt completion, which is otherwise difficult to explain in mechanistic terms. And here we have a very handy, neurobiologically plausible mechanism for it. I could go on for an hour explaining the necessity and benefits of discete representations, but I think do not need to. I think there are lots of reasons, at both the cognitive and neurobiological level, in favour of discrete representations and discrete neural processing units. Now, why do we not see them in current connectionist parallel-distributed processing models? Because these models are toy networks that do not have what the cortex is most famous for, namely massive local connections between neighbouring neurons. Many neural networks avoid strong next-neighbour connections within a particular layer because the network doesn't behave so nicely in this case for certain purposes. We are emphasizing the importance of these next-neighbour, within-layer, within-area connections; these are, as we believe, important for setting up cell assemblies. A similar point applies to rules or rule representations, if you want to speak about such entities. Here, I can speak for Andreas too. Art said that our approach is 'embodied from top to bottom and from bottom to top' and I think he is right. We document that the embodied perspective can even explain the emergence of abstract rules, a claim based on biologically inspired network simulations, and certainly gives rise to other discrete representations.

Glenberg: So, I just want to press this one more time and then I'll give up on it. Suppose we have an experiment where we have people listening to two sentences, and the first sentence is either 'There's a ball in front of you' or 'There's a pen in front of you.' And the next sentence is 'Pick it up.' Do you think activation to the word 'pick' would be the same in both of those contexts?

Pulvermüller: No, absolutely not. Veronique Boulenger, Olaf Hauk, and I are doing a more extreme experiment about 'grasping ideas'. We have action words like 'grasp' presented in sentences like 'He grasps the ball' and 'He grasps the idea'. And, of course, in the idea-grasping case there is a different type of grasping and it seems that the cortical activation pattern changes, although semantic somatotopy is present even in the abstract idiomatic case. Different activations by the same symbol presented in different contexts is not an argument against discreteness, but can be attributed to pre-activation through the context network. It would not be realistic to assume that the discrete representation activates each time in exactly the same way. Rather, its particular activation state at any point in time does not only depend on the internal properties of the network, but also on the type of priming or pre-activation it receives from other networks. So that's the brief answer to a complex question. Context dependence and discreteness are independent features.

Glenberg: Thank you.

Lawrence Barsalou: I'd like to comment on this. Like, one solution to this might be to say that a symbol is not a single state of activation but it's actually something more like an attractor state that has many different possible states of activation.

Pulvermüller: Yes indeed. The word would be understood in an all-or-none discrete manner each time, but in different contexts a set of its possible semantic features are emphasized or even added; for example, referential features related to specific hand motor acts or related mental operations.

Barsalou: So, it may be that all the different instances of 'pick' that get processed are attracted to the same general processing system, but that system can take different states on different occasions for different contexts. So a symbol is really a space of applications not a single application?

Unidentified person: This reminds me of the debate in physics as to the nature of light – the particle/wave debate – which was solved in terms of a field theory of physics. But I wanted to emphasize what you told us about the structure of the column in which, if I understood you correctly, each column in the cortex has a layer especially, functionally, dedicated to perception and another layer to action. Is that right? Because this seems not to be in clear convergence with the standard view of the cortex as specialized in regional terms, not being sensory and motor all the way around.

Andreas Knoblauch: Yes, I think cortical layer 5 has a lot of pyramidal cells that actually project to subcortical structures that are related to actions.

Unidentified person [DEB ROY?]: So I'd like to actually just push a little more on the same question that Art was asking. 'Grasp the idea' versus 'grasp the ball' is, I think, a different point. It's a sort of different word sense, because that's an idiomatic use of 'grasp'. Another way to ask the point I think Art is trying to make, or maybe to make a suggestion, is that if you take a very simple model of how words work – which I think is consistent with the story you've told us – that you have something like, say, a conceptual layer versus a symbolic layer, so you've got words and you've got some sort of sensorimotor something or other connected. Clearly, at the word level, I think we'd all agree that's a discrete system. It's not meaningful to talk about halfway between 'kick' and 'pick' in the lexicon. Phonetically you can talk about that, but you impose a phonological discretization and on it goes morphologically and into the lexicon. But that discrete system is carving up perhaps a continuous conceptual space. So there's no reason for example to believe that 'pick' doesn't bind to a sensorimotor representation that actually has parametric variation, right? So when you kick a ball versus kick a chair versus some other argument of 'kick', there'll be some continuum. So I think there's room to think about both a continuous system *and* a discrete system. And one projection, the projection into language, you can treat as discrete. But it's not clear at all that the back end is discrete.

Pulvermüller: Whether or not the semantic system as such is entirely continuous, I cannot say. At least there is some evidence that word meanings dissect an otherwise continuous conceptual space, for example, in the case of colour concepts. But we are in fact talking here about two different things now. Let me take up the specification. Before, Larry made the point about attractor states. We can say that a cell assembly is a discrete unit – it either becomes active or not – but that its activation states may differ, may change over time and between contexts? So it has an activation minimum, or maybe several activation minima. And depending on where the pre-activation comes from, where it is primed and by what it is primed, certain parts of it can, in a given context, be emphasized, so to speak. And therefore it is, in each case, either active or not, but the particular neurons active in each situation can be different. So its states of activation are not identical. However, it's still in each case a discrete state, it's either active or inactive. So, this would be my explicit answer. There is room for variablitiy among the different activation states of the discrete network

Unidentified person: Yeah, I guess the question is, given the precision of the best imaging technologies, we're reading a lot into whether it is the same network or not, or would you disagree? An average brain has millions of neurons, so there's a lot of room … it's unclear.

Pulvermüller: Well, fMRI is not the best method to address these questions. Multiple-unit recordings in monkeys – especially the work done by Moshe Abeles' group in Israel – have produced very strong data exactly on this issue, that there are specific networks, serving specific cognitive functions, that are active or inactive and exhibit similar activity states and patterns when active. That is, I think, very important work. Also, complex dynamics approaches to EEG and imaging analysis are, maybe, to a degree telling. One wouldn't get that information from fMRI, according to my feeling, because it misses out on the time aspect. Some imaging data show a surprising degree of local specificity, and this is certainly something to look into much more in the future; but we are clearly looking at neuronal mass activity with all neuroimaging methods. Trouble is, the relevant circuits may be large, may include millions of neurons, so a multiple-unit approach alone cannot succeed. Bridging across methods may help. Maybe you want to comment?

Unidentified person: Yeah, I'd say something. It seems to me that several issues are being conflated. First, it seems to me there is likely to be contextual sensitivity of the representation of meaning. Second, I think there's ample evidence that meaning or representations are distributed, and that there are different components of it. I like to think of the different components of meaning as corresponding to different brain locations. That's not the only way in which they might be distributed and it's conceivable that different components have different attributes, that some are much more symbolic in nature and others are much less symbolic. It seems to me that motor representations would be less amenable to, I don't know, a symbolic characterization. Not that you can't say it's 'grasp' – it's still 'grasp'. And I think that's what Deb [Roy] is in agreement with. If you look at the activation carefully, as you say, even without the idiomatic meaning of 'grasp', you would expect to see different representations there. So, I think you can have contextual sensitivity, you could have differential degrees of symbolic-ness for different components. It seems to me that all of this could work quite nicely. And really our job might be to specify, you know, how is it you say or grasp a tennis ball – how does it magically come to be that we're going to get different activation? It seems to me that's a very interesting question. I'd love to be able to answer it. But the contextual sensitivity occurs and we've been doing context effects in experimental psychology probably for a century now. Maybe now we can finally come to the point where we can actually identify the mechanisms by which the context has its effect.

Unidentified person: I also wanted to come back to this issue. Context sensitivity seems to be, like, impossible when you have truly discrete symbols like things that are supposed to be kind of immovable in what they represent, right? But when we think about it in terms of a continuum from analogue to discrete … take the instance of children's development where they go from being very context sensitive to being extremely, you know, able to ignore certain irrelevant contexts. And I think about it terms of Larry Barsalou's suggestion here, in terms of attractor states. It's this idea that first you have really shallow attractor states, where, you know, you're able to kind of be bounced around by irrelevant variations in context. But those attractor states become deeper and deeper as you have more experience with those networks, activating in certain orders and correlating in different areas of neural networks, and you know, 'what fires together wires together.' And eventually you get these very deep, kind of more stable attractor states but that doesn't mean you can't shake them. That even adults, when they have a kind of stable attractor state, could still shake them with different forms of context –they just have to be stronger manipulations in something. And so to take these slices of big neural connections and talk about them as if they were whole trajectories, it's a question of how quickly you can or how much context you need to shake people out of those trajectories versus, you know, how little.

Barsalou: I just wanted to point out that even words are not fully discrete. I mean, a given word spoken by a given person will take a different form as a function of the surrounding linguistic context … different people uttering the same word will have different acoustic and articulatory properties, so even for what seem like symbolic kinds of things it's still a space of possibilities that get realized differently in different contexts. But I would agree that just as the way phonetic input, as you were saying, tends to go towards those attractors, I think the same thing happens at the conceptual level. If you look at the world of objects, rarely do you get stuck between categories. I mean it's a winner-takes-all situation; you go into one conceptual attractor pretty much all of the time, and if you get stuck in between it's really a weird conceptual state.

Unidentified person: I would suggest that the question of contextual sensitivity probably has to do with the level of representation of action in the brain. I mean, Rizzolatti's group recently proposed that action representation or action control in the brain involves somewhat different structures or areas. Some of them have to do with low-level motor problems – particular combinations of muscles, motions, and so on – but some of them are much more abstract. They have to do with sort of general plan of the action. So I think it might be relevant, for instance, to try and decide the granularity of embodied representation of action verbs. An action verb in a noncontextual situation, like your experiments for instance, maybe don't activate lower-level motor structures in the brain but instead some sort of higher-order or abstract representation of action. What is your opinion?

Pulvermüller: This is an excellent question. I think this is exactly the way to go forward. And maybe I can just tell you verbally how we tried to go exactly this way in the last 2 years, looking at words that have a more general action relationship compared with the very primitive, concrete body-related, even body part-specific, action word. So instead of using a word like 'kick', which is in its meaning bound to the leg (one of its semantic features could be described as 'typically done with the leg'), we have used other words that are related to action patterns, words that are highly abstract, as you mentioned. And one such category is words that relate to forms, to shapes; for example 'triangle', 'square', and 'rhomb'. Looking into the neuropsychological literature, these form or shape words are usually treated as visually related, but that's not the entire truth. We know that when people look at shapes they follow the shape with their eyes, they can point out the shape with their hand, and, if needed, they can walk a path of that shape with their feet. So, if you like, the action pattern related to the meaning of the word is not bound to a particular part of the body; there's an abstract action scheme associated with these kinds of words and concepts. And what we find for them in the brain is activation that starts in motor and premotor cortex, but extends into prefrontal cortex, spreading anteriorly compared with the 'primitive' action words. It seems that these higher-level, less-concrete action words involve more frontal areas, Brodmann areas 9 and 46. Action execution, action recognition, and action semantics of different types may involve overlapping but nonidentical systems in frontocentral cortex that include mirror neurons.

Francis Quek: Are there any motion verbs, or motion actions, that are uniquely concrete? For example, if you ask somebody how many ways you can kick, there are myriad ways. Another way to ask the question is whether the neural representation is a specific pattern or a set of patterns: what confidence do we have that each time we think of 'kick' that the exact same distributed pattern takes place? Can there be degrees? Talmy, for example, says that for a particular motion verb there are many aspects to consider – there is ground, there is manner, there is direction. So, none of these verbs, none of these grounded descriptors, need to be absolute in their representation.

Pulvermüller: Well, I think this brings us back to Larry Barsalou's comment on attractor states. Again, my view is this: if there's one discrete network, this network can still have different attractor states and, again as a function of the priming of the network, can be in different basins, or activatity minima of the attractor landscape, so to speak. The property of discreteness is very compatible with context effects and variation in the activation patterns. As for your question about how good the imaging methods are at finding out about the variation, well, we are working on that and maybe in 2 years' time I could give you more concrete answers on that. I would hope that imaging could tell us a little about context-specific activation of the networks.

Unidentified person: How would you contrast a 'do' versus 'make' versus 'bake' in terms of the attractor basin? For 'do' I could imagine just about anything goes, whereas 'make' is going to be a little more constrained but it might have that either/or-ness, whereas 'bake' is going to be very little either/or-ness.

Barsalou: I think the standard linguistic response is that 'do' focuses on a process and 'make' focuses on the result.

Unidentified person: There is a lot of developmental evidence saying that verbs like 'read' are at first highly constrained to objects, like kids learn the word 'read' only in the context of books, and so it's a very early verb; it's early acquired, it's very easy for them to learn. Whereas with a verb like 'put,' it's just like, gosh, you 'put' with almost anything. And there's also languages where verbs are much more specific to objects than English perhaps is, verb-heavy languages like Japanese and Korean. They're highly shape-focused and have very spatial verbs, so in that sense I bet you might be able to find linguistic differences with younger children who might be more object sensitive. Very young children, say 12 months old, are actually sensitive to agent changes, like they won't say the same verb is the same verb if a different person does it, even if it's the same sex. There are lots of superficial similarities, and so I think there's a lot of that evidence, at least behaviourally.

Unidentified person: Yeah, I would like to come back to another question. Marcel in his talk gave this metaphor of a team, and the players in a team, which I thought was a very useful way to think about the whole thing, and also in Friedemann's talk we saw that all over the place there are systems involved. So, my question is, is there anything more already, from a neurobiological point of view, about how this team gets coordinated, or how it gets assembled, I should say? I mean, at some point you mentioned the word 'probe' as if one process would send out, you know, a message for another subsystem to probe.

Unidentified person: Yes, I think that's one of the key questions, because as you enter any situation, you know, depending on what kind of sentence I now use, what kind of reference I make, a different set of areas could activate. If I start speaking in a visual metaphor then there may be more intraparietal sulcus activation, and so on. And the question is how does this magic happen? I work with a computational modelling system – it's kind of a hybrid system, but basically the problem is viewed as a resource allocation in a complex system. You have some job about to be done, maybe a job for an athletic team, a sports team, but it could be anything – a job for a university, for a corporation, or whatever; a complex system, a city – and a challenge comes along, for example, a large snowstorm, a plane crash, you know … a set of some product goes down the tubes or whatever. And the question is, how does this system respond? Some people like to think at an executive level; you know, the president of a university calls together all the deans, how are we going to do this? But there are other ways that a complex system sometimes responds, a more distributed response to a resource need. In the model I work with it's really about production systems down at the level of the individual component systems, such that they're all monitoring what's going on in the system at large. So whenever a particular need arises – suppose there's a desire at some point to develop a spatial representation or there's a desire to have a motor representation, or whatever – the area that is most specialized for that need, that has the greatest efficiency, the greatest history of doing that kind of thing, gets to perform that action. And so on a continuous basis there's always a thinking, the team is always sort of selecting itself – is this my chance to go in? Do I do this?

I'm not saying there can't be executive control. I think that in some cases there are. But my view is that there is a sort of self-allocation, a self-selection based on expertise, efficiency, past experience, and so on. Let me give you another argument for this: how is it that Broca's area gets to be Broca's area? For all of us, you know, when you generate either subvocal or actual speech, Broca's area activates. How does that come about? Developmental theorists suggest you know that the different cell types throughout the brain have different characteristics, and the classic example is that cells with particular timing characteristics that are kind of slowish will allow you to do the timing of phoneme distinctions. And maybe those are around the primary auditory cortex, so the auditory cortex, in Liz Bates's terms, gets the contract for processing auditory information. According to Liz, it's not as though you're born with a sort of a label saying 'auditory cortex', but you're born with a bunch of cells in that area that are really good at getting timing distinctions. And so the team members develop their neuroprocessing capabilities by virtue of their capabilities. So the auditory cortex sort of initially wins the contract for processing timing information, and then when the person is 6 months old or 60 years old, when an auditory challenge comes up, that's what activates. It's been doing it since its infancy, and it sort of holds the contract; it's going to try to develop the appropriate response or representation or whatever. That's a long-winded answer to a complex question.

Pulvermüller: I have some reservations with regard to using humanoid metaphors when speaking about brain function. I think considering neurons football players or contractors can be advantageous for educational purposes, but I would still believe that going down to neuroscientific principles might be an alternative option, and in this case what counts is neuronal correlation as it is a major driving force for binding. What becomes bound together is determined by neuronal correlation of activation. So how does Broca's area become Broca's area? Well, Broca's area happens to be close to the motor cortex and this is pre-wired in a way. And the auditory cortex receives auditory information on the basis of very early ontogenetic processes, which are usually finished at the time when the major phases of language learning start. Therefore, the information about motor programming for articulation goes out and the auditory input arrives in these distant areas, and now there is another problem: the motor cortex and the auditory cortex are not linked directly with strong fibre bundles, so if the neural activation patterns should communicate with each other it needs to take a detour. And this detour is exactly through Broca's area and through superior temporal areas in the so-called belt and parabelt region. This is where the long-distance connections emerge. So I have a two-fold answer: one driving force is neuronal correlation, another is the cables built into the brain – the major neuroanatomical tracts that bridge between the sensory- and action-related activations, that connect the areas in which the correlated activation happens. This is the magic that makes Broca's region to act as a binding site for language: correlated activity sets up representations that span across areas linked by long-distance neuroanatomical connections. Thanks very much for the question, A very important question. I think Andreas is more of an expert on this, so maybe I can pass on.

Knoblauch: 20 milliseconds, I think that's not a big problem. The time window for synaptic plasticity lies somewhere in the range of 50 milliseconds or so, for example, the spike timing dependent plasticity, so 20 milliseconds would still mean the synapse is strengthened.

Quek: I have a question here. One of the things I've been interested in is how embodiment in the neuronal structure that we have, for example, for thinking of space and so on, gets implicated into understanding the limit of a mathematics function – these are the kinds of things that Lakoff and Nunez have talked about. So can you suggest, with the pre-wiring and so on, how does repurposing take place, if it does, if we actually repurpose certain neural architectures that are designed, for example, for sequential-izing of events, to understand mathematical functions. I doubt that we are born with mathematical processing in our brain that waits for the first time we take a mathe-matics class to wake up, okay. So if indeed mathematics, for example, or higher level thought, is actually repurposing of certain neural pathways that have other designs, how might you speculate this takes place?

Pulvermüller: It's a very difficult question. No one wants to take it.

Unidentified person [QUEK?]: Our culture is constantly coming up with new skills, so reading is not god given or biologically given … certainly maths and computer programming. Oh, and driving a car – none of these are biologically inherent skills and so the question is how do you acquire them. Others have talked about how that's where grounding tends to play a role. It's sometimes difficult to go straight to what is sort of an abstract skill without some sort of grounded representation, and that's why – I don't know very much about this – but where you have manipulatives in maths, you know, you're playing with arrays and getting addition and multiplication through some visual or manipulative activities at first before you move to a sort of a more symbolic level.

Robert Goldstone: This is sort of a conceptual question. A lot of the ideas that have been going around with Friedemann and Marcel, and what Larry has said, suggest that a large part of the meaning of the words is based upon the activation of percep-tual and motor cortices that would be normally associated when you were using or normally understanding the words. And I guess my conceptual stumbling block is how do you know that these are perceptual and motor cortices if they're so tightly bound to the word side of things? So if you cut a slice of the skull out and you open up, you wouldn't see a Post-It note that says 'This is the motor cortex', right? And so it seems to me that if you had done the word meaning experiments before you had done the finger twiddling experiments then you could have said 'wow isn't it amazing that the finger twiddling causes the same area to light up as the word's semantic area.' So, I mean, I'm totally on board when you talk about correlations or the distributed nature of the representations, but I get a little bit less sure when I think about ascrib-ing singular functions to those areas.

Pulvermüller: Well, the lucky thing is that these areas are in fact labelled. If we open the brain and look intensely at it we see that exactly from which areas the cables lead to the arm or to the leg. Well, not directly: the large pyramidal neurons in motor (and premotor) cortex project to the neurons in the spinal cord that reach arm or leg muscles, respectively. There are these huge pyramidal neurons in Brodmann area 4, and they can have extremely long axons, sometimes a metre long, going down into the spinal cord, which link the motor cortex through one intermediate neuronal step to the respective effectors.

Goldstone: Right, but that doesn't…

Pulvermüller: The motor cortex is labelled and it has been known for hundreds of years. After the French revolution there was sometimes a French neurologist next to the guillotine and he was sticking his needle in the cut-open spinal cord, looking for whether the foot would be twitching or the arm, and therefore, by such truly revolutionary brain mapping, the first motor maps were made.

Goldstone: Yeah, I understand that.

Pulvermüller: The brain is labelled.

Goldstone: Yes, I understand that, but it doesn't exclude the idea that these areas are multifunctional. I mean, I guess I would just make the general point that there's a sense in which if you're talking about separate areas you haven't totally gotten on board with the idea of a fused perceptual–conceptual system; you're still saying this is the perceptual area that is borrowed by the conceptual area. Maybe you don't want to go this far, but you could have a more integrated view where you said that these areas are actually doing multiple functions. So it's not even totally clear-cut that the only thing that this is doing is controlling the motor processes.

Pulvermüller: Absolutely, the implication is that the area is a multifunctional area, not only doing motor programming but also doing the motor–conceptual processing that is elementary for understanding action words. Action verbs, action words generally, action concepts … I would go with that fully. Nevertheless, as we discussed earlier, overlapping but different activation topographies and network circuits may underlie action execution and recognition, and the comprehension of action language and concepts.

References

Abeles M (1991). *Corticonics: Neural Circuits of the Cerebral Cortex*. Cambridge: Cambridge University Press.

Barsalou L (1999). Perceptual symbol systems. *Behavioural and Brain Sciences*, 22, 577–609.

Barsalou L (2003). Grounding conceptual knowledge in modality-specific systems. *Trends in Cognitive Science*, 7, 84–91.

Braitenberg V (1978). Cell assemblies in the cerebral cortex. In R Heim, G Palm, Eds. *Lecture Notes in Biomathematics 21: Theoretical Approaches to Complex Systems* (pp. 171–88). Berlin: Springer-Verlag..

Braitenberg V, Schüz A (1991). *Anatomy of the Cortex. Statistics and Geometry*. Berlin: Springer-Verlag.

Brooks R (1990). Elephants don't play chess. *Robotics and Autonomous Systems*, 6, 3–15.

Chalmers D (1996). *The Conscious Mind*. Oxford: Oxford University Press.

Coradeschi S, Saffioti A (2003). An introduction to the anchoring problem. *Robotics and Autonomous Systems*, 43, 85–96.

Dennett D (1996). *Kinds of Minds: Towards an Understanding of Consciousness*. London: Weidenfeld & Nicolson.

DiVincenzo D (1995). Quantum computation. *Science*, 270, 255–61.

Douglas R, Martin K (2004). Neuronal circuits of the neocortex. *Annual Review in Neuroscience*, 27, 419–51.

Felleman D, Van Essen D (1991). Distributed hierarchical processing in the primate cerebral cortex. *Cerebral Cortex*, 1, 1–47.

Glenberg A, Kaschak M (2002). Grounding language in action. *Psychonomic Bulletin and Review*, 9, 558–65.

Guillery R (2003). Branching thalamic afferents link action and perception. *Journal of Neurophysiology*, 90, 539–48.

Hammerstrom D, Gao C, Zhu S, Butts M (2006). FPGA implementation of very large associative memories. In A Omondi, Rajapakse J, Eds. *FPGA Implementations of Neural Networks* (pp. 167–95). Cambridge, MA: Springer US.

Harnad S (1990). The symbol grounding problem. *Physica D*, 42, 335–46.

Hawkins J, Blakeslee S (2004). *On Intelligence.* New York, NY: Times Books.

Hebb D (1949). *The Organization of Behaviour: A Neuropsychological Theory.* New York, NY: Wiley.

Hopfield J (1982). Neural networks and physical systems with emergent collective computational abilities. *Proceedings of the National Academy of Science USA*, 79, 2554–8.

Hopfield J (1984). Neurons with graded response have collective computational properties like those of two-state neurons. *Proceedings of the National Academy of Science USA*, 81, 3088–92.

Jackendoff R (1987). *Consciousness and the Computational Mind.* Cambridge, MA: MIT Press/Bradford Books.

Knoblauch A (2005). Neural associative memory for brain modeling and information retrieval. *Information Processing Letters*, 95, 537–44.

Knoblauch A, Fay R, Kaufmann U, Markert H, Palm G (2004). Associating words to visually recognized objects. In S Coradeschi, A Saffiotti, Eds. *Anchoring Symbols to Sensor Data. Papers from the AAAI Workshop. Technical Report WS-04-03* (pp. 10–16). Menlo Park, CA: AAAI Press.

Knoblauch A, Kupper R, Gewaltig M-O, Körner U, Körner E (2005). Design and simulation of a cortical control architecture for object recognition and representational learning. In H Tsujino, K Fujimura, B Sendhoff, Eds. *Proceedings of the 3rd HRI International Workshop on Advances in Computational Intelligence.* Wako, Japan: Honda Research Institute.

Knoblauch A, Kupper R, Gewaltig M-O, Körner U, Körner E (2007). A cell assembly based model for the cortical microcircuitry. *Neurocomputing*, 70, 1838–42.

Knoblauch A, Palm G (2001). Pattern separation and synchronization in spiking associative memories and visual areas. *Neural Networks*, 14, 763–80.

Knoblauch A, Palm G (2002). Scene segmentation by spike synchronization in reciprocally connected visual areas. II. Global assemblies and synchronization on larger space and time scales. *Biological Cybernetics*, 87, 168–84.

Körner E, Gewaltig M-O, Körner U, Richter A, Rodemann T (1999). A model of computation in neocortical architecture. *Neural Networks*, 12, 989–1005.

Mel B, Fiser J (2000). Minimizing binding errors using learned conjunctive features. *Neural Computation*, 12, 247–78.

Metzinger T (2004). *Being No One: The Self-Model Theory of Subjectivity.* Cambridge, MA: MIT Press/Bradford Books.

Palm G (1980). On associative memories. *Biological Cybernetics*, 36, 19–31.

Palm G (1982). *Neural Assemblies: An Alternative Approach to Artificial Intelligence.* Berlin: Springer.

Palm G (1990). Cell assemblies as a guideline for brain research. *Concepts in Neuroscience*, 1, 133–48.

Pulvermüller F (1999). Words in the brain's language. *Behavioural and Brain Sciences*, 22, 253–336.

Pulvermüller F (2003). The Neuroscience of Language: On Brain Circuits of Words and Serial Order. Cambridge: Cambridge University Press.

Quiroga R, Reddy L, Kreiman G, Koch C, Fried I (2005). Invariant visual representation by single neurons in the human brain. *Nature*, 435, 1102–7.

Rao R, Ballard D (1999). Predictive coding in the visual cortex: a functional interpretation of some extra-classical receptive-field effects. *Nature Neuroscience*, 2, 79–87.

Riesenhuber M, Poggio T (1999). Hierarchicalmodels of object recognition in cortex. *Nature Neuroscience*, 2, 1019–25.

Ritz R, Gerstner W, Fuentes U, van Hemmen J (1994). A biologically motivated and analytically soluble model of collective oscillations in the cortex. II. Applications to binding and pattern segmentation. *Biological Cybernetics*, 71, 349–58.

Rizzolatti G, Fadiga L, Fogassi L, Gallese V (2001). Neurophysiological mechanisms underlying the understanding and imitation of action. *Nature Reviews Neuroscience*, 2, 661–70.

Searle J (1980). Minds, brains, and programs. *Behavioural and Brain Sciences*, 3, 417–57.

Shastri L, Ajjanagadde V (1993). From simple associations to systematic reasoning: a connectionistic representation of rules, variables and dynamic bindings. *Behavioural and Brain Sciences*, 16, 417–94.

Steinbuch K (1961). Die Lernmatrix. *Kybernetik*, 1, 36–45.

Treisman A (1998). Feature binding, attention and object perception. *Philosophical Transactions of the Royal Society of London B*, 353, 1295–1306.

von der Malsburg C (1999). The what and why of binding: the modeler's perspective. *Neuron*, 24, 95–104.

Wersing H, Körner E (2003). Learning optimized features for hierarchical models of invariant object recognition. *Neural Computation*, 15, 1559–88.

Willshaw D, Buneman O, Longuet-Higgins H (1969). Non-holographic associative memory. *Nature*, 222, 960–2.

Wilson M (2002). Six views of embodied cognition. *Psychonomic Bulletin and Review*, 9, 625–36.

Young M (1993). The organization of neural systems in the primate cerebral cortex. *Proceedings of the Royal Society: Biological Sciences*, 252, 13–18.

Chapter 8

Symbol systems and perceptual representations

Walter Kintsch

8.1 Introduction

The ability to represent the world symbolically, and hence abstract thought, evolved from other, more concrete forms of representation. The ways in which people model the world in their mind follows an orderly sequence, both phylogenetically and ontogenetically. Donald (1991), for instance, distinguishes a sequence of cultures, each characterized by a particular form of mental representation. All animals learn and have procedural memories. At some level (certainly at the level of primates), animals are able to represent the world in terms of generalized records of past experience that allow them to react directly to situations in a way that summarizes their experience with that situation. This is also true for young children (Nelson, 1996, calls this *general event memory*). As biological evolution continues, representational systems based on intentional imitation emerge. Speech appears at this stage, but its purpose is social communication, not representation. Individuals at this level are characterized by representation through action; cultures are characterized by arts, crafts, and ritual. At this point biological evolution is replaced by cultural evolution: at the sensorimotor level humans are simply primates, but the use of language distinguishes them more and more from their animal ancestors.

Donald makes an important distinction between narrative language and theoretical language use.[1] Narrative language is based on mental models that are fundamentally linguistic. Flexible computations can be performed with models of this kind, giving humans a degree of control over the world that is impossible without language; words and thoughts are inseparable at this level. Most human cultures and human individuals are capable of working at the level of narrative culture. However, few cultures, and by no means every individual in these cultures, achieve the level of formal representation that Donald terms 'theoretical culture'. The engrams of memory are aided by the exograms procured by our technology at this level. Most of what we do in schools is trying to bring students up to that level of cognitive functioning.

This is not the place to fill in the details of this conceptual framework (see Donald 1991 and Nelson 1996). However, a few comments are in order. First, it must be pointed out

---

[1] Several authors have made the same or similar distinctions (e.g., Bruner (1986) with respect to language, or Vygotsky (1987) with respect to everyday and academic concepts.)

that when an individual or a society moves from one level of representation to the next, the earlier level does not disappear but becomes embedded in the new forms as a new kind of representation is added to the behavioural repertoire and integrated with the earlier ones. Educated individuals in our society who function at the level of formal reasoning are, at the same time, also capable of direct action, of participating in art and ritual, and sharing the narrative traditions of our culture. No wonder psychologists have a hard time analyzing human behaviour.

Thus, although event memory is not basically linguistic, it may become linguistically encapsulated: what we remember is sometimes not the event itself but our (explicit or implicit) linguistic description of the event. Furthermore, when the brain develops new forms of mental representation, it not only does not discard the older ones, but in fact uses the old structures for the new purposes. Cognition is indeed embodied.

How the different levels of representation interact in detail is not known at present, but much research, including this volume, is directed at this goal. The question that I want to address here is how to model these overlapping, intertwined mental representations, particularly linguistic representations. If we take seriously the levels of representation analysis that was sketched above, there seem to be two kinds of approaches to the question of mental representations. On the one hand, one could try to model a system of mental representation that combines elements from all of these levels. Thus, we would have images, concepts, somatic markers, S–R links, features, schemata, abstractions, words, etc., all linked and somehow organized in one grand system, at least in principle, for I have no idea how one would go about constructing such a model. Alternatively, one could model each level of representation separately. This has the advantage that, at least for linguistic representations, we do have an idea of how to begin to construct such a model: I'll describe latent semantic analysis (LSA) and its variants shortly. But first I must address an obvious problem created by the divide-and-conquer strategy. If we model the level of linguistic presentation separately, how can it be coordinated with other levels of representation? In behaviour and the brain, these levels function all together and are not easily separated.

One can think of the different levels of mental representations as a set of coordinated maps. Consider the relationship between the level of linguistic representation (as captured by LSA, for example) and other levels of representation, which we can lump together here as the 'real world'. The relationship between the symbolic representation and the real world could be a first-order isomorphism, where the symbol in some way captures meaning directly (e.g., by a picture, feature system, or dictionary definition). The meaning of a word would be its referent in the real world, as in the perceptual symbol system of Barsalou (1999) and similar proposals. This is what I believe people are saying when they claim that language must be 'grounded' (Harnard 1990).

An alternative way of thinking about maps and their relation to the real world is in terms of what Shepard (1987) called a second-order isomorphism. In a second-order isomorphism the meaning of a symbol is not defined through its reference to another level of representation, but through its relationship with other symbols. The meaning of a word lies in its relationship with other words. Specifically, in LSA the meaning of a word is a vector of 300 numbers that are entirely vacuous by themselves; they acquire meaning only because they specify how this vector is related to all the other words and texts in the LSA space. In this scheme the real world does not provide the meaning of a symbol, but symbols and referents are nevertheless coordinated as a second-order isomorphism. On a two-dimensional map, I can trace a path from Boulder to Denver that is isomorphic to the trip from here to there in the real world. In the 300-dimensional LSA space, I can compute the semantic distance between these words and expect it to correspond to features of the real world. The cosine between Boulder and Denver is 0.21, way above the cosine between Boulder and farther away locations such as Tenerife (0.01) and Spain (0.02); for similar examples, see the contributions of Zwaan (Chapter 9) and Louwerse and Jeuniaux (Chapter 15) to this volume.

Calling something a second-order isomorphism does not solve the problem of how symbols and the real world are related. That question remains open for research, but it does alter the way one approaches the problem. The meaning of symbols is not to be reduced to perceptual features or actions; rather the question becomes one of correspondence between different levels of representation.

I shall first sketch a proposal for modelling verbal meaning. This proposal has two components: first, we must describe how people induce verbal meaning – the representation of meaning in memory – which is the goal of LSA and related systems. Second, we must model how stored meaning is used to construct contextually appropriate meanings of words, sentences, and texts in general. It is of course possible that all we need for language is already stored in memory and that meanings simply have to be retrieved from the semantic store ready-made. The alternative presented here is that the semantic store only provides the raw material for the construction of meaning, and that meaning emerges when words, sentences, and text are used in context. After this sketch of symbolic verbal meaning, I shall discuss the proposition that, although language is not grounded in the sense of a first-order isomorphism, language mirrors perceptual features of the world with a high degree of fidelity.

8.2 Memory for verbal meaning: LSA and its extensions

Our goal is to construct a representation of the meaning of words and texts that can serve as a model for the kind of representation the human mind builds. People generate such representations while interacting with the world and other people, listening, speaking, and reading, mostly without explicit instruction. A computer simulating this process is restricted to reading texts, but must be able to generate a semantic representation without guidance, as people do.

The kind of representation generated depends in part on the precise mathematical algorithm used to generate it, but more significantly, upon the language input used. We start with a large set of texts (e.g., 50,000 documents, containing 100,000 different words and over 10 million words total) that is analysed into a document-by-word matrix, whose cell entries are the frequencies with which each word appears in every document. Note that this is by no means all the information in the corpus – all we look at is word co-occurrences, neglecting word order, syntax, as well as discourse structure. Our input matrix is huge (50,000 by 100,000) and sparse, that is, most entries are zero. We then reduce the dimensionality of this matrix, down to about three hundred dimensions. That is, we express the meaning of a word by a vector in a space with far fewer dimensions. This has two effects: first, it is a process of abstraction. We are not interested in the particular documents used to construct the semantic space, the particular topics the authors wrote about, their particular word choices; instead we are interested in the general semantic properties of words. Reducing the dimensionality of our corpus has the effect of discarding a great deal of idiosyncratic information about word use in this particular corpus, while retaining the essential semantic information about word meanings. Secondly, dimension reduction is a process of generalization. Whereas the original input matrix was sparse, the reduced matrix is filled in, so that we have a measure of the semantic distance for every word pair, even though most pairs by far have never co-occurred together in any of our documents. Thus, 'doctor' and 'physician' have a high cosine of 0.61, even though they have rarely been used together in a single document.

There are various ways to perform this dimension reduction. LSA starts with a document-by-word matrix M of size $n \times m$ that is decomposed via singular value decomposition (Landauer and Dumais 1997; Landauer *et al.* 2007). Typical values of n and m are 100,000 and 50,000, respectively. All eigenvectors but the ones corresponding to the d (300 or so) largest eigenvalues are discarded. A word or document is thus represented by a vector of 300 real numbers. This vector is not by itself interpretable but its cosine (or other measure) with other vectors defines its position in the semantic space: hence, meaning is defined by its relationship with other vectors.

Independent component analysis (ICA) assumes that each observation (word, document) is a mixture of independent semantic elements (components, topics) (e.g., Stone 2004; for an application of ICA to language analysis see Mangalath 2007). Several constraints are used to unmix the observations: components are chosen so that they are statistically independent (not merely uncorrelated). This is achieved by searching for components that are maximally non-Gaussian, and/or the least complex components (that is, the most compact or predictable ones).

The ICA model is conceptually related to the *topics model* (Steyvers and Griffith 2007), in which meaning is represented as a mixture of topics. The distribution of topics over words and documents is based upon a Bayesian learning algorithm. An appealing feature of both models is that components or topics are individually interpretable. Thus, the linguistic corpus we are working with most of the time can be analysed into a set of topics, such as drugs, colours, or doctor visits, or, alternatively, into a mixture of independent components.

A quite different approach has been taken by Jones and Mewhort (2007). Word meaning can be represented as a composite distributed representation by coding word co-occurrences across millions of sentences in a corpus into a single *holographic vector* per word. This representation is a pattern of elements that stores the word's history of co-occurrence and usage in sentences.

But which one of these methods – LSA, ICA, topics, or holographic vectors – yields the best semantic representation, or the right one? Experience from working with these models suggests that they produce very similar results. Each has certain advantages for certain purposes, but basically they paint the same picture: what is related in one way in one model is similarly related in the other. There have been no formal comparisons among these approaches, however, so one must be careful with this conclusion, but that is the impression I have at this point. I take this as a positive indication that whatever semantic space we generate is not an artefact of the method used, but truly reflects the semantic information in the corpus.

However, what does make a striking difference is enriching the input to the analysis so that it contains not only word co-occurrence information but also word-order information. This is done in the holographic vector model by adding to the item vector all convolutions of the word with the other words in the sentence. This allows the system to use order information and to infer grammatical categories. It also provides a much better account of human priming data (Jones *et al.* 2006). If only context information is used in the holographic model (as well as in all other such models, like LSA) associative priming (e.g. bee–honey) is reasonably well predicted, but not semantic priming (e.g. deer–pony). On the other hand, if only order information is used in the holographic vector model (equally in n-gram or hyperspace analogue to language [HAL] models; Lund and Burgess 1996), the models handle semantic but not associative priming. But if order and context information is combined in the holographic model, a wide range of experimental data involving both associative and semantic priming can be accounted for.

Including order information in a latent semantic structure allows us to model a far greater range of phenomena than before. But the basic unit of analysis in such an expanded model still remains the word, and it has often been argued that the unit of language and cognition is the proposition, not the word (among others, by Kintsch 1974, 1998). Recent advances in machine learning have made it possible to construct LSA-like systems that represent propositional information in the form of dependency trees. Linguists analyse the syntactic structure of a sentence as a phrase structure tree. A *dependency tree* (Yamada and Matsumoto 2003) is a weak form of a phrase structure tree, lacking the phrasal nodes, and thus does not represent all the relevant linguistic distinctions; however, it does retain information about dependencies among words (i.e., about propositional units). I illustrate this relationship with a simple example. Figure 8.1 shows the dependency tree for the sentence 'Rolls-Royce said it expects its US sales to remain steady at about 1,200 cars,' after Yamada and Matsumoto (2003).[2]

In Figure 8.2, I show the proposition list corresponding to that sentence according to Kintsch (1998), and in Figure 8.3 I superimpose the propositional structure on the dependency tree. The dependency tree shows which words in the sentence belong together as a propositional unit. To obtain a dependency tree, a dependency parser is trained on a

---

[2] The actual dependency tree includes parts of speech tags which are neglected here.

```
                                    Said
                                    ╱╲
                            Rolls-royce expects
                                      ╱╲
                                    It    Remain
                                           ↑
                                           To
                                  Sales         Steady      At
                                  ╱╲                        |
                                Its  Us                    Cars
                                                            |
                                                          1,200
                                                            |
                                                          About
```

Fig. 8.1 Dependency tree for the sentence 'Rolls-Royce said it expects its US sales to remain steady at about 1,200 cars' (after Yamada and Matsumoto 2003).

large corpus of sentences with support vector machines (Yamada and Matsumoto 2003; Nivre *et al.* 2006). This approach is a departure from the purist path of unguided learning, but that is the only way we can do this analysis at present. Praful Mangalath, in our lab, has begun to explore how the information provided by this dependency parse can be incorporated into the systems discussed above to generate a semantic structure that represents word information as well as propositional information. Basically, we analyse, in addition to the word-by-document matrix, a word-by-dependency matrix. Normally, a sentence like 'The ferocious lion killed an antelope' would make both lion and antelope a little bit more ferocious; however, since ferocious is dependent on lion, only the lion should become more ferocious, not the antelope.

Thus, Mangalath uses syntax (in the form of dependency trees) to guide what is learnt, but syntax can also constrain interpretation. It is possible to use the syntactic structure of a sentence to guide the construction of its meaning. We no longer have to add up the vectors of all the words in a sentence to arrive at the sentence meaning, but we can use the syntactic information to generate an interpretation of a sentence in much the same way as we use it in the construction of the semantic space in the first place. Thus, we turn to the construction of meaning – word meanings, sentence meanings, and text meanings.

(1) Said [rolls-royce, 2]

(2) Expect [rolls-royce, 3]

(3) Remain [sales, steady]

(4) Of [rolls-royce, sales]

(5) Us [sales]

(6) At [steady, 8]

(7) 1,200 [cars]

(8) About [7]

Fig. 8.2 Propositional analysis of the sentence in Figure 8.1 (after Kintsch 1998).

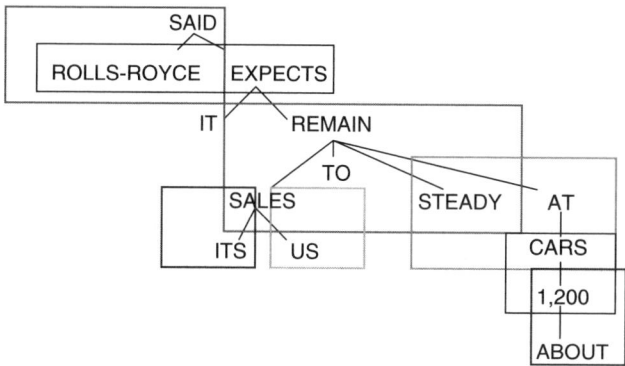

Fig. 8.3 Propositional structure mapped on the dependency tree.

8.3 **The construction of meaning**

Words often have more than one meaning and several different senses. In LSA, as well as in the other methods discussed above, a single vector represents the meaning of each word. It is like a dictionary that lists each word, but is silent about its different meanings and senses. Instead of listing word meanings and senses, as in a conventional dictionary or in a mental lexicon, meanings can be generated in context. While the semantic representation of a word is a single vector that combines all meanings and senses, a context-appropriate meaning is generated every time the word is used in a different context.

I have described such a model for a generative lexicon in Kintsch (2001, 2007, in press). Briefly, it allows the context to modify word vectors in such a way that their context appropriate aspects are strengthened and irrelevant ones suppressed. In the *construction–integration model* of Kintsch (1998), discourse representations are built up via a spreading activation process in a network defined by the concepts and propositions in a text. Meaning construction works in a similar way: a network is constructed containing the word to be modified together with its semantic neighbourhood and linked to the context; spreading activation in that network assures that those elements of the neighborhood most strongly related to the context become activated and are able to modify the original word vector.

Consider the meaning of 'bark' in the context of 'dog' and in the context of 'tree'. The semantic neighborhood of 'bark' includes words related to the dog-meaning of 'bark', such as 'kennel', and words related to the tree-meaning, such as 'lumber'. To generate the meaning of 'bark' in the context of 'dog' ($bark_{dog}$), all neighbours of 'bark' are linked to both 'bark' and 'dog' according to their cosine values. Furthermore, the neighbours themselves inhibit each other in such a way that the total positive and negative link strength balances. As a result of spreading activation in this network, words in the semantic neighbourhood of 'bark' that are related to the context become activated, and words that are unrelated become deactivated. Thus, in the context of 'dog', the activation of 'kennel' increases and the activation of words unrelated to 'dog' decreases; in the context of 'tree', the activation values for 'lumber' increases, while 'kennel' is deactivated. The contextual meaning of $bark_{dog}$ is then the centroid of the 'bark' and its 'dog'-activated neighbours, such as 'kennel'; that of $bark_{tree}$ is the (weighted) centroid of 'bark' and neighbouring words like 'lumber'. $Bark_{dog}$ becomes more 'dog'-like and less 'tree'-like; the opposite happens for $bark_{tree}$.

Two distinct meanings of 'bark' emerge: using the six most highly activated neighbours to modify 'bark' from a neighbourhood of 500, the cosine between $bark_{tree}$ and $bark_{dog}$ is only 0.03. Furthermore, $bark_{dog}$ is no longer related to 'tree' (cosine = −0.04) and $bark_{tree}$ is no longer related to 'dog' (cosine = 0.02).

Figure 8.4 illustrates a different example, where the meaning generated is metaphorical rather than literal (Kintsch and Bowles 2001; Kintsch, in press): the meaning of 'shark' in the context of 'My lawyer is a shark'. The items that 'lawyer' selects from the neighbourhood of 'shark' make $shark_{lawyer}$ more aggressive and vicious, and less fish-like. Thus, context-appropriate word meanings can be generated within a system like LSA, in spite of the fact that the vectors constructed by LSA are context-free word vectors. The use of context-free representations in memory allows us to deal with polysemy. We no longer have to decide

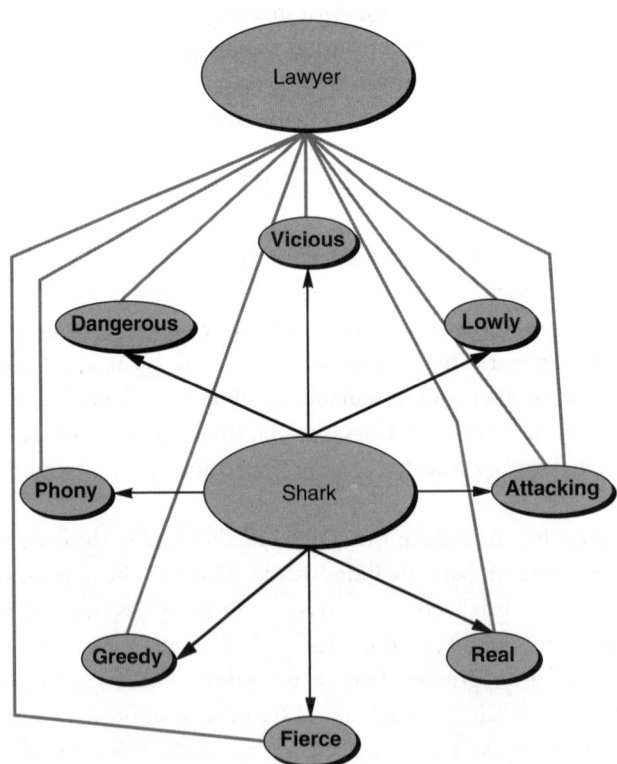

Fig. 8.4 A spreading activation network showing the eight most strongly activated neighbours of 'shark' by 'lawyer' (ICA, TASA corpus).

how many meanings and senses a word has, or how they are to be retrieved when needed. Instead, we only have to deal with a single, context-free word vector plus a process (predication) that generates emergent contextual meanings from that vector.

Predication models how word meanings are constructed in context, but what about sentence meaning, and the meaning of texts? LSA has proven to be so useful because it has a very simple and effective way of constructing the meaning of texts by summing the vectors of all the words involved. Despite neglecting both syntax and text structure, if the texts are long enough it produces excellent results, as attested by the commercial success of Pearson Knowledge Technologies (www.knowledge-technologies.com) and the educational success of *Summary Street* (Wade-Stein and Kintsch 2004; Franzke *et al.* 2005). However, the vector sum of the constituent words is not a useful representation when we are dealing with single sentences or brief texts. The syntax plays a large a role in this case, whereas it apparently averages out over longer text, so that its neglect does not cause fatal problems.

We can use syntax to guide the construction of sentence meanings. As in the construction–integration model, we can use dependency analysis to construct propositional units and then integrate these units as required by the task at hand. I shall illustrate this model with two simple examples.

In a cloze test students are given a sentence with a word missing and are asked to fill in the missing word from a given set of alternatives. For example, the sentence might be 'If you try, you can _____ any problem,' with the alternatives 'solve', 'quack', 'flour', and 'seem'. One way to model this task is to compute the cosine between the sentence fragment and the four response alternatives in LSA. As Table 8.1 shows, this leads to the incorrect choice of 'seem'. However, we get a different prediction if we use dependency analysis of the sentence to focus our comparison on the relevant proposition rather than on the whole sentence. Figure 8.5 shows the dependency tree for this sentence, indicating the proposition containing the target word. This analysis suggests computing cosines between the four response alternatives and 'problem', the relevant content word in the proposition containing the target word. Note that while function words were crucial in the construction phase – they determine what kind of dependency tree is constructed and hence, which content words enter into the integration phase – we neglect function words in this comparison because they would only add noise. Table 8.1

Table 8.1 Cosines between four response alternatives and the sentence fragment '*If you try, you can _____ any problem*'

	LSA (sentence)	LSA (dependency)
solve	0.49	**0.88**
quack	0.18	0.08
flour	**0.15**	0.03
seem	0.58	0.26

LSA = latent semantic analysis.

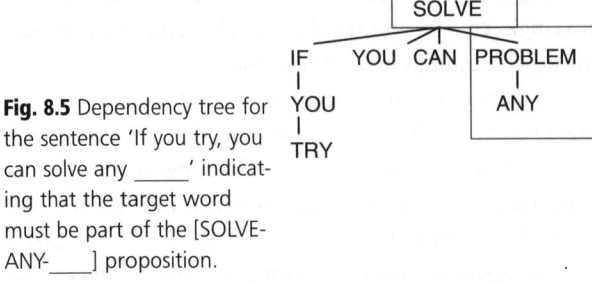

Fig. 8.5 Dependency tree for the sentence 'If you try, you can solve any _____' indicating that the target word must be part of the [SOLVE-ANY-____] proposition.

shows that the correct alternative has the highest cosine with the relevant proposition fragment.[3]

A second example involves the implications that sentences may have. Suppose we have in our corpus the sentence 'Oracle fought to keep the forms from being released' and our task is to determine whether the inferences 'Oracle released a confidential document' and 'Oracle fought to keep the forms released' are true or false. Both inference sentences have a high cosine with the original sentence (cosine = 0.24 and 0.93, respectively), and hence appear to be implied. However, the dependency trees for these sentences suggest otherwise (Figures 8.6 and 8.7). The inference sentence 'Oracle released a confidential document' has the dependencies *Oracle-released*, *released-document*, and *document-confidential*, none of which match any of the dependencies in the original sentence. For the second inference sentence, 'Oracle fought to keep the forms released', the dependencies between oracle, fought, keep, and forms are the same as in the original sentence, but not between the rest: in the original sentence there are dependencies between *keep-from* and *from-released*, which do not match the *forms-released* dependency in the inference sentence. Hence, both inferences can be rejected as invalid. Function words play a crucial role in this process: they do not enter the comparison process (in LSA or ICA, function words like 'from' tend to be uninformative, having been experienced in all kinds of different contexts), but they determine what is compared with what in a sentence.

The point I am making with these examples is this: we have fairly good models of how people store verbal symbolic information in memory, and of how this information is used in constructing the meaning of words, sentences, and texts. The question is, do we have to throw away these models because 'No matter how LSA, HAL, and other ungrounded symbol theories are extended and modified, ungrounded arbitrary symbols cannot be an adequate basis for human meaning' (Glenberg and Robertson 2000, p. 397)? My answer is 'not at all' – this claim, and similar claims by others, is ungrounded and based on a misconception about the relationship between symbolic and nonsymbolic representations.

....

[3] It should be noted that there are other methods that allow LSA to solve cloze problems, using n-gram or word-order information.

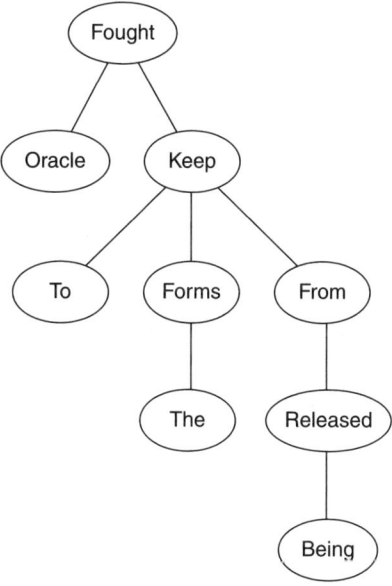

Fig. 8.6 Dependency tree for the sentence 'Oracle fought to keep the forms from being released'.

8.4 **Language as the mirror of the mind**

When people remember a list of words, read a sentence, or listen to a story, they do not function solely at the symbolic level. They use imagery to make words more memorable (Paivio 1969), they construct situation models that are multimedia creations (Bransford and Franks, 1971; Zwaan *et al.* 2004; Zwaan, Chapter 9, this volume); and perceptual and motor areas in the brain become activated (Pulvermüller, Chapter 6, this volume), as was

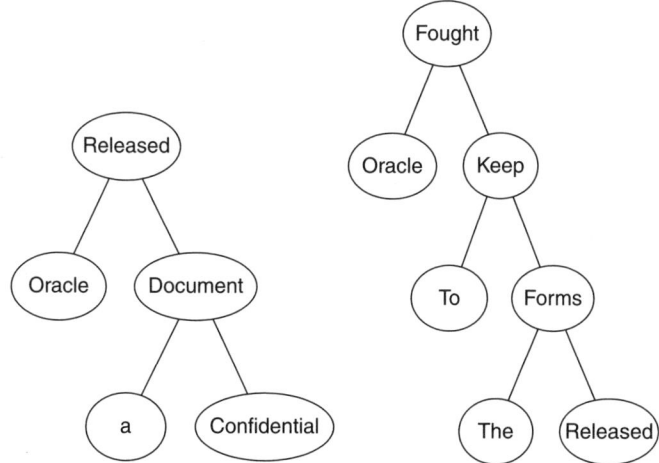

Fig. 8.7 Dependency trees for the inference sentences 'Oracle released a confidential document' and 'Oracle fought to keep the forms released.'

demonstrated by several participants in the debate. All levels of representation that humans are capable of appear to be functioning in a well coordinated chorus. How can a purely symbolic model, such as the one outlined above, do justice to these multilevel processes? The brief answer is that it does not and can not. The more interesting answer is that language has evolved in the service of perception and action, and that it is capable of mirroring aspects of perception and action. Yes, words can distort and bias meaning, but language would not be the great human success story it is if it were incapable of expressing what we experience faithfully.

There have been a large number of claims, such as that of Glenberg and Robertson (2000), that purely verbal (in other words, symbolic) representations of meaning are insufficient for modelling meaning. One the one hand this is a truism; human meaning representations are multilevel and not purely symbolic – no one can doubt that we live and act in a nonsymbolic world. On the other hand, the demonstrations of the short-comings of verbal meaning are often misleading, based on much too crude a conception of verbal meaning. Language does a better job than it is given credit for by the critics of symbolic representations. Louwerse (2007 and Chapter 15, this volume) has argued this point with some compelling examples. I would like to add two more examples, one involving only word meanings, where LSA provides an account that closely mirrors perception, and one involving sentence meanings, where the construction–integration model is shown to mirror affordances.

Salomon and Barsalou (2001) convincingly demonstrated that what things look like is an important factor in verbal tasks of property verification. Concepts – certainly concrete concepts – are represented both perceptually and symbolically, which plays an important role not only in perception and action, but also when we talk and think about a concept. For example, English has only a single word 'wing', but a wasp wing and a butterfly wing look different, and this difference is reflected in the mental representation of these concepts. Solomon and Barsalou showed this in a priming study, where participants first verified either a concept–property pair such as 'wasp wing' or 'butterfly wing', and then a target pair such as 'bee wing'. The perceptually similar 'wasp wing' significantly primed 'bee wing', but the perceptually dissimilar butterfly wing did not. They also showed that perceptual similarity mattered in this case, not merely conceptual similarity, because no priming differences were observed when there was no perceptual discrepancy. In the case of their example, 'butterfly body' and 'wasp body' primed 'bee body' equally well – presumably because the bodies of wasps, bees, and butterflies look more or less alike, unlike their wings.

LSA cannot see wings, but it has indirectly encoded some of the perceptual informa-tion needed to know that a wasp wing and a butterfly wing are really quite different, whereas a wasp body and a butterfly body are not. If we compute a vector for 'wasp wing', 'butterfly wing', and 'bee wing' using the predication procedure described above, we can compute how close to each other these vectors are in the LSA space. The cosine between 'wasp wing' and 'bee wing' is 0.66, whereas the cosine between 'butterfly wing' and 'bee wing' is only 0.48. Hence, 'wasp wing' could be expected to be a better prime for 'bee wing' than 'butterfly wing'. On the other hand, both 'wasp body' and 'butterfly body' are equally close to 'bee body' in the LSA space. The cosines are 0.94 and 0.91, respectively – LSA really can't tell the difference between these concepts (and neither can most of us).[4]

I am not saying that the participants in the Salomon and Barsalou experiment did not employ perceptual representations when verifying properties; they probably did. My claim is simply that even a purely verbal symbolic system like LSA would behave in much the same way, because the verbal information mirrors perception to a considerable degree. This is also the case with my second example from Glenberg and Robertson (2000), who claim that purely symbolic, verbal representations cannot recognize perceptual affordances. Perceptual affordances are obviously based on perceptual and action experience, but these experiences are mirrored in our language, so that even a purely symbolic system like LSA or ICA can detect that some of the sentences below are more sensible than others in the context of a story of someone being caught in the rain. Consider the following sentences:

(a) Being clever, he walked into a store and bought a **newspaper** to cover his face.

(b) Being clever, he walked into a store and bought a **matchbox** to cover his face.

(c) Being clever, he walked into a store and bought a **ski mask** to cover his face.

Both (a) and (c) are possible ways to cover one's face, (c) much better than (a); (b) does not make sense. A ski mask or even a newspaper afford some protection from the rain, whereas a matchbox does not. The affordances inherent in an object are based on experiences, experience in the real world with newspapers and matchboxes, and experience with how the words 'newspaper' and 'matchbox' have been used. Glenberg and Robertson present an LSA analysis that fails to account for the difference between these and related sentences: if one computes the cosine between the bold-faced target words and the rest of the sentences in the examples above, they do not differentiate between sensible and nonsense alternatives (Table 8.2). However, when these sentences are analysed in terms of their propositional constituents, the difference in affordances emerges quite clearly.

Figure 8.8 shows the dependency tree for (a). The figure shows that the relevant proposition involves the target words and '(to) cover (his) face'. If you simply take the cosine between the three targets and the whole sentence, you would conclude that (b) was the best alternative and (c) the worst. Glenberg and Robertson have obtained sensibility ratings on a scale of 1 to 7 which show that (c) is highly sensible, (a) is less sensible, but (b) is clearly not. An ICA (or LSA) analysis that respects the propositional units obtained by a dependency analysis yields similar results, simply because, according to our experience with language, *cover-face* has a high cosine with 'ski mask', not quite as high a cosine with 'newspaper', and a low cosine with 'matchbox'.

---

[4] A more detailed analysis of the Salomon and Barsalou (2001) results is given in Kintsch (2007).

Table 8.2 Three examples from Glenberg and Robertson (2000) with LSA cosines between the sentence frames and the target items, sensibility ratings of the target items for each sentence frame, and ICA cosines between the target words and the relevant portion of the sentence frames

	LSA cosine (sentence)	Sensibility rating	ICA cosine (dependency)
'Being clever, he walked into a store and bought a _____ to cover his face.'			
newspaper	0.35	5.08	0.18
matchbook	0.42	1.16	0.03
ski mask	0.40	6.04	0.53
'She gave him a ——— to play with.'			
plastic spoon	0.48	5.58	0.07
large refrigerator	0.49	1.08	−0.03
red beanbag	0.39	6.29	0.59
'Adam pulled out of his golf bag a _____ and used that to chisel an inch of ice off his windshield.'			
seven iron	0.50	4.50	0.42
ham sandwich	0.56	1.00	0.16
screwdriver	0.60	5.00	0.63

ICA = integrated component analysis; LSA = latent semantic analysis.

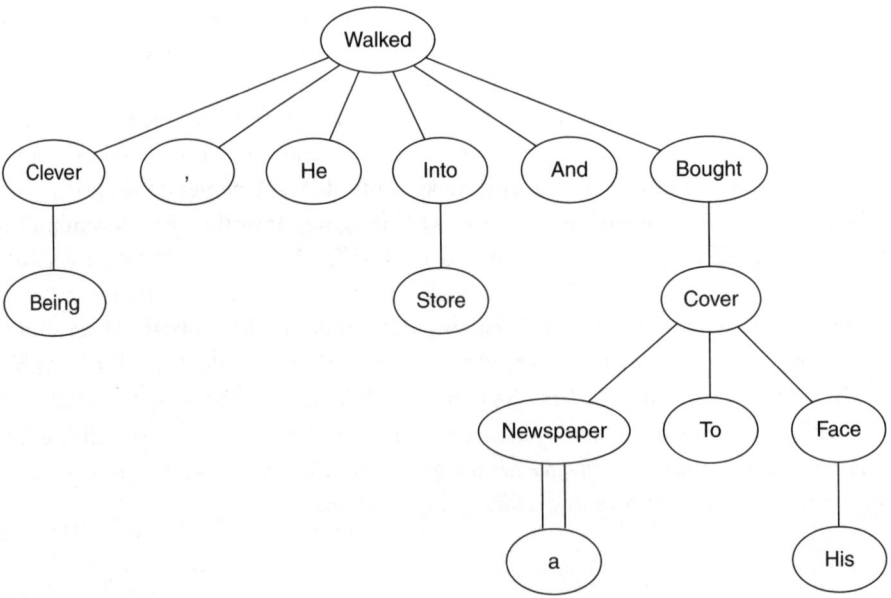

Fig. 8.8 Dependency tree for the sentence 'Being clever, he walked into a store and bought a newspaper to cover his face.'

Table 8.2 also shows two other examples from the Glenberg and Robertson study. For the sentence 'She gave him a _____ to play with,' people accept the target items 'red beanbag' and 'plastic spoon' as sensible, but not 'large refrigerator'. The dependency analysis suggests that the target item is to be compared with '(to) play (with)'. This analysis correctly yields a high cosine for 'red beanbag' and a low cosine for 'large refrigerator', but the cosine for 'plastic spoon' is not as high as it should be according to the sensibility rating participants provided.

For the sentence 'Adam pulled out of his golf bag a _____ and used that to chisel an inch of ice off his windshield,' the target items are 'seven iron', 'ham sandwich', and 'screwdriver'. The dependency analysis shows that 'and' dominates both *pull-[the target item]* and 'used that to chisel an inch of ice off his windshield.' This comparison correctly selects 'screwdriver' as the most sensible completion, and 'seven iron' as also possible, but rejects the 'ham sandwich', much like people do. Overall, ICA correctly selects the completion that is both related and afforded in all of the experimental examples and rejects the unrelated, unafforded alternative. The cosines for the afforded but unrelated alternative are more variable, as we have seen in the examples above. Perceptual representations obviously add something to symbolic representations, but purely symbolic representations do a remarkably good job of discriminating between perceptually afforded and unafforded alternatives.

Let me comment briefly on some other criticisms of symbolic representations that I think are misguided. The objection is not only to ungrounded symbols, but also to arbitrary symbols. Apparently, icons are regarded as somehow better, easier, and more natural than symbols. To comprehend meaning entails understanding the difference between a symbol or icon and the real thing; that is, *dual representation*. This may in fact be easier when the symbol is arbitrary (as words are) than when it is not (as pictures are). Children as old as 2.5 years are apt to mistake photographed objects for the object itself and treat small toys as if they were much larger (DeLoache 2004). They do not confuse an arbitrary sound with the object it represents, but do attempt to sit down on a miniature toy chair. Babies confuse pictures with the objects they depict, as seen in American babies who are familiar with pictures as well as Ivory Coast babies who are not. It is not at all trivial to learn that concrete, nonarbitrary icons represent something that they are not. Teaching math with manipulatives can be counterproductive because the children learn to manipulate objects without ever inferring abstract mathematical principles. The consequences of the confusion between objects and their representation can be serious, as when dolls are used in courtrooms to represent a child's body – but the child is unable to think of the doll as both a doll and a representation of herself (Ceci and Bruck 1995). Icons and analogue presentations are not necessarily easier than digital representations or arbitrary symbols.

The fact that symbolic processes involve the same brain areas as action and perception (e.g. Pulvermüller, Chapter 6, this volume) does not imply that symbolic processes and sensorimotor processes are the same. In the course of evolution nature discards very little, the new is generally built upon the old: fins turn into legs, and sensory brain areas become involved with symbol processing. But symbol processing is still at a different level than sensorimotor processes. Other primates and humans are very similar in their sensorimotor processes, but differ radically in their symbolic processing. Humans operate at multiple levels of representations, from the most primitive ones shared by all animals to the symbolic level. But that does not justify reducing symbolic systems to sensorimotor processes.

Rejecting a reductionism that denies the autonomy of symbolic processes does not imply that we should not make every effort to gain a better understanding of how symbolic and nonsymbolic representations interact. To learn more about how perceptual representations and symbolic representations are coordinated and how they interact, is a very important goal, and considerable progress has been made as the chapters presented in this book attest, both at the level of brain processes, as in the work of Pulvermüller already cited, as well as the behavioural level (e.g., Zwaan, Chapter 9, this volume). This problem has also been addressed at the level of computational modelling. Goldstone *et al.* (2005) explicitly address the problem of connecting words to each other as well as to the world. Howell *et al.* (2005) have described a neural network model that simulates the acquisition of language based on prelinguistic concepts. A group of linguists and computer scientists at Berkeley have developed what they call a 'simulation semantics' to reflect their belief that much of language understanding involves embodied enactment (e.g., Feldman & Narayanan 2004; for more information see http://www.icsi. berkeley.edu/NTL). In simulation semantics, the mind simulates the world while functioning in it. Concrete action schemata, for instance, become symbolically extended. Understanding at the symbolic level may involve this kind of simulation, but it can also remain at the purely symbolic level – the world that LSA describes. Research like this serves as an existence proof that it is possible to model the interface between symbolic and nonsymbolic representations.

8.5 **Discussion**

The sensitivity of language to perceptual information (as well as emotion and action) should surprise no one. Most of the words we know we have learned from reading, which precludes a direct perceptual association. And for words we know at both the verbal level and the action–perception level – the crucial anchor words that link these different levels of representation – language has encoded in its own way the information it needs to mirror the world. I conclude with two quotes that Chomsky used in his discussion of language (Chomsky 1966) and that I have also cited in Kintsch (1998):

Les langues sont le meilleur miroir de l'esprit humain.

(Languages are the best mirror of the human mind.)

Gottfried Wilhelm Leibniz, 1765

Der Mensch lebt mit den Gegenständen hauptsächlich, ja, da Empfinden und Handeln in ihm von seinen Vorstellungen abhängen, sogar ausschliesslich so, wie die Sprache sie ihm zuführt.

(We interact with objects mainly as they are represented by language; indeed, exclusively so, since perception and action depend upon memory images.)

Wilhelm von Humboldt, 1792

Debate

Arthur Glenberg: I agree with you that there's an awful lot of structure that's available in covariance and analysed covariance as you presented it. But, in the spirit of this debate, what I'd like to hear from you is how we might determine whether people are using those sorts of representations; whether people are extracting information from language in similar ways.

Walter Kintsch: I make predictions about what people do, I observe the behaviour and see how well the predictions fit. That's all I can say.

Glenberg: And I agree with you, that the predictions do fit nicely. I think I need to formulate my question more incisively. The question really should be, how can we tell if this is a better theory than perhaps the embodied theory? In other words, where are the predictions that are different from an embodied theory?

Kintsch: Well, I don't think that that's so much the question, what's a better theory? I think most people are agreed that cognition has symbolic aspects and has embodied aspects, and that's worthwhile to study. It's not an either/or kind of thing. This is clearly what I wanted to show.

Friedemann Pulvermüller: Thanks very much for an excellent and very interesting talk. It is a very good thing to have these measures of relatedness of semantic distance between words and concepts, or family resemblance. However, the information that would not be included in this relationship – distance measures – would be, for example, that 'bark' is in one case an action and in the other case an object. And one wouldn't have directions from this very abstract representation about in which brain areas one would have to look. Of course, one could take a very abstract strategy and just say 'I take this description and make a correlation with brain activation.' But on the other hand, it might be a good thing to connect these descriptions, these semantic descriptions, with some information that links the representation to particular cognitive domains and brain systems as well, *a priori*. Obviously, an embodied description might have the advantage in providing that. These measures we are discussing, these covariance measures, they would also incorporate important information which a purely embodied theory might lack. So both sides might learn from each other.

Kintsch: Well, I think that suggestion is a very interesting one, and that's something we can actually look into. Because, if I have a lot of nouns and a lot of verbs, I can get them to cluster in very distinct clusters, so the verbs are here and the nouns are over there. Now, if you have a word like 'bark', which could be either one, I don't know what would happen. But if I can construct, now, the 'dog' meaning of 'bark', which is a verb, and the 'tree' meaning of 'bark', which is a noun, those two vectors should be in different clusters.

Pulvermüller: Sure, yes.

Kintsch: If I'm right. And it might also be interesting from a brain standpoint to see what happens with homonyms that have two different functions, how they are represented in two different places.

Pulvermüller: Right. Two overlapping styles is what we would propose, or have proposed. May I make another little comment on the syntax part? I've tried to argue that, using a model consisting of sequence detectors, it is actually possible to formulate dependency grammar. So I have a feeling that if you have the basic sequencing of words and on top of that a separate grammar dependency processor, there might be some redundancy.

Kintsch: I know, I was thinking whether that couldn't be put together. A very interesting thought.

Author note

Preparation of this report was supported by National Science Foundation grant #153-4712.

References

Barsalou L (1999). Perceptual symbol systems. *Behavioural and Brain Sciences*, 22, 577–660.

Bransford JD, Franks JJ (1971). The abstraction of linguistic ideas. *Cognitive Psychology*, 2, 331–50.

Bruner JS (1986). *Actual Minds, Possible Worlds*. Cambridge, MA: Harvard University Press.

Ceci SJ, Bruck M (1995). *Jeopardy in the Courtroom: The Scientific Analysis of Children's Testimony*. Washington, DC: American Psychological Association.

Chomsky N (1966). *Cartesian Linguistics*. New York, NY: Harper.

DeLoache JS (2004). Becoming symbol-minded. *Trends in Cognitive Sciences*, 8, 66–70.

Donald M (1991). *Origins of the Modern Nind*. Cambridge, MA: Harvard University Press.

Feldman J, Narayanan S (2004). Embodied meaning in a neural theory of language. *Brain and Language*, 89, 385–92.

Franzke M, Kintsch E, Caccamise D, Johnson N, Dooley S (2005). Summary Street®: computer support for comprehension and writing. *Journal of Educational Computing Research*, 33, 53–80.

Glenberg AM, Robertson DA (2000). Symbol grounding and meaning: a comparison of high-dimensional and embodied theories of meaning. *Journal of Memory and Language*, 43, 379–401.

Goldstone RL, Feng Y, Rogosky B (2005). Connecting concepts to the world and each other. In D Pecher, RA Zwaan, Eds. *Grounding Cognition: The Role of Perception and Action in Memory, Language, and Thinking* (pp. 292–314). Cambridge: Cambridge University Press.

Harnad S (1990). The symbol grounding problem. *Physica D*, 42, 335–46.

Howell SR, Jankowicz D, Becker S (2005). A model of grounded language acquisition: Sensorimotor features improve lexical and grammatical learning. *Journal of Memory and Language*, 53, 258–76.

Jones MN, Mewhort DJK (2007). Representing word meaning and order information in a composite holographic lexicon. *Psychological Review*, 114, 1–37.

Jones MN, Kintsch W, Mewhort DJK (2006). High-dimensional semantic space accounts of priming. *Journal of Memory and Language*, 55, 534–52.

Karmiloff-Smith A (1992). *Beyond Modularity*. Cambridge, MA: MIT Press.

Kintsch W (1974). *The Representation of Meaning in Memory*. Hillsdale, NJ: Erlbaum.

Kintsch W (1998). *Comprehension: A Paradigm for Cognition*. New York: Cambridge University Press.

Kintsch W (2001). Predication. *Cognitive Science*, 25, 173–202.

Kintsch W (2007). Meaning in context. In TK Landauer, D McNamara, S Dennis, W Kintsch, Eds. *The Handbook of Latent Semantic Analysis* (pp. 89–106). Mahwah, NJ: Erlbaum.

Kintsch W (in press). How the mind computes the meaning of metaphor: a simulation based on LSA. In R Gibbs, Ed. *Cambridge Handbook of Metaphor and Thought*. New York, NY: Cambridge University Press.

Kintsch W, Bowles AR (2002). Metaphor comprehension: what makes a metaphor difficult to understand? *Metaphor and Symbol*, 17, 249–62.

Landauer TK, Dumais ST (1997). A solution to Plato's problem: the latent semantic analysis theory of acquisition, induction and representation of knowledge. *Psychological Review*, 104, 211–40.

Landauer TK, McNamara D, Dennis S, Kintsch W, Eds (2006). *The Handbook of Latent Semantic Analysis*. Mahwah, NJ: Erlbaum.

Louwerse MM (2007). Symbolic and embodied representations: a case for symbol interdependency. In TK Landauer, D McNamara, S Dennis, W Kintsch, Eds. *The Handbook of Latent Semantic Analysis* (pp. 107–20). Mahwah, NJ: Erlbaum.

Lund K, Burgess C (1996) Producing high-dimensional semantic spaces from lexical co-occurrence. *Behaviour Research Methods, Instrumentation, and Computers*, 28, 203–8.

Mangalath P (2006). *Beyond Latent Semantic Analysis: Cognitive Component Resolution with Independent Component Analysis* [doctoral thesis]. Denver, CO: University of Colorado.

Nelson K (1996). *Language in Cognitive Development: The Emergence of the Mediated Mind*. New York, NY: Cambridge University Press.

J Nivrc, J Hall, J Nilsson (2006). MaltParser: a data-driven parser-generator for dependency parsing. In *Proceedings of the Fifth International Conference on Language Resources and Evaluation (LREC2006)*, May 24–26, 2006, Genoa, Italy.

Paivio A (1969). Mental imagery in associative learning and memory. *Psychological Review*, 76, 241–63.

Salomon KO, Barsalou LW (2001). Representing properties locally. *Cognitive Psychology*, 43, 129–69.

Shepard RN (1987). Toward a universal law of generalization for psychological science. *Science*, 237, 1317–23.

Steyvers M, Griffiths T (2007). Probabilistic topic models. In TK Landauer, D McNamara, S Dennis, W Kintsch, Eds. *The Handbook of Latent Semantic Analysis* (pp. 427–48). Mahwah, NJ: Erlbaum.

Stone JV (2004). *Independent Component Analysis*. Cambridge, MA: MIT Press.

Wade-Stein D, Kintsch E (2004). Summary Street: interactive computer support for writing. *Cognition and Instruction*, 22, 333–62.

Yamada H, Matsumoto Y (2003). Statistical dependency analysis with support vector machines. *Proceedings of IWPT* (pp. 195–206). Nancy, France: IWPT.

Zwaan RA, Madden CJ, Yaxley RH, Aveyard ME (2004). Moving words: dynamic representations in language comprehension. *Cognitive Science*, 28, 611–19.

Chapter 9

Experiential traces and mental simulations in language comprehension

Rolf A Zwaan

9.1 Experiential traces

The goal of this chapter is to put forth a framework for an embodied, experiential view of language comprehension. The claim is not that the elements of this framework are entirely new. The contribution lies more in the integration from insights from the diverse literatures on concepts, episodic and semantic memory, psycholinguistics, computational linguistics, motor control, and action observation. The key notion in the framework is that of the experiential trace. Experimental traces are laid down in multiple modality-specific cortical areas (e.g., Barsalou 1999; Damasio 1999). As such, they are multimodal records of experience. Importantly, linguistic constructions such as words and standard phrases are also multimodal experiential traces (Sadoski and Paivio 2001). For ease of exposition, a distinction will be made between linguistic traces and referential traces, but the key is that the two types of traces are not intrinsically different from one another (Zwaan and Madden 2005).

Associations between co-occurring linguistic traces and referential traces, *L–R associations*, are formed via Hebbian learning. Often, such co-occurrences emerge during *embedded comprehension* (Spivey and Richardson, in press), in which language is used to refer to the immediate environmental context and reference is established via joint attention (e.g., eye gaze, pointing). Developmental research has established that in such contexts young language learners use certain heuristics in resolving reference. For example, they tend to associate new linguistic constructions with whole objects rather than with their parts. They also associate new linguistic constructions with new entities, rather than with items for which they already have a lexical associate.

An additional way in which L–R associations can be formed is via *referential bootstrapping*. For example, a website for elementary school students features the following riddle: "'What has horns like a giraffe, a deer head, a horse neck, and legs like a zebra?' The answer is, of course, an okapi. Although it may not be possible to construct a completely accurate visual representation of an okapi based on the riddle, this representation will probably be sufficiently accurate to enable one to recognize an okapi upon being exposed to this rare mammal, either in the flesh or in the form of a picture.

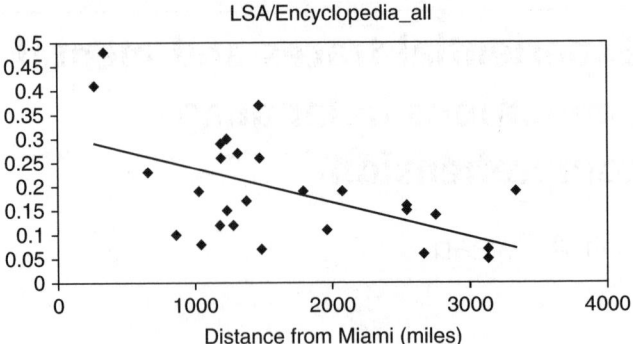

Fig. 9.1a

In addition to L–R associations, the brain forms L–L and R–R associations. Again, Hebbian learning is the mechanism. Computational techniques of large corpora of text, such as *latent semantic analysis* (LSA; Landauer and Dumais 1997), can be used to uncover patterns of L–L co-occurrences. These co-occurrences can be temporal or spatial (linguistic constructions occurring in similar contexts, without necessarily temporarily co-occurring in such contexts). Just as linguistic traces can co-occur in time and space, so referential traces can co-occur in time and space (Zwaan and Madden 2005). For example, tomatoes and lettuce can be found together in salads; staplers and notepads on desktops; and guitars and drums in recording studios. Given that language most often describes situations in the real world, or in realistic environments, L–L co-occurrences tend to reflect R–R co-occurrences.

Techniques such as LSA can capitalize on such second-order correlations between co-occurrences. For example, Figure 9.1a shows the correlation ($r = 0.55$) between the distance from Miami for 27 cities in the United States and the LSA cosine in the

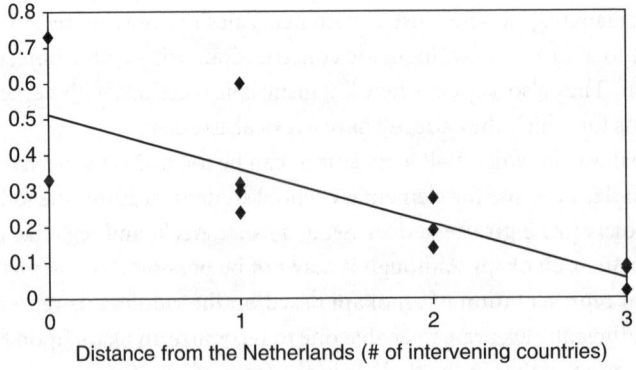

Fig. 9.1b

encyclopaedia corpus (http://lsa.colorado.edu/).[1] Using the same corpus, Figure 9.1b shows the correlation between the ordinal distance from the Netherlands – 0 for bordering countries, 1 for countries separated from the Netherlands by one other country, and so on – and the cosines between them ($r = 0.79$).

Similar substantial correlations can be found for states in the US and South American countries. These correlations presumably reflect the fact that cities or countries in close spatial proximity are more likely to be described in similar contexts than more remote cities or countries. This could be for a variety of reasons including geological, geographical, and historical overlap, and the greater likelihood of interactions via trade, travel, and war: all of these are common topics in encyclopaedias. Although impressive, and potentially useful, such second-order correlations do not allow the inference that meaning is grounded in co-occurrences of linguistic constructions. Rather, they suggest that L–L associations often mirror R–R associations and, due to this correspondence, can be used to enhance language comprehension, for example with regard to enhancing the ability to anticipate upcoming information. The notion of anticipation in comprehension will be developed below.

9.1.1 Cross-modal activations

Given that experiential traces are multimodal representations, it follows that when a trace is activated in one modality, its components in other modalities resonate as well. For example, phonological representations are activated during reading (e.g., Stone *et al.* 1997) and orthographic representations are activated during speech comprehension (e.g., Seidenberg and Tanenhaus 1979) and speech production (Damian and Bowers 2003). Furthermore, there is evidence that auditory representations of words activate the motor programs used to produce these words and viewing letters activates, in expert typists, the motor programs used to type them (Rieger 2004; for related evidence see Beilock, in press; van den Bergh *et al.* 1990). By the same token, aspects of experience in one modality may facilitate retrieval of other aspects of the experience. Classic findings in the memory literature of encoding specificity are cases in point. More recently, it has been shown that proprioceptive cues facilitate retrieval of posture-congruent autobiographical memories (Dijkstra *et al.* 2007). For instance, assuming a reclining posture facilitates retrieval of memories of visits to the dentist.

9.2 Resonance

Linguistic traces resonate with linguistic input, causing associated referential traces to resonate as well. The general idea here is consistent with an instance-based view of memory (e.g., Hintzman 1986), according to which experiential traces resonate to varying degrees with the input.

---

[1] See Kintsch (Chapter 8, this volume) and Louwerse and Jeuniaux (Chapter 15) for similar analyses and somewhat different conclusions.

The extent to which experiential traces resonate is a function of their similarity to the input and their current level of activation. For example, if a referential representation has recently been activated, it is likely to resonate with a later occurring word, thus inserting itself in the mental simulation. In support of this idea, we recently found that incidental exposure to word–picture combinations can affect later reading (Aveyard *et al.*, submitted). The experiment involved two phases: a word–picture verification phase presenting critical objects in specified shapes or conditions (e.g., a perched eagle), and a subsequent reading phase (ostensibly unrelated to the first) in which the subjects' eye movements were tracked. In this phase, texts were presented that implicitly disambiguated shapes or conditions of critical objects (e.g., 'In the sky an eagle …'). Reading times on the target word ('eagle') were longer when the implied shape in the text mismatched the associated shape from the first phase (for example, a perched eagle cannot be found in the sky). Furthermore, total reading times on the prepositional phrase were longer in the mismatch condition than in the match condition, presumably indicating that the activation of the contextually inappropriate shape caused subjects to re-encode the location of the target entity in an attempt to resolve the inconsistency.

Abstractions are an automatic by-product of the resonance process. For example, the word 'not' will activate all instances of having experienced the input " not" (e.g., 'Don't pet the doggie;' 'Do not leave your clothes on the floor;' 'Do not talk in class'). Presumably, the only thing these traces have in common is that a state of the environment has triggered the activation of a goal, the execution of which is subsequently thwarted. This abstraction process (see also Zwaan and Madden 2005), which may involve a different set of traces each time, renders the meaning of 'not' as a sequence of activation patterns. First, a representation of an expected or desired state is activated, subsequently activating the motor programs needed to bring about that state. The execution of these programs is then blocked, which likely results in an emotional response (e.g., surprise, annoyance, anger, frustration). Thus, the difference between concrete and abstract words is that concrete words activate a relatively homogeneous set of traces, whereas abstract words activate a more heterogeneous set of traces, leading to patterns of activation that differ more across and within individuals than traces of concrete concepts (see also Barsalou and Wiemer-Hastings 2005).

9.3 Mental simulation

Activated experiential traces are used in mental simulations of the described situation (Barsalou 1999).[2] A notion of mental simulation that is relevant to language comprehension has been developed in research on motor control and in research on action observation.

Anticipation is a crucial mechanism in motor control. Given the inherent delays in perception – for example, it takes about one hundred milliseconds to process visual feedback – and in action execution, perceptual feedback is not very useful in the control of

---

[2] Elsewhere, I have shown how this view can be viewed as a logical extension of earlier work on situation models (Zwaan 2004; Zwaan and Radvansky 1998).

motor actions. Instead, the central nervous system produces a mental simulation, called a *forward model*, which uses a copy of the efferent signal to the effectors to predict the sensorimotor consequences of motor commands. Discrepancies between the predicted and the observed perceptual input are used to fine-tune the system, by inducing synaptic weight changes, such that future predictions in similar situations will be more accurate. In one model, simulation involves a range of motor programs, which are activated in resonance fashion to the extent that they match the current situation (Wolpert *et al.* 2003).

Similar anticipatory mental simulations are thought to underlie action understanding and social interaction (Gallese *et al.* 2003; Wolpert *et al.* 2003). Empirical evidence is consistent with this idea. For example, during the performance and observation of a block-stacking task, the observer's pattern of eye fixations mirrors that of the performer; both precede the movement of the blocks (Flanagan and Johansson 2003). This experiment suggests that the observer is using the same anticipatory eye movement program to understand the action that the actor is using to perform the action. Importantly, this anticipatory pattern is not found when the actor is hidden from view and the blocks seem to move by themselves. In this case, eye movements trail the blocks. In other words, this experiment shows very elegantly that anticipatory processing occurs during action observation, but only when the observer interprets the actions as being performed by a conspecific.

Single-cell recordings in the macaque monkey ventral premotor cortex provide detailed analyses of the recruitment of motor programs during action observation (e.g., DiPellegrino *et al.* 1992). In this model, some fire when the monkey observes an action being performed that it also has in its own action repertoire (e.g., grasping a food item). These neurons have been termed *mirror neurons*. Mirror neurons have also been shown to fire when the monkey hears a sound associated with an action in its repertoire, for instance cracking a nut (Köhler *et al.* 2002).

Importantly, mirror neurons are responsive to an understanding of the goal of an action. When the monkey knew there was food behind a screen, its mirror neurons responded when the experimenter's hand moved towards the food, even though the hand disappeared behind a screen. The activation pattern was similar to a condition without the screen; some mirror neurons responded equally strongly in both conditions, whereas others responded more strongly in the full vision condition. In contrast, mirror neuron activity did not occur, with or without screen, if there was no food, but the experimenter made the same grasping movement (Umiltà *et al.* 2001). Thus, in the monkey brain, having a mental representation of the goal of a grasping action seems both necessary and sufficient for mirror neuron activation; however, this restriction does not seem to hold for the human brain (Grèzes, Armony *et al.* 2003).

A recent computational approach (Keysers and Perrett 2004) suggests how sensory information becomes associated with motor programs due to the anatomical connections between the superior temporal sulcus area, which responds to visual and auditory stimulation, and areas PF and F5, which receive input from superior temporal sulcus. Because a subset of neurons in these areas show some degree of viewpoint independence, the monkey learns to associate not only the sights and sounds of its own actions with motor programs, but also the sights and sounds of the same actions performed by others. Converging evidence has been provided in brain imaging studies of humans (corresponding human areas are Brodmann areas 44 and 6, posterior parietal lobe, and superior temporal sulcus). When humans observe a facial action that is within their repertoire (e.g., human or monkey lip smacking), blood flow increases in the premotor cortex. However, when a facial action is observed that is outside the human repertoire (e.g., barking), activation of the visual, but not premotor, cortex occurs (Buccino *et al.* 2004). The human mirror system appears to be more flexible than the monkey's in that it responds to a broader range of actions, including mimed ones in which no goal object is present, as well as to the visual presentation of manipulable objects (Grèzes, Armony *et al.* 2003).

If motor resonance occurs during the direct observation of actions, then one might also hypothesize that it is used in the understanding of actions that are conveyed via language (e.g., Gallese and Lakoff 2005; Rizzolatti and Arbib 1998). This hypothesis predicts that exposure to words denoting or sentences describing actions or objects associated with actions would result in motor resonance. Neuroimaging (for a review, see Pulvermüller, Chapter 6, this volume) as well as behavioural research (e.g., Bub *et al.*, 2008; Glenberg and Kaschak 2002; Tucker and Ellis 2003; Zwaan and Taylor 2006) supports this prediction (see Fischer and Zwaan, 2008, for an extensive review of this literature).

9.4 **Presonance**

What is all this activation good for? It allows organisms to anticipate to a certain degree upcoming states of the environment and of the organism itself. Elsewhere, we have called this anticipatory skill, *presonance* (Zwaan and Kaschak, in press). The notion of presonance captures two components of anticipation: prediction and resonance. Experiential traces are activated by way of resonance with a stimulus in a manner that allows the organism to maximize its ability to anticipate future states of the environment and of itself. Presonance should not be thought of as slow and deliberate forecasting. Rather, it should be thought of as *fluency* – it is fast and involuntary. The idea of anticipation as a highly adaptive cognitive mechanism has been proposed by various other researchers (e.g., Berthoz 2000; Knoblich and Wilson 2005; Wolpert *et al.* 2003; Zacks *et al.*, in press).

In language, presonance takes on a dual form. Anticipatory processes operate at two different levels. First, they operate at the linguistic level. For example, co-articulation is an anticipatory process at work in language production. The way in which a phoneme is articulated is influenced by the gesture our articulatory system makes in anticipation of the phonemes that follow. There is a long history in psycholinguistics on research into syntactic strategies which allow the language processing system to anticipate upcoming constituents, thereby minimizing processing effort, except in relatively rare cases of syntactic ambiguity (e.g., Frazier 1979). Similar resource-efficient anticipatory strategies have been identified at the level of connected discourse (e.g., Kintsch and van Dijk 1978; van Dijk and Kintsch 1983). The language processing system uses these strategies to anticipate referents in upcoming clauses, thus facilitating the processing of upcoming clauses.

The idea that language processing involves the rapid use of information to anticipate likely next states in the linguistic input is consistent with constraint-satisfaction views of language processing (e.g., MacDonald *et al.* 1994; McRae *et al.* 1998; Jurafsky 1996). According to constraint satisfaction models, sentence processing proceeds by activating many possible interpretations for the sentence. These interpretations compete for activation on the basis of probabilistic information from the comprehender's experience with language. The likelihood of a particular word being used in a particular syntactic function (MacDonald *et al.* 1994), the likelihood of a particular syntactic structure being used (Jurafsky 1996), the preceding context (e.g., van Berkum *et al.* 2005; Spivey and Tanenhaus 1998), and other such factors simultaneously constrain the unfolding interpretation of the sentence.

The second level at which anticipatory processes operate is that of the referential situation. Comprehenders generate anticipations on their experience with events in the real world, for example, with regard to the chronology and temporal contiguity of the described events (e.g., Zwaan 1996), with regard to the goals and plans of story protagonists (e.g., Graesser *et al.* 1994), and with regard to a number of other situational dimensions (Zwaan *et al.* 1995; Zwaan and Radvansky 1998). Shifts on these dimensions lead to increases in on-line processing load (Zwaan *et al.* 1995; Zwaan *et al.* 1998), presumably because they bring about a decrease in predictability.

Two questions can be asked regarding the relation between presonance and language comprehension: (1) what can presonance do for language comprehension, and (2) what can language comprehension do for presonance? A plausible answer to the first question is: to facilitate the processing of incoming information, thereby enhancing the fluency of the comprehension (or production) process and minimizing the demand on cognitive resources. A plausible answer to the question of what language comprehension can do for presonance might be: to acquire, without any physical danger to the organism, anticipatory mechanisms that can be used in real-world situations. Suppose someone tells you 'Don't eat those red berries or you'll get very sick.' This knowledge allows you to anticipate what will happen when you eat the berries, and will in all likelihood prevent you from trying to eat them in the first place. Importantly, language comprehension can result in anticipatory processes that operate at a much finer level of granularity. This is a promising way of viewing the role of language comprehension because it provides an answer to the question: what is comprehension for?[3]

9.5 **Is comprehension ornamental or instrumental?**

Even if one is convinced by the rapidly growing amount of empirical evidence that perceptual and motor representations are activated during language use, several significant obstacles to embodied theories of language use remain. I will address two of them.

Even if perceptual and motor activation occur routinely during language comprehension, it does not immediately follow that they are *necessary* for comprehension.

---

[3] Glenberg (1997) raised – and proposed an answer to – the same question about memory.

They could be ornamental, or epiphenomenal, rather than instrumental to comprehension. At the theoretical level, the epiphenomenon question can be tackled by making an appeal to parsimony. If a theory based on perceptual and motor representations can account for all the findings that amodal, abstract, and arbitrary (AAA) symbol systems can account for, then why should we postulate the latter, given the wealth of evidence for the former and the paucity of evidence for the latter (Barsalou 1999)? Moreover, if perceptual and motor representations are routinely activated during comprehension, but do not play a meaningful role, we can ask, echoing Wilson and Knoblich (2005): 'What is the purpose of this neurological extravaganza?' It does not seem theoretically parsimonious to assume that there are AAA symbols that are doing all the work and that the embodied representations are just along for the ride. Moreover, no brain region has been found that features as the locus for converting analogue perceptual information into abstract representations (Barsalou 1999).

However, none of this should be taken to imply that lexical representations and the associations between them do not have a role to play in an embodied view of language use. Lexical representations are just another type of experiential trace. Moreover, lexical (and sublexical) associations augment comprehension in at least three ways. First, they enhance our ability to anticipate upcoming information – to presonate – by providing an additional stream of anticipatory information. Not only do we anticipate upcoming information by activating relevant situational information, we also generate anticipations based on lexical associations and typical constituent sequences. Second, lexical associations allow us to create new experiential traces by way of referential bootstrapping. For example, linguistic descriptions help us form experiential traces of unfamiliar animals, such as okapis, or nonexistent animals, such as griffins, by instructing us how to use L–L combinations to create R–R combinations via the already stored L–R associations. Third, L–L associations lower the demands on attention and short-term memory. For example, if we have to memorize a sequence of actions to be performed, it may be efficient to use the associated words as a shorthand.

9.6 **What's next?**

Several questions should be high on the research agenda for the coming years. Among them are the following:

(1) in addition to showing that perceptual and motor representations are activated in language comprehension, we need to specify when and how this occurs,

(2) we need to examine whether and how perceptual and motor simulations interact in language comprehension,

(3) we need to develop theories about different levels of comprehension. For example, deeper comprehension should be attainable if we have the described actions in our repertoire than if we do not, and

(4) we need to examine more closely the role of prior experience (not just declarative knowledge) in language comprehension.

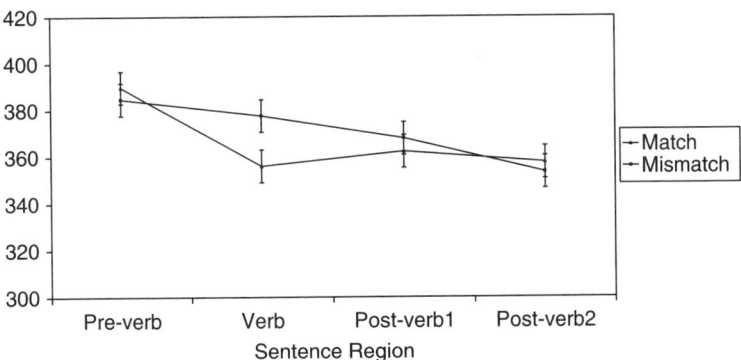

Fig. 9.2

In my laboratory, we have begun addressing some of these questions. I will conclude with a brief description of a series of experiments aimed at examining the waxing and waning of motor resonance during sentence comprehension. Glenberg and Kaschak (2002) showed that motor resonance occurs when subjects make sensibility judgments about sentences ('does this sentence make sense, yes or no?'). However, this does not directly implicate motor resonance in comprehension. In order to address this question, we asked subjects to read sentences one segment (two or three words) at a time by rotating a knob, with every four degrees of rotation resulting in the presentation of a new segment (Zwaan and Taylor, 2006; experiment 4).[4]

The critical sentences implied a manual rotation action, for example, 'After/ lighting/ the candles/ for the/ romantic/ evening,/ he/ dimmed/ the/ lights.' Each sentence had a critical region ('dimmed' in the example). We hypothesized that reading times on the critical region should be longer if the actual and implied rotation were in different directions – for example, turning down the volume requires counterclockwise motion and so should be read more slowly when read via clockwise rotation than when read via counterclockwise rotation. The idea that motor resonance should occur immediately is consistent with constraint-based theories of sentence comprehension and with the view outlined here. Also of interest was the question of what would happen to motor resonance in the regions after the critical region – an article and a noun in all critical sentences. Figure 9.2 shows the reading times per segment for the critical sentences, the regions before the verb having been collapsed into one.

The results show motor resonance first appearing on the critical verb and then disappearing after the verb. The immediate occurrence of motor resonance relative to the semantic content of the sentence was in accordance with our expectations. However, the

---

[4] Reading by turning a knob at first sight may seem awkward. However, as pilot testing revealed, in fact it feels surprisingly natural, especially compared with advancing through the sentence via key presses.

disappearance of the effect after the verb is intriguing. One potential explanation of this finding is that motor resonance only occurs when the action is what is focused on by the sentence: we call this the *linguistic focus hypothesis* (LFH). In a recent series of experiments (Taylor and Zwaan, 2008), we put the LFH to the test. The sentences in the previous experiment all shifted attention away from the performance of the action itself to the acted-upon object (e.g., turning back the clock, dimming the light, screwing in a lightbulb, turning down the volume).[5] According to the LFH, if a sentence maintains focus on the action itself, for example, by modifying it with an adverb, motor resonance should extend beyond the verb. We tested this prediction by using sentence pairs of the following type: 'The runner/ was very/ thirsty./ A fan/ handed him/ a bottle/ of cold/ water/ which he/ opened/ quickly.' According to the LFH, motor resonance should now occur on both the verb and the adverb, and the results support this prediction (Taylor and Zwaan, 2008; experiment 1).

We next performed an even stronger test of the LFH, replacing the action-modifying adverbs from the previous experiment with subject-modifying adverbs. Consider for example: 'The runner/ was very/ thirsty./ A fan/ handed him/ a bottle/ of cold/ water/ which he/ opened/ eagerly.' The adverb in this sentence tells us more about the runner's mental and physical state than about the manner in which the action is performed. Therefore, the LFH predicts that motor resonance should occur on the verb, but not on the adverb. The results support this prediction (Taylor and Zwaan, 2008; experiment 2).

We believe that these, and eventually much more fine-grained analyses, are needed to achieve a better insight into how language orchestrates mental simulations in the comprehender.

9.7 **Discussion**

To summarize, I have tried to outline a view of language comprehension that is decidedly embodied. However, I have also tried to show that this view is not as radical a departure from earlier disembodied views as might be believed at first sight. There still is an important role for lexical representations and their associations. However, these are now conceptualized as multimodal embodied mental representations, or experiential traces. I have identified presonance as a crucial mechanism in motor control, action observation, and language comprehension. Finally, I have outlined several topics that should be high on the research agenda, and have briefly sketched how we are currently attempting to address one of them. These activities lead me to conclude that instead of asking *whether* cognition, and therefore language comprehension, is embodied, it is much more fruitful to ask *how* it is embodied.

Debate

The following is a simulated discussion of this chapter. Because of an unfortunate snafu, Rolf Zwaan was unable to attend the meeting in Garachico. The editors asked

---

[5] See MacWhinney (2005) on grammatical constituents and perspective shifts.

Max Louwerse to prepare questions of the sort he would have asked if the paper had been presented at Garachico, and we asked Rolf Zwaan to respond to those questions.

Max Louwerse: Overview papers like these remind me yet again how innovative the embodied cognition literature is. I am glad that you agree that the question *how* language comprehension is embodied is more fruitful than the question *whether* language comprehension is embodied. I think, in part, thanks to the embodiment paradigm reigning over the last decade, the less fruitful question has already been answered: language comprehension can be embodied. We may, however, disagree on the modal auxiliaries: *can* language comprehension be embodied or *must* it (always) be? You make the argument that (L–R) associations are formed between linguistic traces and referential traces, which establish the link between symbolic processing and embodied processing. You also argue for R–R associations and L–L associations. Computational linguistic techniques like LSA illustrate these L–L associations. These L–L associations are important but merely a by-product of R–R associations. And that's where our points of view start to differ. I would therefore like to raise four questions.

First, if Hebbian learning explains L–L, R–R, and L–R associations, why is it the case that for every linguistic trace a referential trace must be activated in order to assign meaning? We already agreed that, in principle, symbol grounding is needed. What is your view on a hypothesis that would state that some L–R associations are formed, which then allow meaning to be bootstrapped through L–L associations? In other words, could it be the case that some symbol grounding may be needed under certain circumstances, but not all symbol grounding under all circumstances?

Rolf Zwaan: One important question that needs to be addressed first is whether R representations are activated routinely during language processing (production and comprehension, but in my chapter I focus on comprehension). My answer to this question would be 'yes.' Evidence is coming out all of the time that supports this view. I believe you're asking whether this activation is necessary for comprehension. One way to examine this is to see what happens if the relevant brain systems cannot be used for comprehension, for example, because they are otherwise engaged (e.g., in a dual-task paradigm) or because they are incapacitated (either permanently as a result of lesions or temporarily as a result of the administration of transcranial magnetic stimulation). Absolutely key here is to use or develop comprehension tests that are sufficiently sensitive. Researchers are beginning to address this issue, but it is still too early in the game to confidently draw conclusions, although there certainly is some compelling evidence already that provides an affirmative answer to the necessity question. Your hypothesis 'that some symbol grounding may be needed under certain circumstances, but not all symbol grounding under all circumstances' doesn't seem very testable to me. In order to begin answering it, you would have to specify the conditions under which you expect symbol grounding to be necessary and those under which you do not expect symbol grounding to be necessary.

Louwerse: Second, you argue that because perceptual and motor representations are routinely activated during comprehension, they must play a meaningful role ('what would otherwise be the purpose of this neurological extravaganza?'). I would like to pose the same question to you: if language structures show to encode perceptual and motor representations and we routinely use language, they must play a meaningful role in cognition – what would otherwise be the purpose of this linguistic extravaganza? Are these linguistic structures simply a by-product of embodied cognition, or are they in fact used by comprehenders as a mnemonic shortcut to meaning?

Zwaan: I'm not sure I understand what you mean when you say that 'language structures ... encode perceptual and motor representations.' The way I conceptualize it, language structures activate perceptual and motor representations. They therefore play a very important role in comprehension, because they orchestrate the activation and integration of these representations.

Louwerse: Third, you allude to the claim that the 'theory based on perceptual and motor representations can account for all the findings that AAA symbol systems can account for.' We could discuss whether amodal and abstract symbol systems like the language system are truly arbitrary, but I have a more mundane question. I have not found evidence that embodiment theories can account for all the findings that AAA symbol systems can account for. I will not give an overview of the evidence for AAA symbol systems (see Landauer *et al.* 2007), but the two studies I know that attempt to account for abstract words are the ones you refer to: Zwaan and Madden (2005) and Barsalou and Wiemer-Hastings (2005). To call these exploratory studies (I use Barsalou and Wiemer-Hastings' description) evidence that embodiment theories can account for all the findings that AAA symbol systems can account for is a stretch. Moreover, I am tempted to make the opposite argument: in various cases my lab has found evidence that language structures can explain embodiment findings (see Louwerse 2007; Louwerse et al. 2006, Louwerse and Jeuniaux, Chapter 15, this volume). So what would be your view on a position stating that a theory based on perceptual and motor representations can account for some of the findings that AAA symbol systems cannot account for, and that a theory based on amodal and abstract symbolic representations can account for other findings that perceptual and motor representations cannot account for?

Zwaan: The easy answer would be that amodal models account for concrete and abstract representations equally badly. They merely transcode the input. For sure, embodied frameworks are currently doing a much better job accounting for concrete concepts than they do for abstract concepts. But they have made promising ways to address this shortcoming, Barsalou's approach being one of them. But you are obviously right that a lot more work is needed. I strongly disagree with your self-assessment that the research you report in the book chapters you mention can explain embodiment findings that have been published in the literature. What your LSA analyses do, and cleverly so, is mimic the output of selected embodiment experiments. This does not imply that they mimic the underlying cognitive processes. To me, this is the fatal flaw in your approach. Your reasoning amounts to the claim that Deep Blue provides a model of a human chess grandmaster.

Louwerse: Finally, I would like to conclude with a broader, overarching question. Psycholinguistic research was dominated by symbolic theories in the 1960s to the 1990s. It now seems to be dominated by embodiment theories, from the 1990s to 2007. Are we perhaps just dealing with mood swings from our colleagues, who move explanations from one (symbolic) side of the spectrum to the other (embodied) side, while in all truth the answer lies – blatantly obvious – right in the middle?

Zwaan: I'm not sure I agree that psycholinguistic research is 'dominated by embodiment theories.' The situation is quite the opposite, actually. In fact I'd go so far as to say that even the name 'psycholinguistics,' with its focus on syntax, is inherently nonembodied. If we take the area of cognition more generally, however, it is clear that there is an increasing interest in embodiment. In large part this is because of the influence of neuroscience, which keeps on piling up evidence about cognitive processes that seem difficult to explain from a traditional framework, but that can much more easily be explained by embodied theories. However, I'd still hesitate to say that the cognitive literature is dominated by embodiment theories.

    I don't think we're dealing with mood swings. Rather, I think we're dealing with a larger-scale change in the field. Symbolic theories from the 1960s to the 1990s were inspired by the computer metaphor of cognition and by developments in computer science and linguistics. The brain was largely irrelevant to these theories. The advent of neuroimaging techniques, which provide a window on 'the brain in action' as it performs cognitive tasks, has made it increasingly clear that theories of cognition cannot simply ignore the brain and that the brain is not neatly organized in different modules, each with their own jobs to do, as traditional cognitive theories would have it. So I see the change to embodied approaches as a biologically rather than linguistic-philosophically inspired one. I suspect that this trend will continue.

Author note

This research is supported by National Institutes for Health grant MH-63972 and National Science Foundation grant BCS-0446637. I thank Tobias Richter, Larry Taylor, and Rich Yaxley for helpful comments on a previous version. Please address all correspondence regarding this paper to: Rolf A Zwaan, Department of Psychology, Erasmus University, NL-3000 DR Rotterdam, The Netherlands. Email may be sent to: zwaan@fsw.eur.nl.

References

Barsalou LW (1999). Perceptual symbol systems. *Behavioural and Brain Sciences*, 22, 577–660.

Barsalou LW, Wiemer-Hastings K (2005). Situating abstract concepts. In D Pecher, R Zwaan, Eds. *Grounding Cognition: The Role of Perception and Action in Memory, Language, and Thought* (pp. 129–63). New York, NY: Cambridge University Press.

Beilock SL, Holt LE (2006). Embodied preference judgments: can likeability be driven by the motor System? *Psychological Science* **13**, 694–701.

Berthoz A (2000). *The Brain's Sense of Movement*. Cambridge, MA: Harvard University Press.

Bub DN, Masson MEJ, Cree GS (2008). Evocation of functional and volumetric gestural knowledge by objects and words. *Cognition* **106**, 27–58.

Buccino G, Binkofski F, Fink GR *et al.* (2001). Action observation activates premotor and parietal areas in a somatotopic manner: an fMRI study. *European Journal of Neuroscience*, 13, 400–4.

Damasio AR (1999). *The Feeling of What Happens: Body and Emotion in the Making of Consciousness.* New York, NY: Harcourt Brace.

Damian MF, Bowers JS (2003). Effects of orthography on speech production in a form-preparation paradigm. *Journal of Memory and Language*, 49, 119–32.

DiPellegrino G, Fadiga L, Fogassi L, Gallese V, Rizzolatti G (1992). Understanding motor events: a neurophysiological study. *Experimental Brain Research*, 91, 176–80.

Dijkstra K, Kaschak MP, Zwaan RA (2007). Body posture affects retrieval from autobiographical memory. *Cognition*, 102, 139–49.

Fischer MH, Zwaan RA (2008). Embodied language: a review of the role of the motor system in language comprehension. *Quarterly Journal of Experimental Psychology* **61**, 825–850.

Flanagan JR, Johansson RS (2003). Action plans used in action observation. *Nature*, 424, 769–71.

Frazier L (1979). *On Comprehending Sentences: Syntactic Parsing Strategies* [doctoral thesis]. Storrs, CT: University of Connecticut.

Gallese V, Lakoff G (2005) The brain's concepts: the role of the sensory-motor system in reason and language. *Cognitive Neuropsychology*, 22, 455–79

Glenberg AM (1997). What memory is for. *Behavioural and Brain Sciences*, 20, 1–55.

Glenberg AM, Kaschak MP (2002). Grounding language in action. *Psychonomic Bulletin and Review*, 9, 558–65.

Graesser AC, Singer M, Trabasso T (1994). Constructing inferences during narrative text comprehension. *Psychological Review*, 101, 371–95.

Grezes J, Armony JL, Rowe J, Passingham RE (2003) Activations related to 'mirror' and 'canonical' neurones in the human brain. *NeuroImage*, 18, 928–37.

Grezes J, Tucker M, Armony J, Ellis R, Passingham RE (2003). Objects automatically potentiate action: an fMRI study of implicit processing. *European Journal of Neuroscience*, 17, 2735–74.

Hintzman DL (1986). Schema-abstraction in a multiple trace model. *Psychological Review*, 93, 411–28.

Jurafsky D (1996). A probabilistic model of lexical and syntactic access and disambiguation. *Cognitive Science*, 20, 137–94.

Keysers C, Perrett DI (2004). Demystifying social cognition: a Hebbian perspective. *Trends in Cognitive Sciences*, 8, 501–7.

Kintsch W, Van Dijk TA (1978). Toward a model of text comprehension and production. *Psychological Review*, 85, 363–94.

Köhler E, Keysers C, Umiltà MA, Fogassi L, Gallese V, Rizzolatti G (2002). Hearing sounds, understanding actions: action representation in mirror neurons. *Science*, 297, 846–8.

Landauer TK, Dumais ST (1997). A solution to Plato's problem: the latent semantic analysis theory of the acquisition, induction, and representation of knowledge. *Psychological Review*, 104, 211–40.

MacDonald MC, Pearlmutter NJ, Seidenberg MS (1994). The lexical nature of syntactic ambiguity resolution. *Psychological Review*, 89, 483–506.

MacDonald MC (1994). Probabilistic constraints and syntactic ambiguity resolution. *Language and Cognitive Processes*, 9, 157–201.

MacWhinney B (2005). The emergence of grammar from perspective taking. In D Pecher, RA Zwaan, Eds. *Grounding Cognition: The Role of Perception and Action in Memory, Language, and Thinking* (pp. 198–223). Cambridge: Cambridge University Press.

McRae K, Spivey-Knowlton MJ, Tanenhaus MK (1998). Modeling the influence of thematic fit (and other constraints) in on-line sentence comprehension. *Journal of Memory and Language*, 38, 283–312.

Rieger M (2004). Automatic keypress activation in skilled typists. *Journal of Experimental Psychology: Human Perception and Performance*, 3, 555–65

Rizzolatti G, Arbib MA (1998). Language within our grasp. *Trends in Neurosciences*, 21, 188–94.

Sadoski M, Paivio A (2001). *Imagery and Text: A Dual Coding Theory of Reading and Writing*. Mahwah, NJ: Erlbaum.

Seidenberg MS, Tanenhaus MK (1979). Orthographic effects on rhyme monitoring. *Journal of Experimental Psychology: Human Learning and Memory*, 5, 546–54.

Spivey MJ, Tanenhaus MK (1998). Syntactic ambiguity resolution in discourse: modeling the effects of referential context and lexical frequency. *Journal of Experimental Psychology: Learning, Memory, and Cognition*, 24, 1521–43.

Spivey MJ, Richardson DE (in press). Language embedded in the environment. In P Robbins, M Aydede, Eds. *The Cambridge Handbook of Situated Cognition*. Cambridge: Cambridge University Press.

Stone GO, Vanhoy M, Van Orden GC (1997). Perception is a two-way street: feedforward and feedback phonology in visual word recognition. *Journal of Memory and Language*, 36, 337–59.

Taylor LJ, Zwaan RA (2008). Linguistic focus and motor resonance. *Quarterly Journal of Experimental Psychology* **61**, 896–904.

Umiltà MA, Köhler E, Gallese V *et al.* (2001). 'I know what you are doing': a neurophysiologycal study. *Neuron*, 32, 91–101.

Van Berkum JJA, Brown CM, Zwitserlood P, Kooijman V, Hagoort P (2005). Anticipating upcoming words in discourse: evidence from ERPs and reading times. *Journal of Experimental Psychology: Learning, Memory, and Cognition*, 31, 443–67.

Van den Bergh O, Vrana S, Eelen P (1990). Letters from the heart: affective categorization of letter combinations in typists and nontypists. *Journal of Experimental Psychology: Learning, Memory, & Cognition*, 16, 1153–61.

Van Dijk TA, Kintsch W (1983). *Strategies of Discourse Comprehension*. New York, NY: Academic Press.

Wilson M, Knoblich G (2005). The case for motor involvement in perceiving conspecifics. *Psychological Bulletin*, 131, 460–73.

Zwaan RA (1994). Effect of genre expectations on text comprehension. *Journal of Experimental Psychology: Learning, Memory, and Cognition*, 20, 920–33.

Zwaan RA (1996). Processing narrative time shifts. *Journal of Experimental Psychology: Learning, Memory, and Cognition*, 22, 1196–207.

Zwaan RA (2004). The immersed experiencer: toward an embodied theory of language comprehension. In BH Ross, Ed. *The Psychology of Learning and Motivation, Vol. 44* (pp. 35–62). New York, NY: Academic Press.

Zwaan RA, Kaschak MP (in press). Language comprehension as a means of 're-situating' oneself. In P Robbins, M Aydede, Eds. *The Cambridge Handbook of Situated Cognition*. Cambridge: Cambridge University Press.

Zwaan RA, Langston MC, Graesser AC (1995). The construction of situation models in narrative comprehension: an event-indexing model. *Psychological Science*, 6, 292–7.

Zwaan RA, Madden CJ (2005). Embodied sentence comprehension. In D Pecher, RA Zwaan, Eds. *Grounding Cognition: The Role of Perception and Action in Memory, Language, and Thinking* (pp. 224–45). New York, NY: Cambridge University Press.

Zwaan RA, Magliano JP, Graesser AC (1995) Dimensions of situation-model construction in narrative comprehension. *Journal of Experimental Psychology: Learning, Memory, and Cognition*, 21, 386–97.

Zwaan RA, Radvansky GA (1998). Situation models in language comprehension and memory. *Psychological Bulletin*, 123, 162–85.

Zwaan RA, Radvansky GA, Hilliard AE, Curiel JM (1998). Constructing multidimensional situation models during reading. *Scientific Studies of Reading*, 2, 199–220.

Zwaan RA, Taylor LJ (2006). Seeing, acting, understanding: motor resonance in language comprehension. *Journal of Experimental Psychology: General*, 135, 1–11.

Chapter 10

Defining embodiment in understanding

Anthony J Sanford

10.1 Introduction

The present debate is cast in terms of an opposition between purely symbolic and purely embodied notions of meaning and cognition. In this paper, I suggest that the meaning of words is grounded in action and perception. Further, I suggest that knowledge is largely, if not completely, organised in situation-relevant packages, and thus is 'proxy-situated', meaning that it, too, is grounded. More difficult is the notion of embodiment entering into real-time processing, and what constitutes embodied real-time processing. I have tried to examine three levels at which this might take place, and to open up questions of what these levels might imply. In fact, it is likely that abstraction occurs as a continuum, and that these three levels are simply landmarks in the continuum. I have also tried to establish how some of the existing literature relates to these ideas, including sketches of further empirical questions. I end with some remarks about how embodied cognition might relate to the portrayal of emotion and related aspects of writing.

My interim conclusion is that some level of abstraction (which may be modality-free) is necessary for processing, and that some level of embodiment in some circumstances is also likely. However, in general, existing experiments and theoretical frameworks underspecify how processing is supposed to work.

10.2 Grounding and situations

Because I shall consider representations and procedures at different levels of abstraction from actual bodily events, I shall start with some observations about abstract notions (such as features) that have a recognised place in the 'abstract' treatment of semantics, and show how the origins of these ideas were in fact grounded in situatedness. I shall then discuss the notion of *scenario* (Sanford and Garrod 1981, 1998), showing how it relates to grounding and situatedness. The message is the same for both: meaning is grounded in our interactions with the world, and so is significance. I shall use the notion of scenario in the remainder of the paper, which is an attempt to look at some of the trickier issues encountered in trying to characterize embodment.

10.2.1 **Words**

The idea that the meaning of things, including sentences, can be represented in a way that has nothing to do with situatedness and embodiment is essentially nonsensical. In some of the early work in linguistic approaches to semantics, the idea was to represent the meaning of words by decomposing those words into other, simpler expressions that when added together gave the meaning of the original expression. The way to do this was to compare words with each other using that method to pinpoint distinguishing features between words. By iterating this procedure, in principle a complete set of features for everything could be found. But, in practice, a very important restriction of the technique was used: only words within so-called *semantic fields* (see Miller and Johnson-Laird 1976 for examples and a discussion) were compared.

The notion of semantic fields is interesting, referring to terms that have a similarity in meaning (e.g., sets of minimal pairs). But the fact is that a semantic field – like that of kinship terms (Goodenough 1965), verbs of motion (Miller 1969) and terms for containers (Lehrer 1970) – refers to a set of individuals among whom it is important to make distinctions for social and living purposes. In brief, a semantic field is structured to represent a useful set of labels to refer to important and useful social and biological distinctions. In this way, semantic fields exist because of our need as humans to divide up useful aspects of nature into distinguishable entities and actions. They represent a kind of situated knowledge (examined in more detail below). Semantic fields, the origin of feature notions, are intrinsically grounded in real-life, concrete situations. Of course, this is very different from a more willy-nilly approach of ascribing attributes to representations of entities without making principled distinctions.

Note also that a semantic field such as 'kinship' really requires a certain amount of *interdefinition* – of one word in terms of others – simply so that the structure of the semantic field can be made clear. If 'McDonald is the son of Donald', then 'Donald is the father of McDonald.' If we don't know that, then we don't understand how to use the terms 'father' and 'son' at all. The need for interdefinition is ubiquitous: for adequate communication, we need to know that left is the opposite of right, and that north is the opposite of south. But we do not normally need to engage in on-line embodied cognition to use these words or these facts. Our two points are:

(a) Semantic fields are socially useful clusters of words that are typically learnt as part of the process of differentiating aspects of the world. They are thus 'situated' in some sense.

(b) There is a need for interdefinition in order to use words appropriately for communication. Interdefinition is not embodied. It is purely conceptual.

Of course, we are not saying that this is anything like adequate for capturing meaning. Although 'dog' and 'cat' may belong to a semantic field (such as 'animals small children encounter in reality or in media'), and differentiation of them by habits and nature is important knowledge, what they are must be rooted in perceptual representations of them, still and in motion. Even the dictionary definitions of many words show that the ultimate test of a word requires contact with how our bodies interact with the outside world.

For instance, the Oxford English Dictionary defines 'north' as 'Towards, or in the direction of that part of the earth which is most remote from the midday sun' (*Authors note: this is only true in the northern hemisphere*), and 'left' as: 'The distinctive epithet of the hand which is normally the weaker of the two,, and of the other parts on the same side of the human body. Hence also what pertains to the corresponding side of any other body or object.'

So, meaning has to be rooted in our physical (and social) interactions with the world. This is not a new position for psychology; however, the claim that (lexical) meaning is exclusively embodied must be false, since relationships between terms need to be known, and these are essentially abstractions.

10.2.2 Situation representations (scenarios)

Theories of comprehension that do not assume the existence of knowledge organised more generally along situations from everyday life have problems in accommodating many aspects of language comprehension. Sanford and Garrod (1981, 1998; Sanford and Moxey 1995, 1999) have presented arguments for the utilization of situation-specific packages of background knowledge (scenarios) in the service of comprehension. This is a schema-based notion, and is therefore related to other schema-based ideas, such as Schank and Abelson's (1977) theory of scripts as information packages that may be used in comprehension. In comparing our theory with other accounts, we particularly paid attention to how it differed from the idea that people parse texts into propositions as a fundamental building block of the product of comprehension. For us, the fundamental operation was relating an incoming utterance to situation-specific background knowledge (where possible) in order to facilitate comprehension. Our principal arguments are laid out in the references cited above, and will not be repeated here.

Here, we want to stress some characteristics of the scenario notion that have not been aired very much, but that are relevant to the present debate. First, the notion of situation-specific knowledge is directly related to the notion of *situated cognition*. Here, we use Wilson's (2002) definition of situated cognition, which excludes a large amount of human cognition (e.g., Grush 1997). Situated cognition is taken to mean cognition that takes place when one is actually in a particular situation.

For example, reading is situated in the act of reading itself. What we read about does not directly constitute situated cognition. However, we need our interpretations of what we are reading to relate to situated cognition: this we will call *proxy-situatedness*.[1] If we are reading about a familiar situation, such as moving a large piece of furniture up some stairs, then our understanding is determined not just by the meanings of the words, but also by what we know about the situation itself, for instance, that weight and size, and our own strength, are crucial elements in understanding what is entailed by the exercise. Indeed, it has been amply demonstrated that different aspects of an entity, such as piano,

---

[1] Semantic fields, to the extent that they are thought of as networks of inter-related concepts, also support proxy-situated cognition.

are emphasized in situations where they occur (e.g., Tabossi 1988). The idea of a scenario is that it is an organizational structure that deals in the idea currency that is relevant to the situation that it models, and nothing else at all. This automatically gives rise to another important property in that scenarios also specify levels of granularity of representation that are appropriate for understanding the essentials of a situation (Sanford and Garrod 2005). For instance, 'slipping on ice' is a familiar and undesirable winter situation, as is 'a car with no snow-tyres skidding into something.'

Our knowledge of these particular situations means that we don't (typically) see the explanation of someone slipping on the sidewalk in winter as being due to a reduction in friction through a thin layer of water forming between a shoe or a tyre and the ice underneath, which leads to aquaplaning. Rather, we see the explanation as slipping on ice because it's winter. The global physics explanation is something that is not appropriate to our understanding, being at a much finer grain of explanation. Rather, our normal understanding is typically about taking care, choosing the right footwear, and being aware of the consequences of a fall or skidding. Scenarios, by reflecting what is key in situations, keep explanatory possibilities from exploding in every direction, from the 'Will of God' to the physics of ice.[2] Of course, a natural part of learning about this situation is one's own experiences of slipping on ice, and so part of the situation representation almost certainly includes a pointer to this experiential element. Our claims here are:

(a) Knowledge is organized around situations, because situations are what we live in and act out our lives in.

(b) Scenarios constitute proxy-situated cognition. That is, on the basis of a mapping between a language description and a scenario, proxy-situated cognition can occur. This is because the constraints of reasoning employed in real situations apply by proxy in reasoning with scenarios.

At this stage, we are pretty easygoing about what a situation actually is. Some situations, like being in the presence of danger, although abstracted as knowledge will still be directly linked to physiological flight/fight mechanisms, so although they are abstracted, they should still give rise to bodily activity and emotions. Others, like the ice example, may trigger retrievable bodily experience when they are encountered, but they don't have to. Further, while some scenarios have an abstract character, others may well involve mappings onto sense-specific data (e.g., what a car looks like as it recedes into the distance, versus when it is coming towards you, or motor-plan data (e.g., riding a bike). A scenario's primary data structure may be to give you an emotional signal that a situation is dangerous (and these are situations that you really need representations of, for survival!).

---

2 One problem sometimes put forward about scenarios is that it seems that there simply needs to be too many of them for the account to be plausible. But how many is too many? One might just as well complain that the problem with object recognition is that there are too many objects to claim that we somehow have knowledge of all these objects. Learning situations is simply the most useful thing we can do, because next time around, the situation will be recognized.

To summarize, according to the scenario theory, basic comprehension takes place when we relate language input to proxy-situated information, and thus allow appropriate proxy-situated inferential activity and reasoning to take place. Sanford and Garrod (1981, 1998) have always assumed that the mapping process takes place as rapidly as possible, and that the whole of the situational information becomes easily available (in implicit focus) for use. This process of mapping is assumed to be a real-time component of comprehension, and a successful mapping is assumed to be the basis of feeling of understanding (at the most basic level).

So, comprehension on Sanford and Garrod's view is situated by proxy. Things that are important for situations are represented in scenarios, and addressing scenarios is a real-time component of comprehension.

10.3 What happens when we comprehend?

The remainder of this paper is concerned with one question: what real-time embodied processes are necessary for understanding? Three specific ways in which embodied processing could take place are examined, each one of them capable of finer definition. Since the picture we present is one of increasing levels of abstraction, the final one we discuss is scarcely distinguishable from the scenario notion outlined above. The examples below are divided into three classes, in which embodied activity is implicated in different ways, and where (proxy-situated) symbolic activity is implicated to different extents. In each example, we consider how bodily action is implicated as a real-time processing component of language understanding.

Let us spell out the issue. We have reserved the term 'situated cognition' for cognition undertaken in some situation, in real-time. For instance, the cognition involved in climbing up a rock buttress while doing it is situated, and takes place over the time of the climb. Reading about rock climbing is not the same as doing it, though to the extent that we can understand what we read, the language processing involved is proxy-situated. The question for embodiment is: do we make (at least some) use of the same mechanisms in reading about a rock climb as we do in executing it?

What does it mean to use embodied processing as a stage in cognitive or comprehension processes? What is on-line body-based processing? This is a tough issue, right at the core of the present debate, as I understand it. Here are some possibilities for relating embodied processes to cognition and understanding.

10.3.1 Full, real-time embodied program running

This case is the most extreme, in which to execute some cognitive task, bodily activity must be carried out in its entirety. It is hard to think of cases where this unequivocally applies to language, but here is one possibility. Consider the sentence, uttered to a car driver, 'Turn right at the next intersection.' How might this be understood? I shall call on personal experience here, because it is in fact the case that I have a problem with left and right. I simply do not know, without real-time embodied processing, which of my hands is my right hand. I simply do not possess this knowledge directly. When I took my driving test, I wrote a large 'R' on my right hand (thus externalizing my cognition). When I was told to turn right, I went towards the side where the large 'R' was. Normally, I just work out which hand I write with, by simulating a writing movement, and that's my right hand, and that defines which way I move the car. This is real-time embodied cognition, because I use up time doing the physical activity that enables comprehension. (Being given left/right commands in a salsa class is really disheartening).

As a second example, consider a problem that is not about language comprehension but is clearly real-time embodied cognition. Try answering the following: 'What letter precedes U in the alphabet?' If you are like most other people, you will find yourself mentally reciting or running through a portion of the alphabet (e.g., PQRSTU), and then discovering that T is the answer. Small children confronted with questions like this recite the alphabet out loud, and often recite right past the part where the relevant information could be found.

The same sort of problem arises when deciding whether two letters are in the correct order, as in: 'Are the following items in the correct order?: O, R.' When we first learn the alphabet, we learn to produce it as a string, and although later on we may know some orders more directly (e.g., abc, xyz, pq), for some portions of the string we never get beyond the basic skill of sequence production. In fact, inner speech is rate-limited to about eight syllables per second, which equates well with the maximum rate of external syllable production (Landauer 1962; Sanford 1971). When run-through occurs, it precisely determines the time to make a decision concerning the order of alphabetic items (Hamilton and Sanford 1978). This is an example of real-time embodied processing, because the bodily skill involved (saying the alphabet, albeit subvocally) takes up part of the total time to perform the cognitive operation in question. The phenomenon comes about because the main way we have the alphabet stored is as a motor program to say the alphabet.[3]

10.3.2 Real-time simulation running

Running a real program like the ones described above seem to me the most extreme examples of embodied cognitive processing. However, the idea of running a simulation of a real program poses similar issues for real-time resource consumption. Examples include mental rotation (Shepard and Metzler 1971) and, for most people, the task of answering the question 'How many windows does your house have (openings, not frames)?' One way of solving this is to simulate a trip around the inside of the house, counting as you 'go'. With this procedure, the simulation takes less time to run than does the actual act of walking around the house. So, a simulation can be run in less time than the execution of the activity on which it is based.

For language, the question is whether such simulation occurs during comprehension. For instance, when we understand 'John was cutting the hedge', do we simulate the action of cutting the hedge for a while – and if so, how long does that take, and what action

---

[3] Saying the alphabet is the same as other skills requiring continuous output that ends up being distributed discretely, such as playing the piano/guitar/clarinet. It is almost impossible to fix an error in the middle of the phrase without 'reaching it' by starting at the beginning of the phrase.

simulations are actually involved? To what extent is the simulation a stage in processing the sentence? There does not seem to be any really strong evidence for this position with respect to language comprehension. For instance, while experiments on interference between direction of movement in answering yes/no questions, and directions of movement implied by actions or perception seems to implicate some sort of embodied representation being activated, it is not clear to what extent these can be called real simulations.

Three important issues for characterizing simulation are the *locus*, *kinematics*, and *dynamics* of the action. Superficially, the evidence looks favourable for a view that embodiment and simulation may have a role to play in understanding the meaning of sentences, particularly the research into sentence-response incompatibility (e.g., Glenberg and Kaschak 2002; Zwaan and Taylor, 2006). For instance, Glenberg and Kashak (2002) found that sentences depicting movement away from the body (e.g., 'passing a book to someone') interferes with a response requiring a hand movement towards the body. This is taken to be evidence of embodied processing. Also, reading verbs that depict actions of using the hands cause increased activity in those areas of the brain controlling movements of the hands, while reading verbs that depict movements of facial muscles causes increased activity in the areas controlling those movements (e.g., Pulvermüller *et al.* 2001). This is compatible with the idea that language understanding has an embodied component, possibly related to simulation of movements by specific effectors. However, whether data like that of Glenberg and Kaschak shows that simulation is part of the act of understanding, in the senses of being a simulation that is a component of comprehension, is a very different question. Consider the following pair of sentences:

(a) John (or you)[4] kicked the ball to Bill.

(b) John (or you) threw the ball to Bill.

Now suppose that we carried out a study in which responses were made either by moving the feet (away or towards) or the hands (away or towards). If the incompatibility effect only occurs when the relevant body part is used (e.g., when the foot is used as the response effector when the verb is 'kick'), and not when the irrelevant part (hand) is used, then this would suggest that comprehension is indeed embodied in the sense of depending on specific effector systems. If, in contrast, the critical thing is the generalized direction of movement, a more conceptual notion of embodiment is required. The notion of direction of movement is only real-time embodied if it is tied to the particulars of the action (the effector and its mechanics). Lest it sound far-fetched that such a notion of embodiment would ever be correct, recall the example of the alphabet where specific effector systems underlie the ability to determine serial order.

Even if there is localization of possible simulation to the appropriate effector system, two further issues are easily seen as important in deciding just what simulation might be useful, these being the details of the kinematics and dynamics that might be captured

---

[4] It is unclear under what conditions a reader-centric representation would be appropriate, and hence a reader-centric representation (or simulation) of the movement involved, be that at an embodied or a more conceptual level.

in a simulation. Consider the simulations required to capture the actions 'gently push the drawer closed,' and 'forcefully slam the drawer closed.' These actions differ not only in velocity, acceleration, and duration (kinematics), but also in the force of the movement (dynamics). So, the question is, do simulations capture the kinematics and dynamics of the situation, or are they less embodied and more abstract? The same issue arises in the problem of simulating motor movement in general.[5]

If simulations genuinely capture the kinematics and dynamics, then several outcomes may ensue. For instance, if simulating a gentle movement, the requirement to push a lever hard may be an incompatibility, while the requirement to give a lever a delicate push may be incompatible with comprehending the notion of slamming. Furthermore, if there are differences in the time to simulate some gentle, slow movement than a fast movement, then perhaps a relationship between the speed of the movement and the speed of comprehension might be expected if the simulation is a real-time component of the act of understanding. This may only be demonstrable in principle in appropriately controlled eye-tracking studies, but we are not making really hard predictions here. What we are doing is trying to unravel what it might be necessary to specify if we want a theory of embodied understanding that takes simulation as a real-time component of the act of understanding. If the specific argument given above is not right, then we need a theory of what specific simulations are.

Our mention of kinematics and dynamics above raises the question of what constitutes an appropriate level of *granularity* in the simulation. For instance, maybe the crudest level of granularity is used for the two drawer examples given above, such that only the away-from-body directionality is captured and the manner is lost. This is certainly consistent with Glenberg and Kaschak's (2002) demonstration that even conceptual transfers show a directionality effect in the action compatibility effect (ACE) paradigm. If the level of granularity is only as crude as directionality, then this would be a very restricted form of simulation indeed. Furthermore, it still remains necessary to demonstrate that simulation at any level of granularity is a real-time component of comprehension.

10.3.3 Activating the start sites of potential actions

A third, and weaker, possibility is that when a sentence involving an action is understood, mechanisms are identified which might enable the action to be either simulated or carried out. This is like knowing which button to press to start a plane, but not actually having to press it. For instance, when hearing the sentence 'You throw a ball to Mary,' it might be the case that motor areas for executing and controlling throwing a ball are identified though the act of throwing may not be simulated. It is unlikely that the act of

---

[5] See commentaries on Grush's (1997) theory of emulation for a similar issue. Also, present evidence suggests, for instance, that the premotor cortex represents the goals of movements independent of their kinematics, while parietal regions encode information about the movement kinematics and the goal (e.g., Lestou *et al.* 2003; Lestou 2005). Furthermore, note that the original demonstration of the sentence-movement compatibility effect by Glenberg and Kaschak was very clear where abstract transfer was involved (e.g., 'You read a story to Mary' / 'Mary read a story to you').

throwing would be carried out. (However, children sometimes perform bodily actions of simulating flying when they hear a sentence like 'You'll be flying to America next week' – this seems to be support for activating potential actions, but it is not a property of normal adults).

This idea is one realization for the results of an experiment of de Vega *et al.* (2004). They showed that people find it difficult to 'understand' a sentence like 'While John painted the fence he cut the grass.' Their argument was that this is because we attempt to simulate (or visually simulate the perception of the action depicted) the two actions, and discover that they cannot be done together. Simulation would require at least identifying potential effector use, while visualization would entail 'seeing' that the two things cannot be done at once, presumably through simulating scenes in which both are tried and found to fail. Which of these process, if any, might be used is unclear. Can the results be explained without some sort of on-line embodied simulation? Certainly, the data fit with the present model, but the evidence is not really definitive since a proxy-situated representation (scenario) would specify the actions required. Moreover, in principle, the clash in the idea of using effectors for one action that are in use in another could be picked up directly. The evidence is simply not sufficiently well developed at the present stage. If on-line embodied cognition is part of the process of understanding, then it should take up part of the time of comprehension. To demonstrate this, it is necessary to specify the kind of bodily activity involved, and that comprehension is not 'complete' until this has happened. There is no data that I know of that does this.

It is necessary to specify what constitutes a simulation if that is part of the proposed process of comprehension. There is also the important question of whether a simulation is sufficiently fine-grained to merit being called 'embodied' rather than being some sort of an abstraction, even if that abstraction is originally grounded in a specific action or situation. If it is an abstraction, it may still be modality-specific, of course, but it is nonetheless an abstraction. Another important question relating to abstraction is whether the simulation is restricted to the body parts involved in carrying out the real thing. Even identifying the possible bodily effectors involved in an action may constitute a weak version of embodied processing. However, such a view is very close to the scenario theory view of (abstracted) proxy-situatedness, and Sanford and Garrod have always assumed that some scenarios may directly tap into movement or emotion.

10.4 **Balancing up: the role of simulation**

We suggest that most of the demonstrations of the role of embodiment in language processing are convincing in that they demonstrate that there is a role. However, we claim that there is no demonstration at the time of writing that the indicators of embodiment show that they are necessary, real-time components for comprehension. Rather, they show that much of cognition is grounded in embodied actions and perceptions. Indeed, we would argue that this is absolutely necessary, in that there is little point in having a cognitive system that has not evolved (both phylogenetically and ontogenetically) to deal in the currency of the situations and actions that make up our lives. The symbol grounding problem is not at issue, but it seems more needs to be done to specify just how embodiment enters into processing. For instance, it is a distinct possibility that embodied aspects of understanding are not real-time components of comprehension, but in some sense effects correlated with necessary processing.[6] As things stand, provided cognition is proxy-situated there appears to be no obvious reason, from a processing perspective, to bring on-line embodied processing into it generally, attractive though such an idea might be.

And why should bringing in embodied processing be at all attractive? In a nutshell, because our real understanding would be so much poorer without it, and often inadequate, inappropriate, and misleading. It could be the case that the embodied effects observed in many studies are like elaborative inferences – they add to the richness of our experience of understanding, and make that understanding deeper in the sense of a fuller interpretation. (Of course, in some cases they may be more central, but that needs to be demonstrated.) In some cases, the elaboration may be a truly vital process.

Consider the case of reading about an emotional event. Emotion is an excellent example of something where an embodied response is not only possible, but is the *raison d'être* of much writing (including film, TV, and drama) be it literary, popular pulp fiction, or political writing. Here, one would expect the emotion induced by an emotive sentence to interfere with incompatible subsequent input (as shown by Glenberg *et al.* 2005). Furthermore, work on brain responses to emotional written input seems quite a plausible line to pursue (e.g., Moll *et al.* 2005). However, while obtaining emotional reactions is a key aim of much writing, how to actually achieve it is not that straightforward, as witnessed by a large volume of internet literature on the topic for writers.

It is in the literary domain that embodiment takes on a whole different face. For instance, does 'calming' writing slow down reading? Does exciting reading speed it up? We still have to bite the following important bullet in our models of comprehension:

> When reading a story, we may 'experience' cold wind blowing in our face, the smell of stale beer, a kiss on our lips ... (Zwaan 1999, p.83).

Exploring this possibility is really exciting. But we suspect the word 'may' in the above quote is significant. The writer's art is to bring in these bodily experiences only when appropriate. It is now increasingly recognized that not all aspects of discourse are processed equally deeply. Sanford and Sturt (2002) report several situations in which semantic processing, in the sense of utilizing word meaning, can be shown to be shallow (see also Sanford and Graesser 2006). In a similar vein, Ferreira *et al.* (2002) have argued that representations may only be 'good enough' (i.e., detailed enough) for the purpose of communication.

Work in my laboratory has established that semantic analysis seems to be enhanced by certain features, for instance, formatting devices such as truncated 'mini-sentences' (Emmott *et al.* 2006) and italics (Sanford AJS *et al.* 2006), the use of linguistic and discourse focus (Sturt *et al.* 2004), and putting information into main rather than

---

6 This would at least explain the weak effects often observed with this paradigm, including the manifestations of effects in different measures in different studies. There is still much groundwork to be done.

subordinate clauses (Paul 2005). It has also been shown that high sentential processing loads appear to reduce depth of semantic analysis (Sanford AJS *et al.* 2005). It seems very plausible that the use of embodied representations during processing may be a function of these variables. It also seems very likely that forms of descriptions, and the artful use of adverbs and adjectives, would increase embodied processing.[7]

10.5 **Discussion**

While the case for knowledge being essentially situated and for word meaning being grounded in perception and action seems to be inescapable, the use of embodied cognition as a real-time component of comprehension ('embodied processing') is fraught with problems that have not yet been tackled. Some level of abstraction seems to be required. Clarification of researchers' ideas about this is clearly necessary, and the casual and enthusiastic use of the term 'simulation' must now be replaced by a more circumspect approach if we are to continue talking in terms of process models for comprehension. However, there is little doubt that, as users of language, a totally disembodied cognition is impossible. Even if many of the phenomena of embodiment that are currently being uncovered remain outside of the sequential process necessary for the basic understanding of a sentence, much writing and speaking is to do with engendering some sort of simulation of experience, and without it our lives would be poorer; survival might not even be possible. However, the ways in which embodied processing might enrich our ideas of understanding is only just being explored. Doubtless the questions raise here will soon be answered.

Debate

Friedemann Pulvermüller: I have a brief answer to one of your questions: what happens when we look at brain activation during sentences, such as 'He kicked the ball' or 'He threw the ball,' is the same as with action verbs in isolation. It has been recently published by our colleagues in Parma (in the *Journal of Cognitive Neuroscience*), that they found very similar areas of activation as we reported previously. So, it seems they've had primarily premotor activation whereas ours spread into primary motor cortex too. So there might be a little activation difference. Also, theirs wasn't as pronounced; it was like a little spot somewhere, but this is easy to explain by the more complex design and by the presence of more noise in the data. At least the data point in the direction that words in context do the same as those in isolation, but you are right, there might be contexts in which these mechanisms can be switched off.

Anthony Sanford: Yeah, OK.

Arthur Graesser: I agree with your position and, for the record, that a lot of the activity is epiphenomenal when you look at real text, because often you want to read and want

---

[7] Indeed, Zwaan and Taylor (2007) have very recently shown that motor compatibility effects can be absent when a verb is not part of the focus of a sentence, but is present when the verb is put into focus through adjectival modification.

to know why are people doing things, what the social significance is, what's the point? I mean, does it really matter, the exact dynamics of moving through space of these characters? Or do you really want to know who's killing who? But I do think it can become important when you select verbs of a certain type, because there's a big difference between saying 'they got there' and 'they bicycled there,' because why they choose bicycle as opposed to just a change in state I think is nontrivial. So the selection of the lexical items, I think, can be cues to when they are epiphenomal in guiding the level of detail of some of these action dynamics.

Sanford: Yes, I'm sure you're right. I mean, another case where embodied processing of some sort might necessarily enter into comprehension might be manners of walking. I don't know how this fits into LSA research, I'm sure, but consider the verb 'trudge' – 'trudge' as in walking. To me, I would find that very difficult to describe, but it's quite easy to do, it's quite easy to produce as a movement. And there are several different verbs for walking, and it's known that these are easily discriminated in terms of bodily movements. So if you put point sources on the joints of an actor and that's all an observer can see, and the actors trudge, or skip, for instance, or walk, the discrimination is very good on the basis of that information. It's very hard to put it into words, but it's easy to express it algebraically.

Graesser: But why an author would bother using those one of those verbs over another reflects an intention of trying to communicate something, so there's still an issues of intentionality underlying the use of embodiment…

Sanford: Yes, absolutely, of course there is.

Manuel de Vega: You mentioned some of our work, Art's and mine, concerning action sentences like 'cutting the grass and digging a hole,' and you put to us a counterexample, like Plato driving a car, and we say OK, you can verify this as impossible and you don't need any embodied representation. I see that you are coming from someplace quite different. First, you can go to your general knowledge store, your long-term memory and realize that in Plato's time, in classical days, there were no cars, OK. But in our case, both actions, cutting the grass and digging a hole, they go to the same search area, they are stored very close. In a purely symbolic representation, there would be highly correlated vectors directly corresponding to these actions. So it is much more subtle than just going to our general knowledge and realizing that the concepts are semantically related or not.

Sanford: I don't know, I guess I've learned that there are lots of things I can't do at the same time, and I probably learned them when I was quite little. I mean, there may have been a time when as an organism I didn't know that I couldn't crawl forwards and backwards at the same time, but I would probably have made that discovery in some sense.

de Vega: Did your parents make you cut the grass?

Sanford: I guess my primary point is that these areas are so underspecified in terms of how these discrepancies might be discovered that it needs a lot more evidence to make a claim for one way of doing it or another.

Author note

Preparation supported by Arts & Humanities Research Council grant MRG-AN8799/APN19525.

References

de Vega M, Robertson D, Glenberg A, Kashack MP, Rinck M (2004). On doing two things at once: temporal constraints on actions in language comprehension. *Memory and Cognition*, 32, 1033–43.

Emmott C, Sanford AJ, Morrow LI (2006). Capturing the attention of readers? Stylistic and psychological perspectives on the use and effect of text fragmentation in narratives. *Journal of Literary Semantics*, 35, 1–30.

Ferreira F, Ferrara V, Bailey KGD (2002). Good-enough representations in language comprehension. *Current Directions in Psychological Science*, 11, 11–15.

Glenberg A, Havas D, Becker R, Rinck M (2005). Grounding langauge in bodily states: the case of emotion. In D Pecher, RA Zwaan, Eds. *Grounding Cognition: The Role of Perception and Action in Memory, Language, and Thinking* (pp. 115–28). Cambridge: Cambridge University Press.

Glenberg AM, Kaschak MP (2002). Grounding language in action. *Psychonomic Bulletin and Review*, 9, 558–65.

Goodenough WH (1965). Yankee kinship terminology: a problem in componential analysis. *American Anthropologist*, 67, 259–87.

Grush R (2004). The emulation theory of representation: motor control, imagery, and perception. *Behavioural and Brain Sciences*, 27, 377–442.

Hamilton JME, Sanford AJ (1978). The symbolic distance effect in alphabetic order judgements: a subjective report and reaction time analysis. *Quarterly Journal of Experimental Psychology*, 30, 33–43.

Landauer T (1962). Rate of implicit speech. *Perceptual and Motor Skills*, 15, 646.

Lehrere A (1970). Indeterminacy in semantic description. *Glossa*, 4, 87–110.

Lestou V (2005). *Resonance Mechanisms in the Imitation of Human Movement: Behavioural and fMRI Studies* [doctoral thesis]. Glasgow: University of Glasgow.

Lestou V, Pollick FE, Kourtzi Z (2003). A differential involvement of prefrontal and parietal areas revealed by fMRI adaptation. Abstract presented at the 33rd Annual Meeting of the Society for Neuroscience, New Orleans, LA, USA, 7-11 November.

Miller GA (1969). English verbs of motion: a case study in semantics and lexical memory. In AW Melton and E Martin, Eds. *Coding Processes in Human Memory*. Washington, DC: Winston.

Miller GA, Johnson-Laird PN (1976). *Language and Perception*. Cambridge: Cambridge University Press.

Moll J, de Oleveira-Souza R, Moll FT *et al.* (2005). The moral affiliations of disgust: A functional MRI study. *Cognitive and Behavioural Neurology*, 18, 68–78.

Paul A (2005). The effects of linguistic subordination, semantic distance, and recency in discourse processing. Unpublished master's Thesis, Department of Psychology, University of Glasgow, Glasgow, Scotland.

Pulvermueller F, Harle M, Hummel F (2001). Walking or talking? Behavioural and electrophysiological correlates of action verb processing. *Brain and Language*, 78, 143–68.

Sanford AJ (1971). A periodic basis for perception and action. In WP Colquhoun, Ed. *Biological Rhythms and Human Performance*. London: Academic Press.

Sanford AJ, Garrod SC (1981). *Understanding Written Language: Explorations of Comprehension Beyond the Sentence*. Chichester: John Wiley & Sons.

Sanford AJ, Garrod SC (1998). The role of scenario mapping in text comprehension. *Discourse Processes*, 26, 159–90.

Sanford AJ, Garrod SC (2005). Memory-based processing and beyond. *Discourse Processes*, 39, 205–24.

Sanford AJ, Graesser A (2006) Shallow processing and underspecification. *Discourse Processes*, 42, 99–108.

Sanford AJ, Moxey LM (1995) Notes on plural reference and the scenario-mapping principle in comprehension. In C Habel, G Rickheit, Eds. *Focus and Cohesion in Discourse*. Berlin: de Gruyter.

Sanford AJ, Moxey LM (1999). What are mental models made of? In G Rickheit, C Habel, Eds. *Mental Models in Discourse Processing and Reasoning*. North Holland: Elsevier.

Sanford AJ, Sturt P (2002). Depth of processing in language comprehension: not noticing the evidence. *Trends in Cognitive Sciences*, 6, 382–6.

Sanford AJS, Sanford AJ, Filik R, Molle J (2005). Depth of lexical-semantic processing and sentential load. *Journal of Memory and Language*, 53, 378–96.

Sanford AJS, Sanford AJ, Molle J, Emmott C (2006). Shallow processing and attention capture in written and spoken discourse. *Discourse Processes*, 42, 109–30.

Schank R, Abelson R (1977). *Scripts, Plans, Goals and Understanding: An Enquiry into Human Knowledge Structures*. Hillsdale, NJ: Erlbaum.

Shepard RN, Metzler J (1971). Mental rotation of three-dimensional objects. *Science*, 171, 701–3.

Sturt P, Sanford AJ, Stewart AJ, Dawydiak E (2004). Linguistic focus and good-enough representations: an application of the change-detection paradigm. *Psychonomic Bulletin and Review*, 11, 882–8.

Tabossi P (1988). Effects of context on the immediate interpretation of unambiguous nouns. *Journal of Experimental Psychology: Learning, Memory, and Language*, 14, 153–62.

Wilson M (2002). Six views of embodied cognition. *Psychonomic Bulletin and Review*, 9: 625–36.

Zwaan R (1999). Embodied cognition, perceptual symbols, and situation models. *Discourse Processes*, 28, 81–8.

Zwaan R, Taylor LJ (2006). Seeing, acting, understanding: motor resonance in language comprehension. *Journal of Experimental Psychology: General*, 135, 1–11.

Zwaan R, Taylor LJ (2007). Linguistic focus and motor resonance. Paper presented at the meting of the Experimental Psychology and Psychonomic Societies, Edinburgh, July 4–7, 2007.

Chapter 11

A mechanistic model of three facets of meaning

Deb Roy

11.1 Introduction

This chapter presents a physical-computational model of sensorimotor grounded language interpretation for simple speech acts. The model is based on an implemented conversational robot and combines a cybernetic closed-loop control architecture with structured conceptual schemas. The interpretation of directive and descriptive speech acts consists of translating utterances into updates of memory systems in the controller. The same memory systems also mediate sensorimotor interactions, thus serving as a cross-modal bridge between language, perception, and action. The referential, functional, and connotative meanings of speech acts emerge from the effects of memory updates on the future dynamics of the controller as it physically interacts with its environment.

This book is the result of a meeting which was organized around a contrast between two approaches for analyzing and modelling semantics. In the first camp are cognitive scientists who model linguistic meaning as structural relations between symbols. The term 'symbol' in this context is taken to mean, roughly, discrete information elements that may be given word-like labels that make sense to humans (e.g., 'dog', 'is-a'). Examples of this approach include semantic networks such as WordNet (Miller 1995) and Cyc (Lenat 1995) and statistical methods such as *latent semantic analysis* (LSA; Landauer *et al.* 1998).

In contrast to this 'ungrounded' approach, the grounded camp treats language as part of a larger cognitive system in which semantics depends in part on nonsymbolic structures and processes, including those related to perception and motor planning. Examples include Bates' grounding of language in sensorimotor schemas (Bates 1979) and Barsalou's proposal of a *'perceptual symbol system'* that grounds symbolic structures in sensorimotor simulation (Barsalou 1999). Thus, the grounded camp pays more attention to how symbols arise from, and are connected to interactions with the physical and social environment of the symbol user.

Although the ungrounded camp has the advantage of mature computational modelling tools and formalisms, and these have proven to be immensely useful, there are clear limits to the explanatory power of any approach that deals strictly with relations among word-like symbols. Many aspects of linguistic phenomena that depend on conceptual and physical processes, many of which are centrally involved in child language acquisition, are beyond the scope of a purely symbolic analysis. For example, to understand the meaning of the assertion that 'there is a cup on the table' includes the ability to translate this utterance into expectations of how the physical environment will look, and the sorts of things that can be done to and with the environment if the utterance is true. Presumably, those working with ungrounded models would readily acknowledge the limits of their approach in dealing with such issues. Thus, I am not sure whether there is really a debate to be had between the two camps, but rather a difference of opinion in choosing which parts of an immensely complicated overall problem to focus on.

My contribution will be to offer a new model that I believe makes some progress towards understanding interactions between linguistic, conceptual, and physical levels in strictly mechanistic terms. I chose the phrase 'mechanistic model' in the title of this chapter to indicate that the model encompasses both physical processes of sensory and motor transduction and computational processes. My hope is that this model will illuminate some aspects of linguistic meaning related to intentionality, reference, connotations, and illocutionary force that are difficult, if not impossible, to address using conventional symbolic modelling methods alone, but emerge naturally by emphasizing the underlying processes of grounding.

The model bridges the symbolic realm of language processing with the control-theoretic and machine perception realms of robotics. I will present the model in stages, starting with a simple cybernetic control loop that provides the foundation for goal-directed systems. Memory systems and processes of this core model are enriched with schema structures and refined planning mechanisms, yielding a prelinguistic cognitive architecture that is able to support processing of language about an agent's here-and-now environment. Once linguistic interpretation processes are grafted onto this prelinguistic substrate, the result is a model of linguistic meaning with unique explanatory power. One novel aspect of the model is an explanation of connotative meaning that is derived directly from the design of the agent control architecture.

The model may be useful for guiding the construction of embodied/situated natural language processing systems (e.g., conversational robots, video game characters, etc.), and to shed light on aspects of human language processing, especially children's language acquisition. My emphasis on cognitive architectures and processes in this paper complements an earlier paper (Roy 2005) that focuses on schema structures underlying language use.

The remainder of the paper is structured as follows. I begin by suggesting that a viable strategy for modelling language use is to focus on simple language use, such as that of young children. I define three facets of 'meaning' that need to be explained, leading to functional specifications for a model of language use framed as semiotic processing. The next section describes the embodiment and behaviour of Ripley, a conversational robot built in our lab that serves as a concrete launching point for a more general model that is developed in the next sections. The final sections discuss the meaning of 'meaning' suggested by this model.

11.2 **Simple languages**

Given the immense complexity of language use, it is critical to focus on a manageable subset of the phenomena if we are to gain traction. Rather than partition the problem along traditional pragmatic–semantic–syntactic–phonological boundaries, we can instead take a 'vertical' slice through all of these levels by modelling simple but complete language use. Language use by young children is a paradigm case.

Children acquire language by hearing and using words embedded in the rich context of everyday physical and social interaction. Words have meaning for children not because they have memorized dictionary definitions but rather because they have learned to connect words to experiences in their environment. Language directed to, and produced by, young children tends to be tightly bound to the immediate situation. Children talk about the people, objects, and events in the here and now. The foundations of linguistic meaning reside, at least in large part, in the cognitive representations and physical processes that enable this kind of situated language use.

Consider the language production capabilities of a normally developing toddler at the two word phase (e.g., 'more milk'). The child's lexicon is tiny; her grammar is trivial compared to that of an adult. She will mainly refer to the here and now – more complex reference to the past, future, and to distant and imagined worlds will develop later. She will ignore most social roles in choosing her words to express herself. sensitivity to social roles will also develop in time. In other words, toddlers use simple language, simple along dimensions of lexicon size, syntactic complexity, extent of reference, social/cultural sensitivities, and so forth.

For all its simplicity, a toddler's use of language nonetheless demonstrates many of the hallmarks of mature adult language: descriptive and directive speech acts consisting of compositions of symbols that relate to the child's environment and goals. How is it that young children produce and interpret simple speech acts that simultaneously refer (are about something) and serve social functions? How do the child's mental lexicon and grammar interact? How do symbolic (word-level) and subsymbolic (concept-level) processes interact? A detailed mechanistic analysis of speech acts used by children has yet to be offered in the cognitive sciences. Any such model must explain aspects of perception, memory, motor control, planning, inference, and reasoning capacities that play a role in situated language use.

In this chapter, I will describe a model of embodied, situated speech act interpretation that is motivated by these questions. The goal of this model is to make progress towards a mechanistic explanation of meaning, so first we need a working definition of 'meaning'.

11.2.1 **Three facets of meaning**

To analyse and model the meaning of speech acts produced or interpreted by humans there are (at least) three main facets we need to consider. First, speech acts refer to objects, actions, properties, and relations in the physical environment – words have *referential meaning*. Second, speech acts are produced intentionally to achieve effects on the interpreter – speech acts have *functional meaning*. Finally, words and larger linguistic

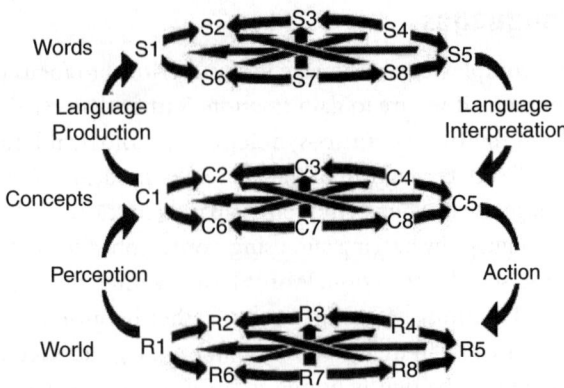

Fig.11.1 Semiotic processes.

structures have *connotative meaning* where connotation is defined as 'an idea or feeling that a word invokes in a person in addition to its literal or primary meaning' (McKean 2005).

Consider first referential meaning. Taking a semiotic perspective,[1] we can view the symbolic constituents of language (words, clauses, etc.) as a layer of interconnected structures that interacts with a conceptual layer via language interpretation and production processes. The conceptual layer in turn interacts with the physical environment of the language user via perception and motor action. These three layers are essentially an expanded view of Odgen and Richards' classic semiotic triangle (Odgen and Richards 1923) in which the link from words to the environment are mediated by concepts (or 'ideas'). In contrast to the static structural view intimated by the original semiotic triangle, Figure 11.1 identifies the dynamic processes that link each layer. The semiotic relation brings our attention to referential meaning: language can be used to refer to (and to depict) objects, events, and relations in the physical environment. To ensure that referential meaning is truly grasped by language users, we can require that language users have the ability to verify referential assertions against the physical environment for themselves.[2]

Figure 11.2 shows how all three facets of meaning may be analysed in the context of a conversational turn. Two people are at a table and there is a cup on the table. The speaker says, 'This coffee is cold.' What do these words mean to the listener? The referential content of the utterance regards the temperature of the liquid in a particular container. 'This' and 'is' bind the referential meaning of the utterance to the here-and-now object that is physically accessible to both speakers. The meaning of 'coffee' and 'cold' have

---

[1] I use the term 'semiotic' based on my interpretation of CS Peirce, which emphasizes the process of interpreting sensory patterns as indications of the future possibilities for action.

[2] This is a strong requirement, one that we certainly would want to relax when we consider reference to referents distant in space, time, and more obviously for fictive/hypothetical referents. But, in the simplest case of talk about the here and now, it is a reasonable requirement, one that rules out any approach that relies merely on perceptual associations without the ability to act.

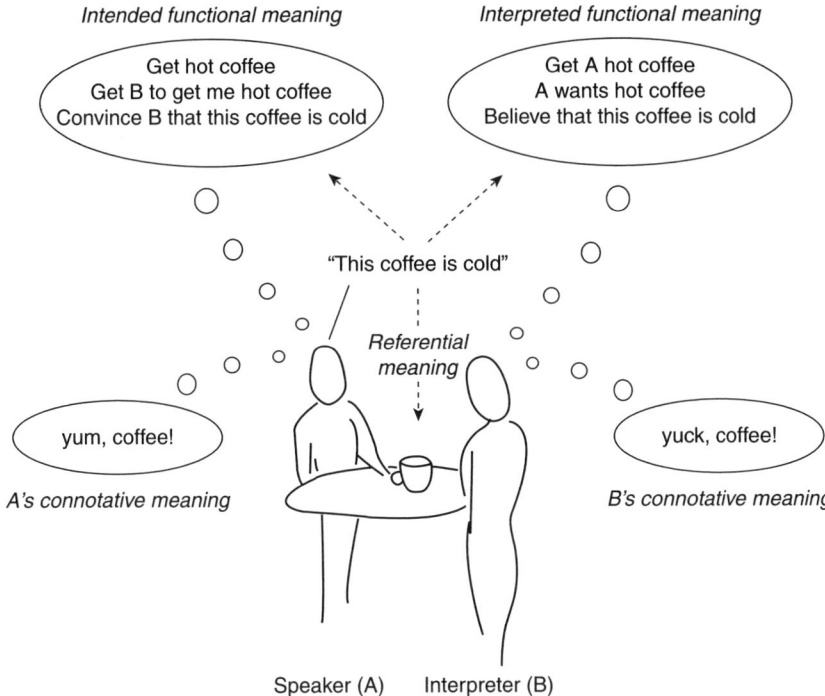

Fig.11.2 Three facets of linguistic meaning.

shared meaning for the speaker and interpreter to the extent that they have associated similar personal experiences to the terms. The connotative meanings of 'coffee' might differ markedly and depend on personal tastes. The speaker's purpose in producing this speech act (Figure 11.1, upper left thought bubble) is layered from higher- to lower-level intended functions. If communication succeeds as intended, the cooperative interpreter should infer a set of layered beliefs and goals that are aligned with those of the speaker.

The degree and kind of alignment of speaker and interpreter meaning is different for each facet of meaning. Referential meaning in the depicted situation should overlap substantially. Shared sensorimotor access to a common physical environment forces alignment of referential meaning. Functional meaning should be complementary at matched levels (e.g., A's attempt to convince B that X is true should translate to B believing that X; A wanting Y should translate to B planning to get Y for A). Connotative meaning may in general diverge significantly as in this example, or alignment might depend on similarities in cultural background or personal experiences.

Within the framework of conventional computational linguistics typified by WordNet and Cyc, my choice of three facets of meaning seems to miss an obvious one: semantic relations. These are diagrammed as arrows within the top layer of Figure 11.1. Virtually all computational models of semantics, from early semantic networks (Quillian 1968) to electronic thesauri such as WordNet (Miller 1995) and commonsense databases such as Cyc (Lenat 1995) all the way to purely statistical approaches such as LSA (Landauer *et al.* 1998) share one fundamental attribute: they model relations between words. Modelling semantic relations can be traced back to Saussarian tradition of structural analysis and in modern semantic analysis in terms of lexical relations (Cruse 1986). Without doubt, semantic relations are essential for a mature language user. We would never learn from texts (e.g., textbooks, dictionaries, news articles, etc.) without them. These relations, however, must ground out in nonlinguistic relations. The meaning of semantic relations such as *is-a*, *has*, or *opposite* must be fleshed out in terms of nonlinguistic structures and processes. My assumption is that semantic relations are derived from the three core facets of meaning and my focus will therefore be on these foundations. Ultimately, semantic relations (and for that matter, syntactic word classes) will gain some degree of autonomy from underlying conceptual structures.[3] The model I discuss here does not address these more advanced aspects of linguistic structure but prepares a way for them.[4]

11.2.2 Modelling approaches: modularity versus holism

A grand challenge for the cognitive sciences, then, is to develop a mechanistic model that simultaneously addresses all three facets of meaning. This is clearly a tall order that will not be met any time soon but, as I hope to show, tangible progress can be made.

A potential objection to taking the grand challenge seriously is the overwhelming number of factors that must be considered together to analyse and model situated language use. When we don't understand how any one 'module' works, how can we possibly consider such a challenge? This objection leads to a natural alternative strategy: divide and conquer. Kintsch conveys just this sentiment in Chapter 8 of this volume:

> … we would have images, concepts, somatic markers, S–R links, features, schemata, abstractions, words, etc., all linked somehow organized in one grand system, at least in principle, for I have no idea how one would go about constructing such a model. Alternatively, one could *model each level of representation separately*. (emphasis added)

This is in fact a widely held attitude by those interested in developing computationally precise models of language. For example, Chomsky expresses a similar position when considering the prospects for explaining language use (as opposed to mere syntactic competence) (Chomsky 2000):

> It would … be a mistake, in considering the nature of performance systems, to move at once to a vacuous 'study of everything'… [instead we should] try to *isolate coherent systems* that are amenable to naturalistic inquiry and that yield some aspects of the full complexity. (emphasis added).

---

[3] For example, although we might assume that syntactic word classes are grounded in conceptual categories (nouns derived from objects, verbs from actions, etc.), a fully autonomous syntax moves beyond conceptual groundings (e.g., 'the flight' or 'cup the ball').

[4] The failure to realize the age-old AI dream of building machines that learn by reading books is perhaps due to an underestimation of the value of first working out the nonlinguistic underpinnings of semantic interpretation.

The idea of focusing on a single level or component is, of course, a reasonable strategy. The problem is that we might miss the forest for the trees. For example, although presumably the reason for studying any aspect of language is to understand how words are used to communicate, as research into syntax delves deeper and deeper into the subtleties of 'rules' of word order, it is unclear how progress towards under-standing these rules in any way furthers our understanding of the process of social communication. I believe that holistic models of language use can be developed if we start with simple kinds of language use and slowly expand the boundaries of analysis to more complex forms. This strategy was anticipated by Wittgenstein (Wittgenstein 1958):

> If we want to study the problems of truth and falsehood, of the agreement and disagreement of propositions with reality, of the nature of assertion, assumption, and question, we shall with great advantage *look at primitive forms of language* in which these forms of thinking appear without the confusing background of highly complicated processes of thought. When we look at such simple forms of language the *mental mist which seems to enshroud our ordinary use of language disappears.* We see activities, reactions, which are clear-cut and transparent. On the other hand we recognize in these simple processes forms of language not separated by a break from our more complicated ones. We see that *we can build up the complicated forms from the primitive ones by gradually adding new forms* (emphasis added).

11.2.3 Functional specifications for a semiotic processor

From a designer's perspective, we can translate the task of here-and-now speech act interpretation into a set of functional specifications that must be met by a 'semiotic processor'. There are two basic kinds of speech acts, descriptives and directives, that I will be concerned with here. Descriptives are used by speakers to make assertions about the current state of affairs, providing the same function as perception. Directives are used to make requests/demands for actions upon the environment. In contrast to descriptives and perceptual input that inform the interpreter of how things *are*, directives suggest how things *should be*.

Imagine a robot that can see its environment through cameras and manipulate objects using its arms. A fly lands next to the robot and the robot happens to be looking in the fly's direction. The fly's presence gives rise to a pattern of activation in the camera. If successfully interpreted, the robot can come to correctly believe that there is a fly out there at some location relative to the robot, with certain properties of size, shape, colour, and so forth. If instead of seeing the fly, the robot is told (by a trustworthy source) that 'there is a fly next to you,' the robot must translate these words into beliefs that are of the same format as those caused by the fly's image. If the robot is asked to 'swat the fly,' those words must be translated into appropriate motor actions selected to make appropriate changes to the environment, and that can be verified by sensoriotor interaction. The robot must translate directives into goals that are in a format compatible with its perceptually-derived beliefs. This set of cross-modal (perception–action–language) requirements of compatible beliefs and goals constitute the functional specifications of a here-and-now semiotic processor at an abstract level of description.

I will now turn to a specific robotic implementation to make some of these ideas concrete, leading to a generalized model of speech act interpretation.

11.3 **Ripley, a conversational robot**

Ripley is an interactive robot that integrates visual and haptic perception, manipulation skills, and spoken language understanding. The robot uses a situation model, a kind of working memory, to hold beliefs about its here-and-now physical environment (Roy *et al.* 2004). The mental model supports interpretation of primitive speech acts (Mavridis and Roy 2006). The robot's behaviours are governed by a motivation system that balances three top-level drives: curiosity about the environment, keeping its motors cool, and doing as told (which includes responding to directives and descriptives) (Hsiao and Roy 2005). The following sections briefly describe Ripley's embodiment and behaviour, which are then translated into a general model that Ripley partially instantiates.

11.3.1 **Ripley's embodiment**

Ripley's body consists of a five degree-of-freedom (DOF) arm (or torso) with a gripper at its end. Thought of as an arm, the five DOFs provide a shoulder (two DOFs), elbow (one DOF), and wrist (two DOFs). The gripper has one DOF allowing it to open and close. A pair of miniature video cameras are mounted on either side of the gripper. The torso is anchored to a table. The table top serves as Ripley's primary physical environment. The robot can examine objects on the tabletop via vision and touch, and move objects around using its gripper. Each DOF of the torso includes a position and a force sensor so that the robot can sense both its body pose and the forces that are acting upon it (e.g., due to gravity or if something resists the robot's movements). The gripper tips house pressure sensors that detect contact with objects.

11.3.2 **Ripley's behaviour**

The robot is designed for physical and verbal interaction with a human partner who sits across the table from the robot. Using a visual face detector, the robot is able to detect and track the location of the person, using this information to guide its motor movements (e.g., when handing the person an object) and to guide its interpretation of speech (e.g., to differentiate the meaning of 'my left' versus 'your left') (Figure 11.3). The following is a typical interaction between Ripley and a human partner:

Ripley: Looks at left side of table, detects a new object that the human has just placed on the table, updates its mental model to encode this new information.

Ripley: Looks to the right, finds that the table is empty there, which matches its expectations based on its current mental model, so no further changes to the model are needed.

Ripley: Goes to sleep by positioning its gripper to a resting position on the table and cutting off electrical power to its motors in order to cool them.

Human: 'Hand me the blue one'

Fig.11.3 Ripley hands an apple to its human partner in response to a spoken directive.

Ripley: Wakes up, leans down, picks up the blue cup, lifts it and holds it out towards the human.

Human: Takes the cup from Ripley.

Ripley: Goes back to sleep.

Ripley: Wakes up by lifting its torso back to an upright position, then looks up towards the human to verify the human's location based on face detection. The person has shifted slightly, so an appropriate update is made to the mental model.

Human: 'There is a red cup on your left'

Ripley: Updates its mental model to integrate a new belief, that there is a cup on the table but currently not in view, says 'OK'

Human: 'Where is the red cup?'

Ripley: 'On the left'

Human: 'How big is the red cup?'

Ripley: 'I have no idea' (Ripley has only been told about the colour of the cup, but has no information about its size)

Human: 'Look to the left'

Ripley: Ripley looks to the left, detects the cup, consolidates new visually derived information (precise location, precise colour, size) into mental model.

Human: 'How big is the red cup?'

Ripley: 'It is small'

Human: 'Pick up the red cup'

Ripley: Attempts three times to grasp object but fails because it is too slippery, updates mental model to record this failure, says 'sorry'.

The following sections explain Ripley's behaviour in terms of cognitive architecture, information structures, and processes. This model yields an analysis of all three facets of meaning, from Ripley's point of view. The model focuses on interpretation of speech acts. Although Ripley does generate limited speech output, speech production is not yet addressed by the generalized model.

11.3.3 A model of situated language interpretation

In a previous paper (Roy 2005) I developed the concept of *semiotic schemas* which 'serve as structured beliefs that are grounded in an agent's physical environment through a causal–predictive cycle of action and perception.' The focus of that paper was on conceptual structures underlying lexical semantics. I chose the term 'semiotic' to highlight the cross-modal (language–action–perception) interpretive and control issues at stake in grasping physically situated language. The following sections build on these ideas but turn attention to the architecture and processes required for semiotic processing. I will treat Ripley as a concrete instance of a class of cognitive architectures. My goal is to develop a model with minimal complexity that yields the capacity for interpreting (i.e., acting successfully in response to) descriptive and directive speech acts about the physical here and now.

Goal-directed behaviour

As we shall see, to account for any of the three facets of meaning, the language interpreter must have its own autonomous interests/goals/purposes. A framework for goal-directed behaviour is therefore a natural starting point. In cybernetic terms, at the core of the simplest goal driven system is a comparator that drives the agent's actions (Rosenblueth *et al.* 1943). A comparator, or difference engine, is shown in Figure 11.4. Rectangles indicate memory systems and ovals indicate processes. The difference engine compares target and actual situations and generates a plan of action designed to eliminate differences. The specification of the target situation must be in a format compatible with the contents of the situation model, although the levels of specificity may differ (e.g., the target might specify a constraint on acceptable situations but not specify a particular target situation).

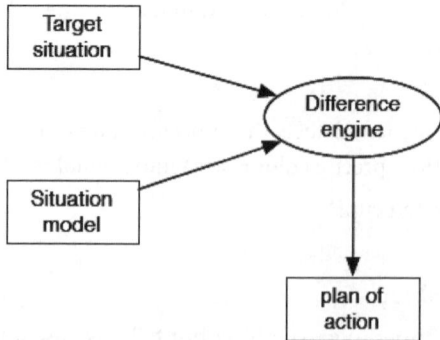

Fig. 11.4 Core elements of a goal driven system.

Since we are interested in physically situated here-and-now semantics, the difference engine must be embodied and embedded in a physical environment. Figure 11.5 introduces interaction processes that couple the difference engine to the agent's environment. A simple (and classic) example of an embodied, situated difference engine is a thermostat coupled with heating and cooling equipment. Consider how parts of the thermostat system map onto Figure 11.5. The target situation is a desired temperature, perhaps dialled in by a human. The situation model is a reading of the current air temperature. The body of the thermostat includes a temperature sensor. Bodily interaction includes mapping the sensor reading into a format compatible with the encoding of the target temperature stored in the target situation memory system. The difference engine must compare inputs and decide whether to turn on a furnace, air conditioner, or neither. The output of the difference engine which encodes this three-way decision constitutes its plan of action. The bodily controller (which is part of the bodily interaction processes) executes actions by activating appropriate bodily actuators (furnace, air conditioner). Acting upon and sensing the environment form a cycle, enabling the system to adapt to changes in either the environment or the target temperature.

Semiotic schemas for conceptual representation and control

To begin scaling up the capabilities of the basic architecture of Figure 11.5, we will need to endow our system with richer situation models. Rather than elaborate on specific

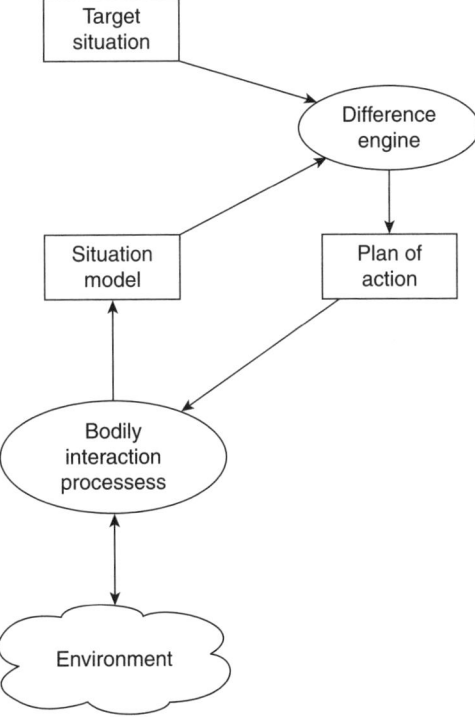

Fig.11.5 Embodied, situated goal driven system.

implementation details, a previous paper (Roy 2005) describing Ripley's software system at the level of semiotic schemas provides a clearer functional explanation suited to our current purposes. Several key attributes of semiotic schemas are:

- Actions, objects, properties, and spatial relations are all represented using a common set of primitives. In other words, rather than treat objects and actions as distinct basic ontological types, both are constructed from the same elementary structures.

- Schemas are used both to encode representations and guide control. Beliefs, goals, and plans are created by instantiating schemas in the situation model, target situation, and plan memory systems respectively.

- Schema types encode concepts and are stored in long-term memory. Schema tokens are instantiations of schema types. Tokens can be created in target, situation, or plan memory systems. Schemas are instantiated in response to (as an interpretation of) bodily and linguistic interactions with the environment as described below.

- Actions are chains of precondition–movement–result triplets. The result of each action provides the precondition of the next. An action chain may be represented compactly by suppressing all but the initial precondition and final result, providing a causal representation of what the chain achieves without specifying the means by which the change is achieved (Drescher 1991).

- Objects schemas are collections of possible actions that are licensed by the existence of external objects. As a result, the composition of actions and objects is seamless. To lift a cup is to activate the appropriate action schema from the collection of schemas that represents the cup.[5]

- The properties of an object are parameters of schemas which may be measured when those schemas are executed (e.g., the weight of an object is represented in terms of expected forces on joints when that object is lifted).

- Schemas may be used to represent both topological concepts (e.g., that a triangle has three sides corresponding to three connected actions that are spatially mutually constrained) and graded effects (e.g., that blue occupies a certain part of a colour space with certain parts of the space that are most typical).

An example of an object schema in Ripley is shown in Figure 11.6. A detailed explanation of the elements of this schema are provided elsewhere (Roy 2005). The main idea is that for Ripley to believe that there is a cup at location L is to believe that two sensorimotor control loops will execute successfully. The upper control loop involves visually detecting a region at location L. The lower loop involves grasping, touching, and moving the object. Both the visual and the haptic loop can effect change in the value of L. Shape and colour parameters may optionally be measured while the visual loop is executed. Although not shown in this figure, touch information may also inform the shape parameter (this was

---

[5] In contrast, other models of verb grounding that use schemas (e.g. Narayanan 1999; Siskind 2001) do not provide a direct path for modelling composition with objects that are acted upon. I view this as a serious shortcoming of these alternative approaches.

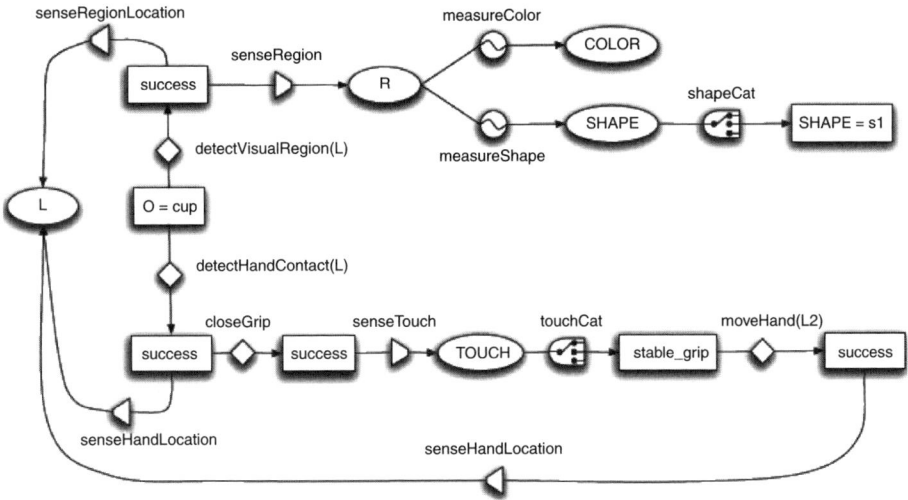

Fig.11.6 Cup schema.

not implemented in Ripley since Ripley's touch sensors are too primitive to characterize object shapes).

An example of an action schema corresponding to the verb lift is shown in Figure 11.7. Executing this action causes the lifted object to change from location from *L1* to *L2*. Note that the lift schema is embedded within the cup schema. This embedding relation enables composition of schemas corresponding to the compositional semantics of 'lift the cup.'

Equipped with schemas, Ripley can move beyond the stimulus-response world of thermostats and represent its physical environment in an object-oriented fashion. In Figure 11.8, bodily interpretation processes gain access to a long-term memory store of schema types (concepts). Ripley's implemented schema store includes representations of simple handle-able objects, their colour, size, weight, and compliance (i.e., soft, hard) properties, and a few actions that may be performed on objects to move them about the table top and to give them to the human partner.

Schema use is goal driven. The first of three top-level drives implemented in Ripley is its curiosity drive. The purpose of the curiosity drive is to maintain an up-to-date set of beliefs about the objects in Ripley's environment. Belief maintenance is challenging for Ripley due to its limited visual field of view. Ripley's cameras allow it to see only a portion of the table top at a time (recall that the cameras are mounted at the end of Ripley's gripper so its visual perspective shifts whenever the robot moves). Also, when

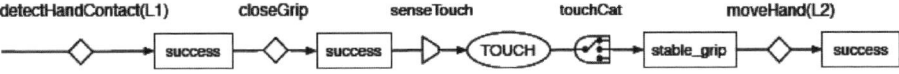

Fig.11.7 The lift schema is embedded in the cup schema.

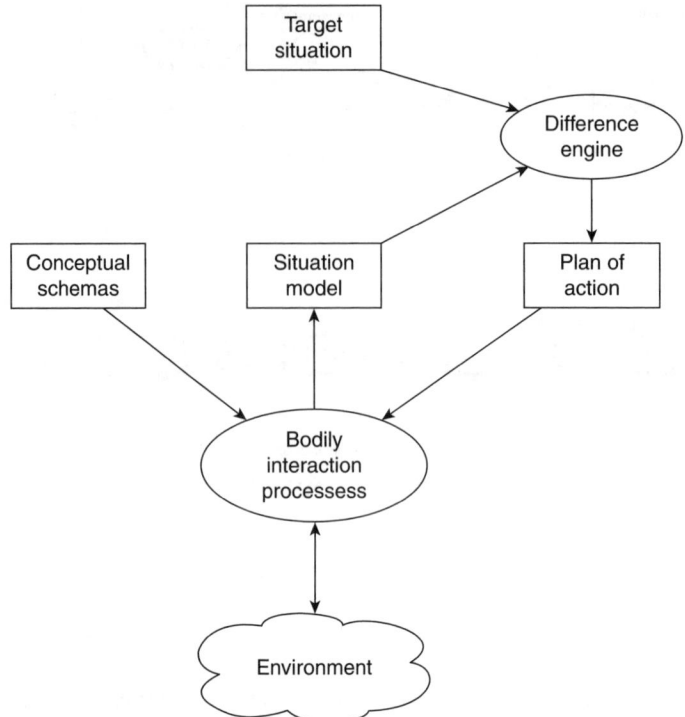

Fig.11.8 Conceptual schemas in long term memory constrain and accelerate interpretation of bodily interactions with the environment.

Ripley looks up to detect and track humans, its field of view only covers part of the space where the human might stand and loses sight of the table altogether.

Ripley's environment is dynamic since the human partner may move around, and may introduce, remove, and move objects on the table. To keep track of changes, Ripley must continuously visually scan its environment. The curiosity drive is implemented as follows (for more details see Hsiao and Roy (2005). The visible environment (defined by the full extent of motions of the robot) is partitioned into a number of visual zones. A scalar weight that ranges from 0 to 1 is associated with each zone and tracks when that zone was last updated based on visual input. The longer it has been since Ripley looks at a zone, the higher its associated weight grows. After a sufficiently long period without a visual update, the weight for the zone will hit the ceiling value of 1 and not change further. The collection of weights constitutes Ripley's curiosity and is stored in the situation memory.

A corresponding set of target curiosity values that are stored in the target situation memory (in the current implementation all target curiosity values are set to 0). The difference engine strives to satisfy curiosity by executing a basic control loop in which each iteration consists of selecting the zone with the highest difference between target and actual curiosity levels, and instantiating a plan to eliminate this difference. The plan is constructed by selecting a target pose for the robot that corresponds to the targeted visual zone, computing a sequence of body poses that will guide the body to that pose, and then pausing at that pose for a sufficient period (about 0.5 seconds) for Ripley's visual processes to incorporate new input into the situation model.

Bodily interaction processes using object schemas

Object schemas guide Ripley's interactions with the environment via a set of object schema management processes:

Create: Instantiates a new object schema token in situation memory when a visual region is detected and no existing token corresponds to (i.e., explains away) the region. Correspondence is determined by comparing location, colour, and size properties of the region to those predicted by each object already in situation memory.

Destroy: Deletes an object schema token from situation memory if sensorimotor evidence does not support existence of an object. This may happen, for example, if an object is removed from the table while Ripley is not looking or if the schema was instantiated due to perceptual errors.

Update: Modifies parameters of an existing object schema (e.g., size, colour, position) based on current perceptual input. Moving objects are tracked by continuous updates of the position parameter based on visual sensing. Although not yet implemented, changes in parameters over time may be attributed to causes within the object (self-driven or agentive objects) or due to external causes (e.g., naïve physics).

Temporal merge: Implements object permanence by tracking objects over perceptual discontinuities. If Ripley looks at an object on the table at time T1, looks away at time *T2*, and then looks again at the same object at *T3*, using only the processes listed so far would lead to creation of an object schema (call it token 1) at time *T1*, then destruction of token 1 at *T2*, and then creation of a new token 2 at *T3*. The temporal merge process detects such sequences and merges token 2 (and all subsequent tokens) into token 1. This process provides the basis for individuating objects, and grounding proper names.

Similar processes are used to maintain beliefs about the human's position using a special-purpose face detector rather than generic region detectors. The collection of object schemas in situation memory provides referential content for both linguistic communication and motor actions. Various details regarding real-time operation of these bodily control and interpretation processes may be found in (Roy *et al.*, 2004).

Several object schema management processes that have not been implemented in Ripley but a more complete maintenance system for a situation model should include are:

Temporal split: The complement of temporal merge, this process splits tokens that have been mistakenly merged in the past back into separate tokens.

Spatial merge type 1: Merges object schema tokens in situation memory that were erroneously instantiated by a single distal object. An object that is visually split by another object in front due to occlusion of it might cause such errors.

Spatial split type 1: Splits an object schema token into two to correct for a previous interpretation error. Two similarly coloured objects that are next to one another might initially be mistaken for a single object, but when one object is moved, this process would account for the unexpected split in visual regions by creating a new object token.

Spatial merge type 2: Merges object schema tokens in situation memory that were instantiated in response to two objects that have since become connected. This process should be executed, for example, if the robot observes one object being attached to another.

Spatial split type 2: Splits an object schema token into two to account for an actual split in what was originally a single object. For example, if an object is cut into two, the token corresponding to the original object is split into two tokens. This process is needed to ground conceptual actions (and ground corresponding verbs) for actions such as cutting or tearing.

Although the processes have been described with an emphasis on visual perception, touch plays a parallel role for all processes (e.g., creating an object schema instance upon touching it).

Schema types guide and limit bodily interpretation process. Since object schemas are collections of possible actions, the situation memory encodes possible actions that the robot expects would succeed if they were to be executed. Which of these possible actions is actually taken depends on the active drive(s) of the agent. When an action is performed on the basis of situated beliefs, failure of the action may lead to belief revision (e.g., removal of a object schema if the robot fails to touch it, update of properties such as weight or slipperiness if the robot fails to lift it).

Schemas in situation memory have intentionality ('aboutness') with respect to their referents through causal histories and future-directed expectations. Physical objects cause patterns to impinge on sensors of the agent, which are interpreted by the agent by creating object schema tokens (causal relationship). Instantiated schemas are the agent's beliefs about the here and now, and guide future actions upon the environment (predictive relationship). Interpretation errors may be discovered when predicted outcomes of bodily interactions diverge from actual experiences, triggering belief revision.

With only the curiosity drive in place, Ripley's motor behaviour is straightforward: it scans visual zones in a predictable sequence, always choosing the zone with highest curiosity value. We now consider a second top-level drive that competes with curiosity and enriches Ripley's autonomous behaviour.

Multiple top-level drives and autonomy of behaviour

For practical reasons, Ripley has been designed to keep its motor temperatures within a target operating range. In early versions of Ripley's controller, the robot would sometimes be accidentally left operating over extended periods, causing motor burnout from over-exertion. Since introducing the self-maintenance drive, Ripley has never lost another motor. This drive is implemented in a similar fashion to the curiosity drive. The situation model maintains the current estimated temperature of three motor zones. Each zone is the average temperature of a group of motors. The target situation memory is programmed to hold target temperatures for each zone. The difference engine compares actual and target temperatures; if any zone level of 'tiredness' is too large, it generates a plan for the robot to bring its body down to a resting position on the table, and power to all motors is shut off for a cool-down period.

The drive for self-maintenance conflicts with the drive for curiosity. If both drives are active at the same time the robot will not effectively satisfy either goal since it cannot rest at the same time that it moves to update its situation model. Ripley's architecture adds a layer of control beyond the components of a thermostat, as shown in Figure 11.9. The drive memory system maintains long-term motivations of the robot (e.g., target temperatures of motor zones, target curiosity values for visual zones) along with priority weights for each drive. The target selector compares the contents of situation memory and drive memory to select which drive 'takes control' of the body. Once a new drive is selected, the target selector copies the associated drive variables into the target situation memory, and then the difference engine goes into action.

The interaction of self-maintenance and curiosity drives in Ripley often leads to surprisingly rich behaviour. From a designer's point of view, it is difficult to predict when the robot will tire of visual scanning and decide to rest. Once it rests sufficiently to cool off its motors, it is also difficult to predict which visual zone it will first update in its situation memory.

Of course, the most unpredictable element of Ripley's environment is the human. For the human to affect Ripley's behaviour using words, Ripley must be equipped with

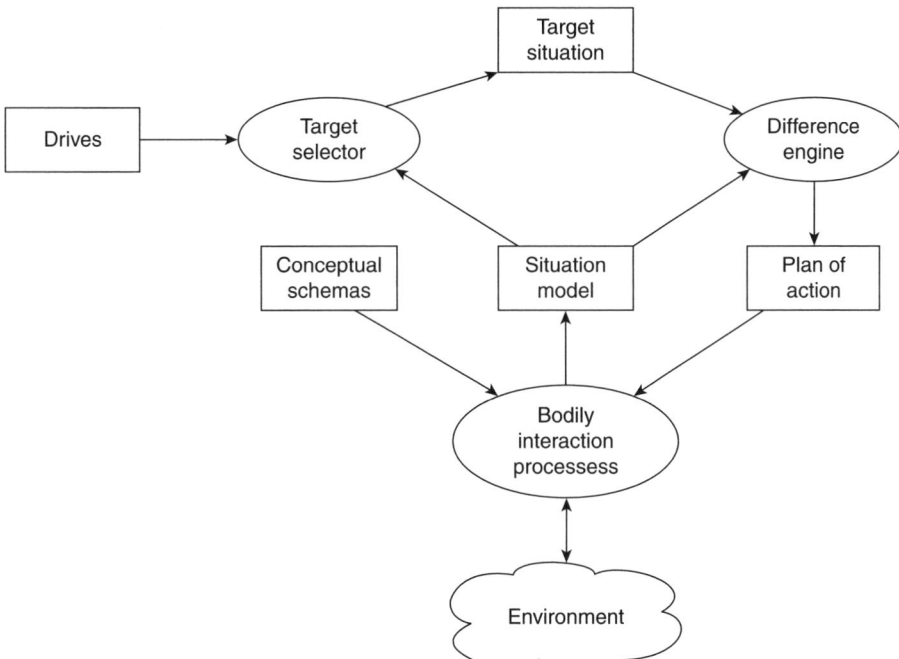

Fig.11.9 Autonomy of goal selection based on multiple top-level drives.

language interpretation skills. To get there, let us briefly reconsider our interpretation of sensoriotor interaction as a kind of semiotic process, or sign interpretation. This view leads directly to a path for language interpretation.

Bodily interpretation: signs and interpretive frames

Recall the situation memory processes that make sense of an object when it first comes into Ripley's view. The visual presence of the object causes a two-dimensional pattern to impinge on the robot's camera sensor, and this pattern, when properly interpreted, causes the creation of an object schema token in situation memory. The same is true of touch patterns. Patterns may signify the presence of an object in the here and now. The robot interprets patterns as signs of objects that caused them.

The proper treatment of patterns as signs requires that the interpreter have contextual knowledge about where, when, and how the pattern was received. For example, if Ripley feels no sensation of pressure at its fingertips, how should this touch pattern be interpreted? In a contextual vacuum, this pattern is meaningless. The robot must integrate contextual knowledge of its body pose (which determines the position of its gripper) and whether the gripper is in motion, and which way it is moving. By combining these bits of knowledge, a touch sensation can be interpreted. Suppose Ripley holds the belief that there is an object within its grasp (i.e., an object schema is in situation memory that predicts grasp-ability), and that Ripley reaches to the location of this object and closes its gripper. The sensation at its fingertips now has clear meaning: either a tangible object is present or it is not. I will call the collection of contextual information necessary to interpret a pattern its *frame of interpretation*. Streams of patterns that emerge as a result of bodily interactions are interpreted using the current frame of interpretation. Over time, the meaning of signs are integrated into situation memory.

The foundation is now laid for grounding language use. Words serve as signs, and syntax provides frames of interpretation for words. Sensorimotor signs embedded in their interpretive frames provide the equivalent of descriptive speech acts – the environment makes 'assertions' about how things are. Like linguistic assertions, beliefs derived from the environment may be faulty and therefore always susceptible to revision. The meanings of speech acts are created in the process of translating word strings into updates on situation memory (in the case of descriptive speech acts) and drive memory (directive speech acts).

Grounding the lexicon: words as signs

Open-class words are defined by three components:

- **Spoken form:** In Ripley the sensorimotor schema for generating and recognizing the spoken form of the word are implemented using standard speech recognition methods based on hidden Markov models combined with model-based speech synthesis.
- **Conceptual association:** Associated schema types in conceptual schema memory.
- **Protosyntactic class:** This is the word class on which syntactic rules will operate. In contrast to a fully autonomous syntax in which word classes are independent of conceptual categories, the protosyntactic model instantiated by Ripley assumes that word classes are fully determined by their conceptual groundings.

Open-class words that Ripley uses include names for objects, actions, object properties, and spatial relations. Examples of schema types for objects and actions were mentioned above. Property schemas specify expected values on measurable parameters of action schemas. Since object schemas embed action schemas, properties can be specified for either actions or objects. Spatial schemas specify expected results of movements between objects.

Grounding protosyntax: frames of interpretation for words

Just as sign can only be interpreted within the context of a frame, words too require an interpretive frame which is provided by syntactic structure. In Ripley, a small set of context-free grammar (CFG) rules are used to encode allowable word sequences. CFG rules are augmented with interpretation rules that specify the semantic roles and argument structure assigned to words based on syntactic structure.[6] Parsing according to these augmented CFG rules generates a semantic analysis of an utterance in terms of its propositional structure and speech act class. Closed class words (e.g., 'there', 'is') used in the CFG rules have meaning purely in their role in determining speech act classes and propositional structure. Of course, other cues beyond syntax (e.g., prosody) may also be integrated to classify speech acts. The propositional structure specifies the role relations between the open class words of the utterance. The structure guides interpretive processes that translate word sequences into memory updates. The speech act class determines whether updates are made to situation or drive memory corresponding to descriptive and directive speech acts respectively.

Figure 11.10 introduces three final components of the model. The lexicon and rules of syntax are stored in dedicated memory systems, and the speech interpreter applies these memory structures to parse incoming spoken utterances and interpret them. As the figure shows, the effects of interpretation are to modify either situation memory in the case of descriptive speech acts, or drive memory in the case of directive speech acts.

Interpretation of descriptive speech acts

A descriptive speech act is an assertion by the speaker about how the world is. Assuming a trustworthy relationship between speaker and interpreter, the literal meaning of a descriptive is constructed by making appropriate changes to the situation memory in order to translate the speaker's assertion into beliefs about the here-and-now environment.

When the syntactic parser determines that the utterance is a descriptive speech act, interpretation consists of translating the propositional structure of the utterance into execution of appropriate schema management processes applied to situation memory. Recall that these processes have already been defined and are also used for interpreting bodily interactions with the environment. The open-class lexical items in the spoken utterance serve as indices into the schema types.

Let us work through an example to clarify the process of descriptive speech act interpretation. Suppose Ripley's human partner says 'there is a bean bag on your left' while

---

6 Ripley's parser is adapted from Gorniak and Roy (2004).

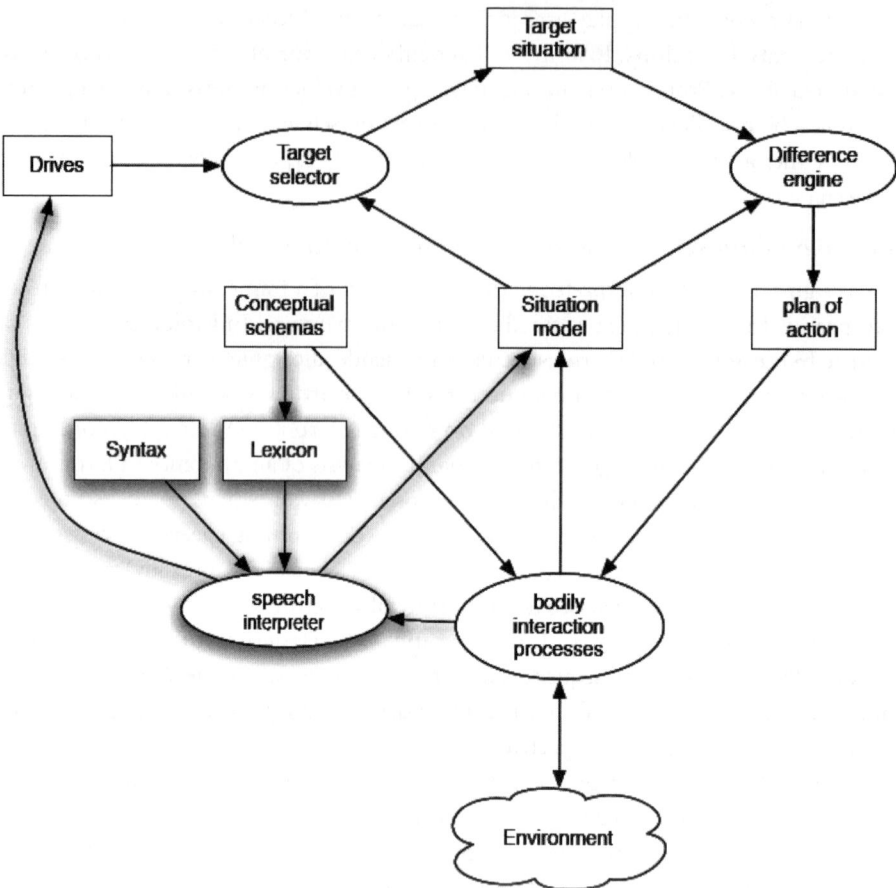

Fig.11.10 Interpretation of descriptive speech acts about the here-and-now environment parallel interpretation of bodily interactions with environment. The semantics of descriptives are fully constituted by their effects on the contents of the situation model.

Ripley is looking up at the person and there are currently no object schemas instantiated in Ripley's situation memory. Interpretation of this descriptive speech act results in the creation of an object schema in situation memory. Here are the stages that lead to this result:

- **Speech recognition:** Speech is detected and translated into a sequence of lexical entries. The acoustic form of words are retrieved from the lexicon and constrain the recognition process.

- **Parsing:** The output of the speech recognizer is parsed using the augmented CFG rules. The parse will result in classification of this utterance as a descriptive speech act, which routes the resulting memory update to the situation memory. The parse also generates a propositional structure with the form *there-exists(bean-bag (location–left(landmark=ego), colour=red))*.

- **Create schema token:** The top level of the propositional structure is interpreted by creating a new bean bag object schema in situation memory. The lexical item with form 'bean bag' provides a link into the conceptual memory store for the corresponding object schema type. By default, the parameters of the schema (location, colour, size, etc.) are set to be unknown. Parameters are encoded using probability density functions (pdfs). To specify a parameter as unknown, the corresponding pdf is set to a uniform distribution over all possible values ('maximum entropy').

- **Update schema location parameter:** The location parameter of the object schema token is updated by replacing its location pdf with a new distributionthat concentrates the probability density to the left region of the table top. The location distribution corresponding to 'left' is stored in the concept memory as a property schema type. The meaning of 'I' and 'your' in the current implementation of Ripley is limited to the impact of these words in interpreting spatial semantics (Roy et al. 2004).

- **Update schema colour parameter:** The colour parameter of the object schema token is updated by replacing its colour pdf with a new distribution that concentrates the probability density in the region corresponding to the English colour term 'red'. This idealized colour distribution is stored in the concept memory as a property schema type.

This example demonstrates the semiotic underpinnings of the model. To process language that depicts an object triggers the same processes that would be used to interpret bodily interactions with an object. As a result, the semiotic requirement of a consistent cross-modal format is met. Words are translated into the same format as sensory patterns since in both interpretive processes, the results are encoded in a common form in situation memory. This enables Ripley to come to believe in the existence of an object via language, and later verify, modify, or even discredit this belief via embodied sensorimotor interaction.

There are differences in the level of ambiguity provided by sensory patterns versus words which are handled naturally in the model through the schema parameter update process. In some cases language is more ambiguous, in other cases less. Suppose Ripley is told that 'there is a cup on the table.' This assertion provides no constraints on the location of the cup beyond the fact that it is resting somewhere on the surface of the table. It also says nothing about the cup's size, colour, orientation, and so forth. In contrast, one clear view of the cup would provide far more precise information about all of these aspects of the object. Although the interpreter might not choose to actually extract and encode all information available in the visual pattern, the information is visually available as opposed to the less informative verbal evidence. This is a case where language is more ambiguous than perception.

In contrast, suppose Ripley knows about (has conceptual schemas of) three-dimensional balls and cylinders.[7] If Ripley were to view an object from above and see a circular form, this pattern would be ambiguous since it might signify either a ball or a

---

[7] Ripley does not encode shapes as part of its concept schemas, but other robots we have built do (Roy 2003).

cylinder (which would also appear round when upright). On the other hand, the utterance 'there is a ball on the table' would not harbour the same ambiguity. More generally, of course, language is a powerful medium for conveying information that is not perceptible at all (i.e., historical facts, theoretical assertions, etc.).

Interpretation of directive speech acts

A directive speech act transmits a request for action (or in the case of a negative directive, a constraint on actions). The directive might specify an action (e.g., 'hand me the cup'), a desired change in situation ('clear the table'), a request for information ('is that box heavy?'), and so forth. I will focus here on the directives that specify actions on objects by working through another example, 'hand me the blue cup', which triggers the following interpretive processes:

- **Speech recognition:** Speech is detected and translated into a sequence of lexical entries using the same processes as descriptives.
- **Parsing:** The parser will classify the utterance as a directive, which routes the resulting memory update to the drive memory. The parser also generates a propositional structure with the form *there-exists(cup(colour=blue), targetlocation=partner)*.
- **Create target structure:** The propositional structure is converted into a situation target linked to a top-level 'do-as-told' drive. The situation target specifies a desired action (lift object and offer to human partner) and a target object specified by selecting an active object schema in situation memory which best matches the contents of the propositional structure. If no object schema in situation memory matches the requested object, this stage will fail.

Once the requested action is placed into drive memory, it will remain there until the target selector determines that the do-as-told drive should gain control. When this happens, the verbally triggered target situation is copied into target memory, which in turn triggers planning processes that carry out the requested motor action.

11.3.4 **Summary**

The model begins with the classic cybernetic structure of a closed loop controller. This skeletal structure provides scaffolding for introducing richer memory structures and processes for control and interpretation. A prelinguistic architecture is developed that integrates drive arbitration with conceptual schemas that guide the interpretation of sensorimotor interactions with the environment. Interpretation of bodily interactions can be viewed as a semiotic process of sign interpretation, where the results of interpretation lead to modifications of situation memory. We can then graft machinery for linguistic interpretation onto this underlying architecture. Interpretation of descriptive speech acts parallels the process of making sense of sensory-motor patterns by effecting situation memory. The meaning of directive speech acts emerges from the modification of drive memory structures.

This model addresses only a sliver of the linguistic phenomena exhibited by a young child. The list of what is missing is long and includes language production, understanding any kind of non-literal meanings, anaphoric resolution, reference beyond the here and now, and so forth. At a prelinguistic level, the conceptual representations I have sketched are also very incomplete. Object schemas as presented here only encode *do-to* affordances, i.e., the kinds of actions the agent can expect to take on an object such as grasping it or pushing it. Eventually we must also model *do-with* affordances (i.e., one can carry things in a cup) and *reacts-to* affordances (i.e., what a ball does when pushed). Type hierarchies for actions, objects, properties, and relations are also surely important, as is goal and belief attribution to communication partners. I believe all of these issues may eventually be addressed by building on the approach developed here, an approach that might be called embodied interactionist semantics.[8]

This model makes some progress in explaining language use in holistic yet mechanistic terms. In contrast to typical computational models of language that focus on one aspect (e.g., syntax, semantic networks, word-to-percept associations, etc.), this model strives to encompass a broad range of cognitive structures and processes in a coherent framework to yield explanatory power.

11.4 Meaning, from Ripley's point of view

In Section 11.2.1, I suggested three facets of meaning that are in need of explanation by a unified mechanistic model. In the model presented here, meaning arises as a consequence of the effects of words on an agent's memory systems, and the ensuing effects on the dynamics of the agent's control processes that are affected by these memory systems. Let us consider how the model relates to each facet of meaning.

11.4.1 Referential meaning

The propositional content of descriptives and directives are bound to the environment via situation memory. Schema instances in situation memory are cashed out in how they make predictions of, and are shaped by referents in the environment. Individual words provide constraints on the establishment of reference. For example, the meaning of 'blue' constrains parameters of visual schemas embedded in whatever object is described to be blue. The model suggests that reference is the result of dynamic interactive processes that cause schemas to come into being and that keep them attuned to specific objects.

11.4.2 Functional meaning

Functional semantics of speech acts emerge from their embeddings in the overall behaviour system of the agent. The functional class of a speech act is determined by word order and the location of specific closed-class words. The meaning of the classification is determined by the memory system that is effected. Descriptive speech acts are translated into updates on situation memory, effectively committing the agent to new beliefs that have potential 'cash value' in terms of future bodily interactions with the environment.

---

[8] In philosophical terms, the approach is closely connected to Peirce's pragmatism (Peirce 1878) and in contemporary philosophy to Millikan's functional semantics (Millikan 2004) and Bickhard's interactivism (Bickhard 1993).

Directive speech acts are translated into updates on drive memory, effectively biasing the agent to act in new ways in order to satisfy the new goals expressed by the speech acts.

11.4.3 Connotative meaning

Connotation refers to the implied, subjective meaning of a word or phrase as opposed to its conventional, literal meaning. In the model, connotative meaning of a word is the accumulated effect of the word's conceptual structure with respect to the motivational drives of the language user. For example, consider the connotation of 'heavy' for Ripley, given that the robot is driven by curiosity, the need to stay cool, and to do as it's told. In terms of doing what it is told to do, the meaning of 'heavy', like any property name, is useful since it lets Ripley pick out the object that the human wants. In this respect, the connotation of heavy is positive, or rather 'positive with respect to the verbal response drive.'

In contrast, any object that is heavy will be negative with respect to Ripley's other top-level drives, since manipulating heavy objects accelerates motor heating, and heavy objects often slip. Although Ripley's curiosity drive is currently based on visual evidence, it would be possible to extend curiosity to include the weight of objects in which case Ripley would be compelled to lift all novel objects to characterize their weights. In such a setting, the negative relation of heaviness to motor heat would also emerge during the activity of satisfying curiosity, which would now involve lifting all novel objects. Overall, we might say that the meaning of 'heavy' is neither totally negative nor positive, but instead is bittersweet. This mixed attitude is due to the interaction of Ripley's multidimensional drives, the particular conceptual structure underlying 'heavy', and the physical laws of force and heat that govern Ripley's bodily interactions with its environment. The model yields an analysis of connotative meanings as the agent's embodied experiences 'projected' through the lens of the particular concept. Even with Ripley's simple drive system we obtain a three-dimensional space for connotative meanings. More generally, the richness of connotative meanings grows with the richness of the drive system of the language user. As a child grows and acquires culturally rooted interests and drives, the effective dimensionality and subtleties of connotative meaning become correspondingly more intricate.

11.5 Discussion

To summarize, I believe the model presented here provides a new way of looking at how each facet of meaning emerges from common causes. Rather than propose three models of meaning, a single model gives rise to all three. The partial implementation of the model in the form of an embodied, interactive, autonomous robot provides a tangible example of a machine that 'grasps' linguistic meanings in a way that goes beyond any symbol manipulation model of meaning.

Debate

Reviewers of an earlier draft of this paper raised numerous questions and concerns, some of which I have tried to address in revisions to the paper. However, I felt that a subset of the questions was better answered in a question and answer format.

Manuel de Vega: 'Create' is a process of object schema token instantiation. If I understood, this means that the corresponding type schema must be stored in memory. My question: is it the case that only objects whose type schema was precompiled by the programmers can be perceived, manipulated, and understood? In other words, if you provide Ripley with a banana and it does not have any previous banana-type schema, could Ripley respond to descriptions and commands about it? My question is also related to Ripley's learning capabilities (or limitations).

Deb Roy: Only precompiled schemas may be instantiated in the current model. Schemas have parameters such as shape and colour which can accommodate new forms of objects. But the topological structure of schemas, as defined by the network of possible actions, is assumed to be fixed and preprogrammed. An important topic for future work is to explore how a machine could learn new schemas (i.e., concept learning), perhaps through a process of discovering new clusters of interaction patterns.

de Vega: 'Destroy' seems, psychologically, to be a radical procedure because object permanence takes place even if an object is hidden. The robot should distinguish between a hidden object that could reappear later on and an actually destroyed object that will never come back. Of course you know that, but then what is the motivation for including destroy? Maybe it is convenient to clean up the working memory in a purely on-line processor such as Ripley, but what about long term memory? For instance, let's suppose that you want to implement an improved version of Ripley with episodic long-term memory. We should ask Ripley displaced reference questions like: where was object X before (now)? To answer this, the destroy routine would be quite inconvenient, because the system should be able to track of past events and organize them into episodic memory traces.

Roy: In psychological terms, Ripley has working memory (the situation model, plan of action system, target situation memory, drive encodings) and long-term semantic memory (for storing conceptual schemas) but no long-term episodic memory. A promising future direction is to extend the model to include episodic memory by selectively encoding chains of activity of the various working memory systems. Furthermore, we might explore learning processes that compress multiple similar episodic traces to form new conceptual schemas.

Arthur Glenberg: Section 11.3.1 begins with the statement 'to account for any of the three facets of meaning, the language interpreter must have its own autonomous interest/goals/purposes.' This is a really interesting claim, but it is given little justification in section 11.3.1, and it is not until section 11.4.3 that it is cashed out. I would really like to see you justify the claim more completely in 11.3.1, or even in 11.4.3.

Roy: Schemas are structured bundles of possible actions. In other words, schemas are collections of potential plan fragments. As such, all instantiations of schemas (beliefs) and their meanings are inextricably imbued with the interests/goals/purposes of the beholder.

Arthur Graesser: The question of whether a system is embodied is another layer beyond the matter of a model being grounded. Systems can be grounded in perception and action, but not be embodied, whereas it would not make sense for an embodied model to be ungrounded. In my view, what makes a model embodied is that its representations and processes are constrained by the properties of the body. So I wonder where Ripley stands on the question of whether it is embodied.

Roy: The terms 'grounded' and 'embodied' have unfortunately been used in many overlapping ways by authors and are therefore bound to be confusing, and any specific definition contentious. That being said, I find it difficult to make sense of the assertion that 'Systems can be grounded in perception and action, but not be embodied' if we are talking about perception of, and action in, the physical world since that would be impossible without a physical body, and I think most definitions of embodiment involve bodies in one way or another. Ripley is a physically instantiated machine that senses and acts upon its physical environment so I would call it embodied, but I don't think there is much significance to being classified as such. The point of the model I have presented is not merely that it was implemented in an embodied system (robots have been around for a long time). Rather, the important idea here is to develop a continuous path from sensorimotor interactions to symbolic communication.

Graesser: In Ripley, schemas are organized around objects (e.g., the lift schema is embedded within the cup schema). These are essentially affordances. I worry that such a position is too bottom-up. A more goal-driven system would presumably also be needed.

Roy: I agree that schemas are closely related to the idea of affordances. Affordances as originally concieved by Gibson (1979) refer to the confluence of top-down and bottom-up factors, so there is no reason to 'worry that such a position is too bottom-up' anymore than one should worry that such a position is too top-down. Ripley's architecture is designed to blend top-down and bottom-up influences. On one hand the drive memory, target selection, and planning mechanisms drive top-down activity, whereas interpretation of bodily interactions with the environment via belief revision adapt to bottom-up influences of the environment.

Graesser: Ripley's continuous scanning of the environment may be a bit different than humans. Human attention is primarily drawn by changes in the environment.

Roy: Ripley's architecture is trivial in richness and capability when compared with humans. Thus, there are vast differences between its scanning processes and analogous functions in humans. That said, it would be inaccurate to say that human attention is 'primarily drawn by changes in the environment.' Attention depends critically on active goals and interests, not just changes in the environment (as shown, for example, by experiments in change blindness).

Graesser: Ripley lives in a simplified toy world, much like Winograd's Blocksworld in the early 1970s. There is a small finite number of goals, a limited number of agents and objects, a small spatial layout, assumed common ground in person–system interaction, and so on. One reason that it is necessary to work with toy worlds is because of combinatorial explosion problems and modules that otherwise would be impossible to implement (such as goal recognition in the wild, and assumed rather than computed common ground). It would be good to justify Ripley's being in the toy world for these and any other reasons.

Roy: As anyone who has tried to build an autonomous robot knows, operating on every-day objects in the physical world – even in the safe haven of Ripley's table-top world – is radically different from operating on symbolically described blocks in a virtual world of the kind that was popular in early artficial intelligence systems. This question appears to be based on a failure to grasp this difference. Since error and ambiguity do not arise in symbolic descriptions, there is no reason to make the fundamental distinction between belief and knowledge when dealing with symbolic microworlds. In contrast, all that Ripley can ever strive for are beliefs that translate into expecta-tions with respect to self-action. The world may of course not satisfy those expecta-tions, leading to belief revision. The model of language interpretation and meaning I have presented are grounded in this conception of belief. Issues of error, ambiguity, and categorization are at the core of the model. None of these issues even arise in clas-sic AI microworlds such as that of Winograd's SHRDLU (Winograd 1973). A better way to think of Ripley's simplified environment is that Ripley exists in a carefully controlled subworld in the sense defined by Dreyfus and Dreyfus (1988). Like the controlled subworld of an infant, Ripley's world opens up into our world as opposed to microworlds that are physically disconnected from the rest of reality. Justifications for keeping Ripley's subworld simple can be found in Sections 11.2.1 and 11.2.3.

Author note

Ripley was built in collaboration with Kai-yuh Hsiao, Nick Mavridis, and Peter Gorniak. I would like to acknowledge valuable feedback on earlier drafts of this paper from Kai-yuh Hsiao, Stefanie Tellex, Michael Fleischman, Rony Kubat, Art Glenberg, Art Graesser, and Manuel de Vega.

References

Barsalou L (1999). Perceptual symbol systems. *Behavioural and Brain Sciences*, 22, 577–609.

Bates E (1979). *The Emergence of Symbols*. New York, NY: Academic Press.

Bickhard MH (1993). Representational content in humans and machines. *Journal of Experimental and Theoretical Artificial Intelligence*, 5, 285–333.

Chomsky N (2000). New horizons in the study of language and mind. Cambridge: Cambridge University Press.

Cruse D (1986). *Lexical Semantics*. Cambridge: Cambridge University Press.

Drescher G (1991). *Made-up Minds*. Cambridge, MA: MIT Press.

Dreyfus H, Dreyfus S (1988). Making a mind versus modeling the brain: intelligence back at a branchpoint. *Daedalus*, 117, 15–43.

Gibson JJ (1979). *The Ecological Approach to Visual Perception*. Mahwah, NJ: Erlbaum.

Gorniak P, Roy D (2004). Grounded semantic composition for visual scenes. *Journal of Artificial Intelligence Research*, 21, 429–70.

Hsiao K, Roy D (2005). A habit system for an interactive robot. Presented at the AAAI Fall Symposium 2005, 'From reactive to anticipatory cognitive embodied systems.'

Landauer TK, Foltz P, Laham D (1998). Introduction to latent semantic analysis. *Discourse Processes*, 255, 259–84.

Lenat D (1995). Cyc: A large-scale investment in knowledge infrastructure. *Communications of the ACM*, 38, 33–8.

Mavridis N, Roy D (2006). Grounded situation models for robots: where words and percepts meet. Presented at the 2006 IEEE/RSJ International Conference on Intelligent Robots and Systems (IROS).

McKean E, Ed (2005). *The New Oxford American Dictionary*. New York, NY: Oxford Univeristy Press.

Miller G (1995). Wordnet: a lexical database for English. *Communications of the ACM*, 38, 39–41.

Millikan RG (2004). *Varieties of Meaning*. Cambridge, MA: MIT Press.

Narayanan S (1999). Moving right along: a computational model of metaphoric reasoning about events. Presented at the 1999 AAAI National Conference on Artificial Intelligence, Orlando, FL, USA.

Odgen C, Richards I (1923). *The Meaning of Meaning*. Harcourt.

Peirce CS (1878). How to make our ideas clear. *Popular Science Monthly*, 12, 286–302.

Quillian R (1968). Semantic memory. In M Minsky, Ed. *Semantic Information Processing*. Cambridge, MA: MIT Press.

Rosenblueth A, Wiener N, Bigelow J (1943). Behaviour, purpose and teleology. *Philosophy of Science*, 10, 18–24.

Roy D (2003). Grounded spoken language acquisition: experiments in word learning. *IEEE Transactions on Multimedia*, 5, 197–209.

Roy D (2005). Semiotic schemas: a framework for grounding language in action and perception. *Artificial Intelligence*, 167, 170–205.

Roy D, Hsiao K, Mavridis N (2004). Mental imagery for a conversational robot. *IEEE Transactions on Systems, Man, and Cybernetics, Part B*, 34, 1374–83.

Siskind J (2001). Grounding the lexical semantics of verbs in visual perception using force dynamics and event logic. *Journal of Artificial Intelligence Research*, 15, 31–90.

Winograd T (1973). A process model of language understanding. In *Computer Models of Thought and Language* (pp. 152–86). Freeman.

Wittgenstein L (1958). *The Blue and Brown Books*. Oxford: Basil Blackwell.

Chapter 12

The symbol grounding problem has been solved, so what's next?

Luc Steels

12.1 Introduction

In the 1980s, a lot of ink was spent on the question of symbol grounding, largely triggered by Searle's Chinese room theory (Searle 1980). Searle's article had the advantage of stirring up discussion about when and how symbols could be about things in the world, whether intelligence involves representations or not, what embodiment means, and under what conditions cognition is embodied. But almost 25 years of philosophical discussion have shed little light on the issue, partly because the discussion has been mixed up with emotional arguments whether artificial intelligence (AI) is possible or not. However, today I believe that sufficient progress has been made in cognitive science and AI so that we can say that the symbol grounding problem has been solved. This chapter briefly discusses the issues of symbols, meanings, and embodiment (the main themes of the workshop), why I claim the symbol grounding problem has been solved, and what we should do next.

12.2 Symbols

As suggested in Chapter 2, let us start from Peirce and the (much longer) semiotic tradition which makes a distinction between a symbol, the objects in the world with which the symbol is associated (for example, for purposes of reference), and the concept associated with the symbol (see Figure 12.1). For example, we could have the symbol 'ball', a concrete round spherical object in the world with which a child is playing, and the concept *ball* which applies to this spherical object so that we can refer to the object using the symbol 'ball'.

In some cases, there is a method that constrains the use of a symbol for the objects with which it is associated. The method could, for example, be a classifier – a perceptual/pattern recognition process that operates over sensorimotor data to decide whether the object 'fits' with the concept. If such an effective method is available, then we call the symbol grounded. There are a lot of symbols which are not about the real world but about abstractions of various sorts, like the word 'serendipity', or about cultural meanings, like the word 'holy water', and so they will never be grounded through perceptual processes. In this chapter I focus on groundable symbols.

12.2.1 **Semiotic networks**

Together with the basic semiotic relations in Figure 12.1, there are some additional semantic relations that provide additional pathways for navigation between concepts, objects, and symbols.

- Objects occur in a context and may have other domain relationships with each other, for example, hierarchical or spatial and temporal relations.

- Symbols co-occur with other symbols in texts and speech, and this statistical structure can be picked up using statistical methods (as in the latent semantic analysis proposal put forward by Landauer and Dumais [1997]).

- Concepts may have semantic relations among each other, for example, because they tend to apply to the same objects.

- There are also relations between methods, for example, because they both use the same feature of the environment or use the same technique for classification.

Humans effortlessly use all these links and make jumps far beyond what would be logically warranted. This is nicely illustrated with social tagging, a new web technology whose usage exploded in recent years (Golder and Huberman 2006; Steels 2006). On sites like Flickr or last.fm, users can associate tags (i.e. symbols) with pictures or music files (i.e. objects). Some of these tags are clearly related to the content of the picture and hence could, in principle, be grounded using a particular method. For example, a picture containing a significant red object may be tagged 'red', a piano concerto may be tagged 'piano'. But most tags are purely associative; for example, a picture containing a dog may be tagged 'New York' because the picture was taken in New York.

Tagging sites compute and display the various semantic relations between tags and objects: they display the co-occurrence relations between tags so that users can navigate between the most widely used tags, as well as the contexts in which objects occur, for example, all pictures taken by the same user. The enormous success of these sites shows that this kind of system resonates strongly with large numbers of people, and I believe this is the case because it reflects and externalizes the same sort of semiotic relationships and navigation strategies that our brains are using in episodic memory or language.

I will call the huge set of links between objects, symbols, concepts, and their methods, a *semiotic network*. Every individual maintains such a network which is entirely his or her own.

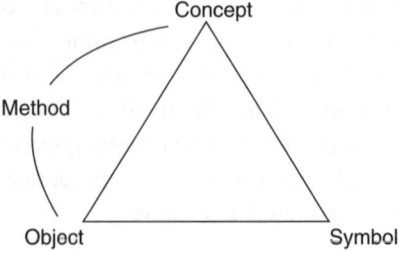

Fig. 12.1 The semiotic triad relates a symbol, an object, and a concept applicable to the object. The method is a procedure to decide whether the concept applies or not.

This semiotic network gets dynamically expanded and reshuffled every time you experience, think, or interact with the world or with others. Individuals navigate through this network for purposes of communication: when a speaker wants to draw the attention of an addressee to an object, he can use a concept whose method applies to the object, then choose the symbol associated with this concept and render it in speech or some other medium. The listener gets the symbol, uses his own vocabulary to retrieve the concept, and hence the method, and applies the method to decide which object might be intended.

Thus, if the speaker wants a bottle of wine on the table he is sitting at, he has to categorize this object as belonging to a certain class, then he has to look up the symbol associated with this class and use it to refer to the bottle. He might say, for example, 'Could you pass me the wine please?' Clearly, grounded symbols play a crucial role in communication about the real world, but the other links may also play important roles. For example, notice that the speaker said 'Could you pass me the wine please?,' whereas in fact he wanted the bottle that contained the wine. So the speaker navigated from container to content in his conceptualization of this situation, a move we commonly make. The speaker could also have said 'the Bordeaux please' and nobody would think that he requested the city of Bordeaux. Here, he navigated from container (bottle) to content (wine), from content (wine) to specific type of content (Bordeaux wine) characterized by the location where the content was made. Note that 'bottle of wine' might potentially be grounded in sensorimotor interaction with the world by some embodied pattern recognition method, but deciding whether a bottle contains Bordeaux wine is already a hopeless case, unless you can read and parse the label or are a highly expert connoisseur of wine.

There is apparently a debate in cognitive psychology between those emphasizing the grounded use of symbols (e.g. Barsalou 1999) and those emphasizing extraction and navigation across nongrounded semantic networks, and the current book reflects some of this debate. However, I think the opposition is a bit of a red herring. Both aspects of semantic processing are clearly extremely important and they interact with each other in normal human cognition. In this chapter I focus on grounded symbols not because I think other kinds of semantic processing are not relevant or important, but because in the Chinese room debate it is accepted that semantic processing can be done by computational systems whereas symbol grounding cannot.

12.2.2 Collective semiotic dynamics

Symbols can play a significant role in the development of an individual (see the example of children's drawings later in this chapter), but most of the time symbols are part of social interaction – such as in communication through language – and partners get feedback on how their own semiotic networks are similar or divergent from those of others. The semiotic networks that each individual builds up and maintains are therefore coupled to those of others and they get progressively coordinated in a group, based on feedback about their usage. If I ask for the wine and you give me the bottle of vinegar, both of us then learn that sometimes a bottle of vinegar looks like a bottle of wine. So we need to expand our methods for grounding 'wine' and 'vinegar' by tightening up the methods associated with the concepts that they express.

Psychological evidence for this progressive and continuous adaptation of semiotic networks is now beginning to come from many sources. First of all there are the studies of natural dialogue (Pickering and Garrod 2004; Clark and Brennan 1991) which show convincingly that speakers and hearers adopt and align their communication systems at all levels within the course of a single conversation. Their sound systems and gestures become similar, they adopt and negotiate new word meanings, they settle on certain grammatical constructions, and align their conceptualizations of the world. There are studies of emergent communication in laboratory conditions (Galantucci 2005) which show that the main skill required to establish a shared communication is the ability to detect miscommunication and repair them by introducing new symbols, changing the meaning of symbols, or adjusting your own conceptualizations based on taking the perspective of your partner. Growing evidence from linguistics (Francis and Michaelis 2005) show how human speakers and hearers engage in intense problem solving to repair inconsistencies or gaps in their grammars, and thus expand and align their linguistic systems with each other.

I call the set of all semiotic networks of a population of interacting individuals a *semiotic landscape*. Such a landscape is undergoing continuous change as every interaction may introduce, expand, or enforce certain relationships in the networks of individuals. Nevertheless there are general tendencies in the semiotic networks of a population, otherwise communication would not be possible. For example, it can be expected that individuals belonging to the same language community will have a similar network of concepts associated with the symbol 'red', and that some of these are grounded in a sufficiently similar way into the hue and brightness sensations so that if one says 'give me the red ball' she gets the red ball and not the blue one. Some tendencies also appear in other semantic relations between concepts. For example, in the Western world, the concept of 'red' is associated with danger, hence it is used in stop signs or traffic lights, whereas in some Asian cultures (e.g., China) 'red' is associated with joy.

Despite strong tendencies towards convergence, individual semiotic networks will never be exactly the same, even between two people interacting every day, as they are so much tied into personal histories and experiences. Psychological data confirms this enormous variation in human populations, even within the same language community (Webster and Kay 2005). The best one can hope for is that our semiotic networks are sufficiently coordinated to make joint action and communication possible.

12.2.3 The symbol grounding problem

Let me return now to the question originally posed by Searle (1980): can a robot deal with grounded symbols? More precisely, is it possible to build an artificial system that has a body, sensors and actuators, signal and image processing, pattern recognition processes, and information structures to store and use semiotic networks, and uses all that for communicating about the world or representing information about the world.

My first reaction is to say 'yes.' As far back as the early 1970s, AI experiments like Shakey the robot achieved this (Nilsson 1984). Shakey was a robot moving around and accepting commands to go to certain places in its environment. To act appropriately upon a command like 'go to the next room and approach the big pyramid standing in the corner,' Shakey had to perceive the world, construct a world model, parse sentences and interpret them in terms of this world model, and then make a plan and execute it. Shakey could do all this. It was slow, but then it used a computer with roughly the same power as the processors we find in modern-day hotel doorknobs, and about as much memory as we find in yesterday's mobile phones.

So what was all the fuss created by Searle about? Was Searle's paper (and subsequent philosophical discussion) based on ignorance or on a lack of understanding of what was going on in these experiments? Probably partly. It has always been popular to bash AI because that puts one in the glorious position of defending humanity. But one part of the criticism was justified: all was programmed by human designers. The semiotic relations were not autonomously established by the artificial agent but carefully mapped out and then coded by human programmers. The semantics therefore came from us, humans. Nils Nilsson and the other designers of Shakey carved up the world, they thought deeply how the semantics of 'pyramid' and 'big' could be operationalized, and they programmed the mechanisms associating words and sentences with their meanings. So the Chinese room argument, if it is to make sense at all, needs to be taken differently, namely that computational system cannot generate their own semantics whereas natural systems (e.g., human brains) can. Indeed the mind/brain is capable to develop autonomously an enormous repertoire of concepts to deal with the environment and to associate them with symbols that are invented, adopted, and negotiated with others.

So the key question for symbol grounding is not whether a robot can be programmed to engage in some interaction which involves the use of symbols grounded in reality through his sensorimotor embodiment: that question has been solved. It is actually another question, well formulated by Harnad (1990): if someone claims that a robot can deal with grounded symbols, we expect that this robot autonomously establishes the semiotic networks that it is going to use to relate symbols with the world.

AI researchers had independently already come to the conclusion that autonomous grounding was necessary. By the late 1970s it was already clear that the methods needed to ground concepts, and hence symbols, had to be hugely complicated, domain-specific, and context-sensitive if they were to work at all. Continuing to program these methods by hand was therefore out of the question and that route was more or less abandoned. Instead, all effort was put into the machine learning of concepts, partly building further on the rapidly expanding field of pattern recognition, and partly by using 'neural network'-like structures such as perceptrons.

The main approach is based on supervised learning: the artificial learning system is shown examples and counterexamples of situations where a particular symbol is appropriate and it is assumed to learn progressively the grounding of the concepts that underlie the symbol in question, and hence to learn how to use the concept appropriately. We now have a wide range of systems and experiments demonstrating that this is entirely feasible. Probably the most impressive recent demonstration is in the work of Deb Roy and his collaborators (see Chapter 11, this volume). They provided example sentences and example situations to a vision-based robotic system and the robot was shown to acquire progressively effective methods to use these symbols in subsequent real world interaction.

So does this mean that the symbol grounding problem is solved? I do not believe so. Even though these artificial systems now autonomously acquire their own methods for grounding concepts (and hence also symbols), it is still the human who sets up the world for the robot, carefully selects the examples and counterexamples, and supplies the symbol systems and conceptualizations of the world by drawing from an existing human language. So the semantics are still coming from us humans. Autonomous concept and symbol acquisition mimicking that of early child language acquisition is a very important step, but it cannot be the only one. The symbol grounding problem is not yet solved by taking this route and I believe it never will.

12.2.4 **Symbols in computer science**

Before continuing, I want to clear up a widespread confusion. The term 'symbol' has a venerable history in philosophy, linguistics, and other cognitive sciences, and I have tried to sketch its usage. However, it was hijacked in the late 1950s by computer scientists – more specifically AI researchers – and adopted in the context of programming language research. The notion of a symbol in so-called symbolic programming languages like LISP or Prolog is quite precise: it is a pointer (i.e., an address in computer memory) to a list structure containing a string known as the 'print name', which is used to read and write the symbol, possibly in addition to a value temporarily bound to this symbol, a definition of a function associated with this symbol, and an open-ended list of further properties and values associated with the symbol.

Thus, while the function of symbols can be recreated in other programming languages (like C++), the programmer then has to take care of allocating and deallocating memory for the internal pointer, reading and writing strings, and turning them into internal pointers, and he has to introduce more data structures for the other information items typically associated with symbols. In a symbolic programming language all that is done automatically. So a symbol is a very useful computational abstraction (like the notion of an array) and a typical LISP program might involve hundreds of thousands of symbols, created on the fly and reclaimed for memory ('garbage collected') when the need arises. Almost all sophisticated AI technology, as well as a lot of web technology, rests on the elegant but enormously powerful concept of symbolic programming.

Clearly this notion of symbol is not related to anything I discussed in the previous paragraphs, so I propose to make a distinction between *c-symbols* (the symbols of computer science)and *m-symbols* (meaning-oriented symbols in the tradition of the arts, humanities, and social and cognitive sciences). Unfortunately, when philosophers and cognitive scientists who are not knowledgable about computer programming read the AI literature they naturally apply the baggage of their own field, immediately assuming that if one uses a symbolic programming language, you must be talking about m-symbols and you possibly subscribe to all sorts of philosophical doctrines about symbols. Sadly, all this has given rise to what is probably the greatest terminological confusion in the history of science. The debate about the role of symbols in cognition or intelligence must be totally decoupled to whether one uses a symbolic programming language or not. The rejection of 'traditional AI' by some cognitive scientists or philosophers seems mostly based on this misunderstanding. Thus, it is perfectly possible to implement a neural network using symbolic programming techniques, but these symbols are then c-symbols and not the fully-fledged 'semiotic' m-symbols I discussed earlier.

12.3 Meaning

The debate on symbol grounding is tied up with another discussion concerning the nature and importance of representations for cognition, and the relation between representations and meanings. The notion of representation also has a venerable history in philosophy, art history, etc., before it was hijacked by computer scientists to become much more narrow in scope. Since then, however, the computational notion of representation has returned to cognitive science through the back door of AI. Neuroscientists now talk without fear about representations and try to find their neural correlates. Let me try to trace these shifts, starting with the original precomputational notion of representation, as we find for example in the famous essay on the hobby horse by Gombrich (1969) or in the more recent writings of Bruner (1990).

12.3.1 Representations and meanings

In traditional usage, a representation is a stand-in for something else, so that it can be made present again (i.e. re-present-ed). Anything can be a representation of anything. For example, a pen can be a representation of a boat, a person, or an upward movement; a broomstrick can be a representation of a hobby horse. The magic of representations happens because one person decides to establish that x is a representation for y, and others either agree with this or accept that this representational relation holds. There is nothing in the nature of an object that makes it a representation or not; it is rather the role the object plays in subsequent interaction. Of course, it helps a lot if the representation has some properties that help others to guess what it might be a representation of. But a representation is seldom a picture-like copy of what is represented.

It is clear that m-symbols are a particular type of representations, one where the physical object being used has a rather arbitrary relation to what is represented. It follows that all the remarks made earlier about m-symbols are also valid for representations. However, the notion of representation is broader. Following Peirce, we can make a distinction between an icon, an index, and a symbol. An icon looks like the thing it signifies, so meaning arises by perceptual processes and assocations, the same way we perceive the objects themselves. An example of an icon is a statue of a saint which looks like the saint, or how people imagine the saint to be. An index does not look like the thing it signifies, but there is nevertheless some causal or associative relation. For example, smoke is an index of fire. A symbol on the other hand has no direct or indirect relation to the thing it signifies. The meaning of a symbol is established purely by convention, and hence you have to know the convention in order to figure out what the symbol is about.

Human representations have some additional features. First of all they seldom represent physical things, but rather meanings. A meaning is a feature that is relevant in the interaction between the person and the thing being represented. For example, a child may represent a fire engine by a square painted red. Why red? Because this is a distinctive important feature of the fire engine for the child. There is a tendency to confuse meanings and representations. A representation 're-presents' meaning but should not be equated with meaning, just like an ambassador may represent a country but is not equal to that country. Something is meaningful if it is important in one way or another: for survival, maintaining a job, social relations, navigating in the world, etc. For example, the differences in colour between different mushrooms may be relevant to me because they help to distinguish those that are poisonous from those that are not. If colour is irrelevant and shape instead is distinctive, then the shape features of mushroom would be meaningful (Cangelosi *et al.* 2000).

Conceiving a representation requires selecting a number of relevant meanings and deciding how these meanings can be invoked in ourselves or others. The invocation process is always indirect and most often requires inference. Without knowing the context, personal history, prior use of representations, etc., it is often very difficult, if not impossible, to decipher representations. Second, human representations typically involve perspective. Things are seen and represented from particular points of view and these perspectives are intermixed.

A nice example of a representation is shown in Figure 12.2. Perhaps you thought that this drawing represents a garden, with the person in the middle watering the plants. Some people report that they interpret this as a kitchen with pots and pans and somebody cooking a meal. As a matter of fact, the drawing represents a British double-decker bus (the word 'baz' is written on the drawing). Once you adopt this interpretation, it is easy to guess that the bus driver must be sitting on the right-hand side in his enclosure.

Fig. 12.2 Drawing by a child called Monica. Can you guess what meanings this representation expresses?

The conductor is shown in the middle of the picture. The top shows a set of windows of increasing size, and the bottom a set of wheels. Everything is enclosed in a big box which covers the whole page. This drawing was made by a child, Monica, aged 4.2 years.

Clearly this drawing is not a realistic depiction of a bus. Instead it expresses some of the important meanings of a bus from the viewpoint of Monica. Some of the meanings have to do with recognizing objects, in this case, recognizing whether a bus is approaching so that you can get ready to get on it. A bus is huge. That is perhaps why the drawing fills the whole page, and why the windows at the top and the wheels at the bottom are drawn as far apart as possible. A bus has many more windows than an ordinary car, so Monica has drawn many windows, and the same thing for the wheels: a bus has many more wheels than an ordinary car so many wheels are drawn. The exact number of wheels or windows does not matter, there must only be enough of them to express the concept of 'many'. The shape of the windows and the wheels has been chosen by analogy with their shape in the real world. They are positioned at the top and the bottom as in a normal bus viewed sideways. The concept of 'many' is expressed in the visual grammar invented by the child.

Showing your ticket to the conductor or buying a ticket must have been an important event for Monica. It is of such importance that the conductor, who plays a central role in this interaction, is put in the middle of the picture and drawn prominently to make them stand out as foreground. Once again, features of the conductor have been selected that are meaningful, in the sense of meaningful for recognizing the conductor. The human figure is schematically represented by drawing essential body parts (head, torso, arms, legs). Nobody fails to recognize that this is a human figure, which cannot be said for the way the driver has been drawn. The conductor carries a ticketing machine. Monica's mother, with whom I corresponded about this picture, wrote that this machine makes a loud 'ping' and would be impossible not to notice. There is also something drawn on the right side of the head, which is most probably a hat, another characteristic feature of the conductor. The activity of the conductor is represented, too: the right arm is extended as if ready to accept money or check a ticket.

Composite objects are superimposed and there is no hesitation to mix different perspectives. This happens quite often in children's drawings. If they draw a table viewed from the side, they typically draw all the forks, knives, and plates as separate objects 'floating above' the table. In Figure 12.2, a bird's eye perspective is adopted so that the driver is located on the right-hand side in a separate box and the conductor in the middle. At the same time, the bus is viewed from the side so that the wheels are at the bottom and the windows near the top. Then there is the third perspective of a sideways view (as if seated inside the bus), which is used to draw the conductor as a standing up figure. Even for drawing the conductor, two perspectives are adopted: the sideways view for the figure itself and the bird's eye view for the ticketing machine so that we can see what is inside. Multiple perspectives in a single drawing are later abandoned as children try to make their drawings more 'realistic', and hence less creative. But it may reappear again in artists' drawings. For example, many of Picasso's paintings play around with different perspectives on the same object in the same image.

The fact that the windows get bigger towards the front expresses another aspect of a bus which is most probably very important to Monica. At the front of a double-decker bus there is a huge window and it is great fun for a child to sit and watch the streets go by. The increasing size of the windows reflects the desirability to be near the front. Size is a general representational tool used by many children (and also in mediaeval art) to emphasize what is considered to be important. Most representations express not only facts but above all attitudes and interpretations of facts. Though the representations of very young children may seem to be random marks on a piece of paper, for them they are almost purely emotional expressions of attitudes and feelings, like anger or love.

Human representations are clearly incredibly rich and serve multiple purposes. They help us to engage with the world and share world views and feelings with others. Symbols or symbolic representations traditionally have this rich connotation. And I believe the symbol grounding problem can only be said to be solved if we understand, at least in principle, how individuals originate and choose the meanings that they find worthwhile to use as basis for their (symbolic) representations, how perspective may arise, and how the expression of different meanings can be combined to create compositional representations.

12.3.2 Representations in computer science

In the 1950s, when higher-level computer programming started to develop, computer scientists began to adopt the term 'representation' for data structures that held information for an ongoing computational process. For example, in order to calculate the amount of water left in a tank with a hole in it, you have to represent the initial amount of water, the flow rate, the amount of water after a certain time, etc., so that calculations can be done over these representations. Information processing came to be understood as designing representations and orchestrating the processes for the creation and manipulation of these representations.

Here, computer scientists (and *ipso facto* AI researchers) are clearly adopting only one aspect of representations: the creation of a 'stand-in' for something into the physical world of the computer so that it could be transformed, stored, transmitted, etc. They are not trying to operationalize the much more complex process of meaning selection, representational choice, composition, perspective taking, inferential interpretation, etc., all of which are an important part of human representation making. I will use the term *c-representations* to mean representations as used in computer science and *m-representations* for the original, meaning-oriented use of representations in social science and humanities.

In building AI systems there can be no doubt that c-representations are needed to do anything at all. It is also now accepted among computational neuroscience researchers that c-representations must be used in the brain (i.e., information structures for visual processing, motor control, planning, language, memory, etc.). Mirror neuron networks are a clear example of c-representations. If the same circuits become active both in the perception of an action and the execution of it, it means that these circuits act as representations of actions, usable as such by other neural circuits. So the question today is no longer whether or not c-representations are necessary but rather what their nature might be.

As AI research progressed, a debate arose between advocates of 'symbolic' c-representations and 'nonsymbolic' c-representations. Simplified, symbolic c-representations are representations for categories, classes, individual objects, events, or anything else that is relevant for reasoning, planning, language, or memory. Nonsymbolic c-representations, on the other hand, use continuous values and are therefore much closer to the sensory and motor streams. They have also been called analogue c-representations because their values change by analogy with a state of the world, like a temperature measure which goes up and down when it is hotter or colder. Thus, we could have on the one hand a sensory channel for the wavelength of light with numerical values (a nonsymbolic c-representation), and on the other hand (symbolic) c-representations for colour categories like 'red', 'green', etc. Similarly, we could have an infrared or sonar channel that numerically reflects the distance of the robot to obstacles (a nonsymbolic c-representation), or we could have a c-symbol representing 'obstacle seen' (a symbolic c-representation).

Sometimes the term 'subsymbolic' c-representation is used to mean either nonsymbolic c-representation or a distributed symbolic c-representation, as employed in connectionist systems (Rumelhart and McClelland 1986; Smolensky 1988). The latter type assumes that the representation (i.e., the physical object representing a concept) is actually a cloud of more or less pertinent features, themselves represented as primitive nodes in the network.

When the neural network movement became more prominent again in the 1980s, they de-emphasized symbolic c-representations in favour of the propagation of (continuous) activation values in neural networks (although some researchers used these networks later in an entirely symbolic way, as pointed out in the discussion of Rogers *et al.* in Chapter 2). When Brooks wrote a paper on 'Intelligence without representation' (Brooks 1991), he argued that real-time robotic behaviour could often be better achieved without symbolic c-representations. For example, instead of having a rule that says 'if obstacle then move back and turn away from obstacle,' we could have a dynamical system that directly couples the change in sensory values to a change on the actuators. The more infrared reflection from obstacles is picked up by the right infrared sensor on a robot, the slower its left motor is made to move, and hence the robot starts to veer away from the obstacle.

But does this mean that all symbolic c-representations have to be rejected? This does not seem to be warranted, particularly not if we are considering language processing, expert problem solving, conceptual memory, etc. At least from a practical point of view it makes much more sense to design and implement such systems using symbolic c-representations and mix symbolic, nonsymbolic, and subsymbolic c-representations whenever appropriate. The debate between symbolic or nonsymbolic representations seems to be based on an unnecessary opposition, just as the debate between grounded and non-grounded symbols. The truth is that we need both.

At the same time, the simple fact of using a c-representation (symbolic or otherwise) does not yet mean that an artificial system is able to come up or interpret the meanings that are represented by the representation. Meaning and representation are different things. In order to see the emergence of meaning, we need a minimum of a task, an environment, and an interaction between the agent and the environment that works towards an achievement of the task.

12.4 Embodiment

We have seen that the notions of symbol and representation are used quite differently by computer scientists than by cognitive scientists, muddling the debate about symbol grounding. Computer science symbols and representations capture only a highly limited aspect of what social and cognitive scientists mean by symbols and representations. This has also happened, I think, with a third notion the debate has considered: embodiment.

12.4.1 Embodiment as implementation

In the technical literature (e.g., in patents), the term embodiment refers to implementation (i.e., the physical realization) of a method or idea. Computer scientists more commonly use the word *implementation*. An implementation of an algorithm is a definition of that algorithm in a form such that it can be physically instantiated on a computer and actually run (i.e., that the various steps of the algorithm find their analogue in physical operations). It is absolute dogma in computer science that an algorithm can be implemented in many different physical media, on many different types of computer architectures, and in different programming languages, even though, of course, it may take much longer (either to implement or to execute) in one medium versus another.

Computer scientists naturally take a systems perspective when they try to understand a phenomenon. In contrast, most physical natural scientists seek material explanations, they seek an explanation in the materials and material properties involved in a phenomenon. For example, their explanation of why oil floats on top of water is in terms of the attraction properties of molecules, the density of oil and water, and the upward buoyancy forces. System explanations, on the other hand, are in terms of elements and processes between the elements. For example, the explanation for money is not in terms of the material properties of coins or bank notes, but in terms of legal agreements, conventions, central banks, trust relations, etc. The franc, lire, or Deutschmark were replaced overnight by the Euro and everything kept going.

It is therefore natural that computer scientists apply system thinking to cognition and the brain. Their goal is to identify the algorithms (in other words the systems) underlying cognitive activities like symbol grounding and study them through various kinds of implementations which may not at all be brain-like. They might also talk about neural implementation or neural embodiment, meaning the physical instantiation of a particular algorithm with the hardware components and processes available to the brain. But most computer scientists know that the distance between an algorithm (even a simple one) and its embodiment, for example in an electronic circuit, is huge, with many layers of complexity intervening. Moreover, the translation through all these layers is never done by hand but by compilers and assemblers. It is very possible that we will have to follow the same strategy to bridge the gap between high-level models of cognition and low-level neural implementations.

There has been considerable resistance both from biologists and philosophers alike for accepting the systems perspective, i.e., the idea that the same process (or algorithm) can be instantiated (embodied) in many different media. They seem to believe that the (bio-)physics of the brain must have unique characteristics with unique causal powers. For example, Penrose (1989) has argued that intelligence and consciousness is due to certain (unknown) quantum processes which are unique to the brain, and therefore any other type of implementation can never obtain the same functionality. This turns out to be also the fundamental criticism of Searle. The reason why he argues that artificial systems, even if they are housed in robotic bodies, cannot deal with symbols is because they will for ever lack *intentionality*. Intentionality is 'that feature of certain mental states by which they are directed at or about objects and states of affairs in the world' (Searle 1980, footnote 2), and it is of course essential for generating and using symbols about the world. According to Searle, an adequate explanation for intentionality can only be a material one: 'Whatever else intentionality is, it is a biological phenomenon, and it is as likely to be as causally dependent on the specific biochemistry of its origins as lactation, photosynthesis, or any other biological phenomena.' If that is indeed true, investigating the symbol grounding problem through experiments with artificial robotic agents is totally futile.

Searle invokes biology, but most biologists today accept that the critical features of living systems point to the need to adopt a system perspective instead of a material one; even photosynthesis can be done through many different materials. For example, leading evolutionary biologist John Maynard Smith (2000) has been arguing clearly that the genetic system is best understood in information processing terms and not (just) in molecular terms. Neurobiologists Gerald Edelman and Giulilo Tononi (2000) have proposed system explanations for phenomena-like consciousness in terms of re-entrant processing and coordination of networks of neural maps, rather than specific biosubstances or quantum processes. Thus, the system viewpoint is gaining more and more prominence in biology rather than diminishing in productivity.

12.4.2 Embodiment as having a physical body

There is another notion of embodiment that is also relevant in this discussion. This refers quite literally to 'having a body' for interacting with the world, i.e. having a physical structure, dotted with sensors and actuators, and the necessary signal processing and pattern recognition to bridge the gap from reality to symbol use. Embodiment in this sense is a clear precondition to symbol grounding. As soon as we step outside the realm of computer simulations and embed computers in physical robotic systems, we achieve some form of embodiment (even if the embodiment is not the same as human embodiment). It is entirely possible, and in fact quite likely, that the human body and its biophysical properties for interacting with the world make certain behaviours possible and allow certain forms of interaction that are unique. This puts a limit on how similar AI systems can be to human intelligence. But it only means that embodied AI is necessarily of a different nature due to the differences in embodiment, not that embodying AI is impossible.

12.5 A solution to the symbol grounding problem?

Over the past decade I have been working with a team of a dozen graduate students at the University of Brussels (VUB AI Lab) and the Sony Computer Science Lab in Paris on various experiments in language emergence (Steels 2003). They take place in the context of a broader field of study concerned with modelling language evolution (see Minett and Wang 2005 or Vogt 2006 for recent overviews; see Vogt 2002 for other examples). We are carrying them out to investigate many aspects of language and cognition, but here I focus on their relevance for the symbol grounding problem. I will illustrate the ingredients that we put in these experiments using a specific example of a colour guessing game, discussed in much more detail in Steels and Belpaeme (2005).

12.5.1 Embodiment

A first prerequisite for solving the symbol grounding problem is that we can work with physically embodied autonomous agents, autonomous in the sense that they have their own source of energy and computing power, they are physically present in the world through a body, they have a variety of sensors and actuators to interact with the world, and, most importantly, they move and behave without any remote control or further human intervention once the experiment starts. All these conditions are satisfied in our experiments.

We have been using progressively more complex embodiments, starting from pan-tilt cameras in our earlier 'Talking Heads' experiments, moving to Sony AIBO dog-like robots, and, more recently, fully-fledged humanoid QRIO robots (see Figure 12.2). These robots are among the most complex physical robots currently available. They have a humanoid shape with two legs, a head, and two arms with hands and fingers. The robots are 0.6 metres (2 feet) tall and weigh 7.3 kilograms (16 pounds). They have cameras for visual input, microphones for audio input, touch and infrared sensors, and a large collection of motors at various joints, with sensors at each motor. The robots have a huge amount of computing power on board with general- and special-purpose processors. In addition they are wirelessly connected to off-board computers which can be harnassed to increase computing power, up to the level of supercomputers. Even extremely computation-intensive image processing and motor control is possible in real time. The robots have been programmed using a behaviour-based approach (Brooks 1991) so that obstacle avoidance, locomotion, tracking, grasping, etc., are all available as solid smooth behaviours to build upon.

By using these robots, we achieve the first prerequisite for embodied cognition and symbol grounding, namely that there is a rich embodiment. Often we use the same robot bodies for different agents by uploading and downloading the complete state of an agent after and before a game. This way we can do experiments with larger population sizes.

12.5.2 Sources of meaning

If we want to solve the symbol grounding problem, we next need a mechanism by which an (artificial) agent can autonomously generate its own meanings. This means that there must be distinctions that are relevant to the agent in its agent–environment interaction. The agent must therefore have a way to introduce new distinctions based on the needs of the task. As also argued by Cangelosi *et al.* (2000), this implies that there must be a task setting in which some distinctions become relevant and others do not.

One could imagine a variety of activities that generate meaning, but we have focused on language games. A language game is a routinized situated interaction between two embodied agents who have a cooperative goal (e.g., one agent wants to draw the attention of another agent to an object in the environment, or one agent wants the other one to perform a particular action) and who use some form of symbolic interaction to achieve that goal (e.g., by exchanging language-like symbols or sentences, augmented with nonverbal interactions such as pointing or bodily movement towards objects).

In order to play a language game the agents need a script with which they can establish a joint attention frame, in the sense that the context becomes restricted and it is possible to guess meaning and interpret feedback about the outcome of a game. Specifically, in the colour guessing game shown in Plate 12.1 (the so-called Mondriaan experiment), agents walk towards a table on which there are colour samples. One agent randomly becomes speaker, chooses one sample as topic, uses his available grounding methods to categorize the colour of this sample as distinctive from the other colour samples, and names the category. The other agent is hearer; it decodes the name to retrieve the colour category, uses the method associated with this category to see to which sample it applies, and then points to the sample. The game is a success if the speaker agrees that the hearer pointed to the sample it had originally chosen as topic.

Needless to say that the robots have no idea which colours they are going to encounter. We introduce different samples and can therefore manipulate the environment driving the categories and symbols that the agents need. For example, we can only introduce samples with different shades of blue (as in Plate 12.1) which will lead to very fine-grained distinctions in the blue range of the spectrum, or we can spread the colours far apart.

12.5.3 Grounding of categories

Next we need a mechanism by which agents can internally represent and ground their relevant meanings. In the experiment, agents start with no prior inventory of categories and no inventory of methods (classifiers) that apply categories to the features (sensory experience) they extracted from the visual sensation they received through their cameras. In the experiments, the classifiers use a prototype-based approach implemented with radial basis function networks (Steels and Belpaeme 2005). A category is distinctive for a chosen topic if the colour of the topic falls within the region around a particular prototype and all other samples fall outside of it. For example, if there is a red, green, and blue sample and the red one is the topic, then if the topic's colour falls within the region around the red prototype and the others do not, red is a valid distinctive category for the topic. If an agent cannot make a distinction between the colour of the topic and the colour of the other samples in the context, it introduces a new prototype and will later progressively adjust or tighten the boundaries of the region around the prototype by changing weights.

12.5.4 Self-organization of symbols

The next requirement is that agents autonomously can establish and negotiate symbols to express the meanings that they need to express. In the experiment, agents generate new symbols by combining randomly a number of syllables into a word, like 'wabado' or 'bolima'. The meaning of a word is a perceptually grounded category. No prior lexicon is given to the agents, and there is no central control that will determine by remote control how each agent has to use a word. Instead, a speaker invents a new word when it does not have a word yet to name a particular category; the hearer will try to guess the meaning of the unknown word based on feedback after a failed game, and thus new words enter into the lexicons of the agents and propagate through the group.

Coordination process

If every agent generates his own meanings, perceptually grounded categories, and symbols, then no communication is possible, so we need a process of coordination that creates the right kind of semiotic dynamics so that the semiotic networks of the individual agents become sufficiently coordinated to form a relatively organized semiotic landscape.

This is achieved in two ways. Firstly, speakers and hearers continue to adjust the score of form-meaning associations in their lexicon based on the outcome of a game: When the game is a success they increase the score and dampen the score of competing associations; when the game is a failure the score is diminished. The net effect of this update mechanism is that a positive feedback arises: words that are successful are used more often and hence become even more successful. After a while the population settles on a shared lexicon (see Figure 12.3).

Speakers and hearers also maintain scores about the success of perceptually grounded categories in the language game, and adjust these scores based on the outcome. As a consequence, the perceptually grounded categories also get coordinated in the sense that they become more similar, even though they will never be absolutely identical. This is shown for the colour guessing game in Plate 12.2. We note that if agents just play discrimination games but do not use their perceptually grounded categories as part of language games, they succeed in discrimination but their categorical repertoires will show much more variation.

We have been carrying out many more experiments addressing issues related to the origins of more complex meanings, the more complex use of embodiments (e.g., in action), and the emergence of more complex human language-like symbolic representations, even with grammatical structure (Steels 2005). For example, our 'perspective reversal experiment' has demonstrated how agents self-organize an inventory of spatial categories and spatial language symbolizing these categories, as well as the ability to use perspective reversal and the marking of perspective (Steels and Loetzsch 2007).

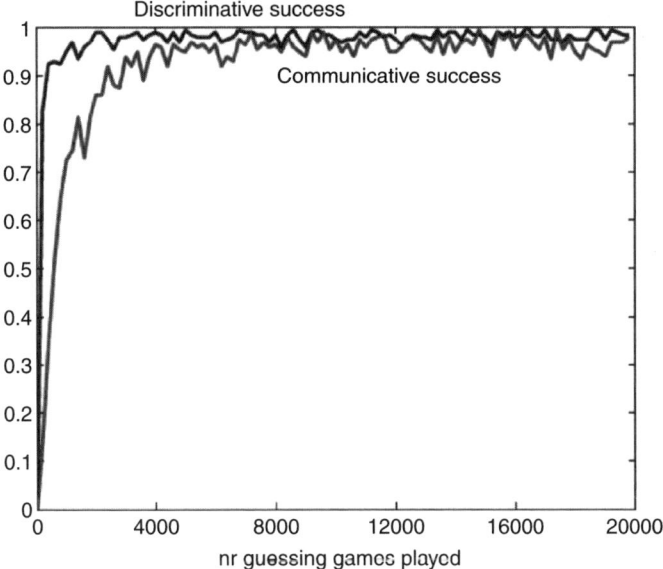

Fig. 12.3 Results of experiments showing the coordination of colour categories in a cultural evo-lution process. A population of 20 agents play guessing games naming colour distinctions and coordinating both their lexicon and colour categories through the game. We see progressively that discriminative success and communicative success (y-axis) increase as the number of games progresses (x-axis).

12.5.5 **So, have we solved the grounding problem?**

I argue that these experiments show that we have an effective solution to the symbol grounding problem, if there is ever going to be one: we have identified the right mecha-nisms and interaction patterns so that the agents autonomously generate meaning, autonomously ground meaning in the world through a sensorimotor embodiment and perceptually grounded categorization methods, and autonomously introduce and nego-tiate symbols for invoking these meanings. The objective test for this claim is in the increased success of agents in the language games. Clearly, if the agents do not manage to generate meanings and coordinate their perceptually grounded categories and symbols, they will have only a random chance of succeeding, whereas we see that they reach almost total success in the game. There is no human prior design to supply the symbols or their semantics, neither by direct programming nor by supervised learning.

The explanatory power of these experiments does not come from the identification of some biochemical substance for intentionality of the sort Searle is looking for, but it is a system explanation in terms of semiotic networks and semiotic dynamics operating over these networks. Each agent builds up a semiotic network relating sensations and sensory experiences to perceptually grounded categories and symbols for these cate-gories (see Plate 12.3). All the links in these networks have a continuously valued strength or score that is continually adjusted as part of the interactions of the agent with the environment and with other agents. Links in the network may at any time be added or changed as a side effect of a game. Although each agent does this locally, an overall coordinated semiotic landscape arises.

12.6 Discussion

The main goal of this chapter was to clarify some terminology and attempt to indicate where we are with respect to one of the most fundamental questions in cognition, namely the symbol grounding problem. From the viewpoint of AI, the question is whether it is possible to ever conceive of an artificial system that is able to invent and use grounded symbols in its sensorimotor interactions with the world and others. Several discussants, most notably Searle, have argued that this will never be possible because artificial systems will forever lack the critical biomaterial substance. I now boldly state that the symbol grounding problem is solved, by that I mean we now understand enough to create experiments in which groups of agents self-organize symbolic system that are grounded in their interactions with the world and others.

So where do we go from here? Clearly there is still a lot to learn. First of all, we can do many more experiments with artificial robotic agents to progressively understand many more aspects of meaning, conceptualization, symbolization, and the dynamical interactions between them. These experiments might focus, for example, on issues of time and its conceptualization in terms of tense, mood, modality, and the roles of objects in events, as well as its expression in case grammars, the categorization of physical actions and their expression, raising issues in the mirror neuron debate, etc. I see this as the job of AI researchers and (computational) linguists.

Second, we can investigate whether there are any neural correlates in the brain for the semiotic networks being discussed here, which would be impossible to find with local cell recordings but would require looking at global connections and long-distance correlations between neural group firings (Sporns and Tononi 2005). The work of Pulvermüller (Chapter 6, this volume) has already been able to identify some of these networks and shown that they span the whole brain. For example, smell words activate areas for olfactory processing or action words activate cells in motor circuitry. We can also investigate whether we can find neural correlates for the dynamical mechanisms that allow individuals to participate in the semiotic dynamics going on in their community. All this is going to be the job of cognitive neuroscientists.

Finally, there is a need for new types of psychological observations and experiments investigating representation-making in action (for example in dialogue or drawing) and investigating group dynamics. The experiments I have been discussing provide models to understand how humans manage to self-organize grounded communication systems, although I have produced no evidence in this chapter for such a claim: this is the job of the experimental psychologists.

All too often psychological experiments have focused on the single individual in isolated circumstances, whereas enormously exciting new discoveries can be expected if we track the semiotic dynamics in populations, something that is becoming more feasible thanks to internet technologies. Psychological experiments have often assumed as well that symbol use and categorization is static, whereas the interesting feature is precisely in its dynamics.

I believe there has been clear progress on the issue of symbol grounding and that there will be much more progress in the coming decade, provided we keep an open mind, engage in interdisciplinary curiosity, and avoid false debates.

Debate

Arthur Graesser: I was curious whether you tried to have your robots handle 'my,' 'your,' or 'our' with a modifier, which would be very diagnostic of perspective and point of view, like 'on *my* left,' 'on *your* left,' 'on *the* left,' 'on *our* left,' and things like that. And that's where you also get syntax coming in, and ordering. Is there any emergent way that was helping them figure out these different points of view?

Luc Steels: Well, right now in the experiment there is just my point of view and your point of view. But we are currently doing one in which there would be several points of view, like a third robot who also sees the situation. We needed to set up a communicative challenge to see the thing emerge. I think syntax is not necessarily because of that. You know, we can go into that, but that's a bigger topic. I have precise ideas about this, which I'm willing to share, but maybe not now.

Mitchell Nathan: So Luc, in one of your very first slides, you made this pure convention claim, but what if we did the following experiment, if we ask people to decide which of the labels 'tanak' or 'oblio' should be assigned to a sawtooth image and which to a smoothly curving image? Would it come out as purely arbitrary or would we find some agreement, and if we did, what does that imply?

Steels: Well we could do the experiment later, maybe during the coffee break. But personally, I think the assignment would be arbitrary. If it's really names.

Deb Roy: I had a question. I noticed several times you emphasized that things were being learnt, that they weren't innate. But I thought you are modelling evolutionary processes, so I was wondering if you could clarify. Because typically when people make that distinction of learnt versus innate, they mean these are not evolved structures; these are lifetime learnt. And yet all of the experiments, as I understood them, were models of evolutionary processes. So are you seeing a relationship here between what we call language acquisition in a lifetime versus evolution of language, or are you treating this process as sort of contiguous of both? What do you mean by innate versus learnt?

Steels: This is a very good point, which I didn't clarify. So, these things are cultural processes, it's not evolution over centuries. I mean this happens very quickly, right? And I think what psychological dialogue studies have shown us in the last 10 years is in fact that human beings, when they sit together and go into a dialogue, they continuously adapt their communication system at all levels. And this is going on here, too. Now, the model we should adopt for understanding symbolization is not that at one time somebody symbolized. It's a continuously adaptive process, you know, all the time changing. It's true that here we start from scratch – this is because I wanted to emphasize that we didn't put anything in, and also to find an explanation for the origin of these things; how it is possible that language has emerged? But in fact if there's already a system, then you can throw in new agents and they will learn preferably the system that's already there. This is because they are exposed with a higher frequency to existing conventions and so they absorb them. But, in that sense, it's a cultural process. Also, I briefly mentioned recruitment. This is obviously happening in developmental time, so when you pull in new systems – an emotional system, perspective transform, or what have you – they are all pulled into the language faculty in developmental time. So in terms of what is innate I would say it's the fundamental hardware of all these mechanisms, like bidirectional associative memory, etc., dealing with sequences – all of that stuff. You have to pull them in to be able to participate in the collective construction of language.

Roy: We talked earlier about the process of symbolization, building on things you hinted at. I was wondering what might be uniquely human about that process. Sort of the possibility of looking for what is maximally distinctive between two things and having the motor repertoire to mimic it, or to indicate it.

Lawrence Shapiro: I was wondering what role embodiment plays in the learning of these categories. It occurred to me that it depends on the sort of categories that you want to learn. We could play this game just sitting around the table here and that doesn't seem to require embodiment in the senses that I've heard others discussing.

Steels: Well, if you play around the table we need still the perceptual system, we need the pointing for gestures. But this is, I think, where this perspective reversal experiment is relevant. Because Chomsky, for example, would say language has nothing to do with communication, right? It evolved for internal representation and then, accidentally, it became used externally through some translation. Now then, how would you explain that perspective marking is so common in languages, right? 'Your left,' 'my left,' etc. So, the problem of perspective is a direct consequence of the problem of being embodied in the world. We had to put that in these different, unpredictable perspectives to get this feature of language going. That's part of the answer.

Friedemann Pulvermüller: Thanks for an excellent talk, very interesting. I'm not sure whether I fully followed and so allow me a potentially very stupid question. What does it buy the agents if they align their word usage?

Steels: Well, if I say 'ba' and this means left for you but right for me, or move forward for you and move left for me, we're going to have failure in communication, right? The whole process is tremendously difficult because every situation can be conceptualized from many points of view. You get a search process. You're dealing with noisy sensing, you know, multiple world models. So you're swimming in a sea of uncertainty and you try to constrain the degrees of freedom that you need to explore. Now the more you are aligned, the higher your chances are to succeed. And so that's why I think you see in many experiments now, of Simon Garrod and others, that people engaged in dialogue on the spot very quickly invent new words and new grammatical constructions.

Author note

This research was funded and carried out at the Sony Computer Science Laboratory in Paris, with additional funding from the European Union Future Emerging Technologies EC Agents Project IST-1940, and at the University of Brussels VUB AI Lab. I am indebted to the participants of the Garachico workshop on symbol grounding for their tremendously valuable feedback and to the organisers Manuel de Vega, Art Glenberg, and Arthur Graesser for orchestrating such a remarkably fruitful workshop. Experiments and graphs discussed in this paper were based on the work of many of my collaborators, but specifically Tony Belpaeme, Joris Bleys, and Martin Loetzsch.

References

Barsalou L (1999). Perceptual symbol systems. *Behavioural and Brain Sciences*, 22, 577–609.

Brooks R (1991). Intelligence without representation. *Artificial Intelligence Journal*, 47, 139–59.

Bruner J (1990). *Acts of Meaning*. Cambridge, MA: Harvard University Press.

Cangelosi A, Greco A, Harnad S (2002). Symbol grounding and the symbolic theft hypothesis. In A Cangelosi, D Parisi, Eds. (2000). *Simulating the Evolution of Language*. Berlin: Springer Verlag.

Clark H, Brennan S (1991). Grounding in communication. In: L Resnick, S Levine, S Teasley, Eds. *Perspectives on Socially Shared Cognition* (pp. 127–49). Washington, DC: APA Books.

Davidoff J (2001). Language and perceptual categorisation. *Trends in Cognitive Sciences*, 5, 382–7.

Edelman G (1999). *Bright Air, Brilliant Fire: On the Matter of the Mind*. New York, NY: Basic Books.

Edelman G, Tononi G (2000). *A Universe of Consciousness. How Matter Becomes Imagination*. New York, NY: Basic Books.

Francis E, Michaelis L (2002). *Mismatch: Form-Function Incongruity and the Architecture of Grammar*. Stanford, CA: CSLI Publications.

Galantucci B (2005). An experimental study of the emergence of human communication systems. *Cognitive Science*, 29, 737–67.

Golder S, B Huberman (2006). The structure of collaborative tagging. *Journal of Information Science*, 32, 198–208.

Gombrich EH (1969). *Art and Illusion: A Study in the Psychology of Pictorial Representation*. Princeton, NJ: Princeton University Press.

Harnad S (1990). The symbol grounding problem. *Physica D*, 42, 335–46.

Kay P, Regier T (2003). Resolving the question of color naming universals. *Proceedings of the National Academy of Sciences USA*, 100, 9085–9.

Landauer TK, ST Dumais (1997). A solution to Plato's problem: the latent semantic Analysis theory of acquisition, induction and representation of knowledge. *Psychological Review*, 104, 211–40.

Maynard Smith J (2000). The concept of information in biology. *Philosophy of Science*, 67, 177–94.

Minett JW, Wang WS-Y (2005). *Language Acquisition, Change and Emergence: Essays in Evolutionary Linguistics*. Hong Kong: City University of Hong Kong Press.

Nilsson NJ (1984). Shakey the Robot, Technical Note 323. Menlo Park, CA: AI Center, SRI International.

Penrose R (1989). *The Emperor's New Mind: Concerning Computers, Minds, and the Laws of Physics*. Oxford: Oxford University Press.

Pickering MJ, Garrod S (2004). Toward a mechanistic psychology of dialogue. *Behavioural and Brain Sciences*, 27, 169–225.

Rumelhart DE, McClelland JL; PDP Research Group (1986). *Parallel Distributed Processing*. Cambridge, MA: MIT Press.

Searle JR (1980). Minds, brains, and programs. *Behavioural and Brain Sciences*, 3, 417–57

Smolensky P (1988). On the proper treatment of connectionism. *Behavioural and Brain Sciences*, 11, 1–74.

Sporns O, Tononi G, Kotter R (2005). The human connectome: a structural description of the human brain. *PLoS Computational Biology*, 1, e42.

Steels L (2003). Evolving grounded communication for robots. *Trends in Cognitive Science*, 7, 308–12.

Steels L (2004). Intelligence with representation. *Philosophical Transactions of the Royal Society of London A*, 361, 2381–95.

Steels L (2005). The emergence and evolution of linguistic structure: from minimal to grammatical communication systems. *Connection Science*, 17, 213–30.

Steels L (2006). Collaborative tagging as distributed cognition. *Pragmatics and Cognition*, 14, 287–92.

Steels L, Belpaeme T (2005). Coordinating perceptually grounded categories through language: a case study for colour. *Behavioural and Brain Sciences*, 24, 469–89.

Steels L, Loetzsch M (2007). Perspective alignment in spatial language. In KR Coventry, T Tenbrink, JA Bateman, Eds. *Spatial Language and Dialogue*. Oxford: Oxford University Press.

Vogt P (2002). Physical symbol grounding. *Cognitive Systems Research*, 3, 429–57.

Vogt P, Sugita Y, Tuci E, Nehaniv C, Eds (2006). *Symbol Grounding and Beyond*. Berlin: Springer Verlag.

Webster MA, Kay P (2005). Variations in color naming within and across populations. *Behavioural and Brain Sciences*, 28, 512–13.

Chapter 13

Language and simulation in conceptual processing

Lawrence W Barsalou, Ava Santos, W Kyle Simmons, and Christine D Wilson

13.1 Introduction

Theories of cognition often assume that a single type of representation underlies knowledge. Traditionally, most theories have assumed that amodal symbols provide uniform knowledge representation (e.g., Collins and Loftus 1975; Fodor 1975; Newell and Simon 1972; Pylyshyn 1984). More recently, theories have adopted statistical representations (e.g., McClelland *et al.* 1986; O'Reilly and Munakata, 2000; Rumelhart *et al.* 1986). Most recently, theories have proposed that knowledge is grounded in modal simulations, embodiments, and situations (e.g., Allport 1985; Barsalou 1999, 2008a; Damasio 1989; Glenberg 1997; Martin 2001, 2007; Thompson-Schill 2003), while other theories have proposed that knowledge is grounded in linguistic context-vectors (e.g., Burgess and Lund 1997; Landauer and Dumais 1997).

Our theme in this chapter is that multiple systems—not just one—represent knowledge. We focus on two sources of knowledge that we believe have strong empirical support: linguistic forms in the brain's language systems, and situated simulations in the brain's modal systems. Although we focus on these two sources of knowledge, we do not exclude the possibility that other types are important as well. In particular, we believe that statistical representations play central roles throughout the brain, and that they underlie linguistic forms and situated simulations. At this point, we are somewhat skeptical that completely amodal representations exist in the brain, for both theoretical and empirical reasons (Barsalou 1999, 2008a; Simmons and Barsalou, 2003), but we are open to compelling arguments otherwise.

We begin by reviewing linguistic and modal approaches to the representation of knowledge. We then propose the *language and situated simulation (LASS) theory* as a preliminary framework for integrating these approaches. We then turn to empirical evidence for the LASS theory, including evidence for *dual code theory* (Paivio 1971, 1986), evidence for Glaser's (1992) revision of dual code theory (the *lexical hypothesis*), evidence from our laboratory, and evidence from other laboratories. Finally, we address future issues that research from the LASS perspective could address.

13.2 **Linguistically-motivated representations of knowledge**

Traditional theories of amodal symbols are closely related to language (e.g., Collins and Loftus 1975; Fodor 1975; Newell and Simon 1972; Pylyshyn 1984). Although these theories typically assume that the amodal symbols underlying knowledge differ from linguistic forms (e.g., words), close correspondences exist. Predicates that represent object, property, and event concepts in amodal theories often hold a rough one-to-one correspondence with words that refer to them. For example, the predicates $bird(X)$, $red(X)$, and $buy(X,Y)$ represent concepts that correspond roughly to the words 'bird,' 'red,' and 'buy'.[1] Similarly, the propositional representations constructed to represent the meanings of sentences and texts conform closely to their linguistic forms (e.g., Kintsch and van Dijk 1978; van Dijk and Kintsch 1983). For example, the propositional structure $sing(Elizabeth,aria)$ corresponds to the sentence, 'Elizabeth sings the aria.'

As these examples illustrate, amodal approaches to representing knowledge are grounded in language. Although these approaches assume that amodal symbols differ from linguistic forms, they adopt an approach to knowledge representation that mirrors language structure.

13.2.1 **Theories of linguistic context**

More recently, researchers have argued that linguistic forms per se represent knowledge (e.g., Burgess and Lund 1997; Landauer and Dumais 1997). According to this approach, there is no underlying system of amodal symbols that correspond to language – there are only linguistic forms (i.e., words). The intriguing proposal is that statistical distributions of linguistic forms represent knowledge. For example, the representation of *bird* is not an amodal symbol but is instead the distribution of words that co-occur with 'bird' in natural language. According to this approach, two concepts become increasingly similar as their distributions of co-occuring words become increasingly similar.

Thus, this approach to representing knowledge is even more linguistic than traditional amodal approaches. Rather than linguistically-inspired symbols representing knowledge, linguistic forms themselves represent knowledge.

13.2.2 **Situated simulation**

A very different and much older account assumes that knowledge is grounded in the brain's modal systems. Philosophical theories from ancient philosophers to later empiricist and nativist philosophers assumed that images of experience play central roles in knowledge representation (Barsalou 1999, 2008a; Prinz 2002). Only in the 20th Century have linguistically-inspired theories dominated, especially since the cognitive revolution, when the computer metaphor transformed theories of cognition.

---

[1] Italics will be used to indicate concepts, and quotes will be used to indicate linguistic forms (words, sentences). Thus, *bird* indicates a concept, and 'bird' indicates the corresponding word.

In recent times, these older theories have been reinvented in the modern contexts of cognitive science (e.g., Barsalou 1999, 2003a, 2005a; Glenberg 1997) and neuroscience (e.g., Allport 1985; Damasio 1989; Pulvermüller 1999; Simmons and Barsalou 2003). According to these theories, the brain captures modal states during perception, action, and introspection, and then later simulates these states to represent knowledge. On perceiving dogs, for example, the brain captures modal states in the visual, auditory, and somatosensory systems about how dogs look, sound, and feel, respectively. On interacting with dogs, the brain similarly captures modal states in the motor and proprioceptive systems about appropriate actions. During these interactions, the brain also captures introspective states associated with affect and mental operations. On later occasions, when representing knowledge about dogs, the brain attempts to reactivate these multimodal states, typically only succeeding partially. The resultant simulation of the brain states associated with experiencing dogs can then be used for a wide variety of purposes, including inference, recollection, language, and thought.

Much empirical evidence has accumulated for this view across disciplines. Reviews of supporting evidence from cognitive psychology can be found in Barsalou (2003b, 2008), Barsalou, Simmons *et al.* (2003), and Pecher and Zwaan (2005). Reviews of results from cognitive neuroscience can be found in Martin (2001, 2007) and Thompson-Schill (2003). Reviews of results from social psychology can be found in Barsalou, Niedenthal *et al.* (2003) and Niedenthal *et al.* (2005). Reviews of developmental evidence can be found in Thelen (2000) and Smith and Gasser (2005). In general, much evidence exists that modal representations play central roles in the knowledge that pervades cognition.

A related theme is that knowledge representations are situated. Rather than being abstract and detached, knowledge about something is simulated in the context of likely background situations. Instead of simulating knowledge in a vacuum, people simulate it in the context of relevant settings, actions, events, and introspections. For example, knowledge about chairs might be simulated in the context of a kitchen, with someone sitting in a chair, feeling comfortable. The presence of situational information prepares agents for situated action. Rather than only representing the focal knowledge of interest, as in a dictionary or encyclopaedia entry, a category representation prepares agents for interacting effectively with its members.

Much evidence documents the importance of situational information in the representation and processing of knowledge. Furthermore, much of this evidence indicates that simulations in the respective modal systems represent situational information. For reviews, see Barsalou (2003b, 2005b, 2008a, in press), Barsalou, Niedenthal *et al.* (2003), and Yeh and Barsalou (2006).

13.3 The LASS theory of conceptual processing

Research on knowledge typically focuses on categories of things in the world and on concepts in the cognitive system that represent them (for a broad review, see Murphy 2002). Rather than representing knowledge as holistic images (as a camera does), humans use a powerful attentional system to focus on components of multimodal experience and form concepts that represent knowledge about them. As people focus attention on objects, properties, settings, actions, events, mental states, affect, relations, etc., concepts develop over time to represent the corresponding categories of exemplars experienced. After focusing attention on robins, for example, a concept develops to represent this category. Focusing attention on hands, valleys, waving, storms, hopes, etc., similarly produces concepts of these categories.

The theme of this chapter is that the representation and processing of concepts relies heavily on both language and situated simulation. The linguistic representations that we believe important are linguistic forms, as in theories of linguistic context, not amodal symbols. In general, we assume that linguistic forms and situated simulations interact continuously in varying mixtures to produce conceptual processing.

Given that most research has addressed conceptual processing when words are presented as stimuli, we focus on word-based tasks. In our opinion, however, research has suffered considerably from an over-reliance on words. We suspect that the conceptual system evolved primarily to process nonlinguistic stimuli, including perceptual, motor, and intro-spective aspects of experience. We further suspect that the processing of experience contin-ues to be more central in human cognition than the processing of words. At later points, we address implications of the LASS framework for processing experience. Nevertheless, because most research has focused on words, we focus our treatment here on them. In the following sections, we address four aspects of the LASS framework: (1) linguistic process-ing, (2) situated simulation, (3) mixtures and interactions of language and situated simula-tion, and (4) the statistical underpinnings of language and situated simulation.

13.3.1 Linguistic processing

We assume that when a word is perceived, the linguistic system becomes engaged imme-diately to categorize the linguistic form (which could be auditory, visual, tactile, etc.). As Figure 13.1 illustrates, we assume that the linguistic system and the simulation system both become active initially, but that activation for the word form peaks before activation

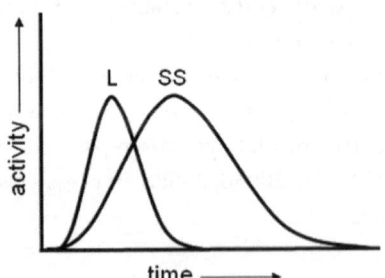

Fig. 13.1 Initial contributions from the linguistic system (L) and the situated simulation system (SS) during conceptual processing. When the cue is a word, contributions from the linguistic system precede those from the simulation system. The height, width, shape, and offset of the two distributions are not assumed to be fixed. In response to different words in different task contexts, all these parameters are expected to change (e.g., SS activity could be more intense than L activity). Thus, the two distributions in this figure illustrate one of infinitely many differ-ent forms that activations of the L and SS systems could take.

for the simulation. Following the content addressability and encoding specificity principles, we assume that information in memory most similar to the cue becomes active
most rapidly (e.g., Tulving and Thomson, 1973). Because representations of linguistic
forms are more similar to presented words than are simulations of experience, representations of linguistic forms peak first.

Once the word has been recognized, we assume that associated linguistic forms are
generated as inferences, and as pointers to associated conceptual information. In many of
the experiments to follow, the generation of linguistic forms is realized as the simple
process of word association, where a cue word elicits other words associated with it
(e.g., 'cat' elicits 'fur,' 'purr,' and 'pet'). As we will see, word association plays a central role
in the early stages of conceptual processing when words are presented as cues. We hasten
to add that word association is the simplest possible form of the linguistic processing that
occurs during conceptual processing. Much more complex processing occurs, as
compounds, phrases, and syntactic structures are generated and processed.

Once associated linguistic forms are generated, they support a variety of superficial strategies (e.g., Glaser 1992). As we will see later, associations between words can be
sufficient to produce correct responses on conceptual tasks – use of deeper conceptual
information is not necessary (Kan *et al.* 2003; Solomon and Barsalou 2004). Consistent
with linguistic context theory (e.g., Burgess and Lund 1997; Landauer and Dumais
1997), we assume that samples of associated words are generated that provide linguistic
context for the presented word. Once a sample becomes active, it supports a wide variety
of tasks and implements many basic effects.[2]

We assume that these linguistic strategies are relatively superficial (following Glaser
1992). Rather than providing deep conceptual information, these strategies provide
shallow heuristics that make correct performance easily possible. When the retrieval of
linguistic forms and associated statistical information is sufficient for adequate performance, no retrieval of deeper conceptual information is necessary.

Much work on lexical processing is consistent with this proposal. In the lexical decision
task, activation of a word's meaning is shallow when read in the context of nonwords that
lack acceptable phonology and orthography. Because discriminating words from
nonwords can be based on linguistic form alone, there is no need to access and consider
meaning. Conversely, when nonwords satisfy rules of phonology and orthography, words
access meaning more deeply. Because linguistic form no longer discriminates words from
nonwords, meaning must be retrieved to verify that a stimulus is a word (e.g., James,
1975; Joordens and Becker, 1997; Shulman and Davidson, 1977; Stone and Van Orden,
1993; Yap, Balota, Cortese, and Watson, 2006). Similarly, the depth of processing literature

---

[2] It is important to note, however, that it is not always clear whether linguistic contexts cause these
effects or are merely correlated with them. Because linguistic contexts are correlated with conceptual
information, such as the information contained in situated simulations, apparent effects of linguistic
context could actually be due to correlated conceptual information. More work is needed to resolve
this issue. A likely possibility – the one pursued here – is that both factors contribute to conceptual
processing effects.

shows that phonemic orienting tasks produce relatively less activation of meaning than semantic orienting tasks (e.g., Craik 2002; Craik and Lockhart 1972; Craik and Tulving 1975; Lockhart 2002; Morris *et al.* 1977). Broadly speaking, many findings are consistent with the proposal that linguistic forms can be processed superficially.

By no means does superficial processing imply lack of utility. As described later, superficial linguistic strategies can often be highly effective in producing accurate performance, similar to the heuristic value of many other superficial strategies (Gigerenzer 2000). Nevertheless, attributing more conceptual depth to these heuristics than actually exists may mischaracterize them and obscure other important mechanisms that provide deeper conceptual processing.

We further realize that proponents of linguistic and amodal theories are likely to disagree strongly with our claim that purely linguistic processing is superficial. Such theorists are likely to believe instead that the language system contains semantic content and produces deep conceptual understanding. Resolution of this issue depends on further empirical evidence. If future empirical evidence indicates that the linguistic system contains its own semantics, our position will need revision.

13.3.2 Situated simulation

As the linguistic system begins to recognize the presented word, the word immediately begins to activate associated simulations. Linguistic forms associated with the presented word also become active and may begin to activate simulations as well. Thus, activated linguistic forms serve as pointers to simulations that are potentially useful for representing the cue word's meaning. As described earlier, we assume that these simulations tend to be situated, preparing agents for situated action. Specifically, correlated information in perceptual, motor, and introspective brain areas becomes active to represent the concept in a likely situation. We also assume that these simulations are often activated automatically and quickly (e.g., within 200 milliseconds of word onset; Pulvermüller *et al.* 2005; Pulvermüller, Chapter 6, this volume).

Although simulations may become active quickly, they may not dominate conscious deliberate cognition immediately. If executive processing selects a processing strategy that utilizes another system, that system may control behaviour initially. As we will see later, executive processing can focus on the linguistic system as its primary source of information for at least several seconds, before simulations begin to have effects on behaviour. One possibility is that executive processes focus attention on the linguistic system as a source of information for producing responses until this system stops being useful. At that point, executive processes shift attention to the simulation system as an alternative source of information. Notably, simulations are likely to be activated simultaneously while the executive system is producing responses from the linguistic system. This account reconciles the fast access of simulations that Pulvermüller reports with our later findings that the linguistic system can dominate processing for several seconds. We return to this issue when presenting evidence from a functional magnetic resonance imaging (fMRI) study that assessed the production of conceptual information from the linguistic and simulation systems.

Finally, we assume that simulations represent deep conceptual information, unlike linguistic representations, which we view as more superficial. Specifically, we assume that conceptual content about properties and relations resides in simulations. We further assume that basic symbolic processes such as predication, conceptual combination, and recursion, result from operations on simulations. Barsalou (1999, 2003a, 2005a) describes how simulation mechanisms can implement symbolic operations. Barsalou (2008b), reviews relevant evidence. We assume that linguistic forms are not capable of implementing these operations in the absence of simulations. Attempting to perform symbolic operations on linguistic forms alone would be like manipulating symbols in an unfamiliar language, with no true comprehension (Searle 1980). Because simulations provide the meanings of linguistic forms, they are required for implementing symbolic operations. As we will see later, human participants cannot perform the symbolic operation of predication on linguistic forms alone (Solomon and Barsalou 2004). Only when simulations are constructed do such operations become possible. We hasten to add that linguistic forms almost certainly play central roles in symbolic operations as well. As suggested later, symbolic operations probably operate most effectively when both linguistic forms and simulations contribute.

13.3.3 Mixtures and interactions of language and situated simulation

We assume that different mixtures of the two systems underlie a wide variety of tasks. When superficial linguistic processing is sufficient to support adequate task performance, processing may rely mostly on the linguistic system and little on simulation (Glaser 1992; Kan *et al.* 2003; Solomon and Barsalou 2004). Conversely, when linguistic processing is unable to support adequate performance, the simulation system must be consulted for the required conceptual information. Depending on task conditions, conceptual processing may mostly consist of linguistic processing or simulation. Under many conditions, both may contribute equally. We assume that both processes are typically engaged to some extent. As described later, simulation appears to become the dominant processing strategy, at least initially, when non-linguistic stimuli are processed.

When linguistic processing occurs in mixtures, we do not assume that the two systems operate independently. Instead we assume that extensive interactions occur between them. As linguistic forms become active initially, they activate simulations. Once a simulation becomes active, words that refer to its space–time regions become active. As these words become active, they activate simulators that interpret these regions conceptually. Following Barsalou (1999, 2003a, 2005a), we assume that simulators are roughly equivalent to concepts in traditional theories of knowledge.

We further assume that complex linguistic interactions arise from the interplay of these two systems. When a speaker has something to say, for example, a simulation represents it initially (e.g., a speaker simulating anticipated enjoyment while hearing a concert later that evening). Putting the simulation into words requires control of attention across the simulation. When attention focuses on a region, a simulator categorizes it. Linguistic forms associated with the simulator, such as words and syntactic structures, become active, which are then integrated into the evolving motor program for an utterance (e.g., the speaker stating, 'I'm looking forward to the concert this evening'). In turn, when a listener comprehends the utterance, its words and syntactic structures function as cues to assemble a simulation compositionally that should, ideally, correspond to the speaker's simulation (e.g., simulating the speaker's anticipated enjoyment at the concert).

In reasoning, we assume that similar interactions occur extensively. As people try to figure things out during decision making, planning, and problem solving, they simultaneously engage in simulating the relevant situation and verbalizing about it (e.g., deciding whether it would be better to answer email or review a paper during a 2-hour break between meetings). Whereas simulations represent the content of thought, words provide tools for indexing and manipulating this content, as possibilities are evaluated and decisions made.

In general, we assume that linguistic forms provide a powerful means of indexing simulations (via simulators), and for manipulating simulations in language and thought. As the two systems interact, one may dominate momentarily, followed by the other, perhaps cycling many times, with both systems being active simultaneously at many points.

13.3.4 Statistical underpinnings of language and situated simulation

We further assume that both systems are exquisitely sensitive to the statistical structure of their respective domains. In the simulation system, simulators capture the statistical frequencies of properties and the relations between them in experience. In the linguistic system, it is well established that the frequency of words, the associations between them, and their relations to syntactic structures are coded statistically.

We further assume that the statistical structures in the two systems roughly mirror each other (cf. Louwerse and Jeuniaux, Chapter 15, this volume). One reason is that people constantly hear language that corresponds to perceived situations. As a result, frequencies and correlations in perceived situations are mirrored in frequencies and correlations of words used to describe them. Similarly, when people use language to describe non-present situations, statistical correspondences occur between the situated simulations in memory and the linguistic forms used. As a result of these correspondences, statistical information in each system mirrors experience and each other. For this reason, each system can be useful in providing relevant statistical information under appropriate task conditions. We assume that neural architecture naturally stores extensive amounts of statistical information in this manner.

13.3.5 Caveats about the 'linguistic' and 'simulation' systems

Discussion of the LASS theory so far has assumed simplifications of the 'linguistic' and 'simulation' systems that must be qualified. First, we do not mean to imply that these are modular systems. Clearly, each system is highly complex and draws on many systems distributed throughout the brain. Furthermore, many of these systems probably contribute to other processes besides language and simulation (e.g., the vision and motor systems contribute to perception and action, respectively, not just to language).

Second, we do not mean to imply that that each system takes the same rigid form in every situation. To the contrary, we assume that each system is dynamical such that it draws on different configurations of processes in different situations (Barsalou, Breazeal *et al.* 2007). Furthermore, we do not assume that there is only a single form of simulation in the brain. To the contrary, we believe that the brain implements diverse forms of simulation across different cognitive processes (Barsalou 2008a).

Third, when we refer to the linguistic system here, we are referring to the system that processes linguistic forms, not to the system that represents linguistic meaning. As described earlier, we assume that meaning is largely represented in the simulation system. Clearly in other contexts, the linguistic system would include the representation of meaning, thereby including the simulation system. Because we wish to contrast linguistic forms and linguistic meaning here, we use the 'linguistic system' for the former and 'simulation system' for the latter.

In summary, we use 'linguistic system' and 'simulation system' as simplifications so that we can focus on mechanisms of interest, in particular, linguistic forms versus situated simulations. This usage, however, should not be taken as a commitment to rigid modular systems, nor to the view that the linguistic system is unrelated to the simulation system.

13.4 **Previous Evidence Consistent with the LASS theory**

We turn to empirical evidence for the LASS theory, beginning with evidence for two related views: Paivio's (1971, 1986) dual code theory, and Glaser's (1992) lexical hypothesis, an extension of dual code theory. We then turn to recent findings from our own laboratory, and conclude with potentially relevant findings from other laboratories.

13.4.1 **Paivio's dual code theory**

Dual code theory and LASS have much in common. Both assume two basic systems (among others), one linguistic and the other grounded in modalities. Both assume that the two processes underlie a broad spectrum of cognitive activities. Both assume that the two processes operate interactively in different mixtures across different task conditions. Many other deep similarities exist between the two approaches.

Differences exist as well. Whereas LASS assumes that the simulation system performs slower and deeper conceptual processing than does the linguistic system, dual code theory assumes that deep conceptual processing occurs in both systems. Whereas LASS assumes that the simulation system is central to the representation of abstract concepts, dual code theory assumes that the linguistic system is central. In general, LASS places less computational power in the linguistic system than does dual code theory and more in the simulation system.

Over the past 40 years, dual code theory has generated an impressive body of empirical support (for reviews see Paivio 1971, 1986). Much evidence indicates compellingly that cognition relies on two systems, one that processes linguistic representations, and another that processes modal representations. Evidence for two systems has accumulated in developmental psychology, where modal systems develop faster than the linguistic system. Evidence for two systems has accrued in the individual differences literature, with different individuals relying more on one system than the other. Evidence for two systems has accrued in the literatures addressing episodic memory, semantic memory, and language comprehension. Because of the substantial empirical support that dual code theory has accumulated, the central assumption of LASS that cognition relies on the constant interplay between a linguistic system and a simulation system appears on solid ground.

13.4.2 Glaser's lexical hypothesis

Glaser (1992) reviewed evidence consistent with the view that the linguistic system has less computational power than the simulation system. Glaser starts with dual code theory and modifies it in two ways. First, consistent with the theoretical Zeitgeist of the time, he considers the possibility that Paivio's imagery system might be better viewed as a conceptual system that contains amodal representations. As we will see, however, the evidence that Glaser reviews suggests that his conceptual system might actually be populated with modal representations, not amodal ones (he often appears to approach this conclusion himself). Second, Glaser modifies dual code theory by proposing that the linguistic system can perform relatively superficial processing independently of the conceptual system—what he calls the 'lexical hypothesis'.

Glaser adopts the lexical hypothesis based on findings across many literatures. On the one hand, he addresses the ability of pictures vs. words to access the conceptual system during verification tasks. On the other hand, he addresses the ability of pictures versus words to produce conceptual effects in priming and interference tasks. Each set of studies is addressed in turn.

In verification tasks, pictures are faster than words in accessing the conceptual system. When verifying whether something belongs to a category (e.g., *living things* versus *artefacts*), pictures are verified as category members faster than the corresponding words (e.g., a picture of a cat versus the word 'cat'). This finding suggests to Glaser that pictures provide the fastest access to the conceptual system. It further suggests that the conceptual system may have a perceptual character, rather than an amodal one. Additionally, Glaser concludes that words access the linguistic system first – not the conceptual system – which explains why they take longer to verify than pictures. This latter assumption is a key component of the lexical hypothesis: words can bypass the conceptual system and be processed solely by the linguistic system.

Pictures also produce stronger conceptual effects on priming and interference tasks than do words. In priming tasks, picture primes tend to produce priming effects that are two to three times as large as those produced by word primes. In a representative task, one concept appears as a prime (e.g., *chair*) for a target concept (e.g., *table*), whose superordinate must be produced (e.g., *furniture*). When the prime and target are both pictures (e.g., pictures of a chair and table), priming effects are much larger than when they are both words (e.g., 'chair' and 'table'). The analogous pattern occurs in interference tasks. In a representative task, one concept (e.g., *table*) appears as a target to be categorized (e.g., into the superordinate *furniture*), but is immediately followed by a distracter that belongs to a different superordinate (e.g., *horse* from *mammals*). When the target and distracter are both pictures, interference effects are much larger than when they are both words.

Glaser explains larger conceptual effects for pictures than for words in the same way that he explains the faster access of pictures to the conceptual system. Pictures access the conceptual system directly, perhaps because of its perceptual character. As a result of this direct access, large priming and interference effects result, given that relations between concepts in the conceptual system are retrieved, producing these effects.

Conversely, words bypass the conceptual system initially, such that the conceptual relations responsible for priming and interference are not retrieved. If the conceptual system is accessed occasionally, or becomes slightly activated, this could produce the small priming and interference effects observed for words. Alternatively, statistical relations between words that mirror conceptual relations between the respective concepts could be responsible, as suggested earlier (evidence for this mechanism will also be provided later when reviewing Solomon and Barsalou 2004). Both mechanisms could contribute to priming and interference effects for words. Conceptual effects could be relatively weak for words either because the conceptual system is accessed occasionally, and/or because statistical relations in the linguistic system tend to be weaker than conceptual relations in the conceptual system.

Glaser's revision of dual code theory with the lexical hypothesis yields an account that is highly similar to the LASS theory. Both assume that there is a linguistic system and a conceptual system. Both assume that the linguistic system performs relatively superficial processing, whereas the conceptual system performs deeper processing. Both assume that superficial linguistic processing can be sufficient under certain conditions for adequate task performance. The primary difference is that Glaser is less inclined than we are to view the conceptual system as containing modal representations, even though he leans in that direction on many occasions when noting the powerful ability of pictures to access conceptual knowledge.

13.5 Evidence for mixtures of two systems in conceptual processing

We next turn to evidence for the LASS theory from our laboratory. We begin with evidence for the presence of two systems – language and simulation – in conceptual processing, and show that the linguistic system tends to provide information before the simulation system. We review evidence from three lines of research: (1) word association and property generation, (2) property verification, and (3) abstract concepts. In all these experiments, we focus on conceptual processing in response to words. As Glaser's review suggested, however, conceptual processing to pictures may operate quite differently.

13.5.1 Word Association and Property Generation

In a series of experiments (Santos *et al.* 2008), participants received a word for a concept and generated related information verbally. In experiment 1, participants generated word associates. In experiment 2, participants generated properties typically true of a concept's instances. In both experiments, LASS predicts that the linguistic system and the simulation system should both contribute to the responses that participants produce verbally over the course of conceptual processing. Initially, the earliest responses should come from the linguistic system (Figure 13.1). As the simulation system becomes increasingly active, however, responses should increasingly be produced from it as well. Thus, we predicted that the responses produced in both tasks would reflect mixtures of responses from the two systems, with the first responses tending to come from the linguistic system. We also predicted that the linguistic system would contribute a larger amount of information in the word association task than in the property generation task, given that word association focuses attention on the linguistic system.

Word association

In Experiment 1, participants received a word on each trial and were asked to generate associated words (Santos *et al.* 2008). Specifically, participants were asked, 'For the following word, what other words come to mind immediately?' The experimenter recorded the participant's responses on tape. Typically, participants produced 1–3 responses in less than 5 seconds for each cue word. As soon as the participant paused, the experimenter ended the trial. Thus, the experiment aimed to capture the dominant word associates associated with the cue. Each participant produced word associates to 16 cues, drawn from a larger set of 64 cues. The 64 cues referred to highly diverse concepts, including objects (e.g., *car, bee*), actions (e.g., *throw, calculate*), abstract concepts (e.g., *self, fashion*), properties (e.g., *good, heavy*), and proper names (e.g., *Jupiter, Nike*).

All responses to a given cue word across participants were merged into a single master list, with minor lexical variants combined into a single response (e.g., 'flower' and 'flowers'). As described next, two judges then coded all responses in the master list using a hierarchical coding scheme applied sequentially.

If a response was linguistically related to the cue, it was automatically coded as a linguistically-related response. Consideration of other possible coding categories proceeded no further. For example, the response 'hive' to the cue 'bee' was coded as a linguistically-related response, because 'bee-hive' is a common compound phrase. Participants could have generated 'hive' in response to 'bee' after 'bee' activated the compound linguistic form, 'bee hive,' in the lexical system, which in turn produced 'hive' as a response. Possible linguistic responses included forward compound continuations (e.g., 'bee' → 'hive), backward compound continuations (e.g., 'bee' → 'honey' from 'honey-bee'), synonyms (e.g., 'car' → 'automobile'), antonyms (e.g., 'good' → 'bad'), root similarity (e.g., 'self' → 'selfish'), and sound similarity (e.g., 'bumpy' → 'lumpy').[3] In each case, some type of linguistic relation could have related the cue and response.

If a response did not fall into one of these linguistic response categories, it was then evaluated for being a taxonomic response (e.g., 'dog' → 'animal'). If a response was taxonomically related (but not linguistically related), it was automatically coded as such. Consideration of other possible coding categories proceeded no further. Taxonomic responses included superordinate categories (e.g., 'dog' → 'animal'), coordinate categories (e.g., 'dog' → 'cat'), and subordinate categories (e.g., 'dog' ' → 'terrier').

---

[3] The syntax of the examples shown here is 'cue' → 'response'.

If a response did not fall into either a linguistic or taxonomic coding category, it was automatically coded as an object–situation response. Interestingly, every valid response that was not a linguistic or taxonomic response, always described either a property of the cue concept or a thematic associate of the cue concept that could co-occur with it in a situation. For example, 'bee' produced *bee* properties (e.g., 'wings') and situational associates (e.g., 'flowers'). Similarly, 'golf' produced *golf* properties (e.g., 'boring') and situational associates (e.g., 'sunshine').

The LASS theory makes predictions about the three general coding categories during the word association task. First, LASS predicts that linguistically-related responses should tend to come from the linguistic system. As described earlier, the response 'hive' to the cue 'bee' could result from 'bee' activating the compound linguistic form, 'bee-hive,' in the lexical system, which in turn produces 'hive' as a response. Importantly, however, 'hive' could also result from describing a simulation of a situation containing a bee and a hive. Although this is possible, and probably occurred to some extent, we assume that this possibility is statistically less likely than 'hive' originating in the linguistic system. Thus, the prediction is that linguistic responses should be statistically more likely to originate from linguistic processing than from simulation. As a result, linguistic responses should tend to occur early in participants' protocols, given our assumption that the linguistic system produces responses faster than the simulation system (Figure 13.1).

Conversely, LASS predicts that object–situation responses should be statistically more likely to originate from describing situated simulations than from retrieving linguistic forms. Although the response 'flowers' could be associated with 'bee' in the linguistic system, we predicted that it would be more likely to arise from the simulation of a situation containing a bee and flowers. In general, we assume that object–situation responses are statistically more likely to result from describing simulations than to result from linguistic retrieval. As a result, object–situation responses should tend to occur relatively late in participants' protocols, given our assumption that simulations become active more slowly than linguistic forms.

The LASS theory's predictions for taxonomic responses are less clear than its predictions for linguistic and object–situation responses. On the one hand, taxonomic categories are generally viewed as residing in conceptual systems. On the other hand, people memorize phrases for taxonomic relations during childhood, such as 'a dog is an animal.' Thus, taxonomic responses could result from retrieving linguistic forms. Furthermore, it is not clear how taxonomic categories are realized in simulations. How is the superordinate *animal* evident in a situated simulation of a dog? *Animal* is not a concrete property of a dog that is simulated, nor is it a thematic associate that co-occurs with dogs in situations. These observations suggest that taxonomic responses such as 'animal' could largely originate in the linguistic system (especially superordinates). As we will see in experiment 2, coordinates and subordinates may often occur as thematic associates in situated simulations (e.g., a simulation of a dog chasing a cat, or of a dog simulated as a collie).

The results of experiment 1 supported the LASS theory. Linguistically-related responses were produced significantly earlier than object–situation responses. Responses that were more likely to originate in the linguistic system occurred earlier than responses that were more likely to originate in the simulation system. Taxonomic responses fell halfway in between, significantly later than linguistic responses, and significantly earlier than object–situation responses. This suggests that taxonomic responses were sometimes retrieved as memorized lexical phrases, but on other occasions described the content of simulations.

Property generation

Experiment 2 from Santos *et al.* (2008) offered similar evidence using the property generation task. Of the 64 concepts from the word association experiment, 60 were used and, were again highly diverse. Each participant received 30 of the 60 concepts and had to generate typical properties of each. For example, participants were asked, 'What characteristics are typically true of dogs?' Participants typically produced 6–7 responses to each cue in the 15-second period allowed for responding. As in experiment 1, participants produced responses verbally, and responses were coded sequentially into the same linguistic, taxonomic, and object–situation coding categories.

One prediction was that participants would produce fewer linguistic responses and more object–situation responses than in the word association experiment. Because the task is more conceptual in nature, and because participants produced responses for longer periods, more responses should originate in the simulation system. This prediction was strongly confirmed. A second prediction was that, again, linguistic responses should precede object–situation responses. As described earlier, linguistic responses should tend to originate in the faster linguistic system, whereas object–situation responses should tend to originate in the slower simulation system. Again, the results confirmed this prediction. Consistent with the LASS theory, linguistic responses occurred significantly earlier than object–situation responses.

Taxonomic responses did not differ from linguistic responses overall, with both tending to be produced early. Importantly, however, different kinds of taxonomic responses varied considerably in how early they occurred. Superordinates were one of the earliest type of responses produced, occurring earlier than all but one type of linguistic response. This finding suggests that superordinates may often be stored linguistically and be generated from the linguistic system. In contrast, coordinates and subordinates were as slow as object–situation responses. This finding suggests that participants may have been simulating coordinates and subordinates in situations, such that these taxonomic categories were reported at the same time as other situational content.

In summary, experiments 1 and 2 from Santos *et al.* (2008) confirmed predictions of the LASS theory: linguistic responses tended to occur earlier than object-situation responses in both experiments, consistent with the theory's assumptions that responses are produced from a faster linguistic system and a slower simulation system.

Property generation with fMRI

Simmons *et al.* (2008) performed experiment 2 from Santos *et al.* (2008) in a 3-Tesla fMRI scanner. Each participant was scanned twice. In the first scanning session, participants received 30 of the 60 concepts from Santos *et al.'s* experiment 2. As the word for each concept was presented visually, participants generated the typical properties of the concept to themselves for 15 seconds. Participants practiced generating properties for other concepts out loud outside the scanner before the scanning session, so that covert generation would be similar to overt generation.

In a second scanning session a week later, participants performed two localizer tasks that allowed us to test the LASS theory's predictions about conceptual processing. Participants received the other 30 concepts from Santos *et al.*'s Eeperiment 2 that they had not received in the first session. For 24 of these concepts, participants were asked to generate word associates for 5 seconds each. For the other six concepts, participants were asked to spontaneously imagine a situation that contained the concept for 15 seconds each (e.g., for *bee*, a participant might imagine a garden with a bee buzzing around a flower, a hive with bees in it, etc.). Concepts were counterbalanced so that each concept occurred in all three generation conditions (i.e., property generation, word association, situation simulation), with a given participant receiving each concept once. Participants received concepts for all three conditions in a blocked design.

Our predictions for the two localizer tasks were as follows. First, we predicted that the word association task would primarily activate left-hemisphere language areas, especially Broca's area. Second, we predicted that the situation simulation task would activate bilateral posterior areas that are typically involved in the generation of mental imagery. Our predictions for conceptual processing during the critical property generation task were as follows. First, we predicted that conceptual processing would contain activations found in both localizer tasks. Second, we predicted that activations found in the word association localizer would occur earlier than activations found in the situation localizer.

Panel A of Figure 13.2 illustrates these predictions. As can be seen, we assume that both linguistic processing and simulation begin immediately. Linguistic processing, however, peaks during the first half of the generation period, whereas simulation processing peaks in the second half. We further assume that the executive system focuses initially on information in the linguistic system, because linguistic information becomes available initially (due to encoding specificity) and/or because verbal responses are requested. As responses from the linguistic system decrease, the executive system then turns to the simulation system as a source of responses. Consequently, the linguistic system is more active during the first half of the generation than during the second, where the simulation system is more active during the second half. Because the executive system extends the activity of each system in time, using it as a source of responses, differences in the processing activity during the two halves are large enough for fMRI to detect (given its relatively low temporal resolution).

Second, consider an alternative account that the linguistic and simulation systems operate fully in parallel from the onset of the cue word, with properties being generated at equal rates from both (Figure 13.2, panel B). If this account is correct, then linguistic processing activity should not be greater in the first 7.5-second generation period than in the second 7.5-second period, and simulation activity should also not differ between the two periods. Note that the predictions in panel B also hold for an additional account that only one system – not two –generates properties. If only one system generates properties, then early versus late processing should not be differentially associated with brain activations that reflect language versus simulation.

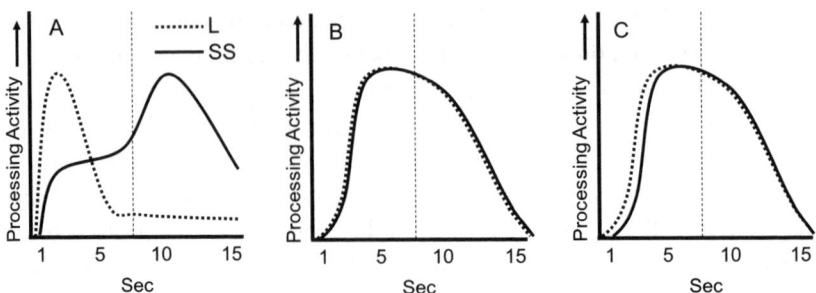

Fig. 13.2 Possible predictions for contributions from the linguistic system (L) and the situated simulation system (SS) during the 15-second property generation periods in Simmons *et al.* (2008). *Panel A*: predictions for the view that the executive system primarily produces responses from the L system for the first 7.5 seconds of the production period and then produces responses from the SS system for the second 7.5-second period. *Panel B*: Predictions for the view that the L and SS systems operate completely in parallel (and also for the view that only one system – not two – produces properties). *Panel C*: Predictions for the view that contributions from the L system only precede contributions from the SS system by about one second or so. The height, width, shape, and offset of the two distributions are not assumed to be fixed. In response to different words in different task contexts, all these parameters are expected to change (e.g., SS activity could be more intense than L activity). Thus, the distributions in this figure illustrate one of infinitely many different forms that activations of the L and SS systems could take.

Finally, consider another alternative account that the linguistic system produces more properties for the first second or so, but that both systems produce properties at equal rates for the remainder of the 15-second period (Figure 13.2, panel C). If this account is correct, then again linguistic processing should not be greater in the first 7.5-second period than in the second. Although simulation does not start quite as early as in panel A, the difference in simulation activity across the two periods should also probably not differ (given the low temporal resolution of fMRI). If there were a significant difference in simulation, this account still predicts no difference for linguistic processing across the two periods.

Turning to the results, first consider activation in the localizer tasks for word association and situated simulation. Activations during these two tasks occurred in the expected areas. Areas that were more active for the word association localizer than for the situation localizer included a large activation in left inferior frontal gyrus (Broca's area), along with large activations in left inferior temporal gyrus, and right cerebellum. All of these areas have been reported previously in research on word processing, especially word generation. Areas that were more active for the situation localizer than for the word association localizer included a large activation in the precuneus, along with a large activation in right middle temporal gyrus. An area in right middle frontal gyrus was also active, but at a lower significance level. These areas are generally not associated with linguistic processing. The precuneus, in particular, is associated with the generation of mental imagery.

Of interest was whether the patterns of activity in the property generation task conformed to the predictions in panel A of Figure 13.2. If so, then the pattern of activation for the word association localizer should have been more active during the first half of the property period than during the second, whereas the pattern of activation for the situation localizer should have been more active during the second half than during the first. Alternatively, if the linguistic and simulation systems generated properties at equal rates either simultaneously or staggered in time slightly, the patterns of the two localizer tasks should not have been more prevalent in either half of the production period (panels B and C of Figure 13.2).

To test these hypotheses, we divided each 15-second property generation block for a single concept into two smaller 7.5-second blocks for the early versus late phases. We then identified brain areas that were more active in the early phase of property generation than in the late phase, and vice versa. Areas that were more active in the early phase of property generation than in the late phase included left inferior frontal gyrus and right cerebellum. A conjunction analysis showed that these activations lay directly within the same areas observed in the word association localizer. Thus, the linguistic system appeared responsible for responses produced during the early phase of property generation. Areas that were more active in the late phase of property generation included precuneus and right middle temporal gyrus. A conjunction analysis showed that these activations lay directly within the same areas observed in the situation localizer. Thus, the simulation system appeared responsible for responses produced during the late phase of property generation.

Because this experiment used a blocked design, assessing the detailed time course of activation in these brain areas was not possible. An important goal for future research is to assess these time courses in greater detail using event-related fMRI designs, and using imaging techniques that have higher temporal resolution (electroencephalography, magnetoencephalography). Nevertheless, these large predicted differences in activation over a 15-second interval provide strong evidence for the LASS theory. If responses from the linguistic system are not produced for an extended duration first, followed by responses from the simulation system for an extended duration, we would have not have observed different activations in the two 7.5-second periods that fell within the localizer activations. The fact that we observed such large differences in the respective localizer areas suggests that the linguistic and simulation systems make large extended contributions to conceptual processing over long periods of time (panel A of Figure 13.2). Simmons *et al.* (2008) discuss this issue in further detail.

In summary, these fMRI findings corroborate findings from the behavioural experiments in Santos *et al.* (2008). Two systems appear responsible for producing conceptual information: the linguistic system and the simulation system. The linguistic system appears to produce responses earlier than the simulation system. Together, the findings from these three experiments support the LASS theory.

13.5.2 **Property verification**

We next show that task conditions within a single experiment can modulate the specific mixture of linguistic and simulation information that represents a concept on a given occasion. The experiments described here used the property verification task. On a given trial, participants first received an object name on a computer screen (e.g., 'horse'), and then verified whether a subsequently presented property was a *part* of the respective object (e.g., 'mane'). Of interest were the times to verify properties, and the accuracy of doing so.

Solomon and Barsalou (2004) proposed that participants can verify properties using either of the two LASS systems. When task conditions allow, participants use a superficial linguistic strategy (following Glaser 1992). When deeper conceptual processing is required, however, participants use simulation. Each approach to verifying properties is addressed in turn.

Linguistic strategy

What conditions might lead participants to adopt a superficial linguistic strategy when verifying properties? Solomon and Barsalou (2004) proposed that participants adopt this strategy spontaneously when information in the linguistic system is sufficient for adequate task performance. One situation where linguistic information is sufficient occurs when the words for the true properties are related to the words for target objects, and when the words for false properties are unrelated to the words for their target objects. For example, the words for the following true object–property pairs are all related: 'bathtub–drain,' 'beaver–teeth,' 'elephant–tusk,' 'sailboat–mast,' 'taxi–meter,' and 'watermelon–seed.' Conversely, the words for the following false object–property pairs are all unrelated: 'pliers–river,' 'airplane–cake,' 'bus–fruit,' 'asparagus–furniture,' 'briefcase–wick,' and 'lion–wire.'

When participants receive true and false pairs like these, they can rely solely on the linguistic system for statistical information that is sufficient for adequate task performance. As described earlier, extensive statistical information resides in the language system, including the associative strength between words. After the object and property words on a trial have been read, participants can simply assess whether they are associated. When the two words are associated, participants respond 'true'; when the two words are not associated, participants respond 'false'. Because linguistic associativeness is highly correlated with correct responses, it can be used to produce correct performance. Importantly, participants need not retrieve any conceptual information. They simply need to detect if the two words are associated, which can be assessed quickly by consulting the linguistic system. Thus, Solomon and Barsalou (2004) predicted that under these task conditions participants would show evidence of using a fast linguistic strategy.

Simulation strategy

Conversely, imagine that a second group of participants receives the same true trials but receives false trials on which the object and property words are related. For example, all the words for the following false object–property pairs are associated (as verified by independent scaling): 'banana–monkey,' 'otter–river,' 'donkey–mule,' 'table–furniture,' 'guitar–keyboard,' 'flashlight–wick.' Note that all of these object–property pairs are false, because the property is not a *part* of the object, which was what participants were asked to verify. Because the properties on these trials were not parts of their respective objects, participants had to respond 'false'.

When participants receive true and false pairs like these, they cannot rely on the linguistic system, because superficial information about word associativeness is not sufficient for adequate task performance. Because both true and false properties are associated with their respective objects, statistical information about word associativeness from the linguistic system is not diagnostic for correct responding. Instead, participants must retrieve conceptual information that specifies whether the property is a part of the object. A *part relation* linking the object and property concepts must be found.[4]

Solomon and Barsalou (2004) proposed that simulations provide the requisite information for making these deeper decisions. Based on Solomon and Barsalou (2001), they argued that on reading the object word, participants first simulate the object. Then, on reading the property word, participants simulate the property. Once both simulations are active, participants assess whether the property simulation can be found in the object simulation. If it can, participants respond 'true'; if it cannot, they respond 'false'.

In summary, Solomon and Barsalou (2004) manipulated whether two different groups of participants received 100 false trials in which the object and property words were either unrelated or related. The object and property words were identical in the two conditions but were paired differently to manipulate relatedness. Both groups received the same 100 true trials, mixed randomly with the 100 false trials.

Results

Solomon and Barsalou (2004) obtained evidence that the false-trial manipulation modulated the extent to which participants used the language or simulation system for verifying properties. When the false trials were unrelated, participants drew more heavily on the linguistic system. When the false trials were related, participants drew more heavily on the simulation system.

First, participants were over 100 milliseconds faster to verify the true trials when the false trials were unrelated than when they were related. This is consistent with the LASS prediction that participants used the faster linguistic system first, when associative strength between words was adequate for task performance in the unrelated false-trials condition. When associative strength between words was not adequate in the related false-trials condition, participants had to use the slower simulation system to find the simulated properties in the simulated objects. This finding is consistent with the analogous findings

---

[4] As discussed earlier for Santos *et al.* (2008), people may often store linguistic phrases that describe taxonomic relations (e.g., 'dogs are animals'). In principle, people could similarly store linguistic phrases that describe part relations (e.g., 'elephants have tusks'). If so, then manipulating the relatedness of false trials should have no effect, because people could always assess whether a property is a part by simply consulting the linguistic system – it should never be necessary to consult the conceptual system, even when false trials are related. As will be seen, however, the false-trial manipulation has large effects. When related false trials block the superficial linguistic strategy, conceptual knowledge must be consulted to assess whether a property is actually a part. Thus, people do not appear to typically store linguistic phrases for part relations as they do for taxonomic relations, or at least to the same extent, perhaps because verbal part descriptions are encountered less frequently.

in Santos *et al.* (2008) and Simmons *et al.* (2008) that information is available earlier from the linguistic system than from the simulation system when words are presented.

Regression analyses provided evidence for a qualitative shift in verification strategies across the false-trial manipulation. Solomon and Barsalou (2004) scaled objects and properties on a wide variety of variables that could potentially predict variance in reaction times and errors (see Solomon, 1997, for additional details). Whereas some properties were verified quickly and with high accuracy, others were verified slowly and with lower accuracy. Of interest was identifying variables that explained this variance.

Three groups of variables seemed potentially important: linguistic, perceptual, and expectation variables. The linguistic variables included the associative strength from the object words to the property words, the word frequency of the property words, the length of the property words, etc. The perceptual variables included the size of the properties relative to the objects, the salience of the properties, whether the properties are occluded, whether they are handled, etc. The expectation variables included the variability of property forms, whether properties could be separate objects, etc.

When the false trials were unrelated, Solomon and Barsalou (2004) found that the linguistic variables best predicted verification performance. In particular, associative strength from the object word to the property word was the best predictor. As the associative strength between an object word and a property word increased, participants verified properties faster and with higher accuracy. Thus, these participants appeared to be using superficial statistical information from the linguistic system to verify properties. The stronger the associative strength between the concept and property words, the easier it was to verify the properties. Interestingly, as associative strength became weak between two words, participants relied increasingly on the simulation system, as described later.

A different pattern of prediction emerged in the related false-trials condition. The importance of linguistic variables in explaining performance decreased significantly, and the importance of perceptual variables increased significantly. Indeed, perceptual variables became the strongest predictors, with the size of a property being the best predictor. As properties became larger, they took longer to verify and produced more errors. Because larger properties take longer to simulate and to match against a simulated object, verifying them led to longer response times and produced more errors (more errors resulted from participants responding before they had taken sufficient time to simulate the larger properties). Kosslyn et al. (1983, 1988) provide related evidence that simulating large objects takes longer than simulating small ones.

Thus, this experiment offers evidence that conceptual processing relies on both language and simulation, as the LASS theory predicts. Under conditions that allowed the use of word associations, participants relied on the linguistic system. When the presence of related false trials blocked this strategy, participants used simulation to assess whether properties were parts of objects.

An additional finding from the unrelated false-trial condition provides evidence that these participants drew on the language and simulation systems dynamically. When participants in the unrelated false-trial condition responded quickly on true trials, the linguistic variables best predicted their performance. When these same participants responded slowly on true trials, however, their performance was best predicted by the perceptual variables. This pattern suggests that the language and simulation systems were operating in parallel. When a strong linguistic association was readily available between the object and property words, participants used it to respond quickly, given that the false trials were unrelated. Conversely, when a strong linguistic association was not available, participants could not respond on the basis of linguistic information and relied instead on the simulation system. Thus, in the unrelated false trials condition, performance relied dynamically on the two systems, depending on whether the linguistic system could provide the requisite information quickly for using the linguistic strategy.

Notably, definitive information for performing the task came from the simulation system, not from the linguistic system, consistent with the LASS theory. If the linguistic system contained deep conceptual information, then it should have been sufficient to produce the information required for correct decisions even when the false trials were related. If this system contained classic amodal propositions, it should have been unnecessary to access the simulation system. Instead, participants had to shift from the linguistic system to the simulation system to find definitive conceptual information.

Solomon (1997) reports a replication of this experiment using different materials and procedures. Solomon and Barsalou (2001) provide further evidence that participants rely on perceptual information to verify properties when the superficial linguistic strategy is blocked, as do Pecher et al. (2003, 2004; for a review, see Barsalou, Pecher et al. 2005).

Neural corroboration

Kan et al. (2003) performed the Solomon and Barsalou (2004) experiment in an fMRI scanner. They predicted that if the false-trial manipulation had modulated verification strategies in Solomon and Barsalou's experiment, then neural evidence should corroborate this modulation. Specifically, they predicted that when related false trials forced participants to use the simulation strategy, brain areas that process visual images should become active. Conversely, when unrelated false trials allowed participants to use the linguistic strategy, activation in these visual imagery areas should not occur.

Kan et al. predicted that the false-trial manipulation should modulate neural activity in the left fusiform gyrus, given that previous research on generating visual images from concrete words activated this area (e.g., D'Esposito et al. 1997; Thompson-Schill et al. 1999). When participants in those studies generated images of concrete objects from names, left fusiform areas became active, suggesting that these areas should similarly be active when simulating objects and verifying their properties.

Kan et al.'s fMRI findings corroborated the behavioural findings of Solomon and Barsalou (2004). When false trials were related, the left fusiform area observed in the previous imagery studies was active. When false trials were unrelated, this area was not active. Like the results of Solomon and Barsalou, this pattern indicates that two systems support conceptual processing. When task conditions allow, participants use linguistic information, such that the simulation system does not play a central role. When task conditions block the use of linguistic information, the simulation system becomes necessary for adequate performance. Again, these results are consistent with the conclusion that deep conceptual information resides in the simulation system, not in the linguistic system.

13.5.3 **Abstract concepts**

Based on the finding that memory tends to be better for concrete concepts than for abstract concepts, dual code theory proposed that the linguistic system represents abstract concepts for the following reasons (Paivio 1971, 1986). Because both the linguistic and simulation systems represent concrete concepts (two systems), memory for concrete concepts is good. Because only the linguistic system represents abstract concepts (one system), memory for abstract concepts is inferior. Based on a wide variety of compatible findings, many researchers have since echoed this view. In particular, nearly all neuroimaging researchers who have assessed the neural bases of concrete and abstract concepts have concurred with dual code theory (for a review, see Sabsevitz *et al.* 2005). Because these studies have generally found left-hemisphere language areas more active for abstract than for concrete concepts (especially Broca's area), they, too, have concluded that language represents abstract concepts.

A logical problem with this account is that language *per se* cannot represent a concept. If people use an unfamiliar foreign language to describe the meaning of an abstract concept, they do not understand the concept (cf. Searle 1980). They only understand the concept once they can ground the language in experience. This suggests to us that simulations of situations should be central to the representations of abstract concepts (Barsalou 1999). Evidence for the *context availability theory* of abstract concepts supports this conclusion (Schwanenflugel 1991).

Findings from Barsalou and Wiemer-Hastings (2005) also support this conclusion. When participants generated properties of concrete and abstract concepts, the properties generated for both showed more similarities than differences. For each type of concept, participants tended to describe the situations in which a concept occurred, including relevant information about agents, objects, settings, events, and mental states. Participants produced all these different kinds of properties for both concrete and abstract concepts, and produced them in roughly the same distributions.

Although strong similarities existed between these distributions, differences existed in content and complexity. Regarding content, abstract concepts focused on mental states and events significantly more than concrete concepts, whereas concrete concepts focused more on objects and settings. Regarding complexity, abstract concepts included more information, deeper hierarchical structures, and more contingency relations. Regardless, situations appeared equally important for both abstract and concrete concepts. Rather than only depending on language, abstract concepts appear to include extensive situational information as well (Schwanenflugel 1991). As much work has shown, however, concrete words tend to access situations faster than do abstract concepts, thereby giving concrete concepts an advantage in superficial processing tasks.

Wilson *et al.* (2008) assessed this conclusion further in an fMRI experiment. These researchers argued that previous neuroimaging experiments had only found evidence for linguistic representations of abstract concepts because they used tasks that allowed and encouraged superficial linguistic processing (e.g., lexical decision, synonym judgments). Because the tasks in these experiments can be performed using information from the lexical system, they did not engage the simulation system that represents deeper conceptual information (Glaser 1992; Solomon and Barsalou 2004; Kan *et al.* 2003).

To engage the simulation system, Wilson *et al.* gave participants the word for an abstract concept for 5 seconds and then had them verify whether the concept applied to a subsequent picture. For example, participants received the word 'convince' and assessed whether the concept *convince* applied to a picture of a politician speaking to a crowd. Wilson *et al.* argued that participants had to activate deep conceptual representations of the abstract concept during the 5-second priming period to determine whether its associated concept applied to the subsequent picture. If so, then areas involved in simulation should become active as participants prime conceptual representations of the abstract concept.

Wilson *et al.* confirmed this prediction. When participants received the abstract concept *convince* and prepared to assess whether it applied to a subsequent picture (across many trials), they activated brain areas involved in representing mental states and social interaction (e.g., medial prefrontal cortex). Similarly, when people prepared to assess whether the abstract concept *arithmetic* applied to a subsequent picture (again across many trials), they activated brain areas involved in performing arithmetic operations (e.g., intraparietal sulcus). For both concepts, participants simulated relevant situations to represent the respective concept prior to receiving a picture. Notably, the linguistic system was not more active for abstract concepts than for concrete concepts under these task conditions.[5]

These results suggest that the representation of abstract concepts can differentially recruit the language and simulation systems. When task conditions allow, as in previous experiments, participants rely only on the language system, because it is adequate for task performance (e.g., in lexical decision and synonym tasks). When task conditions require deeper conceptual processing, participants rely on the simulation system, because it provides the necessary information for performing the task (e.g., verifying that an abstract concept applies to a picture). Similar to Glaser's (1992) conclusion, processing pictures tends to produce deeper conceptual processing. Consistent with findings in previous sections, different mixtures of the language and simulation systems support the processing of abstract concepts under different task conditions.

13.6 **Potential relevance of the LASS theory to other phenomena**

The previous sections offered direct evidence for the LASS theory. Here we turn to more speculative evidence from a *post hoc* perspective. We next review phenomena where the two LASS systems – language and simulation – could potentially play important roles. We hasten to add, however, that the researchers studying these phenomena typically offered

---

[5] In these analyses, only activations for the words were analysed, not activations for the subsequent pictures. Activations for the words and pictures were deconvolved so that activations for word meaning could be examined in isolation.

alternative interpretations of them based on different pairs of systems, not the LASS systems. Nevertheless, we believe that the LASS systems could be part of the story. Although the accounts proposed originally for these phenomena may be correct to some extent, interplay between the language and simulation systems may be important as well. We review the phenomena in this section simply to raise this alternative interpretation. Future research will be necessary to resolve which account is correct. We would not be surprised if multiple accounts, invoking multiple systems, are necessary. In general, however, we suspect that the interplay between language and simulation is a central theme across the spectrum of cognitive activities (again, see Paivio 1971, 1986, for a similar view).

13.6.1 **Language comprehension**

Much work in language comprehension is consistent with the view that one system processes linguistic forms, whereas another system uses simulation to represent meanings. In classic work, Sachs (1967) showed that surface memory for sentences lasts around 20 seconds and is then replaced by gist representations; much subsequent research confirmed this phenomenon. Typically, surface memory is viewed as residing in working memory, whereas gist is viewed as residing in long-term memory, represented by amodal symbols.

Although the working versus long-term memory distinction is probably an important part of the story, so may be the distinction between language and simulation. Whereas surface memory reflects linguistic structures in the linguistic system, gist memory may reflect simulations in the simulation system. Once the linguistic form of a sentence is lost from working memory, the only information remaining is a simulation in long-term memory. Because this simulation is not linguistic, it does not enable direct recovery of the sentence's linguistic form but is nevertheless consistent with its meaning (i.e., gist).

Based on these findings, Bransford and Franks (1971) further demonstrated that gist memory loses information about the specific form of linguistic input. Bransford and Franks presented participants with a series of sentences about a situation and showed that the meanings of these sentences were integrated into a coherent semantic representation. As a result, participants could no longer remember the actual sentences studied. Instead, the more that a test sentence corresponded to the integrated gist, the more participants believed that they had seen the sentence, even when they had not. Other researchers showed, however, that under various conditions, participants could remember the surface forms of the input sentences (e.g., Katz 1973; Flagg et al. 1975). A large literature has continued to develop around this issue (e.g., Brainerd and Reyna 2004).

Again, the distinction between surface form and gist can be aligned with the LASS distinction between language and simulation. Whereas the surface forms of the input sentences are stored in the language system, the integrated gist is stored in the simulation system. As participants hear a sentence sequence, they incrementally construct a simulation to represent the situation being described, losing linguistic forms in the process.

More recent work on comprehension echoes these older themes. McKoon and Ratcliff (1992) argued that when people read texts superficially, they do not compute many inferences that go beyond the linguistic forms mentioned explicitly.

Conversely, Graesser *et al.* (1994) argued that when people process texts deeply, they compute a wide variety of inferences. This distinction between minimal versus rich inference can again be aligned with the language and simulations systems in LASS. When minimal inferencing occurs, people primarily process linguistic forms in the linguistic system. If they construct simulations, they may primarily simulate the meanings of individual words without integrating them into a coherent global simulation – instead, simulations of the individual word meanings are relatively fragmented. Conversely, when rich inferencing occurs, people may perform much more simulation and, in particular, integrate simulations for individual words into a global simulation (as in Bransford and Franks 1971). During the integration process, deep comprehenders may add additional information into the global simulation to make it coherent. As a result of greater integration and coherence, these simulations contain inferences that go considerably beyond words mentioned in the text.

Individual differences in text comprehension can similarly be tied to differential use of the two LASS systems. Poor comprehenders may have to expend so much effort processing linguistic forms that they have minimal capacity left to simulate and integrate word meanings. Because good comprehenders are superior at processing linguistic forms, they spend more time simulating and integrating meaning, and thus exhibit higher comprehension. Van Petten *et al.* (1997) offer evidence for this account. They found that readers with low working memory capacity readily produced linguistic inferences but did not produce meaning inferences. Conversely, readers with high working memory capacity produced both. From the LASS perspective, the poor readers had enough capacity to produce word-level inferences within the linguistic system, but did not have enough capacity to construct rich integrated simulations that represent meaning.

13.6.2 Conceptual processing

Researchers who study concepts have reported related results. Wisniewski and Bassok (1999) found that when participants assess the similarity of two concepts, they inadvertently allow thematic associations to affect their similarity judgements. Participants should have only assessed shared and distinctive properties of the two concepts, ignoring thematic associations between them. For example, when participants judged the similarity of *coffee* and *cup*, they should have only assessed their shared and distinctive properties, ignoring the thematic relation that coffee is drunk from cups. Problematically, however, participants allow thematic relations like these to inflate their similarity judgements.

Gentner and Brem (1999) showed that participants can filter out thematic associations when they receive sufficient time to judge similarity. From the LASS perspective, thematic associations may originate quickly in the linguistic system, similar to the linguistic associations between concepts and properties in Solomon and Barsalou (2004). Conversely, similarity judgements may operate on simulations, as people compare simulations of the two concepts for shared and distinctive properties. When fast responses are possible, thematic information from the linguistic system dominates, and similarity information from the simulation system has less effect. When participants must take more time, they can suppress thematic responses from the linguistic system and focus attention more on the assessment of simulations.

Chaffin (1997) had participants produce word associates to high- versus low-frequency words. In general, the high frequency words often produced semantic responses that described events. In contrast, the low-frequency words often produced linguistic responses, such as synonyms and sound similarities. Low-frequency words also often produced definitions. According to Chaffin, the high-frequency words produced deeper processing associated with the pragmatics of using these words, whereas the low-frequency words produced shallower processing associated with trying to establish their meanings.

A complementary explanation is that high-frequency words readily activate situated simulations in the simulation system, whereas low-frequency words primarily activate linguistic forms in the language system. High-frequency words are associated with pragmatic information because they activate well established event simulations from experience that support situated action. Low-frequency words are associated with linguistic information because they have not been associated with enough experience to activate familiar situations. As a result, low-frequency words activate synonyms and definitions in the linguistic system, because this is what people have primarily learned about them from hearsay.

13.6.3 Social processes

Smith and DeCoster (2000) proposed that two systems underlie a wide variety of social processes. One system provides fast associative information; the other provides slower rule-based information. On some occasions, social processing results from quickly accessing relatively superficial information that is statistically likely (e.g., stereotypes). On other occasions, social processing results from more thoughtful processing that relies on careful reasoning about particular situations.

Although the distinction between associations versus rules is probably central to these two forms of processing, we suspect that the distinction between the linguistic and simulation systems may be central as well. Following Glaser (1992), we suspect that fast superficial processing in social situations often draws on the linguistic system, with linguistic structures being sufficient for task performance (as in Solomon and Barsalou 2004; Kan *et al.* 2003). Conversely, we suspect that slower, more careful processing often operates on simulations of social situations. Simulating how a social situation developed and how it may evolve over time may often underlie deliberate social reasoning. For a related account, see Barsalou, Niedenthal *et al.* (2003).

13.6.4 Clinical phenomena

A recent study by Schlamann *et al.* (2006) suggests that language and simulation underlie treatments in medical and psychological settings. The authors performed an fMRI study with stroke patients who had undergone therapy to help them simulate helpful motor activity. Interestingly, these patients activated the simulation system when asked to think about various motor actions. Conversely, control patients who had not received the therapy did not activate the simulation system, suggesting that they processed the motor actions more superficially.

This pattern suggests that patients in medical and psychological settings may vary in how deeply they understand their illnesses and treatments. Whereas some patients may only have superficial understandings of their situations as described in verbal descriptions, other patients may have deeper understandings grounded in simulation (analogous to the earlier distinction between shallow and rich comprehension). If so, then one important question is whether understanding illnesses and treatments in terms of simulations produces better treatment outcomes. If simulation-based understanding improves outcomes, then inducing such understandings in patients could have significant benefits.

13.6.5 Education

The distinction between superficial linguistic comprehension and deep simulation-based comprehension also appears central in education (cf. Glenberg *et al.*, in press). Students may vary widely in how well their understanding of a particular domain engages both the language and simulation systems. Whereas some students may only be able to regurgitate memorized verbal descriptions about a domain, stronger students may be able to manipulate simulations of the domain, thereby having deep insights about it, along with the ability to go beyond explicit instruction. I suspect that seasoned instructors are familiar with both kinds of students.

13.6.6 Summary

As these speculative examples illustrate, the interplay between the language and simulation systems may be pervasive throughout diverse psychological phenomena, as Paivio (1971) noted originally. We suspect that such interplay is likely to occur in many other areas besides those just covered (for many additional examples see Paivio 1986). Again, further research is necessary to explore these possibilities.

13.7 Discussion

We began by proposing that multiple systems – not one – support conceptual processing. In particular, we have focused on contributions from the linguistic and simulation systems. We saw evidence of both in conceptual processing, and we saw that they play different roles in different concepts and in different task contexts. We also saw that deeper conceptual processing requires the simulation system. When the linguistic system dominates, conceptual processing appears to be relatively superficial, consistent with Glaser's (1992) lexical hypothesis.

Although we have focused on the language and simulation systems, we do not mean to preclude contributions from other systems as well. As described earlier, we believe that statistical representations underlie the processing of both language and simulation. Both the frequency of representations in these systems, along with correlations between them, enter ubiquitously into conceptual processing.

An interesting question is whether stand-alone statistical structures can serve representational purposes in the absence of linguistic and modal representations. Some theorists, such as Damasio (1989), have argued that statistical structures primarily serve to trigger modal simulations. Others have suggested that statistical structures can function as stand-alone representations (e.g., Rogers and McClelland 2004). Still others have suggested that statistical structures primarily serve to trigger simulations, but can function as stand-alone representations in automatic stimulus–response sequences (e.g., Simmons and Barsalou 2003). Regardless of where the empirical findings come down on this particular issue, there is no doubt that statistical representations play central roles throughout conceptual processing.

13.7.1 Complex linguistic processing

Much of the work so far that has assessed interactions between the linguistic and simulation systems has assessed relatively simple forms of linguistic processing. In our experiments, we have primarily assessed word association, property generation, and property verification. Furthermore, in Figure 13.1, we only considered a single cycle of interaction between the language and simulation systems. As Figure 13.3 illustrates, however, we assume that much more complex interactions occur. Over time, both systems cycle through periods of relative activity and inactivity as processing evolves. Rather than operating independently, we assume that the two systems are highly interdependent. The activation of linguistic forms activates simulations. In turn, simulations activate words that describe and manipulate them. We assume that these processes cycle interactively over time in myriad patterns.

Following Barsalou (1999, 2003a, 2005a), we assume that the linguistic system plays central roles in producing the compositional structure of simulations. Specifically, we assume that the syntactic structure of sentences controls the retrieval, assembly, and transformation of the componential simulations that people integrate to represent sentences and texts. Similarly, we assume that interactions between the two systems are responsible for the representation of propositions, conceptual combinations,

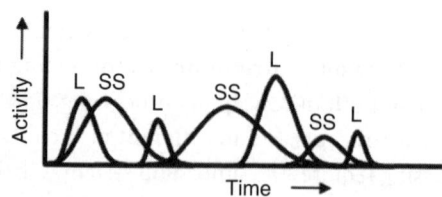

Fig. 13.3 Iterating and interacting contributions between the linguistic system (L) and the situated simulation system (SS) during conceptual processing. When the cue is a word, contributions from the linguistic system precede those from the simulation system. After the initial cycle of processing, both systems cycle through periods of activity and inactivity as they interact with each other. As in Figures 13.1 and 13.2, the height, width, shape, and offset of the distributions are not assumed to be fixed (i.e., the particular distributions in this figure illustrate one of infinitely many different courses that interaction between the L and SS systems could take).

productively produced phrases, recursively embedded structures, etc. In general, we assume that symbolic structures and symbolic operations on these structures emerge from ongoing interactions between the language and simulation systems. In the future, we believe that it will be essential for researchers to explore these more complicated interactions between language and simulation.

Psycholinguistics research increasingly explores these complex interactions. For example, Glenberg and colleagues have explored relations between syntactic structures and the affordances available from simulations (e.g., Glenberg and Roberston 2000; Kaschak and Glenberg 2000). Zwaan and colleagues have explored how sentences activate corresponding simulations and operations on them (Zwaan and Madden 2005), as have de Vega and colleagues (e.g., de Vega *et al.* 2004; de Vega 2005). Other researchers similarly explore complex relations between language and simulation, including Spivey *et al.* (2000), Richardson et al. (2003), Matlock (2004), and Richardson and Matlock (in press).

In the future, we believe that it will be increasingly productive to explore detailed relations between compositional linguistic forms and compositional simulations. Theories in cognitive linguistics offer many intriguing ideas about the corresponding compositional structures of language and experience that researchers could explore in rigorous empirical experiments (e.g., Coulson 2000; Fauconnier 1997; Goldberg 1995; Kemmerer 2006; Lakoff 1987; Langacker 1987; Talmy 1983).

13.7.2 **Superficial conceptual processing in the linguistic system**

According to traditional views, deep conceptual processing results from processing language-like propositional structures, with modal representations playing peripheral roles. As we have seen, however, the opposite may be true. Deep conceptual processing may require the simulation system. When only the linguistic system is engaged, conceptual processing appears relatively superficial (Glaser 1992). As described earlier, Solomon and Barsalou (2004) found that participants could not determine whether a property was a part of an object only using linguistic information (also Kan *et al.* 2003). When word associations did not distinguish true from false trials, participants switched from using the linguistic system to the simulation system. This suggests that processing simulations was required to establish that a property was part of an object. Rather than this relation being stored amodally – '$part(X,Y)$' – or linguistically – 'object Y has property X' – this relation appeared to be computed by simulating the property and determining whether it could be found in the simulated object (see Solomon and Barsalou 2001 for further evidence).

Why would the human conceptual system have evolved this way? Why would the linguistic system only provide superficial information relevant to conceptual processing? Why would processing simulations produce deeper conceptual information? Following evolutionary theorists, we believe that the basic architecture of the human conceptual system existed in previous species (e.g., Donald 1993). Specifically, we believe that simulation systems existed long before humans, so that multimodal experience could be captured to inform situated action (e.g., Barsalou 2005b). By storing memories of multimodal experience, these memories could later be simulated to generate anticipatory inferences that supported feeding, reproduction, etc. Language evolved later for controlling the simulation system to a much greater extent. Adding language increased the ability of the simulation system to represent non-present situations (past, future, counter-factual). Adding language increased the ability to construct simulations compositionally and the ability to coordinate simulations between agents, yielding more powerful forms of social organization.

From this perspective, it is perhaps not surprising that the linguistic system by itself is only capable of superficial conceptual processing. If it primarily serves as a control system for manipulating simulations, then it would be unlikely to contain the most central conceptual information. If it did, it would not be a control system – it would be the main conceptual system (i.e., there would be no need for a simulation system). As we have seen, however, the simulation system appears necessary for deep conceptual processing. Furthermore, its evolutionary precedence suggests that it has had longer to evolve sophisticated mechanisms than the language system.

Most importantly, both systems are probably essential for achieving the powerful symbolic abilities of the human cognitive system. Neither system alone is likely to be sufficient for symbolic behaviour. Indeed, adding a linguistic system to the simulation system almost certainly enhanced symbolic behaviour considerably. Across many abilities, the two systems work together to achieve the power and distinctive properties of human intelligence.

Interestingly, the linguistic system appears to contain considerable amounts of statisti-cal information that mirrors the content of the simulation system, which in turn mirrors the content of experience. As a result, when linguistic cues are received, they initially trigger statistically-related linguistic forms that provide fast heuristic processing. The success of linguistic context theories to explain diverse cognitive phenomena may reflect this use of the linguistic system (e.g., Burgess and Lund 1997; Landauer and Dumais 1997; Louwerse and Jeuniaux, Chapter 15, this volume). Again, however, the ability of the linguistic system to play a heuristic role should probably not be equated with deep conceptual processing (e.g., Glaser 1992; Glenberg and Robertson 2000; Solomon and Barsalou 2004).

13.7.3 The time course of processing simulations

Across the literatures we have reviewed, superficial linguistic processing preceded deep simulation processing temporally. As Pulvermüller and colleagues have found, however, simulations can become activated automatically and quickly, within 200 milliseconds of word onset (Pulvermüller *et al.*, 2005; Pulvermüller, Chapter 6, this volume). As suggested earlier, however, simulations may not dominate executive processing immediately (Figure 13.2, panel A). When the executive system focuses attention on another system as a source of information, this system may control responses, while simulations run unat-tended in parallel.

Consider the examples reviewed here. In Santos *et al.* (2008), we found that linguistic responses tended to initially dominate word association and property generation for at least a second or two. In Solomon and Barsalou (2004), we found that accessing the simulation system required at least another 100 milliseconds of processing than accessing the linguistic system. In Simmons *et al.* (2008), we found that language areas dominated the first 7.5 seconds of brain activation.

Why does executive processing select the linguistic system first as a source of relevant information on these tasks? One potential factor is that the cues on all these tasks are words. When participants receive words as cues for conceptual processing, the information available first may be other words, following the principles of content-addressable memory and encoding specificity (e.g., Tulving and Thomson 1973). Furthermore, because participants must respond with words in many of these tasks, this may further orient executive processing towards the linguistic system. When the linguistic system is capable of generating a correct response (as in the unrelated false condition of Solomon and Barsalou 2004), there is no need to go outside this system. When the linguistic system cannot generate responses on its own, however, attention must shift to the simulation system, which takes extra time. Although the simulation system may produce simulations all along, it may only be consulted when necessary.

One important goal for future research is to further document parallel streams of activity in the two systems, along with interactions between them. Another important goal is to articulate the executive processing strategies that draw on these two processing streams. How do executive strategies determine which stream to process under what conditions? When do executive strategies shift attention to a different stream? How do executive strategies make decisions based on the content of the stream(s) processed?

13.7.4 Conceptual processing of nonlinguistic stimuli

Throughout this paper, we have focused on paradigms where words serve as cues for conceptual processing. From the perspective of evolution, however, words have only played this role a very short time – nonlinguistic experiential states have played this role much longer.

Prior to humans, other animals certainly possessed conceptual skills. As they experienced motivational states, such as feeling hungry or thirsty, they recognized these states as instances of familiar concepts (e.g., *hungry*). Similarly, as they experienced sensory states, such as seeing or hearing prey, they recognized these states as indicating the presence of familiar categories in the world (e.g., *deer* for a lion). Clearly, nonhumans do not have linguistic labels for these categories, nor may they experience acts of categorization consciously (although perhaps they do). Nevertheless, a system in the brain that represents the concept identifies instances as category members. Furthermore, once an instance has been bound to a concept, the concept produces conceptual inferences about what is likely to happen next. Once *hungry* has been categorized as an internal state, for example, simulations of hunting and finding prey become active. Once *deer* becomes bound to a perceived object, simulations of attacking and eating become active. In these senses, nonhumans have powerful conceptual systems that evolved to process experience, not words.

Because humans evolved from nonlinguistic species, their conceptual systems, too, are probably heavily oriented toward processing nonlinguistic experience. Indeed, it seems obvious that tremendous amounts of categorization and inference take place as we process events in our bodies and in the world. Furthermore, our ability to consciously perceive and categorize introspective states may vastly exceed this ability in other species. In particular, greater introspective abilities may be central to the significant presence of abstract concepts in the human conceptual system (Barsalou 1999; Barsalou and Wiemer-Hastings 2005; Wilson *et al.* 2008).

How might conceptual processing differ as a function of receiving sensorimotor and introspective cues instead of receiving word cues? One obvious possibility is that simulations that situate perceived information may precede the activation of linguistic forms. Again, following the principles of content addressability and encoding specificity, experience may activate situated simulations faster than it activates language because simulations are more similar to cue information. Such simulations may provide the myriad situational inferences documented in the literature (e.g., Barsalou 2003b, 2005b, in press; Barsalou, Niedenthal *et al.* 2003; Yeh and Barsalou, 2006).

Another likely effect of experiential input to the conceptual system is stronger conceptual effects. As Glaser (1992) found, conceptual priming and interference are both much larger for picture cues than for word cues. Again, this suggests that conceptual information is established most strongly in the simulation system, with corresponding statistical structures in the linguistic system being weaker. This further suggests that deeper understandings of situations occur when people receive experiential input than when they receive linguistic input. Experiential input may be more likely to activate the simulations that carry deep conceptual information about a situation than do words that describe it.

Linguistic structures are also likely to become active in response to experiential input, although they may become active more slowly than simulations. Much remains to be learned about the roles that these linguistic activations play in the conceptualization of experience. One possibility is that the linguistic structures activated by an experience activate simulations that are more distant than the simulations activated by experience itself. Whereas experience may tend to activate simulations that map closely onto it, linguistic structures may tend to activate simulations of situations that are likely to follow the perceived situation (i.e., predictions), or that preceded it (i.e., explanations). Another important function of linguistic activations may be to draw attention towards important regions of experience that are relevant for situated action. As a word becomes active, it may name a region of a simulation and shift attention towards it (e.g., Estes *et al.*, 2008).

In general, we believe that the conceptual processing of experience deserves much greater scientific examination than it has received so far. Researchers typically study words because it is much easier to use words as laboratory stimuli than it is to use pictures, sounds, touches, actions, and introspections. In our opinion, however, this has led to distorted views of cognition, in general, and of the conceptual system, in particular. Clearly, language plays central roles in cognition and conceptualization. Nevertheless, experience plays a role that is at least as central. We hope that researchers increasingly study language and conceptualization in the context of experience. More balanced theories of cognition are likely to result.

Debate

Lawrence Shapiro: First let me say that was, I think, just a great series of experiments. I have some questions about how it ties in with the LSA stuff. I have the impression that you want to say that the word association process could be using some mechanism like LSA and then we move to the situation stuff, which is using some other sort of mechanism. But why can't we assume that LSA explains all of it? And also what kind of reaction-time predictions should LSA be making? As an advocate of LSA I don't think we have to be committed to any sort of reaction-time predictions.

Lawrence Barsalou: Yeah, I see no *a priori* reason why you couldn't try to explain all of these results with an LSA-like mechanism. I think where things get tricky is explaining the effects of factors like the size of properties in property verification, and why you see all these activations in modality-specific areas. Not just in our work, but in many other people's work – Friedemann's and Alex Martin's and others. There's just so much evidence now that these other mechanisms are engaged, it seems likely that they play a role. Now they could be epiphenomenal, as various people have suggested, and maybe there is something like an LSA mechanism that's at the core of everything. I think that's a plausible hypothesis and it's an empirical question which of those views is correct. So much evidence now exists though, some of it causal, that I strongly doubt that simulations are epiphenomenal.

Friedemann Pulvermüller: Thanks very much for an excellent, very interesting talk. Let me comment on a minor thing. You started with Paivio's theory, and I think this is an interesting starting point. However his model was not fully appropriate with regard to brain models; for example, he placed his verbal system in the left hemisphere and the imagery system in the right, which we have argued in the past doesn't work out. The imagery system has to be bihemispheric.

Barsalou: Let me just comment and then I'll let you continue. We're not at all arguing that Paivio's entire theory is totally correct. It's just the spirit of the idea that there are two systems for language and modality-specific processing that are both central to conceptual processing. At many detailed levels I agree with your concerns and have several disagreements with dual coding theory.

Pulvermüller: Let me also say something about the imaging experiment you mentioned towards the end. There were significant differences between the methods of the localizer conditions, for example, in one case there were trials of 5 seconds, in the other case trials of 15 seconds. And then there were many word cues in one condition and only a few word cues in the other. So these are all things that could lead to why you observed more activation in Broca's area in one condition than in the other.

Barsalou: Well, again, let me describe the way that those localizer tasks worked. For both the word association and the situation simulation localizer tasks, there were 15-second blocks and every 5 seconds a visual stimulus appeared in both conditions. In the word association condition, a different word occurred during each of those 5 seconds. And in the situated simulation it was the same word every 5 seconds.

Now, it could be that there is a difference between the same word and a different word. But the concept condition was just like the situated simulation condition – the same word occurred every 5 seconds. If that's what was going on, then the results in the concept condition should have looked more like the situated simulation condition, overall, than like word association at any piont. However, the results of the concept condition looked more like word association than like situated simulation early on, and more like situated simulation later. If the minor differences in presentation had been critical, this pattern wouldn't have occurred – the entire concept processing period should have only looked like situation generation. In general, the controls in this experiment were strong; for example, the words were identical in all three conditions. Also, we replicated the results twice.

Pulvermüller: I'm totally happy with your conclusions apart from one, namely that the simulation should be so late, because our data very strongly indicate that the earliest neurophysiological correlates of automatic simulation occur as early as the brain responses we can relate to lexical access – to the word-form processing. But now you might say, well, we are looking only at the first brain responses and what you are perhaps tapping into here are mechanisms that extend later in time. Is this correct?

Barsalou: Yeah, I totally agree. Your results are much more diagnostic on the nature of early activation than ours are for a variety of reasons. I'm totally comfortable with the idea that simulation areas become active immediately. It's an interesting question, though, whether the whole simulation is active initially. My guess is that that takes time, and I think there might be some ERP findings that are consistent with late semantic processing. One reason why I don't think we see, say, activations in the motor area in those first 7.5 seconds is that the semantics of these diverse concepts are probably distributed all over the brain. And one reason we only see, in the second 7.5 seconds, the precunius and the right temporal area is because the semantics of all these words are so diverse that they're activating very different sorts of modality-specific areas, such that we don't get an aggregation of signal for any particular kind of property, such as motor properties. We assume that activations for these properties are there, but because all concepts don't activate the same properties, not enough aggregation occurs for a BOLD signal to be detected. We suspect the precunius is involved in generating the situation, as opposed to particular content of the situation. We think if one did this kind of experiment with a very specific kind of concept, e.g., motor terms, you might find activation in those first 7.5 seconds. It would be consistent with what many people have been finding.

Arthur Glenberg: So you noted the relation of your theory to Paivio and to Glaser, but it seemed like there was another obvious relation and that's to construction–integration theory. It seemed really close. I was wondering if you or Walter had anything to say about that.

Barsalou: Absolutely.

Walter Kintsch: This is exciting work but the ideas and processes need a lot of elaboration. It's not just the initial activation of the meaning but that the meanings get
elaborated in time, which involves not just linguistic processes but also perceptual
simulations. This kind of processing is like what we tried to do with the integration
process, yes.

Unidentified person: What about Rips, Shoben, and Smith? I thought the perceptual
stuff was supposed to come quickly and that it's the more analytic *is-a* relations that
come later. How do you reconcile their model and findings with your approach?

Barsalou: That's a great connection. One of the reasons we got into the false-trial
manipulation in semantic memory experiments is because of experiments like theirs,
which manipulated false trials as well. I think they're right that there are different
phases of processing, but I disagree with the interpretations of what the early versus
late phases are doing. I think that the early processing they observed probably reflects
word association, not perceptual or semantic representations of characteristic
features.

Author note

We are grateful to Arthur Graesser and Marcel Just for helpful comments on an earlier
draft. This work was supported by National Science Foundation grants SBR-9421326,
SBR-9796200, and BCS-0212134 and by Defense Advanced Research Projects Agency
contract BICA FA8650-05-C-7256 to Lawrence Barsalou, by a Graduate Fellowship from
the Starr Foundation to Ava Santos, by a National Research Scholar Award
1F31MH070152-01 from the National Institute of Mental Health to Kyle Simmons, and
by a Graduate Research Fellowship from the National Science Foundation to Christine
Wilson. Address correspondence to Lawrence W Barsalou, Department of Psychology,
Emory University, Atlanta, GA 30322, USA (barsalou@emory.edu, www.psychology.
emory.edu/cognition/barsalou/index.html).

References

Allport DA (1985). Distributed memory, modular subsystems, and dysphasia. In SK Newman,
   R Epstein, Eds. *Current Perspectives in Dysphasia* (pp. 207–44). Edinburgh: Churchill Livingstone.

Barsalou LW (1999). Perceptual symbol systems. *Behavioural and Brain Sciences*, 22, 577–660.

Barsalou LW (2003a). Abstraction in perceptual symbol systems. *Philosophical Transactions of the Royal
   Society of London: Biological Sciences*, 358, 1177–87.

Barsalou LW (2003b). Situated simulation in the human conceptual system. *Language and Cognitive
   Processes*, 18, 513–62.

Barsalou LW (2005a). Abstraction as dynamic interpretation in perceptual symbol systems.
   In L Gershkoff-Stowe, D Rakison, Eds. *Building Object Categories* (pp. 389–431). Mahwah,
   NJ: Erlbaum.

Barsalou LW (2005b). Continuity of the conceptual system across species. *Trends in Cognitive Sciences*,
   9, 309–11.

Barsalou LW (2008a). Grounded cognition. *Annual Review of Psychology*, 59, 617–45.

Barsalou LW (2008b). Grounding symbolic operations in the brain's modal systems. In GR Semin, ER Smith, Eds. *Embodied Grounding: Social, Cognitive, Affective, and Neuroscientific Approaches.* (pp. 9–42) New York, NY: Cambridge University Press.

Barsalou LW (in press). Situating concepts. In P Robbins, M Aydede, Eds. *Cambridge Handbook of Situated Cognition.* New York, NY: Cambridge University Press.

Barsalou LW, Breazeal C, Smith LB (2007). Language as coordinated non-cognition. *Cognitive Processing,* 8, 79–91.

Barsalou LW, Niedenthal PM, Barbey A, Ruppert J (2003). Social embodiment. In B Ross, Ed. *The Psychology of Learning and Motivation, Vol. 43* (pp. 43–92). San Diego, CA: Academic Press.

Barsalou LW, Pecher D, Zeelenberg R, Simmons WK, Hamann SB (2005). Multi-modal simulation in conceptual processing. In W Ahn, R Goldstone, B Love, A Markman, P Wolff, Eds. *Categorization Inside and Outside the Lab: Essays in Honor of Douglas L. Medin* (pp. 249–70). Washington, DC: American Psychological Association.

Barsalou LW, Simmons WK, Barbey AK, Wilson CD (2003). Grounding conceptual knowledge in modality-specific systems. *Trends in Cognitive Sciences,* 7, 84–91.

Barsalou LW, Wiemer-Hastings K (2005). Situating abstract concepts. In D Pecher, R Zwaan, Eds. *Grounding Cognition: The Role of Perception and Action in Memory, Language, and Thought* (pp. 129–63). New York, NY: Cambridge University Press.

Brainerd CJ, Reyna VF (2004). Fuzzy-trace theory and memory development. *Developmental Review,* 24, 396–439.

Bransford JD, Franks JJ (1971). The abstraction of linguistic ideas. *Cognitive Psychology,* 3, 331–50.

Burgess C, Lund K (1997). Modelling parsing constraints with high-dimensional context space. *Language and Cognitive Processes,* 12, 177–210.

Chaffin R (1997). Associations to unfamiliar words: learning the meanings of new words. *Memory and Cognition,* 25, 203-226.

Collins AM, Loftus EF (1975). A spreading activation theory of semantic processing. *Psychological Review,* 82, 407–28.

Coulson S (2000). *Semantic Leaps: Frame-shifting and Conceptual Blending in Meaning Construction.* Cambridge: Cambridge University Press.

Craik FIM, Lockhart RS (1972). Levels of processing: a framework for memory research. *Journal of Verbal Learning and Verbal Behaviour,* 11, 671–84.

Craik FIM, Tulving E (1975). Depth of processing and the retention of words in episodic memory. *Journal of Experimental Psychology: General,* 104, 268–94.

Craik FIM (2002). Levels of processing: past, present … and future? *Memory,* 10, 305–18.

Damasio AR (1989). Time-locked multiregional retroactivation: a systems-level proposal for the neural substrates of recall and recognition. *Cognition,* 33, 25–62.

D'Esposito M, Detre JA, Aguirre GK *et al.* (1997). A functional MRI study of mental image generation. *Neuropsychologia,* 35, 725–30.

de Vega M (2005). Processing of sentences with causal or adversative connectives. *Cognitiva,* 17, 85–108.

de Vega M, Robertson DA, Glenberg AM, Kaschak MP, Rinck M (2004). On doing two things at once: temporal constraints on actions in language comprehension. *Memory and Cognition,* 32, 1033–43.

Donald M (1993). Precis of "Origins of the modern mind: three stages in the evolution of culture and cognition." *Behavioural and Brain Sciences,* 16, 739–791.

Estes Z, Verges M, Barsalou LW (2008). Head up, foot down: object words orient attention to the object's typical location. *Psychological Science* 19, 93–97.

Fauconnier G (1997). *Mappings in Thought and Language.* New York, NY: Cambridge University Press.

Flagg PW, Potts GR, Reynolds AG (1975). Instructions and response strategies in recognition memory for sentences. *Journal of Experimental Psychology: Human Learning and Memory,* 1, 592–8.

Fodor JA (1975). *The Language of Thought*. Cambridge, MA: Harvard University Press.

Gentner D, Brem S (1999). Is snow really like a shovel? Distinguishing similarity from thematic relatedness. In M Hahn, SC Stoness, Eds. *Proceedings of the Twenty-First Annual Meeting of the Cognitive Science Society* (pp. 179–84). Mahwah, NJ: Erlbaum.

Gigerenzer G (2000). *Adaptive Thinking: Rationality in the Real World*. New York, NY: Oxford University Press.

Glaser WR (1992). Picture naming. *Cognition*, 42, 61–105.

Glenberg AM (1997). What memory is for. *Behavioural and Brain Sciences, 20*, 1–55.

Glenberg AM, Jaworski B, Rischal M, Levin JR (in press). What brains are for: action, meaning, and reading comprehension. In D McNamara, Ed. *Reading Comprehension Strategies: Theories, Interventions, and Technologies*. Mahwah, NJ: Erlbaum.

Glenberg AM, Robertson DA (2000). Symbol grounding and meaning: a comparison of high-dimensional and embodied theories of meaning. *Journal of Memory and Language*, 43, 379–401.

Goldberg AE (1995). *Constructions: A Construction Grammar Approach to Argument Structure*. Chicago, IL: University of Chicago Press.

Graesser AC, Singer M, Trabasso T (1994). Constructing inferences during narrative text comprehension. *Psychological Review*, 101, 371–95.

James C (1975). The role of semantic information in lexical decisions. *Journal of Experimental Psychology: Human Perception and Performance*, 104, 130–6.

Joordens S, Becker S (1997). The long and short of semantic priming effects in lexical decision. *Journal of Experimental Psychology: Learning, Memory, and Cognition*, 23, 1083–105.

Kan IP, Barsalou LW, Solomon KO, Minor JK, Thompson-Schill SL (2003). Role of mental imagery in a property verification task: fMRI evidence for perceptual representations of conceptual knowledge. *Cognitive Neuropsychology*, 20, 525–40.

Kaschak MP, Glenberg AM (2000). Constructing meaning: the role of affordances and grammatical constructions in sentence comprehension. *Journal of Memory and Language*, 43, 508–29.

Katz S (1973). Role of instruction in abstraction of linguistic ideas. *Journal of Experimental Psychology*, 98, 79–84.

Kemmerer D (2006). The semantics of space: integrating linguistic typology and cognitive neuroscience. *Neuropsychologia*, 44, 1607–21.

Kintsch W, van Dijk TA (1978). Toward a model of text comprehension and production. *Psychological Review*, 85, 363–94.

Kosslyn SM, Cave CB, Provost DA, von Gierke SM (1988). Sequential processes in image generation. *Cognitive Psychology*, 20, 319–43.

Kosslyn SM, Reiser BJ, Farah MJ, Fliegal SL (1983). Generating visual images: units and relations. *Journal of Experimental Psychology: Learning, Memory, and Cognition*, 112, 278–303.

Lakoff G (1987). *Women, Fire, and Dangerous Things: What Categories Reveal about the Mind*. Chicago, IL: University of Chicago Press.

Landauer TK, Dumais ST (1997). A solution to Plato's problem: the latent semantic analysis theory of acquisition, induction, and representation of knowledge. *Psycyhological Review*, 104, 211–40.

Langacker RW (1987). *Foundations of Cognitive Grammar, Vol. 1. Theoretical Prerequisites*. Stanford, CA: Stanford University Press.

Lockhart RS (2002). Levels of processing, transfer-appropriate processing, and the concept of robust encoding. *Memory*, 10, 397-403.

Martin A (2001). Functional neuroimaging of semantic memory. In R Cabeza, A Kingstone, Eds. *Handbook of Functional Neuroimaging of Cognition* (pp. 153–86). Cambridge, MA: MIT Press.

Martin A (2007). The representation of object concepts in the brain. *Annual Review of Psychology*, 58, 25–45.

Matlock T (2004). Fictive motion as cognitive simulation. *Memory and Cognition*, 32, 1389–400.

McClelland JL, Rumelhart DE,; PDP Research Group (1986). *Parallel Distributed Processing: Explorations in the Microstructure of Cognition, Vol. 2. Psychological and Biological Models.* Cambridge, MA: MIT Press.

McKoon G, Ratcliff R (1992). Inference during reading. *Psychological Review*, 99, 440–66.

Morris CD, Bransford JD, Franks JJ (1977). Levels of processing versus test-appropriate strategies. *Journal of Verbal Learning and Verbal Behaviour*, 16, 519–33.

Murphy GL (2002). *The Big Book of Concepts*. Cambridge, MA: MIT Press.

Newell A, Simon HA (1972). *Human Problem Solving*. Englewood Cliffs, NJ: Prentice-Hall.

Niedenthal PM, Barsalou LW, Winkielman P, Krauth-Gruber S, Ric F (2005). Embodiment in attitudes, social perception, and emotion. *Personality and Social Psychology Review*, 9, 184–211.

O'Reilly RC, Munakata Y (2000). *Computational Explorations in Cognitive Neuroscience: Understanding the Mind by Simulating the Brain*. Cambridge, MA: MIT Press.

Paivio A (1971). *Imagery and Verbal Processes*. New York, NY: Holt, Rinehart, & Winston.

Paivio A (1986). *Mental Representations: A Dual Coding Approach*. New York, NY: Oxford University Press

Pecher D, Zeelenberg R, Barsalou LW (2003). Verifying properties from different modalities for concepts produces switching costs. *Psychological Science*, 14, 119–24.

Pecher D, Zeelenberg R, Barsalou LW (2004). Sensorimotor simulations underlie conceptual representations: modality-specific effects of prior activation. *Psychonomic Bulletin and Review*, 11, 164–7.

Pecher C, Zwaan R, Eds (2005) *Grounding Cognition: The Role of Perception and Action in Memory, Language, and Thought*. New York, NY: Cambridge University Press.

Pylyshyn ZW (1984). *Computation and Cognition*. Cambridge, MA: MIT Press.

Prinz J (2002). *Furnishing the Mind: Concepts and their Perceptual Basis*. Cambridge, MA: MIT Press.

Pulvermüller F (1999). Words in the brain's language. *Behavioural and Brain Sciences*, 22, 253–336.

Pulvermüller F, Shtyrov Y, Ilmoniemi R (2005). Brain signatures of meaning access in action word recognition. *Journal of Cognitive Neuroscience*, 17, 884–92.

Richardson DC, Matlock T (in press). The integration of figurative language and static depictions: an eye movement study of fictive motion. *Cognition*.

Richardson DC, Spivey MJ, Barsalou LW, McRae K (2003). Spatial representations activated during real-time comprehension of verbs. *Cognitive Science*, 27, 767–80.

Rogers TT, McClelland JL (2004). *Semantic Cognition: A Parallel Distributed Processing Approach*. Cambridge, MA: MIT Press.

Rumelhart DE, McClelland JL; PDP Research Group (1986). *Parallel Distributed Processing: Explorations in the Microstructure of Cognition, Vol 1. Foundations*. Cambridge, MA: MIT Press.

Sabsevitz DS, Medler DA, Seidenberg M, Binder JR (2005). Modulation of the semantic system by word imageability. *NeuroImage*, 27, 188–200.

Sachs JDS (1967). Recognition memory for syntactic and semantic aspects of connected discourse. *Perception and Psychophysics*, 2, 437–42.

Santos A, Barsalou LW, Chaigneau SE (2008). Word association and situated simulation in conceptual processing. Manuscript in preparation.

Schlamann M, Naglatzki R, deGreiff A, Forsting M, Gizewski E (2006). Autogenic training and fMRI. Poster presented at the Meeting of the Organization for Human Brain Mapping. Florence, Italy.

Schwanenflugel PJ (1991). Why are abstract concepts hard to undersand? In PJ Schwanenflugel, Ed. *The Psychology of Word Meaning* (pp. 223–50). Mahwah, NJ: Erlbaum.

Searle JR (1980). Minds, brains, and programs. *Behavioural and Brain Sciences*, 3, 417–24.

Shulman HG, Davidson TCB (1977). Control properties of semantic coding in the lexical decision task. *Journal of Verbal Learning and Verbal Behaviour*, 16, 91–8

Simmons WK, Barsalou LW (2003). The similarity-in-topography principle: Reconciling theories of conceptual deficits. *Cognitive Neuropsychology*, 20, 451–86.

Simmons WK, Hamann SB, Harenski CN, Hu XP, Barsalou LW (2008). fMRI evidence for word association and situated simulation in conceptual processing. Journal of Physiology Paris.

Smith ER, DeCoster J (2000). Dual-process models in social and cognitive psychology: Conceptual integration and links to underlying memory systems. *Personality and Social Psychology Review*, 4, 108–31.

Smith LB, Gasser M (2005). The development of embodied cognition: six lessons from babies. *Artificial Life*, 11, 13–30.

Solomon KO (1997). The spontaneous use of perceptual representations during conceptual processing. Doctoral dissertation. Chicago, IL: University of Chicago.

Solomon KO, Barsalou LW (2001). Representing properties locally. *Cognitive Psychology*, 43, 129–69.

Solomon KO, Barsalou LW (2004). Perceptual simulation in property verification. *Memory and Cognition*, 32, 244–59.

Spivey M, Tyler M, Richardson D, Young E (2000). Eye movements during comprehension of spoken scene descriptions. *Proceedings of the 22nd Annual Conference of the Cognitive Science Society* (pp. 487–92). Mahwah, NJ: Erlbaum.

Stone GO, Van Orden GC (1993). Strategic control of processing in word recognition. *Journal of Experimental Psychology: Human Perception and Performance*, 19, 744–74.

Talmy L (1983). How language structures space. In H Pick, L Acredelo, Eds. *Spatial Orientation: Theory, Research, and Application* (pp. 225–82). New York, NY: Plenum Press.

Thelen E (2000). Grounded in the world: developmental origins of the embodied mind. *Infancy*, 1, 3–30.

Thompson-Schill SL (2003). Neuroimaging studies of semantic memory: inferring 'how' from 'where'. *Neurosychologia*, 41, 280–92.

Thompson-Schill SL, Aguirre GK, D'Esposito M, Farah MJ (1999). A neural basis for category and modality specificty in semantic knowledge. *Neuropsychologia*, 37, 671–6.

Tulving E, Thomson DM (1973). Encoding specificity and retrieval processes in episodic memory. *Psychological Review*, 80, 352–73.

van Dijk TA, Kintsch W (1983). *Strategies of Discourse Comprehension*. New York, NY: Academic Press.

Van Petten C, Weckerly J, McIssac HK, Kutas M (1997). Working memory capacity dissociates lexical and sentential context effects. *Psychological Science*, 8, 238–42.

Wilson CD, Simmons WK, Martin A, Barsalou LW (2008). Simulating properties of abstract concepts. Manuscript in preparation.

Wisniewski EJ, Bassok M (1999). What makes a man similar to a tie? Stimulus compatibility with comparison and integration. *Cognitive Psychology*, 39, 208–38.

Yap MJ, Balota DA, Cortese MJ, Watson JM (2006). Single versus dual process models of lexical decision performance: insights from RT distributional analysis. *Journal of Experimental Psychology: Human Perception and Performance*, 32, 1324–44

Yeh W, Barsalou LW (2006). The situated nature of concepts. *American Journal of Psychology*, 119, 349–84.

Zwaan RA, Madden CJ (2005). Embodied sentence comprehension. In D Pecher, R Zwaan, Eds. *Grounding Cognition* (pp. 224–45). New York, NY: Cambridge University Press.

Chapter 14

Levels of embodied meaning: From pointing to counterfactuals

Manuel de Vega

14.1 Introduction

The goal of this chapter is to analyse the nature of meaning (embodied versus symbolic) in the light of linguistic reference. I propose three levels of reference differing in their computational demands, and presumably in the quality or degree of embodied representations. In on-line reference, speakers communicate about the current perceptual environment; words, sentences, and gestures are mapped directly into objects and events. In displaced reference, speakers refer to entities that are not present in the current situation; words and sentences are mapped into embodied simulations of the referred situations. Finally, in decoupled reference, speakers refer to their own or others' mental states, such as their beliefs or intentions, which in turn refers recursively to other predicates. In this case, speakers build complex situation models, which can also be embodied, providing alternative points of view of a situation. The three levels of reference produce different computational burdens and, what is more important, might involve different degrees of embodied cognition.

A general function of language is to provide speakers with tools to establish referents cooperatively (e.g., Clark 1996). The interlocutors share the goal of building equivalent meanings from discourse. For instance, they tend to automatically align their representations at different linguistic levels by making use of each others' choices of words, sounds, grammatical constructions, and situation models (e.g., Garrod and Pickering 2004; Steels, Chapter 12, this volume). Alignment among speakers is a powerful dynamic process that could be the basis for optimizing communication among individuals. In the long term, alignment is perhaps a driving force for the evolution of languages and conceptualization across generations, as suggested some robotic experiments described by Steels (1999, Chapter 12).

Besides the general cooperative alignment processes between interlocutors, we can analyse the computational task of specifying reference, which considerably differs depending on what the interlocutors refer to, as well as on the communication setting. Let's start with a clarification of what 'specifying reference' means in the current context. When a speaker or writer produces a message, he or she must provide enough linguistic and sometimes nonlinguistic cues to allow the addressee to represent a particular object, event, or episode. For instance, suppose that you are with a friend in a library, not surprisingly full of books, and you say to them: 'Please give me the red book on the table.' This example illustrates some features of reference specification. First, the speaker's choice of words and grammar is guided by a pragmatic goal or intention, in this case getting a particular book. Second, the addressee's comprehension determines a selection process of a token (e.g., a book) in a contrasting set of many other tokens (e.g., the other books in the library). Third, as a consequence of comprehension the addressee may have some reaction; for instance, producing a verbal response, changing their mood or beliefs or, in the example, producing a motor action of transferring a book to the speaker.

Specifying reference is not a trivial task, even in a simple case as the one mentioned above. In most circumstances, specifying reference is a more challenging process, as when the speaker refers to a memory-based rather than a current perceptual entity, or when the speaker describes his mental states. For instance, if your friend says to you 'I enjoyed having dinner in the restaurant Roma last night,' your friend wants you to create a representation of a singular episode discriminable from many other similar episodes stored in your memory, coming from your own experiences of happy dinners in restaurants, or second-hand experiences from movies, literature, or previous conversations. In addition, your friend is not only informing you about the specific dinner episode, but also about their emotional reaction of enjoyment.

I propose three levels of reference specification, which differ in their computational demands for the speakers and comprehenders, and that might have implications for establishing a more articulated view of language grounding and embodiment: (1) *on-line reference*, (2) *displaced reference*, and (3) *decoupled reference*.

On-line reference occurs when we communicate about the current environment, as when we use deictic language and we point to objects. In this case, words and gestures are directly indexed to perceptual entities. Language provides quite primitive deictic tools such as demonstrative and personal pronouns (e.g., 'give me that') for on-line reference. Displaced reference occurs when people refer to memory-based entities that are not currently available for perception. Specifying reference in this case requires an embodied model of the situation (a simulation), which is detached from the current perceptual experience but still preserves some perceptual or motor properties of the referred objects. Linguistic constructions for this kind of reference are much more sophisticated, in many cases involving relational expressions with a *figure* and a *ground*. For instance, in the sentence 'John's home is behind the mountain', 'John's home' is the figure (the entity under attentional focus) and the 'mountain' is the ground (the entity used as a framework). Finally, decoupled reference takes place when people refer to their own or others' mental states, such as their wishes, beliefs, and the like. Language provides recursive structures for this level of reference, such as mentalist periphrases that govern a subordinate predicate. For instance, 'John wished to buy a new pair of shoes.' In this example, it is not clear whether John has performed or will perform the intended action, because the sentence focuses on John's mental states rather than on a factual action. At this level, a possibility emerges that symbolic operators play a role in meaning construction, and even that perceptual grounding is diminished or entirely suppressed.

It should be noted that these three levels of reference can be combined into a single utterance. For instance, when you say 'I would like that this was yours', involves deictic terms ('I', 'this', 'yours') whose reference only could be specified by perceptual cues in the current communication setting, but these terms are embedded into a mentalist predicate ('would like that …') that signals to a decoupled representation.

14.2 **On-line and displaced reference**

In this section I will present some evidence of embodied meaning in on-line and displaced reference. But I will also explore the idea that the two kinds of reference may involve different degrees of embodied cognition. Everyone may agree that on-line reference to the here-and-now situation is directly grounded on current perception. The notion of grounding, however, could be less direct for displaced reference, as the referents are retrieved or constructed from memory rather than being here and now.

14.2.1 **On-line reference**

This level of communication corresponds basically to the notion of *indices* in Peirce's theory of signs (see Chapter 1, this volume). An index designates its referent directly by means of a spatial connection between the sign and the object. In a typical on-line communication, the speaker and the listener are co-present and they communicate about their current perceptual experience. All languages have deictic tools to refer to the current perceptual situation; for instance, demonstratives ('this', 'that'), pronouns ('I', 'you'), or adverbs ('now', 'afterwards', 'here', 'there'). Deictic words are semantically very impoverished in that they provide little information about the referent. For instance, if I hear the deictic sentence 'give me that', 'that' could be almost any object that is visible in the surroundings. In spite of this ambiguity, the specification of reference is quite easy in a deictic setting because the speaker frequently uses paralinguistic resources which are strongly embodied themselves. For instance, while saying 'give me that' the speaker could align their body with the referred to entity, use a pointing gesture, or gaze at the object to ensure that the listener identifies it. Concurrently, the listener could track the speaker's motions and gaze to figure out which object he or she is referring to.

I propose that on-line (deictic) reference involves a first-order embodiment, by which the speaker uses his/her body as an egocentric framework to specify reference. This first-order embodiment has been clearly demonstrated for pointing gestures. For instance, Rieser (1989) found that blindfolded participants were faster and more accurate in their pointing responses when they actually rotated to face a different orientation than when they merely imagined facing this new orientation. According to Rieser, participants in the physical rotation condition used proprioceptive feedback from locomotion that facilitates access to knowledge of the spatial structure from novel points of view.

Basically, this first-order embodiment is supported by the sensorimotor system we use when we manipulate objects or navigate in the environment. This system automatically updates spatial relations when we move. Thus, despite our continuous reorientation and navigation in the environment, we manage to keep track of objects' positions in our peripersonal space. Neurological studies confirm the automatic updating of the object-to-body relations after physical motion (Graziano *et al.* 1997; Rizzolatti *et al.* 1997). Thus, some neurons in the ventral premotor cortex of monkeys automatically keep track of the sensorimotor 'here and now,' which means maintaining the constancy of the object positions with respect to the body, even in the darkness or for hidden objects. I will provide evidence that sensorimotor updating is also necessary for deictic communication.

14.2.2 Displaced reference

A remarkable feature of human language is *displacement* (Hockett 1960), namely the capability to communicate beyond the here and now, for example, when we refer to entities and events retrieved or constructed from memory. The computational problem of specifying absent referents is much more complex than the computational problem of on-line reference. Speakers and listeners cannot rely primarily on the current perceptual environment, although they may use their bodies as indices to designate absent referents in an indirect manner. For instance, while talking about an episode retrieved from memory, they can make iconic gestures with a communicative value, referring to absent objects and events (McNeill 1992). Also, they can use the deictic gesture of pointing to set up a region of space in which they metaphorically place an absent target. At a later moment in the conversation, they can point again to this region to refer deictically to the absent entity 'placed' there (Bühler 1982; McNeill and Duncan 2000). However, this so-called 'abstract deixis', as well as ordinary iconic gestures, are 'displaced' reference markers; they refer to memory-based or situation-model entities rather than perceptual entities in the current environment.

The use of gestures and abstract deixis in face-to-face communication is an interesting case of using bodily cues to refer to displaced referents. However, these cues play a relatively minor role in the comprehension of complex discourse. In some communication settings gestures cannot provide any useful information for comprehension, purely because the interlocutors do not see each other (e.g., listening to the radio or talking by phone). In these cases, the speaker has to provide enough linguistic cues to help the listener to elaborate a situation model of the absent referents.

An extreme case of displaced reference occurs in written narratives, in which the writer and the reader do not share a communication setting. Surprisingly, even in this case, deictic terms such as 'I', 'here', or 'that' are frequently used. But unlike in on-line communication, readers make a 'deictic shift' to refer these deictic terms to entities in their situation model rather than to the current environment (Duchan *et al.* 1995). On their side, writers must provide explicit descriptions of the deictic referents, by means of quotations (e.g., '"give me that," said John pointing to the red book'). Notice that using this sort of quotation would be superfluous or even extravagant in an on-line reference setting, but it is perfectly appropriate for displaced reference either in the written or the oral modality.

In addition to reusing deictic terms in a new way, displaced reference also employs new and more sophisticated linguistic constructions. These constructions force the speaker and the listener to pack information in certain ways. For instance, understanding locative sentences like 'the book is behind the telephone' requires (1) to focus on a ground object (the telephone), (2) to segment the space around the ground object into discrete topological regions, focusing on the 'behind' region and (c) to mentally place the figure (the book) in the critical topological region. Similar figure/ground framing can be observed for more abstract information such as temporal relations. Thus, when people describe simultaneous events by means of the adverb 'while', they tend to use a larger duration event as frame and a shorter duration event as figure or focus. For instance, it is more acceptable to say 'While Mary wrote a letter, John knocked on the door' than to say 'While John knocked on the door, Mary wrote a letter' (de Vega Rinck, *et al.* 2007). None of these processes seem necessary when we communicate on-line about the here and now.

Embodiment theories propose that, even for displaced reference situations, meaning is embodied by means of simulation of perceptual experience. I will illustrate how displaced reference activates embodied meaning, by describing experiments on the comprehension of action sentences.

14.2.3 **Is displaced reference embodied?**

I have already mentioned that displaced reference can rely on embodied cues such as gestures. However, in many cases, especially in written language, addressees are constrained to process the linguistic message in its purest symbolic format. The question is whether readers activate perceptual or motor representations when they understand written messages. A basic experimental procedure to answer this question is to ask participants to listen or read sentences describing events while they perform a parallel perceptual or motor task, designed to match or mismatch the meaning of the sentences (Glenberg and Kaschak, 2002; Richardson *et al.* 2003; Kaschak *et al.* 2005; Moreno and de Vega 2005). The rationale of these double-task paradigms is to create conditions in which the semantic meaning interacts with the concurrent perceptual or motor task, especially in the matching conditions. If this interaction takes place it would be a clue that meaning and perception share some cognitive or neural mechanism.

An example of a semantic–perceptual double-task paradigm is a study performed by Kaschak *et al.* (2005). They asked participants to listen to and make semantic judgements on sentences that described motions towards you (e.g. 'The car approached you') or away from you (e.g., 'The horse ran away from you'), and simultaneously viewed dynamic stimuli that produce an illusory motion towards or away from you. Semantic judgments were faster in the mismatching condition (e.g., sentences describing a towards motion presented concurrently with a dynamic visual stimulus producing an illusion of an away motion). This suggests that the processing mechanisms recruited to construct visual simulations during language comprehension are also used during visual perception. For instance, it might be the case that constructing a simulation of motion during language comprehension requires the activation of some specific neurons that are tuned to respond to motions in that direction, and this causes a momentary competition for the same neuronal resources (see Kaschak *et al.* 2005). A recent neuroimaging study confirmed this idea (Rüschemeyer *et al.* submitted). The authors used materials and procedures similar to Kaschak *et al.* (2005) while participants' brain BOLD response was recorded, particularly in the MT/V5 region of the temporal cortex. This region is responsible for processing dynamic visual stimuli. They found that MT/V5 was activated with the visual animations as well as some of the transfer sentences, in comparison with baseline conditions of static visual stimuli or sentences describing static situations.

In other double-task studies, participants listened to sentences describing actions, while simultaneously performing a motion themselves. The critical manipulation is that the performed motion either matches or mismatches the motion conveyed by the sentence meaning. In these cases an action–sentence compatibility effect (ACE) has been reported. Thus, Glenberg and Kaschak (2002) asked people to judge how sensible were sentences describing a transfer motion towards or away from you (e.g., 'you give Liz the toy'). For some participants the response key for 'yes' involved moving the finger toward the body, whereas the response key for 'no' required moving the finger away from the body. For other participants the assignment of yes/no keys was reversed. The judgement time was faster for the matching (for instance, the sentence describing a transfer towards you and the 'yes' response was also towards you) than for the mismatching condition, suggesting a sort of priming effect from the linguistic simulation to the real motor response.

Notice that there is a discrepancy in the above results: some experiments obtained a matching advantage (Glenberg and Kaschak 2002; Zwaan *et al.* 2004), and others reported a mismatching advantage (Richardson *et al.* 2003; Kaschak, et al., 2005). Kaschak *et al.* (2005) suggested that a possible cause of this discrepancy is the different temporal pattern used in the double-task procedures. When the linguistic and the concurrent task temporally overlap, a mismatch advantage was seen, whereas when the linguistic information and the secondary task were consecutive, a match advantage was reported.

To test the temporal overlapping hypothesis, Moreno and de Vega (2005) used a variant of the double-task paradigm. The linguistic materials were sentences with a transfer verb describing a motion 'away from me' or 'towards me', followed by a sentence completion task. For example:[1]

(1) 'I threw the tennis ball to my friend over the ... *net / basket*'

(2) 'My friend threw me the tennis ball over the... *net / basket*'

Each sentence was automatically presented word-by-word on the centre of the screen, and the transfer verb (e.g., 'threw') was animated with a forward or backward motion that either matched or mismatched the sentence meaning. The interval between the verb presentation and its apparent motion was manipulated between participants (200 vs. 350 milliseconds) as to vary the temporal overlapping between the transfer verb and the sensorimotor event. While reading, participants had their right-hand index finger on a resting key, and once they had identified the moving word direction,

---

[1] The examples are free translations from the original Spanish into English. Word order in Spanish is more flexible than in English and we exploited this feature to give the reader the identity of the transferred object before the critical verb. Thus, the original sentence for (1) was: '*La pelota de tenis se la lancé a mi amigo...*' ('The tennis ball [I] it threw to my friend...').

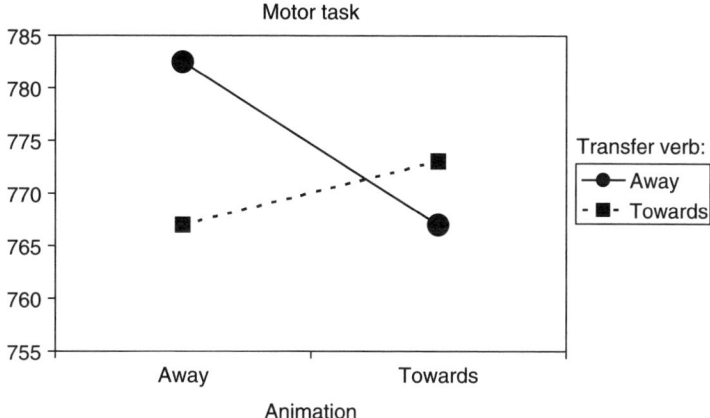

Fig. 14.1 Latencies for the motor task as a function of the transfer verb and the word animation (Moreno and de Vega 2005).

they had to press one of the two alternative keys placed ahead or behind the resting key to match the direction of the word motion (motor task). To ensure participants' comprehension, at the end of the incomplete sentence they had to choose between the two alternative completion words.

When the word animation overlapped the process of comprehending and integrating the meaning of the verb (short interval condition), there was a mismatching advantage. In particular, key pressing time and choice latency were significantly faster for the 'towards transfer–away animation' than for the 'away transfer–away animation'. Notice that the interference was double: from meaning to the perceptual/motor task (Figure 14.1), and from the motor/perceptual task to the processing of meaning (Figure 14.2).

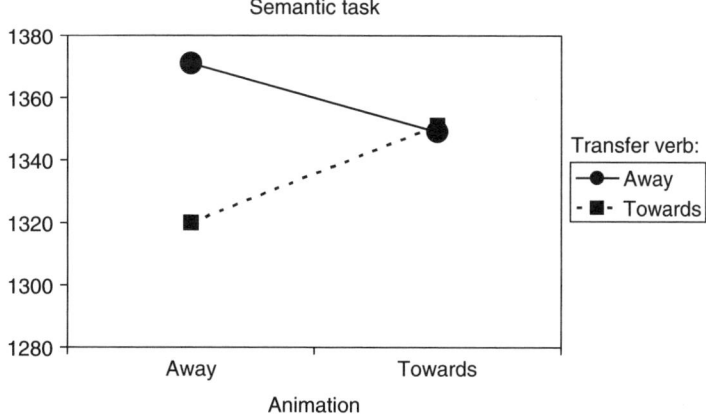

Fig. 14.2 Latencies for the semantic choice task as a function of the transfer verb and the word animation (Moreno and de Vega 2005).

By contrast, when the animation took place after the linguistic processing of the verb had been completed (large interval condition), there was no significant effect either on the motor task or the semantic task, although the observed trend was a matching advantage. Thus, the experiment demonstrated that embodied simulation of action has two stages. The perceptual or motor simulation of events described by language occurs very early and seems to recruit the same brain areas as perception/action. This would explain the disadvantage observed for the matching conditions in the temporal overlapping conditions. However, this motor simulation seems short-lived because 350 milliseconds after the action verb onset the ACE had faded. Short-lived effects of motor simulation have also been observed by Zwaan and Taylor (2006; see also Zwaan, Chapter 9, this volume) with a quite different task. Their participants read sentences describing a manual rotation; for example, 'After/ lighting/ the candles/ for the/ romantic/ evening,/ he/ dimmed/ the/ lights.' Each text segment (separated by /) was self-paced by rotating a knob slightly. They found ACE effects (matching advantage) in the action verb ('dimmed') but not in the next segments ('the/ lights').

In summary, the double-task experiments provide clear evidence of embodied simulations in the comprehension of displaced reference sentences, describing dynamic events. First, these simulations involve the activation of visual and motor features. Second, these simulations are triggered very soon while reading or listening the action or motion verb. Third, these simulations are short-lived activations, rather than long-term situation models.

14.2.4 Dissociating on-line and displaced reference

Grounding meaning on sensorimotor process could be a general feature shared by on-line and displaced reference. However, we can expect some differences between grounding on entities in the current environment (on-line reference) and grounding on entities in a simulated environment (displaced reference). I will explore these differences in this section, changing the focus from language describing events to language describing spatial layouts.

Communicating spatial information is a privileged domain to explore one aspect of embodied meaning, contrasting on-line and displaced reference. English, like other languages, has a repertoire of locative prepositions and adverbs to encode spatial relations among objects such as distances ('near', 'far'), containment ('into', 'out of'), or the canonical directions ('front', 'back', 'right', 'left', 'above', and 'below'), among others. A strong embodiment hypothesis is that when we understand spatial descriptions involving such locative terms, there is a direct interface between language and the spatial system we use to navigate or interact with the physical world. Thus, the representational layout we elaborate for these expressions would be a surrogate for experience.

A different characterization is that language involves a modality specific conceptualization of spatial relations, which differ from perceptual experience in important ways, although it can still be considered embodied. I called this language-specific process *framing* (de Vega and Rodrigo 2005). According to the framing approach, people are forced to parse the environment in a different way when they use locative terms to communicate verbally about object position compared with when they perceive the object or communicate about it deictically. For instance, when speakers communicate the spatial position of an object (i.e., a figure) they are forced to choose another object, or a person's body, as a framework or ground, and this involves different sorts of computations than just pointing to the object or saying 'that'.

Consider a familiar situation: you are asked in the street for directions to a particular place (e.g., a shop) that is close but not visible to the interlocutors. You have to choose a particular ground object as a framework, but you have many potential grounds available such as the town hall, a big bank, a fountain, a church, etc. The choice of a ground does not seem arbitrary, nor based on simple figure/ground co-occurrence in the past. For instance, telephones and ceilings are usually present together in the same situation, but nobody says that the telephone is below the ceiling because it is not informative. Choosing an appropriate ground might depend on assessing several geometric, functional, and informational features of the ground that make it easy for the hearer to locate the shop. Thus, it has been suggested that grounds are usually more static, larger, and more salient than figures (e.g., Talmy 1983; Landau and Jackendoff 1993; de Vega *et al.* 2002). Once you choose the appropriate landmark as ground (e.g., a big fountain), then you must parse space around the ground into gross topological categories ('in front of', 'behind', 'above', etc), and construct a directional expression such as 'the shop is behind the fountain' (see Logan and Sadler 1996 for a computational analysis of this process).

Compare the above situation with another one in which the target object (the shop) is perfectly visible for the interlocutors. In this case you would not need to make a figure/ground computation like the previous one. Instead, you would use a deictic procedure, using an expression such as 'here it is' or 'that building,' while you align your body with the shop and point to it. From the point of view of the listener, the two situations (identifying the referred object of a figure/ground locative expression versus tracking the referent of a deictic utterance accompanied by pointing) also seem quite different.

In several experiments communication using locative language and pointing were compared in order to explore how both modalities demand a different kind of embodied conceptualization (de Vega and Rodrigo 2001; Rodrigo and de Vega, in preparation). For instance, in one experiment (Rodrigo and de Vega, in preparation), people were asked to locate objects in a layout under two task conditions: by pointing to them or by using the directional terms 'front', 'back', 'right', and 'left'. To force participants to update the objects' positions, they were asked to periodically reorient themselves either by turning physically their body to face a given object (physical rotation), or by imagining themselves as turning to face an object while their body remained still (imaginary rotation). In each new reorientation, all participants had to locate the position of several hidden objects, whose position had been previously learned, either by pointing or saying their direction. Physical rotation can be considered an on-line reference task, because participants locate objects in their real position in the layout. In contrast, imaginary rotation is similar to displaced reference, as participants have to rely on a mental representation of the layout that is misaligned with their body. Latency and accuracy of responses were recorded and the results showed striking differences between pointing and verbal labelling.

Pointing was very efficient under physical rotation, in both speed and accuracy of response, but performance dramatically decreased under imaginary rotation. These results replicated others described elsewhere (e.g., Rieser 1989). However, when spatial terms were used to describe object positions, performance was equally efficient under both physical and imaginary rotation (although there was a small but significant impairment in accuracy under physical rotation). These results are robust and have been replicated by using different kind of layouts and pointing procedures, such as pressing directional keys, using a compass, or pointing directly with the index finger (Wang 2005; Avraamides 2003; Wraga 2003).

A possible interpretation relies on how people use different embodied systems in pointing and verbal communication. Pointing employs proprioceptive information of the body, the same sensorimotor information responsible for continuously updating object locations during self-motions in the perceptual environment. This would explain why pointing is so easy under physical rotation, but also why it is so difficult under imaginary rotation. In the former, each time the body rotates the object-to-body relations are automatically updated. In the latter, participants are asked to suppress the dominant proprioceptive information in favour of a conflicting mental framework, produced after imaginary rotation. Most importantly, the verbal system seems immune to this conflicting information, and, in fact, the physical rotation produces interference (de Vega and Rodrigo 2001; Wraga 2003), suggesting that language-based spatial representations are more efficient when they are detached from the current proprioceptive information and exclusively rely on a mental framework.

Spatial mental frameworks seem to be linguistically-guided constructions that rely on a different mechanism than the proprioceptive updating system used for navigation and orientation in the environment. However, mental frameworks are still embodied representations, rather than symbolic or propositional, as suggested by the dimensional pattern of accessibility observed in language-based spatial frameworks (e.g., Franklin and Tversky 1990). Spatial judgements in the front–back dimension are fast and accurate, presumably because these directions are mapped into the morphological and functional asymmetries of the body: the front region is visible, and corresponds to the walking direction, whereas the back region is not visible and is less attended. In contrast, judgements in the right–left dimension are slower because both sides of the body are symmetric and functionally equivalent, and therefore less distinctive.

So, communicating about spatial layouts seems to be governed by different mechanisms when communication is the on-line pointing gesture, and when it is mediated by locative terms, which are the proper linguistic devices for displaced reference. Thus, on-line communication is essentially egocentric, because it is strongly anchored on the body and the proprioceptive mechanisms that update object-to-body spatial relationships. Instead, displaced communication by means of spatial prepositions could be detached from this proprioceptive machinery. These spatial prepositions can eventually be used to refer to on-line situations (e.g. 'give me the book behind you'). However, when locative prepositions are used for displaced reference (like when we describe spatial relations in a memory-based layout) they are not grounded on the body, but in a mental framework that is misaligned with the body.

14.3 Decoupled reference

People communicate not only about current or memory-based entities, but they also communicate their mental states, such as beliefs, wishes, and emotions, observed introspectively in themselves or inferred in others. People also build up and communicate about states of the world that are alternative options about past, current, or future situations. Moreover, all languages have a repertoire of lexical and syntactical resources to express these complex kinds of reference. For instance, sentences involving mental periphrases like 'John thought that the yellow book was on the shelves.' This sentence does not describe a factual situation – it refers to a mental state or propositional attitude. As proof of this, its truth value cannot be easily established just by checking the perceptual environment. If someone pronounces the example sentence, we cannot verify it by looking for the yellow book in the shelves. Whether or not the yellow book is there is quite irrelevant for the reality of John's belief. A better solution would be, for instance, asking John about his belief or evaluating how trustful the speaker is. In addition, mentalist utterances posit the computational problem of dealing with alternative beliefs of a situation at the same time.

For instance, consider the previous sentence embedded in a larger context: 'John thought that the yellow book was on the shelves, but *I* know that it was in the suitcase.' To produce this sentence, the speaker has to dissociate her true belief (she knows that the book was in the case) from John's false belief (that the book was on the shelves). On the listener's side, he also has to be able to decouple the two beliefs implicit in the sentence (and maybe add his own belief!). According to many researchers, we use our mentalist skills or theory of mind to dissociate and deal with alternative beliefs of the situation. Eventually, we have to be able to inhibit our own true belief and take the perspective of others' false beliefs (see Leslie *et al.* 2004 for a review).

Two questions are relevant in our context: first, how do people integrate the meaning of mental and factual sentences, and second, is the meaning of these mental expressions embodied? Some experiments suggest that the content of mental expressions is kept apart in memory and is not integrated in the current situation model. For instance, de Vega *et al.* (2004, experiment 2) presented readers with sentences involving temporal adverbs ('while' or 'after') to indicate a protagonist who performed two actions simultaneously or successively. The two described actions were difficult to perform simultaneously because they involved similar body motions (e.g., 'painting a fence' and 'chopping wood'). Another experimental manipulation was the introduction of a verbal periphrasis in the second clause that was either factual ('started to') or mentalist ('thought of'). Here is an example of the four versions of the materials:

(1) While painting a fence the farmer *started to* chop the wood.

(2) After painting a fence the farmer *started to* chop the wood.

(3) While painting a fence the farmer *thought of* chopping the wood.

(4) After painting a fence the farmer *thought of* chopping the wood.

In sentences with a factual periphrasis, the reading time for the second clause was much slower in sentence (1) than in sentence (2). De Vega *et al.* suggested that readers try to run an embodied (perceptual and motor) simulation, but in the simultaneity condition they fail to accomplish it because the motor incompatibility of the actions (see Sanford, Chapter 10, this volume, for an alternative interpretation of this experiment). However, in sentences with a mental periphrasis, the 'while' (3) and 'after' (4) conditions were read equally fast by participants. One explanation is that the mental periphrasis prompts the reader to build a separate situation model, decoupled from the factual situation model. Thus, the factual content and the mental content are not merged into a single simulation but kept as separate situation models.

A similar case of decoupled meaning has been shown with counterfactual sentences embedded in texts (de Vega, Urrutia and Riffo, 2007). Participants read short stories describing a factual state in the first paragraph (e.g., 'Marta switched on the radio and heard the winning lottery numbers'), followed by additional information that either was factual or counterfactual. For instance:

- 'Since she won the lottery prize, the first thing she did was to buy a new Mercedes car' (factual continuation)
- 'If she had won the lottery prize, the first thing she would have done was to buy a new Mercedes car' (counterfactual continuation).

After reading a story, participants were asked to verify a test probe that either belonged to the beginning of the story ('heard'), or to the final clause shared by the factual and counterfactual version of the story ('buy'). Readers verified the first test probe faster in the counterfactual than in the factual stories (1569 vs. 1703 milliseconds). Concerning the second probe, the results were more complex. Immediately after reading the last sentence, response to the probe was equally fast in the counterfactual and factual versions. Only after reading a filler sentence was the second test probe responded to faster in the factual than in the counterfactual stories (1361 vs. 1409 milliseconds).

If we consider the test probe latency a measure of concept activation or accessibility, the interpretation of these results is straightforward. In the factual stories, the situation model is updated and the attention focus moves on to the new factual content described in the example. Instead, in the counterfactual version updating is cancelled and the attention focus goes back to the factual information in the initial paragraph, but the new information conveyed by the counterfactual sentence is also accessible for a while. Thus, two situation models seem to operate in counterfactuals momentarily: a realistic model in which the counterfactual events did not occur, and an alternative model that considers the counterfactual events 'as if' they had happened. In summary, the above experiments clearly indicate a decoupling process by which readers process mentalist and counterfactual contents apart from factual contents.

At this point, it seems necessary to explore the status of mentalist contents more carefully. I have argued that mentalist and counterfactual information are not integrated with factual information. Readers seem to partition the situation model into functional separate spaces for 'real' and 'epistemic' information and simulations do not merge across spaces. But the data do not tell us anything about our second question: are nonfactual contents still processed as embodied simulations, or are their sensorimotor properties suppressed? To test the embodied (or disembodied) properties of counterfactual information, de Vega *et al.* (2006) used the double-task paradigm described earlier (Moreno and de Vega 2005). The experimental materials were both factual and counterfactual sentences describing a transfer situation either 'away from me' or 'towards me'. These are some examples translated from Spanish:

(1) 'Because I am generous, I've lent you the *Harry Potter* book' (factual away from me).

(2) 'If I had been generous, I would have lent you the *Harry Potter* book' (counterfactual away from me).

(3) 'Because the jeweller is a good friend of mine, he has shown me the imperial diamond' (factual towards me).

(4) 'If the jeweller had been a good friend of mine, he would have shown me the imperial diamond' (counterfactual towards me).

Participants were given each sentence auditorily except for the transfer verb (always a participle form) that was presented visually. After a stimulus-onset asynchrony (SOA) of 100–200 milliseconds, the verb apparently moved towards or away from the participant. The participant was prompted to keep the index finger pressing a resting key (target) while listening to the sentence. Only after the visual word started to move, they released the target key and moved their finger to press the corresponding towards or away key. Both the target-releasing latency and the key-press latency were recorded. To ensure that participants were paying attention to the linguistic task, they were asked to perform a word identification task (test probe) and to answer comprehension questions at the end of the sentence. The main results were as follows.

First, the expected mismatching advantage was found for factual sentences in the 200-millisecond interval, as shown in Figure 14.3. In other words, there was again interference between the processing of the sentence meaning and the perceptual motion

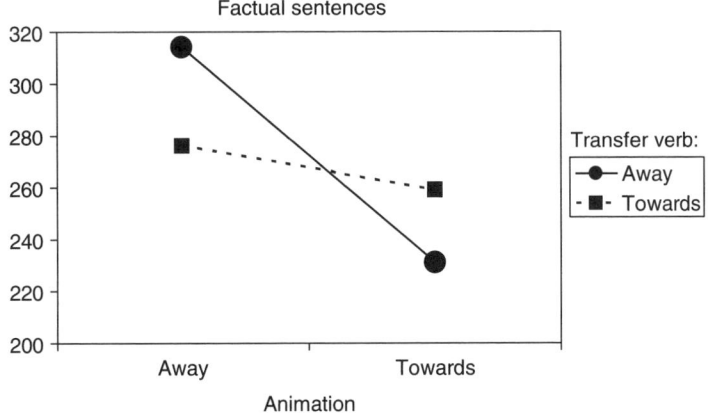

Fig. 14.3 Factual sentences: response time for the animated word (de Vega *et al.* 2006).

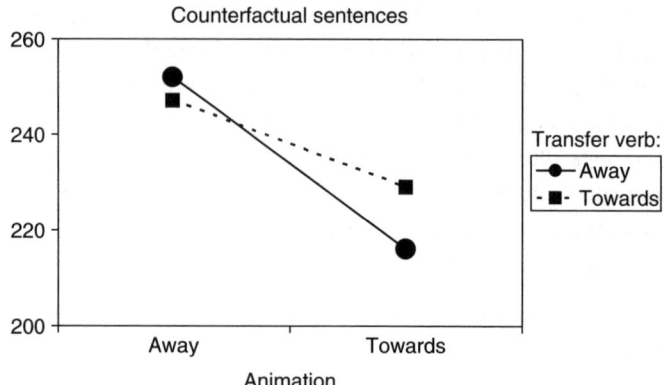

Fig. 14.4 Counterfactual sentences: response time for the animated word (de Vega et al.2006).

for the matching conditions, confirming the previous results (Moreno and de Vega 2005). Given the complete counterbalancing of the materials across participants, the interaction can be interpreted as a genuine interference between semantic and perceptual–motor stimuli.

Second, and more important, a similar interference pattern was obtained with counterfactual sentences in the 100-millisecond interval, as illustrated in Figure 14.4. We can conclude that counterfactual expressions do not initially suppress the sensorimotor anchoring of meaning, at least for transfer verbs. This is quite remarkable, considering that counterfactuals generate separate representations with a hypothetical status that does not merge with factual or 'real' information. I mentioned before that counterfactuals prompt the reader to cancel situation model updating, and that the reader's attention moves backwards to old factual information rather than to the most recent counterfactual contents. In spite of this, a transfer construction embedded in a counterfactual seems to automatically generate an embodied simulation of its literal meaning. This simulation is probably much degraded or abstract and quite short-lived, as in the 200-millisecond SOA, the mismatching advantage disappeared and no other clear pattern was observed on reaction times.

This conclusion converges with other results in the neuroscience literature. For instance, recent studies using functional magnetic resonance imaging (fMRI) and magnetoencephalography (MEG) demonstrate that reading action verbs or sentences that refer to actions performed with the hands, the mouth, or the foot rapidly activate somatotopic regions in the brain that partially overlap those that govern the corresponding motions (see Pulvermüller, Chapter 6, this volume). The activation of a motor representation by transfer verbs like 'give' may be rapid and automatic such that it precludes any impact of counterfactual construction. However, notice that the interference observed in our experiments cannot be considered an exclusively lexical effect. The direction of the transference ('away from me' or 'towards me') does not rely on the isolated verb participle (e.g., 'given'), but requires a phrase or clause integration (e.g., 'I have given you' versus 'You have given me').

Counterfactuals are special cases of negation. For instance, in a sentence like 'If I had been generous, I would have lent you the *Harry Potter* book,' the antecedent is an implicit negation (in fact I am not generous), as it is the consequent (in fact, I did not lend you the book). Thus, it is useful to compare the counterfactual results with those obtained with sentences involving negations. In a classical study, McDonald and Just (1989) have shown that negation automatically reduces the accessibility of the negated concept. Their explanation was that negation is a linguistic operator working within a propositional representation. Thus, in the sentence 'Mary bakes bread but no cookies,' the negated concept 'cookies' becomes less accessible than the concept 'bread'. However, some recent studies demonstrate that situation models modulate the impact of negation on accessibility. Kaup (2001) showed that negations strongly reduce accessibility in sentences describing a constructive activity (like baking in the example by McDonald and Just), but accessibility is much less reduced when the sentences describe a destructive activity. For instance, in 'Sarah was burning the letters but not the photographs,' the negated token 'photographs' is quite accessible because it remains in the situation model. This situation model effect partially counteracts the linguistic effect of negation.

In a new set of experiments, Kaup and Zwaan (2003) established the time course of accessibility of negated concepts. They demonstrated that the impact of negation and of the situation model occur at different moments. When accessibility was tested 500 milliseconds after the target sentence, the negation operator regulates the concept accessibility. After 1500 milliseconds, however, the situation model governs the accessibility of concepts: a concept was accessible when its referent was present in the situation, regardless of whether it was negated in the sentence.

The results obtained with counterfactuals provide a quite different picture than that which is seen with ordinary negations. Using a double-task paradigm, we observed a matching interference – a sign of embodied representations – which occurs very early, less than 200 milliseconds after the verb presentation. The activation measures are consistent with this idea: immediately after reading a counterfactual its information is still very accessible, and only after several hundreds of milliseconds, the information becomes suppressed. Thus, counterfactual constructions seem to initially activate embodied features automatically and, later on (probably at the stage of discourse integration), they promote a reduction of accessibility of concepts under their scope.

14.4 Discussion

In this chapter, I have proposed that the nature of meaning (embodied, symbolic or mixed) could be explored using as guidelines three levels of reference in linguistic communication: on-line, displaced, and decoupled. These levels of reference could help to resolve the debate about meaning because different scholar might be focused on different referential phenomena. For instance, a text-based research program based on latent semantic analysis might relate to displaced reference (see Kintsch, Chapter 8, this volume, and Graesser and Jackson, Chapter 3), whereas the robotics approach is related to on-line reference (see Steels, Chapter 12, and Roy, Chapter 11).

A very basic 'computational theory', in Marr's terminology, underlies the notion of levels of reference.[2] According to this computational theory, a goal of human language is to provide the speakers with tools to specify reference cooperatively. However, the computational demands considerably differ for the three levels of reference, and so must the representations of meaning. Although I assume that embodied representations are activated routinely in most linguistic structures, I am not thinking of a unitary notion of embodiment.

On-line reference involves a first-order embodiment as utterances are strongly grounded on current perception and action. At this level, embodiment is quite a literal notion: utterances are mapped into current perceptual objects and events in the periper-sonal space, and motor programs are executed by speakers to specify reference by guiding their join attention to objects. Thus, the speaker and the addressee align their bodies or part of their bodies with the referred objects, and use gazes to ensure joint attention. This first-order embodiment has important advantages: linguistic expressions can be considerably simplified by using general purpose deictic words, and the anchoring on proprioceptive machinery allows an automatic, low-cost updating of referred entities in space and time.

In contrast, displaced reference relies on a second-order embodiment which is much more detached from current perception and action. For instance, we do not need to align our body with perceptual objects when we hear a description of a distant (or fictitious) spatial layout, such as 'there is a golden sculpture in front of the Rockefeller Center.' People who understand displaced sentences like this rely on a mental framework which is entirely detached from the current perceptual and motor events. An advantage of displaced reference is its flexibility: people can describe not only the current perceptual environment to a co-present interlocutor, but they can communicate about memory-based or fictitious situations to an absent interlocutor, as in written language.

How can we characterize this second-order embodiment for displaced reference? Perceptual detachment has been clearly demonstrated for spatial descriptions (de Vega and Rodrigo 2001; Avraamides 2003; Wraga 2003), but it might seem at odds with the ACE phenomena described elsewhere in the literature, and obtained in our own laboratory. Understanding the meaning of an action sentence momentarily interferes with a simultaneous matching perceptual or motor event. This strongly suggests that motor systems are activated to some extent as part of meaning processes, which is in conflict with our proposal of second-order embodiment. Neurological evidence confirms that motor and perceptual systems are involved in the meaning of actions: the processing of motion verbs produces a partial resonance in somatotopic areas related to their meaning (see Pulvermüller, Chapter 6, this volume).

---

[2] Marr, among others, defends that cognitive sciences must have a 'computational theory', which establishes 'what is the goal of the computation, why it is appropriate, and the logic of the strategy by which it can be carried out' (Marr 1980, p. 25).

However, there are alternative interpretations of the ACE and the neurological data that are compatible with the idea of sensorimotor detachment. A first argument is that action verbs automatically activate high-order motor representations rather than specific motor programs. The very existence of interference in the ACE studies supports this hypothesis. The transfer verbs employed in ACE experiments refer to actions with quite different motor programs (e.g., throwing a ball versus telling a story), and all of them differ considerably from the motor response (e.g., pressing a key or moving a lever) used as the concurrent task. In spite of that interference occurs, suggesting that actions are simulated at a relatively abstract level of motor processing rather than in detail, e.g., at the level of motor programs (see Sanford, Chapter 10, this volume, for a similar argument). A complementary argument is that specific motor programs must be suppressed to avoid enactment of the actions during language comprehension.

It is quite obvious that enactment would be rather inconvenient for most ordinary comprehension tasks. While we listen to or read a sentence about someone performing an action, say 'hammering a nail', resonance of the premotor areas in the brain should not extend to the motor areas responsible of triggering the action of hammering the nail.[3] The early interference observed in some ACE studies for the matching conditions could correspond precisely to the stage of enactment suppression. The interference would result from the dual task in the ACE procedure: while understanding a sentence describing an action, participants have to perform an action similar to that which should be suppressed.

Embodied meaning for displaced reference could occur at the level of image schemas (Lakoff 1987; Mandler 1992), rather than at low-level motor programs. According to Mandler (1992) 'Image schemas are more abstract than images; they consist of dynamic spatial patterns that underlie the spatial relations and movements found in actual concrete images' (p. 592). Thus, understanding transitive sentences, like the ones used in the double-task paradigm with transfer verbs, might activate an abstract image schema of transference. Some recent studies in the neuroscience literature converge with this idea (Nelissen et al. 2005). Understanding others' actions involves multiple levels of representation in the mirror neurons system of the primate brain, from very specific motor programs (area F5C) to relatively abstract action programs (areas F5B and 45B). The authors speculate that language comprehension of described actions activate human areas 44 and 45 homologous to F5B and 45B in the monkey. In the same vein, Pulvermüller (Chapter 6, this volume) found high-order or 'abstract' motor representations in the human brain, in addition to the motor and premotor regions.

I have also provided some preliminary data and arguments concerning embodied meaning in decoupled reference (mentalist or counterfactual expressions), although much more research is needed. It is quite surprising that a linguistic construction as 'abstract' as a counterfactual can trigger embodied representations, even if only momentarily. I can envision

---

[3] An exception to this is when an addressee follows instructions or orders to use an artefact; for instance, following the instruction to hammer a nail on the wall.

for this kind of sentence a two-stage process: embodied representations are automatically activated by default, but later on the counterfactual information is partially deactivated and segregated from the ongoing situation model, preventing its updating. It could be useful in the future to test these hypotheses about counterfactuals and mental predicates with contents other than transfer. The proposed process of segregating an epistemic situation model (e.g., counterfactuals, beliefs, wishes) from the ongoing 'real' situation is quite intriguing and, in any case, it does not seem to be an embodied process itself. Decoupled meanings could be initially grounded on embodied representations, but then they would be put in a mental quarantine, and probably 'disembodied'. Decoupled reference involves grounding but also something else, namely the management of two or more alternative situation models. This representation management is not embodied, although some of its consequences could be (e.g., emotions). The management of alternative situation models is perhaps a genuine symbolic function that goes beyond sensorimotor reference.

The possibility of a symbolic function for decouple reference differs however from other symbolic approaches. For instance, Kintsch (Chapter 8, this volume), develops a purely symbolic theory to explore how much linguistic meaning could be computed by means of symbolic procedures, explaining a variety of psychological phenomena including some that are usually presented as evidence of embodied meaning (see also Louwerse and Jeuniaux, Chapter 15). In addition, Graesser and Jackson (Chapter 3) have developed an artificial intelligence system (AutoTutor) with a partially embodied interface (voice and body position analysers, etc.) using mainly a symbolic approach. By contrast, I assume that meaning is grounded by default, but some linguistic structures such as counterfactuals and mentalist predicates prompt the addressee to go beyond the perceptual and motoric processing of meaning. The function of these structures is to instruct the addressee how to deal with multiple representations of events. They have to do with the semantics of beliefs, wishes, possibilities, and the like, rather than the semantics of facts.

Is this the territory for symbolic processes? Maybe it is, but if so this is a much more constrained view of symbols than the standard one. They are not general-purpose (e.g. 'Turing machines') but special-purpose operators aimed at regulating the flow of multiple representations, and they do not substitute nor suppress the grounding of meaning on perceptual/motor processes.

Debate

Several issues arose in the discussion: (1) to what extent is grounding on perception/action mandatory for language comprehension, (2) how separable are the levels of reference from a functional point of view, (3) which are the optimal temporal parameters for experiments to better reveal the ACE effects, and (4) how abstract are the representations of actions when we understand language.

To what extent is grounding on perception/action mandatory for language comprehension?

Max Louwerse: In your first statement you say that grounding occurs routinely for most language constructions. I'm trying to figure out whether grounding *can* occur routinely (and everyone could agree on that) or *must* occur routinely. The question is *must* it occur routinely?

Manuel de Vega: I think it must!

Arthur Graesser: I think this is an important question. I disagree that it must. There are conditions in which it doesn't happen, and there must be…

Louwerse: So, if the answer is 'it must', then my question is why, given that Walter [Kintsch] has shown that a lot of symbolic representations can be built through symbols only.

de Vega: I suppose that it is a matter of brain wiring. The human brain is a system involving modular devices. Some of them were primarily designed for action, perception, and so forth, but these very structures can also be used opportunistically for other tasks, such as language understanding. Maybe to extend the use of functional mechanisms to new functions is a matter of cognitive economy. I don't dare to say that the system is designed in such and such way for such and such reasons. But my intuition is that the linguistic system is functionally connected to sensorimotor areas in the brain as a way to ground information in the world, and this seems quite useful if we want to explain meaning beyond pure linguistic associations.

Mitchell Nathan: In addition to the removed nature of decouple reference there is also the complexity of the propositional structures. These are more complex structures, in addition to being more removed from the immediate circumstances. So, could it be true that the complexity is also an issue of embodiment, and that we have limits to the degree to which our ability to enact or simulate and ignore the fuzziness of, you know, levels 1 and 2? These processes scale up differently in the face of tremendous amount of complexity because they make enormous demands. One would like to have a system where complexity being one of the properties. Symbol systems, as one might describe them, do scale up to complex structures and complex relational situations, as well as shedding light in the face of complexity.

de Vega: I think that complexity derives from the levels of reference. Namely, complexity is a consequence of the computational task of specifying reference. So the computational task is very simple when you refer to something that is here and now, and consequently you can rely on very gross linguistic devices like demonstratives. When you go to displaced language, referring to something that is not present, you have to help the listener to build up a mental framework, and this involves more complex structures like figure/ground structures that allow the listener to establish a virtual reference by means of simulation. Finally, the third level [decoupled reference] describes a mental state that refers to another predicate. The subordinate predicate could be embodied, as I found in our counterfactuals study, but the superordinate predicate is something different, a sort of operator that instructs the listener how to build alternative representations of the world; for instance, my belief versus your belief, or an event and its possible alternative events for the same situation. So I think that linguistic complexity derives from the computational task of providing cues for different referential tasks, which differ in complexity themselves. Linguistic complexity is not the primary phenomenon but the secondary one.

Graesser: Could I argue with that? The conditions that specify complexity did matter. In the decoupled level you have to create that environment that is not in the here and now, and that's very taxing to do, it takes a lot of mental resources. My claim is that you often don't embody it when you have to construct the whole [discourse meaning], that it takes a lot of resources as opposed to the here and now where it is given. I want to reassert my position that it's going to be in the discourse focus, it's going to be a goal, it's going to need enough time to do it and resources are limited. If those conditions are met then you do not have much of an embodiment experience. This is my extreme position as a contrast. But if you are in the middle of a novel, having spent 4 hours creating this world and it is very concrete and salient, that can be like the here and now, and that's were the embodiment must take place. Because if you build a whole mental world like this, it's as if it's all around you.

How separable are the levels of reference from a functional point of view?

Lawrence Barsalou: Art [Glenberg] has shown (Glenberg *et al.* 1998) that when you ask people to solve problems, the harder the problem gets they seem to shut down the on-line processes system. You begin to see eye aversion, they close their eyes, and so forth... suggesting that they are primarily using their on-line system as part of the computational system for solving a problem. I wonder how you might reconcile these kinds of results with your own data. And if you can predict whether, when people process counterfactuals, they show this trend of shutting down the on-line system.

de Vega: Shutting down the on-line processing is not contradictory to my proposal of displaced reference, in which a mental framework 'detached' from current perception is required. Speakers would avert their gaze or close their eyes to avoid interference between current perception and mental simulation, because both are embodied. Concerning counterfactuals, I don't know whether this perceptual shutdown occurs. This is an empirical question that deserves specific research.

Anthony Gomila: You suggested that the three levels of reference might correspond to three different kinds of linguistic devices for making references. Thus, you suggested that in on-line communications there are pronouns and deictic or indexical devices, and so forth. But it seems that there must be some different way to make the distinction of levels of reference, because you can use deictics in all three levels of reference. I can say a counterfactual including 'I' or 'here'. Therefore, the levels of reference cannot be characterized in terms of the linguistic devices used for creating the reference.

de Vega: I agree. In real life you can use the three levels combined into a single utterance, like in 'if I were thirsty I would drink this.' But I do not think this fact invalidates the basic idea of levels of reference as involving specific linguistic devices and differing in computational demands.

Gomila: In the philosophy of science, counterfactuals were a matter of interest because we have intuitions about their truth. They are used to cash out the notion of natural law. A generalization is a law, for instance, if it supports counterfactuals, namely if the counterfactuals are true. For instance, 'heat dilates metals' is a law because 'if this was a metal and it was heated it would dilate.' Since this seems to be true, that's the law. Could it be interesting to study these intuitions of the truth of counterfactuals? In other words, decoupled reference is not so decoupled from real-life situations of how things work, because to be able to have intuitions about whether a counterfactual is true or not depends on how you think the causal relations are effectively placed.

de Vega: The issue of truth values for counterfactuals is interesting not only from a logical or epistemological point of view, but also in terms of cognitive processes. Notice that counterfactuals involve a dual nature; they could be momentarily considered as a factual expression 'p and q' ('as if'), and also as a negation of such expression: 'not-p and not-q'. Therefore, counterfactuals could be used to make a sort of 'mental experiment' to test the 'p & q' under imaginary conditions.

Which are the optimal temporal parameters for experiments to better reveal the ACE effects?

Arthur Glenberg: One of the points that Tony [Sanford] made is that some of the ACE phenomena are not as robust as we would like. In your talk you provided a hint, namely that you found a mismatch advantage, as did Kaschak in some of his work. One hypothesis is that this mismatch advantage comes when there is a strict simultaneity of processing: your verb was moving on the screen while people tried to understand the meaning of that verb, whereas in the paradigm that I used the action came after people had understood the sentence. Maybe if we are more careful in formulating when the secondary task is simultaneous with the understanding and when it is not, maybe that would help to make the ACE effect more robust.

de Vega: I agree that the temporality of the double task is very important, and our particular paradigm allows us to calibrate very well the amount of temporal overlapping between the secondary (visual/perceptual) task and the semantic task. In some experiments we found that when the SOA between the onset of the verb and the onset of its motion was very short – in the range of 100–200 milliseconds – there was a mismatch advantage. But when the SOA was larger (300–400 milliseconds), the results were the other way around, a sort of priming in the matching condition. These results make sense: the embodiment effects are fairly automatic and short-lived. And this feature is important for language comprehension, because the linguistic input is extended and you cannot continue activating a given motor pattern or any other perceptual feature for long. So activation is very fast, and the temporal overlapping has to be very well calibrated [in your experiments] to obtain the interference effect.

How abstract is the level of representation for actions when we understand language?

Friedemann Pulvermüller: At some stage of your talk you give the impression that interference should always be at a relatively high representational level (e.g., goals or very abstract nonmotoric action representations, at least in the context of displaced reference). However, I should mention that this might be specific to the particular experiment or set of sentences you looked at. From philosophy of action, we know that actions can be described at various levels: (1) basic actions, including all the motor descriptions, the very basic physical features of the movement; (2) high-order description of actions; and (3) goal description. We have sentences that allow us to put the focus on one of these levels. If I have a sentence like 'He moved the nut from here to there.' then there would be a correspondence to Rizzolatti's pincer. It doesn't matter if it is this movement (closing pincer) or this movement (release pincer) that does the gripping. But if we have a basic description such as 'closing the fist around his glass,' then I doubt you have a high order representation. I want to emphasize that there are a multitude of levels of description [for actions] and perhaps of the corresponding neural processes.

de Vega: This is an empirical question related to Sanford's claim on the granularity issue; you can use different linguistic devices to get into more and more details about actions. In our case, sentences were isolated sentences without much context, and at this very simple level a gross motion representation would be enough. Just a topological description such as 'there is a motion away from me to other' or 'from other towards me.' This is a sort of categorical, relatively abstract representation, but still sufficiently embodied as to be directional (transfer differs from me to you and from you to me). I agree that if you use more detailed sentences describing the motor features of the motion you could recruit more specific motor programs in the brain.

Author note

The preparation of this paper was supported by a grant from the Spanish *Ministerio de Educación y Ciencia* (SEJ2004-02360).

References

Avraamides MA (2003). Spatial updating of environments described in texts. *Cognitive Psychology*, 47, 402–31.

Barsalou L (1999). Perceptual symbol systems. *Behavioural and Brain Sciences*, 22, 577–660.

Barsalou L, Wiemer-Hastings K (2005). Situating abstract concepts. In D Pecher, R Zwaan, Eds. *Grounding Cognition. The Role of Perception and Action in Memory, Language, and Thinking.* Cambridge: Cambridge University Press.

Bühler C (1982). The deictic field of language and deictic words. In RJ Jarvella, W Klein, Eds and Trans. *Speech, Place, and Action* (pp. 9–30). London: Wiley.

Clark HH (1996). *Using Language*. New York, NY. Cambridge University Press.

de Vega M, Rinck M, Díaz JM, Leon I (2007). Figure and ground in temporal sentences: the role of the adverbs when and while. *Discourse Processes*, 43, 1–23.

de Vega M, Robertson DA, Glenberg AM, Kaschak MP, Rinck M (2004). On doing two things at once: temporal constraints on actions in language comprehension. *Memory and Cognition*, 32, 1033–43.

de Vega M, Rodrigo MJ (2001). Updating spatial layouts mediated by pointing and labelling under physical and imaginary rotation. *European Journal of Cognitive Psychology*, 13, 369–93.

de Vega M, Rodrigo MJ (2005). The bicycle pedal is in front of the table. Why some objects do not fit into some spatial relations. In L Carlson, E van der Zee, Eds. *Functional Features in Language and Space*. Oxford: Oxford University Press.

de Vega M, Rodrigo MJ, Ato M, Dehn D, Barquero B (2002). How nouns and prepositions fit together. An exploration of the semantics of locative sentences. *Discourse Processes*, 34, 117–43.

de Vega M, Urrutia M, Morera Y (2006). Is counterfactual meaning grounded on sensorimotor cognition? Communication at the 47th Annual Meeting of the Psychonomic Society, Houston, TX.

de Vega M, Urrutia M, Riffo B (2007). Cancelling updating in the comprehension of counterfactuals embedded in narratives. *Memory and Cognition*, 35, 1410–1421.

Duchan JF, Bruder GA, Hevitt LE, Eds (1995). *Deixis in Narrative*. Hillsdale, NJ: Lawrence Erlbaum Associates.

Franklin N, Tversky B (1990). Searching imagined environments. *Journal of Experimental Psychology: General*, 119, 63–76.

Garrod S, Pickering MJ (2004). Why is conversation so easy? *Trends in Cognitive Sciences*, 8, 8–11.

Glenberg AM, Kaschak MP (2002). Grounding language in action. *Psychonomic Bulletin and Review*, 9, 558–65.

Glenberg AM, Schroeder JL, Robertson DA (1998). Averting the gaze disengages the environment and facilitates remembering. *Memory and Cognition*, 26, 651–8.

Graziano MSA, Hu XT, Gross G (1997). Coding the locations of objects in the dark. *Science*, 277, 239–41.

Hockett CF (1960). The origin of speech. *Scientific American*, 203, 88–96.

Kaschak MP, Madden CJ, Therriault DJ *et al.* (2005). Perception of motion affects language processing. *Cognition*, 94, 79–89

Kaup B (2001). Negation and its impact on the accessibility of text information. *Memory and Cognition*, 29, 960–7.

Kaup B, Zwaan B (2003). Effects of negation and situational presence on the accessibility of text information. *Journal of Experimental Psychology: Learning, Memory, and Cognition*, 29, 439–46.

Lakoff G (1987). *Woman, Fire, and Dangerous Things: What Categories Reveal about the Mind.* Chicago, IL: University of Chicago Press.

Landau B, Jackendoff R (1993). 'What' and 'where' in spatial language and spatial cognition. *Behavioural and Brain Sciences*, 16, 217–65.

Leslie AM, Friedman O, German TP (2004). Core mechanisms in 'theory of mind'. *Trends in Cognitive Sciences*, 8, 528–33.

Logan GD, Sadler DD (1996). A computational analysis of the apprehension of spatial relations. In P Bloom, MA Peterson, L Nadel, M Garrett, Eds. *Language and Space* (pp. 493–529). Cambridge, MA: MIT Press.

Marr D (1982). *Vision.* New York, NY: Freeman.

McNeill D (1992). *Hand and Mind: What Gestures Reveal about Thought.* Chicago, IL: University of Chicago Press.

McNeill D, Duncan SD (2000). Growth points in thinking-for-speaking. In D McNeill, Ed. *Language and Gesture* (pp. 141–61). Cambridge, MA: Cambridge University Press.

Mandler JM (1992). How to build a baby: II. Conceptual primitives. *Psychological Review*, 99, 587–604.

McDonald MC, Just MA (1989). Changes in activation levels with negation. *Journal of Experimental Psychology: Learning, Memory, and Cognition*, 15, 633–42.

Moreno V, de Vega M (2005). Animating words during the comprehension of transference sentences. Communication in the XV Annual Meeting of the Society for Text and Discourse, Amsterdam, The Netherlands.

Nelissen K, Luppino G, Vanduffel W, Rizzolatti G, Orban GA (2005). Observing others: multiple action representation in the frontal lobe. *Science*, 310, 332–6.

Richardson DC, Spivey MJ, Barsalou LW, McRae K (2003). Spatial representations activated during real time comprehension of verbs. *Cognitive Science*, 27, 767–80.

Rieser JJ (1989). Access to knowledge of spatial structure at novel points of observation. *Journal of Experimental Psychology: Learning, Memory, and Cognition*, 15, 1157–65.

Rizzolatti G, Fadiga L, Fogassi L, Gallese V (1997). The space around us. *Science*, 277, 190–1.

Rodrigo MJ, de Vega M (in preparation). Dimension effects in spatial localization by pointing and verbal labels.

Steels L (1999). *The Talking Heads Experiment. Words and Meaning, Vol. 1*. Brussels: VUP Artificial Intelligence Laboratory.

Rüschemeyer SA, Glenberg AM, Kaschak MP, Friederici AD (submitted). Listening to sentences describing visual motions activates MT/V5.

Talmy L (1983). How language structures space. In H Pick, L Acredolo, Eds. *Spatial Orientation: Theory, Research, and Application* (pp. 225–82). New York, NY: Plenum.

Wang RF (2005). Dissociation between verbal and pointing responding in perspective change problems. In L Carlson, E van der Zee, Eds. *Functional Features in Language and Space*. Oxford: Oxford University Press.

Wraga M (2003). Thinking outside the body: an advantage for spatial updating during imagined versus physical self-rotation. *Journal of Experimental Psychology: Learning, Memory, and Cognition*, 29, 993–1005.

Zwaan RA, Madden CJ (2005). Embodied sentence comprehension. In D Pecher, RA Zwaan, Eds. *Grounding Cognition: The Role of Perception and Action in Memory, Language, and Thinking*. Cambridge: Cambridge University Press.

Zwaan RA, Madden CJ, Yaxley RH, Aveyard ME (2004). Moving words: dynamic mental representations in language comprehension. *Cognitive Science*, 28, 611–9.

Zwaan RA, Taylor LJ (2006). Seeing, acting, understanding: motor resonance in language comprehension. *Journal of Experimental Psychology: General*, 135, 1–11.

Chapter 15

Language comprehension is both embodied and symbolic

Max Louwerse and Patrick Jeuniaux

15.1 Introduction

Over the last decades a divide seems to have emerged in the cognitive sciences between embodied and symbolic approaches to language understanding. Embodied approaches argue that language comprehension requires activation of our experiences with the *world*, whereas symbolic approaches argue that language comprehension relies on interdependencies of *words*. The current paper argues that language comprehension is both embodied and symbolic. According to the *symbol interdependency hypothesis* comprehenders can ultimately ground symbols, but they also can rely on interdependencies across symbols as a shortcut to the meaning of words. An overview is given of the evidence supporting this hypothesis suggesting that embodied representations are activated under certain conditions and ultimately tend to be encoded in language structures.

Attempts to answer the question to what extent language comprehension is embodied or symbolic are not new. For instance, around 360 BC, Plato describes a discussion between Cratylus, Hermogenes, and Socrates, with Socrates asking:

> Then if you admit that primitive or first nouns are representations of things, is there any better way of framing representations than by assimilating them to the objects as much as you can; or do you prefer the notion of Hermogenes and of many others, who say that names are conventional, and have a meaning to those who have agreed about them, and who have previous knowledge of the things intended by them, and that convention is the only principle; and whether you abide by our present convention, or make a new and opposite one, according to which you call small great and great small – that, they would say, makes no difference, if you are only agreed. Which of these two notions do you prefer? (Plato 360 BC/1902)

Over the last few decades the field of cognitive science seems to be divided between those who prefer one of Socrates's notions over the other. On the one hand, embodied accounts of language processing argue that symbols are fundamentally grounded in embodied experience (whereby the processing of symbols is modal), and indeed there is compelling evidence that symbols can be grounded, subconsciously or consciously (Barsalou 1999; Glenberg 1997; Harnad 1990; Pulvermüller 1999; Searle 1980; Zwaan 2004). For instance, we all know intuitively that we *can* imagine the colour red when reading the word 'rose', the smell when reading the word 'perfume', and the loud noise when reading the word 'thunder'. However, the specific stance of the embodied accounts of language processing, is that all words *must always* activate perceptual representations of their referents (whereby the processing of meaning is modal) in order to activate meaning.

On the other hand, symbolic accounts of language processing argue that symbols retrieve much of their meaning by their association with other symbols (whereby the processing of meaning is amodal). Indeed, there is also compelling evidence that symbols can derive meaning from other symbols (Fodor 1975; Kintsch 1998; Landauer and Dumais 1997; Pylyshyn 1984). For instance, we know that a rose is a flower because 'rose' appears in linguistic contexts similar to the contexts in which 'flower' appears. The same is true for associations between 'perfume' and 'smell' and for 'thunder' and 'loud'.

Though the discussion whether language comprehension is fundamentally symbolic or embodied has become a centuries-old debate (see Ogden and Richards 1923 for an overview), this debate recently received new impetus. For instance, Pecher and Zwaan's (2005) edited book *Grounding Cognition: The Role of Perception and Action in Memory, Language, and Thinking* leaves little room for symbolic representations while showing that language is always grounded (in modal representations), whereas Landauer *et al.'s* (2007) edited book *Handbook of Latent Semantic Analysis* pays little attention to embodied representations while showing the strengths of higher-order relationships between (amodal) language units.

Advocates of embodied versus symbolic accounts of language processing defend their side fiercely, as illustrated by the following quotes.

> ... (amodal) propositions are 'a convenient shorthand' for representing information. Indeed, as these authors and the many researchers inspired by them have shown, impressive results can be obtained by using this shorthand, but as is shown here, it would be a mistake to elevate the shorthand to the status of longhand (Zwaan 2004, p. 56–7).

> Therefore, it is gratuitous to conclude that [latent semantic analysis (LSA)] is wrong in principle from the observation that it is sometimes wrong as implemented and trained ... The use of these [embodiment] ideas to oppose [hyperspace analogue to language (HAL)] and LSA is a case of what Dennett calls an 'intuition pump', pushing the introspective mystery of a mental phenomenon to discredit a mechanistic explanation (Landauer 2002, p. 66–7).

At the same time, however, neither side argues that aspects from the other side should be excluded from the theoretical framework.

> ...Thus, it can be argued that the information captured by amodal propositions forms a subset of the information captured by a perceptual analysis (Zwaan 2004, p. 56).

> ... [Embodiment] is an appealing idea; it offers the beginnings of a way to explain the relation between some important aspects of thought, language, and action that appear to capture analogue properties of the cognition of experience (Landauer 2002, p. 67).

It is noteworthy that the literature on the debate between embodiment and symbolism largely follows the distinction between psycholinguistic and computational linguistic approaches. The lion's share of embodiment publications is experimental in nature, ranging from reaction time to event related-potential experiments (Boroditsky 2000; Fincher-Kiefer 2001; Glenberg and Kaschak 2002; Wiemer-Hastings and Xu 2005; Zwaan *et al.* 2002). Far fewer computational models have been developed that integrate embodied representations (but see Cangelosi and Harnad 2000; Roy and Pentland 2002; Vigliocco *et al.* 2004; Westermann and Miranda 2004).

At the same time, the majority of symbolism publications is computational in nature, showing how data are related to corpus linguistic findings (Burgess 1998; Howard and Kahanna 2002, Kintsch 1998a,b; Landauer 2002; Landauer and Dumais 1997; Lund and Burgess 1996; Steyvers *et al.* 2004). Fewer experimental studies have been conducted arguing that language processing is to a large extent amodal (but see Bransford and Franks 1971; Kintsch and Keenan 1973; Van Dijk and Kintsch 1983). Overall, there seems to be a methodological preference for one side or the other. An interdisciplinary account incorporating psycholinguistics and computational linguistics is probably as hard to find as is a unified account incorporating embodiment and symbolism.

15.2 **Language comprehension is embodied**

There is no doubt that linguistic symbols *can* be grounded in the physical world of the comprehender's experiences. A range of theoretical and experimental evidence shows that, in the process of language comprehension, the comprehender activates embodied representations. Philosophical arguments have posited that language comprehension *must* involve more than translating one symbol into another (Harnad 1990; Searle 1980). Whether or not these thought experiments prove their case remains open for discussion (Bishop and Preston 2001), but at least strong theoretical arguments have been posed against a solely symbolic account of language comprehension.

Indeed, there is overwhelming experimental evidence for embodied representations. For instance, Pecher *et al.* (2003) and Spence *et al.* (2000) found that stimuli from the same modality resulted in faster processing times than stimuli from different modalities (even when the modalities were described in words), suggesting that modalities don't seem to get recoded into an amodal symbolic representation. Furthermore, there is evidence that motor movements of comprehenders match those described in the linguistic input. Klatzky *et al.* (1989) showed that the comprehension of verbally described actions was facilitated by preceding primes specifying the motor movement. Glenberg and Kaschak (2002) found an action-sentence compatibility effect (ACE) with motor imagery described in a sentence effecting motor actions. Zwaan *et al.* (2002) showed that response times for sentences were shorter with pictures matching a sentence than mismatching pictures. Zwaan and Yaxley (2003) showed that the spatial configuration of items presented on a screen affects semantic judgements about them, suggesting that visual representations get activated during language comprehension.

The argument that nonlinguistic representations are tightly coupled to language is also found in eye-tracking studies showing specific patterns of eye gaze influenced by text comprehension or problem solving activities of various degrees of complexity (Altman and Kamide 1999; Chambers *et al.* 2002; Demarais and Cohen 1998; Matlock and Richardson 2004; Spivey and Geng 2001; Spivey *et al.* 2000, 2002). A slightly different kind of embodiment is also found in studies investigating multimodal conversations, whereby gestures iconically support language functions (Louwerse *et al.* 2006). Eye gaze, facial expressions and hand gestures provide a wealth of examples of how language is often grounded in bodily experiences. Louwerse and Bangerter (2005) showed how language and gesture interact, and can even substitute each other. The use of these bodily experiences in communication facilitates thinking (Goldin-Meadow 2003; McNeill 1992). Gesture and speech are thereby co-expressive manifestations of one integrated system and form complementary components of one underlying process. In summary, arguing that language comprehension is not embodied would be analogous to arguing that language comprehension is not, to at least some extent, symbolic.

15.3 **Language comprehension is symbolic**

There is equally compelling evidence that symbols can derive meaning on the sole basis of other amodal symbols (Fodor 1975; Kintsch 1998; Landauer and Dumais 1997; Pylyshyn 1984). These symbolic approaches assume that conceptual representations are typically nonperceptual. That is, they are autonomous symbolic representations. Much of the recent evidence for the symbolic account of language processing comes from computational models. Two symbolic models of language processing have particularly been used in applications and experiments. Models like HAL and LSA are computational models that determine the semantic relatedness by the frequency of word co-occurrences, as well as by the similarity of the contexts in which they co-occur (Kintsch 1998a; Landauer and Dumais 1997). Thus, these models compute higher-order relationships between text units such as words, sentences, paragraphs, and texts.

There is a large body of evidence showing that the output of these models correlate with human performance. For instance, HAL reliably categorized types of nouns (Lund and Burgess 1996), clustered common noun/proper name distinctions (Burgess and Conley 1998) as well as abstract or concrete words (Burgess and Lund 1997). LSA, another amodal symbolic system, does an impressive job on a variety of language tasks. For instance, by comparing student essays with ideal essays, LSA can automatically grade assignments (Landauer et al. 1998) and performs equally well as students on the Test of English as a Foreign Language (Landauer and Dumais 1997). LSA can even measure the coherence of texts (Foltz et al. 1998). Recently, LSA has been used as a measure of coherence in Coh-Metrix (Graesser *et al.* 2004; Louwerse *et al.* 2004), an internet-based tool that analyses texts on over 50 types of cohesion relations and over 200 measures of language, text, and readability.

LSA has also been used in intelligent tutoring systems like AutoTutor and iSTART. AutoTutor engages the student in a conversation on a particular topic like conceptual physics or computer literacy. AutoTutor uses LSA for its world knowledge and determines the semantic associations between a student answer, and ideal good and bad answers (Graesser *et al.* 2000; Graesser, Chapter 3, this volume). iSTART uses LSA in its teaching of reading strategies to students by providing appropriate feedback to students' self-explanations (McNamara *et al.* 2004). LSA is able to 'understand' metaphors (Kintsch 2000), and can assist in problem solving (Quesada *et al.* 2002). These studies suggest that meaning of language can be derived merely through the interrelations of words, sentences, and paragraphs.

15.4 **Language comprehension is both embodied and symbolic**

It seems obvious: the various examples shown so far suggest that neither an embodied nor a symbolic account alone can explain language comprehension, but that language comprehension a convergence of the two accounts. However, if language comprehension is embodied and symbolic, how can this convergence be explained? Based on various theories, most notably Deacon (1997), Louwerse (2007) proposed a symbol interdependency hypothesis, making the argument that symbols can, but do not always have to, be grounded. That is, language is structured in such a way that many relationships that can also be found in the embodied world are structured in language. Language thereby provides a shortcut to the embodied relations in the world.

This claim is illustrated using Deacon's (1997) *language evolution theory* that proposes that icons, indices, and symbols are hierarchically structured. Deacon draws upon Peirce's (1923) theory of signs. A sign can be an icon, index, or symbol. An icon is mediated by a physical similarity between the sign and the object it refers to, as is the case in a picture representing an object. Icons require limited interpretation: we 're-cognize' what the sign stands for. Different from an icon, an index is mediated by spatial or temporal contiguity between sign and object, like smoke indicating the presence of fire. Note that there is no physical resemblance between smoke and fire, and that at least two iconic relationships are needed in order to build the indexical relationship. That is, the indexical link comes about through the iconic relationship between the representation of smoke and its referent, as well as the iconic relationship between the representation of fire and its referent. This indexical link between smoke and fire needs to be learnt at a certain time in a certain space and can be learnt through straightforward learning processes, as in classical and operational conditioning. Finally, a symbol is mediated by a conventional relationship between sign and object, like a yellow wristband as a symbol for hope, courage, and perseverance. Most of language belongs to this latter category. Whether we call an elephant an 'elephant' or a 'bulb' merely depends on what a language community agrees upon, not on a physical resemblance or a temporal and spatial contiguity.

Deacon now argues that icons, indices, and symbols represent levels of representation whereby indices are derived from icons (which correspond to objects in the real world) and symbols are derived from indices. The correspondence between words and objects thereby becomes a secondary relationship. But words do not only (indirectly) refer to objects, they are also linked to each other. Utilizing this network of symbolic relationships is what makes humans the only symbolic species. Whereas most species do not have the neurological architecture to process information beyond iconic and indexical levels, primates are able to recognize symbols, but only humans use the symbol system called language to its full extent.

The symbol interdependency hypothesis follows Deacon's taxonomy of icons, indices, and symbols in the sense that it recognizes that symbols can refer to each other, but are also connected as well to more concrete entities. This interdependency of symbols – to both objects and symbols – allows us to ground some words without the necessity to ground others, providing comprehenders with a convenient mnemonic support for language processing. Comprehenders do not have to activate every single detail of the embodied representations, but they can use a simple shorthand. Using a neural network metaphor we would say that all nodes related to these symbols – including embodied representations – receive some activation. However, by virtue of being located far away from the symbols as implied by Deacon (1997)'s theory – separated by several intermediate layers of neurons – the embodied representations by default don't get enough activation to be in working memory. That is, in a default conversation and reading setting, whereby comprehenders quickly process language, they do not have to perceptually simulate each and every word they encounter. That some particular words trigger visual/motor imagery or other embodied representations is common in certain circumstances, but it is argued that efficiency in language processing simply does not yield a continuous systematic grounding of each and every symbol.

Similar to other accounts, the symbol interdependency hypothesis implies the transduction of information into a new representation. Take for instance Barsalou's (1999) theory of *perceptual symbols*. According to Barsalou, the brain creates neural records during perception. Subsets of these perceptual states are extracted and stored in long-term memory. These subsets of perceptual states are what Barsalou calls perceptual symbols. Thus, knowledge is not transduced into an amodal symbolic representation, but instead into modality-specific states (e.g., visual states). However, during retrieval these perceptual states can function symbolically. Barsalou's point is that a physical stimulus (e.g., seeing a car) is transduced into modality-specific states (e.g., sensory representation of a car), rather than an amodal representation outside the sensorimotor system. It seems more convenient to remember a smell as a smell (with associative sensory representations) and a touch as a touch. Similarly, the symbol interdependency hypothesis states that linguistic expressions by default get transduced into a symbolic representation, because it is generally more efficient to do so. It seems more convenient to remember an amodal symbol as an amodal symbol (with corresponding associative symbolic representations).

15.5 **Activation of embodied representations**

According to the symbol interdependency hypothesis, comprehenders can bootstrap the meaning of amodal symbols through their interrelations and don't need to ground all amodal symbols under all circumstances. There may be theoretical evidence for this claim, but the question needs to be answered how this claim can explain the wealth of evidence for embodied representations. Dozens of psycholinguistic studies have been reported over the last half decade – some mentioned above – all finding evidence for the activation of embodied representations in language comprehension. How can this wealth of evidence warrant the claim that embodied representations are not always sufficiently activated under all circumstances? We have two answers to this question.

First, in some psycholinguistic embodiment experiments participants are typically cued to activate embodied representations like images or motor skills. These experiments demonstrate quite powerfully that cognition cannot be strictly symbolic, because non-linguistic modalities get activated. However, these experiments do not show that no amodal representations get activated. For instance, Stanfield and Zwaan (2001) asked participants to verify whether a picture depicted an object in a sentence. Participants responded more quickly to a picture of an eagle with its wings spread out after reading 'The ranger saw an eagle in the sky' than after 'The ranger saw an eagle in the tree' (Zwaan *et al.* 2002) or to a picture of a horizontally depicted nail after reading 'He pounded the nail into the wall' than to a vertically depicted nail, but *vice versa* after reading 'He pounded the nail into the floor' (Stanfield and Zwaan 2001). According to the authors, understanding the sentence activates pictorial information (i.e., an embodied representation) matching the content of that sentence. Consequently, the reaction times are shorter when the picture matches this pictorial information. Similarly, according to Glenberg and Kaschak's (2002) ACE paradigm there is a compatibility effect between the direction described linguistically in a sentence and the action executed motorically. When subjects were presented with a sentence like 'Courtney handed you the notebook' subjects were faster in making sensibility judgements when the response button was close to them, but when a sentence like 'You handed Courtney the notebook' was presented subjects were faster in making the judgement when the response button was away from them. Both Stanfield and Zwaan (2001) and Glenberg and Kaschak (2002) show that it is more efficient to transduce a physical stimulus in a modality-specific state rather than transducing it into an amodal symbolic representation (Barsalou 1999). That is, these studies show that embodied representations are activated when comprehenders are cued to activate them, but not that amodal symbolic representations do not get activated.

Secondly, in some psycholinguistic embodiment experiments, the experimental instructions activate deep levels of processing, whereby we consider depth in terms of the extent to which meaningfulness is extracted from the stimulus (Lockhart and Craik 1990). Interestingly, as we will show later, most evidence for activation of embodied representations comes from experiments whereby participants are involved in tasks that require cognitive processing deeper than what would typically be required in default language comprehension. For instance, Spivey and Geng (2001) found that subjects acted out the mental image of a passage they read. Subjects listened to a story that had descriptions of upward, downward, leftward, and rightward events, like in the following text:

> Imagine that you are standing across the street from a 40-storey apartment building. At the bottom there is a doorman in blue. On the 10th floor, a woman is hanging her laundry out of the window. On the 29th floor, two kids are sitting on the fire escape, smoking cigarettes. On the very top floor, two people are screaming.

Unbeknownst to participants, their eye movements were recorded as they followed the text. Spivey and Geng found that eye movements were in the direction of the described directions (vertical in this case), suggesting that lower-level motor processes are activated with higher-level cognitive processes. The story presented to participants required considerable cognitive effort. The question could therefore be raised whether a text not cueing the participant to think deeply through the materials, as in the average narrative text, may have led to the same mental imagery. Though this question remains unanswered for now, it is noteworthy that many embodiment experiments require the participant to deeply process the materials, more than what would be needed to assign meaning to language, as we will argue next.

In summary, embodiment experiments clearly show that language comprehension cannot be solely symbolic. However, they do not show that language comprehension is solely embodied. Comprehenders are often explicitly cued to activate nonlinguistic modalities or are involved in a deep semantic analysis task whose nature is such that different kinds of representations – modal and amodal – are activated. Roughly stated, psycholinguistic embodiment experiments can be classified by task and nature of the stimuli. Table 15.1 gives an overview of prominent psycholinguistic embodiment experiments that provide evidence for embodied representations and the associated experimental tasks. Though this may not be a complete overview, we believe it is an adequate representation of psycholinguistic studies in embodied cognition.

The table indicates that the number of embodiment experiments whereby deeper semantic analysis is needed (semantic judgement, sensibility ratings, memory load) or whereby nonlinguistic stimuli are used or motor skills are required (pictorial information, movement to button press) outweigh the studies whereby solely linguistic information is used and shallow language processing suffices. The purpose of this table is not to be exhaustive, to debate whether a study uniquely belongs in a cell, or even whether studies can be classified solely on the basis of these two dimensions. Instead, the table poses the question under what circumstances embodied representations get activated and whether embodied representations supersede symbolic representations in standard (shallow linguistic) language tasks.

Table 15.1 Classification embodiment studies

	Linguistic stimuli only	Linguistic and nonlinguistic stimuli
'Shallow' processing	Pecher et al. (2003); 1,2	Fincher-Kiefer (2001); 2 Matlock and Richardson (2004) Richardson et al. (2001); 1 Richardson et al. (2003); 1,2 Stanfield and Zwaan (2001) Zwaan et al. (2004) Zwaan et al. (2002); 1,2 Zwaan and Yaxley (2003); 1–3
'Deep' processing	Borghi and Barsalou (2001) Borghi (2004); 1–5 Borghi et al. (2004a); 1,2 Demarais and Cohen (1998); 1 Fincher-Kiefer (2001); 1 Glenberg and Robertson (2000); 1–3 Matlock et al. (2005); 1–3 Richardson et al. (2001); 2 Spivey et al. (2000) Spivey and Geng (2001); 1 Wu and Barsalou (in prep.)	Bergen and Wheeler (2005); 1, 2 Demarais and Cohen (1998); 2 Borghi et al. (2004a); 3 Boroditsky (2000); 1–3 Boroditsky and Ramscar (2002); 1–4 Glenberg and Kaschak (2002); 1,2AB Kaschak et al. (2005); 1,2 Spivey and Geng (2001); 2 Tseng and Bergen (2005)

N.B. The number following the author and year refers to the experiment in the study.

15.5.1 **Activating embodied relations under certain conditions**

What is the evidence that in many circumstances embodiment representations are not necessarily activated? Is there evidence that comprehenders can actually rely on the organization of the symbolic system to derive meaning? The symbol interdependency hypothesis makes two predictions. First, because symbols are interdependent on one another, it seems convenient for language users to access language symbolically, unless embodied representations are cued by the task, visual imagery, or motor movements. Second, because language develops through our interaction with the physical world, and because of its efficiency embodied relations are encoded in language structures. In the remainder of this paper we will give an overview of the evidence for these two predictions.

There are various experiments whereby the participant is exposed to linguistic information only, and whereby the task does not explicitly trigger mental imagery. One such study by Zwaan and Yaxley (2003) utilized a spatial Stroop test in which participants saw word-pairs that were either presented on the vertical axis in their iconic order ('attic' above 'basement') or in a reverse-iconic order ('basement' above 'attic'). Zwaan and Yaxley found that participants were faster in making a semantic judgement for the iconic than for the reverse-iconic pairs. They argued that this proves that embodied representations are activated during language processing. According to the symbol interdependency hypothesis, embodied relations are activated under certain conditions, like the 'deeper' processing of the stimuli or a processing such that the activation of embodied representations is desirable.

Intrigued by the interpretation of Zwaan and Yaxley's (2003) findings, Louwerse and Jeuniaux (in press) repeated the experiment by varying a series of factors, one of which was the semantic relation between the two items. Motivated by the idea that something intrinsically symbolic would also have a role to play, we split all pairs in function of the degree of semantic relationship of their items as determined by LSA. This semantic separation was realized by computing the LSA cosine value for each pair. We had therefore items with a strong semantic relation ('airplane'–'runway') and items with a weak semantic relation ('penthouse'–'lobby'). Besides these experimental pairs, we added filler pairs composed of one or two nonwords. Secondly, we varied the task. In the first experiment we used the same semantic judgement task implemented by Zwaan and Yaxley. In the second experiment we used a lexical decision task, where participants decided whether the pair consisted of words only or included nonwords. The idea behind varying the task was that in both cases some semantic processing would be involved but with different degrees of intensity ('deeper' processing in the case of semantic judgement and 'shallower' processing in the case of lexical judgement), and that the activation of embodied representations would be more likely in the former than in the latter, thereby illustrating that these representations are not activated under all circumstances involving semantic processing.

What we found corresponds to this line of reasoning: an effect for iconicity was found in the semantic judgement task but not in the lexical decision task. In the semantic task, as in Zwaan and Yaxley's (2003) experiment, the participants were faster at judging the semantic similarity of two items if they were located in iconic order on the vertical axis. We concluded that for deeper levels of processing – like semantic judgement, where processing times tend to be higher than for shallower levels of processing like lexical decision – embodied relations would be likely more activated. Of course, the argument could be made that in a shallow language processing task like lexical decision where participants merely decide whether a word is a word or a non-word no real language processing takes place. Interestingly, however, this argument does not hold since we found semantic effects stronger than those for iconicity in both experiments (see also Meyer and Schvaneveldt 1971). Indeed, the judgement of semantic relatedness and the response times for lexicality were shorter when the two items were strongly related semantically. Therefore, comprehension *must* have taken place, but at a level shallower (and faster) than in the semantic judgement task.

In a follow-up experiment, we presented the word-pairs horizontally, as did Zwaan and Yaxley (2003). For-word pairs presented horizontally, semanticity but not iconicity effects were found. One explanation is that word-pairs presented horizontally are not perceptually simulated in the same way as word-pairs presented horizontally. An alternative explanation that we pursued is that vertically presented items seem to require the comprehender to process the words independently, whereas with horizontally presented items the words are integrated into one representation (Bradshaw *et al.* 1981). These data do not support the claim that no embodied representations are formed. Instead, they favour the idea that both symbolic and embodied representation are at the core of language processing. The data also suggest that the consideration of embodied relationships slows down the processing of language and therefore may not always occur by default in language processing.

15.6 Language encodes relations in the world

If relations between amodal symbols provide a convenient shortcut to understanding the world described by the language, the structure of the amodal symbol system would be most efficient if it somehow encodes the relations in the physical world. Louwerse (2007) argues that it is indeed likely that language has been shaped by thought. For instance, phrasal structures match perceptual concepts, with adjectives being placed right before or after a noun, so that the 'grey mouse' is easily understood as a mouse that is grey. This even applies to languages that use case suffixes like German, Latin, and Turkish, and would allow for adjectives to be detached from their nouns. Similarly, when we think of actions, we think of subjects (often agents) and objects (often patients), most likely in that order. Indeed, in the vast majority of the world's languages, subjects precede objects (Greenberg 1963).

In a number of other studies we have further explored the extent to which language structures encode perceptual information. In some studies (Louwerse 2007; in press) we used n-grams to compute the frequency with which words occur in a certain order; for others we had to rely on higher-order relations between words using LSA to avoid problems of data sparsity. LSA estimates semantic similarities on a scale of −1 to 1 between the latent semantic structure of words and linguistic contexts. The input to LSA is a set of corpora segmented into contexts like sentences, paragraphs, or whole texts. Mathematical transformations create a large multidimensional words–contexts matrix from the input. By removing dimensions corresponding to small variance and keeping the dimensions corresponding to larger variance, the representation of each term is further reduced as a smaller vector with only k dimensions. Each word now becomes a weighted vector on 300 or so dimensions. The semantic relationship between words can be estimated by taking the dot product (cosine) between two vectors.

Various studies have shown that LSA is able to accurately predict the semantic relations between words. Louwerse *et al.* (2006) have shown that LSA is also able to predict temporal and spatial relations. For instance, LSA in combination with multidimensional scaling (MDS) algorithms was able to accurately order the days of the week as well as the months of the year. Moreover, it was able to determine the order of time distances of relative time units ('second', 'minute', 'hour', 'day', 'week', 'month', 'year') and pointing time adverbs ('ago' and 'later'). LSA was also able to capture world knowledge by predicting the distances between spatial coordinates. Cosine values were higher when comparing 'New York' and 'Boston' than for 'New York' and 'Seattle' because, in natural discourse, place names that are closer in proximity occur more frequently together and, similarly, their contexts are shared across multiple documents. This finding emerged in the comparison of cosines and physical distances of 28 of the largest cities in the United States, the largest cities in the world, and even for different languages (French and English) (see also Louwerse and Zwaan; in press).

We have also conducted a number of studies to determine to what extent symbolic algorithms can explain data found in embodiment experiments. Louwerse (2007) for instance used the same technique as Louwerse *et al.* (2006) computing cosine values from the Glenberg and Robertson (2000) data and analyzing the results using MDS. Glenberg and Robertson presented subjects with a setting sentence ('After wading barefoot in the lake, Erik needed something to get dry'), followed by a related sentence ('He used his towel to dry his feet'), an afforded sentence ('He used his shirt to dry his feet) or a nonafforded sentence ('He used his glasses to dry his feet'). Sensibility and envisioning data from human subjects showed a difference between related/afforded versus nonafforded sentences with lowest scores for the latter. Glenberg and Robertson showed that no such difference was found by comparing the LSA cosine values. However, Louwerse (2007) showed that the combination of LSA with MDS showed no significant differences between the related and the afforded sentences, and an expected difference between the related sentences and the nonafforded sentences, with the related sentences yielding higher scores than nonafforded sentences. Only a marginally significant difference was found between afforded and nonafforded sentences. A correlation between the nonafforded sentence and the setting ('dry'–'glasses'), however, showed a strong negative correlation, suggesting that these are furthest away in terms of Euclidean distance. Furthermore, when the MDS coordinates were compared with Glenberg and Robertson's sensibility and envisioning ratings, a positive correlation was found. These findings show that the embodiment differences could also be obtained using symbolic approaches, suggesting an encoding of embodied relations in language structures.

Similarly, Louwerse (in press; under review) reanalysed data from Borghi *et al.* (2004) and Zwaan and Yaxley (2003) using corpus linguistic methods. For instance, Borghi *et al.* tested concept sets that consisted of an orientation sentence ('There is a doll standing on a table in front of you') followed by a probe word. Two probe words were related to the concept noun of the sentence ('doll'), one describing the upper part of the concept ('hair') the other describing the lower part ('ankle'). Participants were asked to verify whether the probe word was part of the concept by pressing 'yes' or 'no'. The beauty of the experiment lay in the position of the 'yes' and 'no' buttons. In half of the experiments 'yes' was located above 'no' (yes-is-up condition); in the other half the position was reversed (yes-is-down condition). Participants in a movement condition reached out for the buttons, whereas those in a no-movement condition had their hands on the buttons already. Borghi *et al.*'s results showed that response times for the upper parts (e.g., 'hair') were shorter than for the lower parts (e.g., 'ankle') in the yes-is-up condition in the movement condition only. We replicated Borghi *et al.*'s movement results in a no-movement condition by presenting participants with vertically presented word-pairs ('doll'–'hair' and 'doll'–'ankle'). Incidentally, in a corpus linguistic analysis we found that the frequency of concept–upper-part word-pairs is higher than that of concept–lower-part word-pairs, suggesting that upper–lower part distinctions found in the physical world can also be found in language structures. A similar pattern was found when an experiment replicated Zwaan and Yaxley's findings and word-order frequencies were computed for these stimuli. Iconic word order ('attic' preceding 'basement') was more frequent than reverse-iconic word order ('basement' preceding 'attic'). Moreover, word-order frequency explained more of the variance of the reaction time data than did iconicity. These findings do not falsify the results of Borghi *et al.* or Zwaan and Yaxley, but they do suggest that the iconic representations of concepts in the world are also encoded in language structures. The notion of thought shaping language should at least be considered when debating the extent to which language comprehension is embodied and symbolic.

15.7 **Conclusion**

There is evidence that language comprehension is embodied. There is also evidence that language comprehension is symbolic. The symbol interdependency hypothesis argues that it is both: symbols link to other symbols through higher-order relationships and they also refer to objects in the physical world. In many language tasks, comprehenders can draw meaning from symbolic relationships without any extensive grounding being needed. In different circumstances, due to the nature of the instructions or the stimuli; embodied representations become more desirable or useful. The impression may have been given that our account is polemic in nature. It is not. Instead, it argues that debate whether language comprehension is symbolic or embodied should make room for more timely research questions. These include under what circumstances embodied and symbolic representations are activated, and to what extent language shapes thought and thought shapes language.

Debate

Julio Santiago: A really interesting talk, thank you. What I don't really get is the last part in which you compared the horizontal position of word-pairs and argue that the horizontal presentation yields easier processing than vertical presentation. Zwaan and Yaxley would argue that this is exactly the evidence for perceptual simulations.

Max Louwerse: But my point is that when you present items horizontally, you will find shorter reaction times in the first place compared with when you present items vertically. So, there may be different psycholinguistic processes if you process words vertically or horizontally. For instance, it may be that if you process items horizontally, the activation of one word immediately affects the activation of the second word, whereas in vertical processing this carry-over effect is less salient. That is, in vertical processes, the items in the word pairs may be processed more independently. That is at least a possible explanation.

Santiago: But it still doesn't give an explanation for the effect that you find when the words are presented in the vertical dimension.

Louwerse: Well, actually, we ran a multiple regression analysis on the reaction-time data we got from the semantic judgement task and lexical decision experiments, dummy coding for word order. That is, we had two variables in the regression analysis, one (embodied) iconicity and one (symbolic) word order. And what we found was that only 2% of the data in both cases can be explained, but the data are better explained by word order than by iconicity (Louwerse in press).

Robert Goldstone: I'm interested in the experiments that you talked about at the beginning, where you showed that you could reconstruct the spatial configuration of the days of the week, months of the year, or US cities. I mean, I guess, part of my skepticism is wondering whether it's really just a sort of a side effect, That there's something going on in people's minds that would make them talk about Los Angeles at the same time they would talk about San Francisco because of their spatial proximity.

Louwerse: Exactly.

Goldstone: So it makes sense that you could then do some statistical analysis to reconstruct the underlying space that was in the mind of the producer when they were creating it, but I guess I'm still holding out for wanting to have a sort of communal process that explains both the production and the comprehension processes, in order to facilitate communication between them. So, I guess my question to you is, do you see the same thing going on at the production stage? In other words, is there no spatial grounding when they are talking about these cities together?

Louwerse: I'm not arguing against spatial grounding, but, according to what I call the symbol independency hypothesis, you don't have to ground every single item. You only have to ground one or two, and that will bootstrap the grounding of all the others. That's basically what I argue for. So, with regards to what you say about cities, yes we talk about Los Angeles more in the context of San Francisco than New York, for instance. I think that's the whole point. Apparently language has those structures encoded based on the fact that the physical world is structured this way.

Walter Kintsch: So, the series is incomplete. It is a series of comprehension: it does not explain production. And I guess it's embodied. The reason why San Francisco and Los Angeles are closer together in LSA is because people live in the world where they are more likely to talk about those two things. So you can't understand new things by making new combinations of all things. But it doesn't explain production.

Arthur Glenberg: Marvellously creative, thank you. Of course I've got a lot of questions about details but we can do that later. The main question I want to pose is: LSA depends on what's already been done, what's already been written. And, it then seems, if we're going to depend completely on LSA we can never be creative, we can never use language to describe something new. So, do you think that whenever we are using language to describe a new experience we must be embodied?

Louwerse: Again, I'm not arguing against embodiment. With regards to your question about novel situations, I don't know. But LSA could still apply to novel situations. For instance, we could run an experiment where we replaced some existing words with nonwords – that would be a sort of novel scenario I guess – and determine the meaning of the new words. If you replace existing words with nonwords, the nonwords would acquire semantic information from the other real words. I think that sort of relates to the novel situation.

Glenberg: I meant something more along the lines of, well, not learning a novel word but, for instance, telling somebody about my experience in Garachico. And you know, assuming that Garachico is represented in a TASA corpus, would I then be able to describe to you the layout of the town so that somebody could then know how to get around the town?

Louwerse: First of all, when you're talking about a new place you have to at least know the place; information about that place should be somewhere in your corpus. I would then argue that if there is enough contextual information, you would indeed be able to draw a map. In fact, that's what we have done in Louwerse *et al.* (2006) and what we are doing in current studies; (Louwerse and Zwaan in press). Granted, it's true that LSA usually works backwards and needs lots of context. But when it does have that context I don't really see why it would not be able to make predictions for the future, for novel situations.

Author note

This research was supported by National Science Foundation grant NSF-IIS-0416128. Any opinions, findings, and conclusions or recommendations expressed in this material are those of the authors and do not necessarily reflect the views of the funding institution. Correspondence concerning this article should be addressed to Max M Louwerse, Institute for Intelligent Systems / Department of Psychology, University of Memphis, 202 Psychology Building, Memphis, TN 38152-3230, USA.

References

Altmann GTM, Kamide Y (1999). Incremental interpretation at verbs: restricting the domain of subsequent reference. *Cognition*, 73, 247–64.

Barsalou LW (1999). Perceptual symbol systems. *Behaviour and Brain Sciences*, 22, 577–660.

Bergen B, Wheeler K (2005). Sentence understanding engages motor processes. In *Proceedings of the 27th Annual Conference of the Cognitive Science Society* (pp. 238–43). Mahwah, NJ: Erlbaum.

Bishop M, Preston J (2001). *Essays on Searle's Chinese Room Argument*. Oxford: Oxford University Press.

Borghi AM (2004). Objects concepts and action: extracting affordances from objects' parts. *Acta Psychologica*, 115, 69–96.

Borghi AM, Barsalou LW (2001). Perspectives in the conceptualization of categories. *Meeting of the Psychonomic Society*, Orlando, FL, November 2001.

Borghi AM, Caramelli N, Setti A (2004). Conceptual information on objects' location. *Brain and Language*, 93, 140–51

Borghi AM, Glenberg AM, Kaschak MP (2004a). Putting words in perspective. *Memory and Cognition*, 32, 863–73.

Boroditsky L (2000). Metaphoric structuring: understanding time through spatial metaphors. *Cognition*, 75, 1–28.

Boroditsky L and Ramscar M (2002). The roles of body and mind in abstract thought. *Psychological Science*, 13, 185-188.

Bradshaw JL, Nettleton NC, Taylor MJ (1981). The use of laterally presented words in research intocerebral asymmetry: is directional scanning likely to be a source of artifact? *Brain and Language*, 14, 1–14.

Bransford JD, Franks JJ (1971). The abstraction of linguistic ideas. *Cognitive Psychology*, 2, 331–50.

Burgess C (1998). From simple associations to the building blocks of language: Modeling meaning in memory with the HAL model. *Behaviour Research Methods, Instruments, and Computers*, 30, 188–98.

Burgess C, Conley P (1998). Developing semantic representations for proper names. *Proceedings of the Cognitive Science Society* (pp. 185–90). Hillsdale, NJ: Erlbaum.

Burgess C, Lund K (2000). The dynamics of meaning in memory. In E Dietrich, A Markman, Eds. *Cognitive Dynamics: Conceptual Change in Humans* and *Machines* (pp. 295–342). Mahwah, NJ: Erlbaum.

Burgess C, Lund K (1997). Modeling parsing constraints with high-dimensional context space. *Language and Cognitive Processes*, 12, 177–210.

Cangelosi A, Harnad S (2000). The adaptive advantage of symbolic theft over sensorimotor toil: grounding language in perceptual categories. *Evolution of Communication*, 4, 117–42.

Chambers CG, Tanenhaus MK, Eberhard KM, Filip H, Carlson GN (2002). Circumscribing referential domains in real-time language comprehension. *Journal of Memory and Language*, 47, 30–49.

Deacon T (1997). *The Symbolic Species: The Co-evolution of Language and the Human Brain*. London: Allen Lane.

Demarais A, Cohen BH (1998). Evidence for image-scanning eye movements during transitive inference. *Biological Psychology*, 49, 229–47.

Fincher-Kiefer R (2001). Perceptual components of situation models. *Memory and Cognition*, 29, 336–43.

Fodor JA (1975). *The Language of Thought*. Cambridge, MA: Harvard University Press.

Foltz PW, Kintsch W, Landauer TK (1998). The measurement of textual coherence with latent semantic analysis. *Discourse Processes*, 25, 285–307.

Glenberg AM, Kaschak MP (2002). Grounding language in action. *Psychonomic Bulletin and Review*, 9, 558–65.

Glenberg AM, Robertson DA (2000). Symbol grounding and meaning: a comparison of high-dimensional and embodied theories of meaning. *Journal of Memory and Language*, 43, 379–401.

Glenberg AM (1997). What memory is for. *Behavioural and Brain Sciences*, 20, 1–55.

Goldin-Meadow S (2003). *Hearing gesture: How our Hands Help us Think*. Cambridge, MA: Harvard University Press.

Graesser AC, McNamara DS, Louwerse MM, Cai Z (2004). Coh-Metrix: analysis of text on cohesion and language. *Behaviour Research Methods, Instruments, and Computers*, 36, 193–202.

Graesser AC, Wiemer-Hastings P, Wiemer-Hastings K, Harter D, Person N; Tutoring Research Group (2000). Using latent semantic analysis to evaluate the contributions of students in AutoTutor. *Interactive Learning Environments*, 8, 128–48.

Greenberg J, Ed (1963). *Universals of Language*. Cambridge, MA: MIT Press.

Harnad S (1990). The symbol grounding problem. *Physica D*, 42, 335–46.

Howard MW, Kahanna MJ (2002). When does semantic similarity help episodic retrieval? *Journal of Memory and Language*, 46, 85–98.

Kaschak MP, Madden CJ, Therrriault DJ *et al.* (2005). Perception of motion affects language processing. *Cognition*, 94, B79–89.

Kintsch W (2000). Metaphor comprehension: a computational theory. *Psychonomic Bulletin and Review*, 7, 257-266.

Kintsch W (1998a). *Comprehension: A Paradigm for Cognition*. New York, NY: Cambridge University Press.

Kintsch W (1998b). The representation of knowledge in minds and machines. *International Journal of Psychology*, 33, 411–20.

Kintsch W, Keenan J (1973). Reading rate and retention as a function of the number of propositions in the base structure of the text. *Cognitive Psychology*, 5, 257–74.

Klatzky RL, Pellegrino JW, McCloskey DP, Doherty S (1989). Can you squeeze a tomato? The role of motor representations in semantic sensibility judgements. *Journal of Memory and Language*, 28, 56–77

Landauer TK, Dumais ST (1997). A solution to Plato's problem: the latent semantic analysis theory of acquisition, induction, and representation of knowledge. *Psychological Review*, 104, 211–40.

Landauer TK, Foltz PW, Laham D (1998a). An introduction to latent semantic analysis. *Discourse Processes*, 25, 259–84.

Landauer TK, Laham D, Foltz PW (1998b). Learning human-like knowledge by singular value decomposition: a progress report. In MI Jordan, MJ Kearns, SA Solla, Eds. *Advances in Neural Information Processing Systems 10* (pp. 45–51). Cambridge, MA: MIT Press.

Landauer TK, McNamara DS, Dennis S, Kintsch W, Eds. (2007). *Handbook of Latent Semantic Analysis*. Mahwah, NJ: Erlbaum.

Landauer TK (2002). On the computational basis of learning and cognition: arguments from LSA. In N Ross, Ed. *The Psychology of Learning and Motivation*, 41, 43–84.

Louwerse MM (2007). Iconicity in amodal symbolic representations. In T Landauer, D McNamara, S Dennis, W Kintsch, Eds. *Handbook of Latent Semantic Analysis*. Mahwah, NJ: Erlbaum.

Louwerse MM (in press). Embodied representations are encoded in language. Psychonomic Bulletin and Review.

Louwerse MM (under review). Min(d)ing symbols. Embodied representations are encoded in language.

Louwerse MM, Bangerter A (2005). Focusing attention with deictic gestures and linguistic expressions. In B Bara, L Barsalou, M Bucciarelli, Eds. *Proceedings of the Cognitive Science Society* (pp. 1331–6). Mahwah, NJ: Erlbaum.

Louwerse MM, Cai Z, Hu X, Ventura M, Jeuniaux P (2006). Cognitively inspired natural-language based knowledge representations: further explorations of latent semantic analysis. *International Journal of Artificial Intelligence Tools*, 15, 1021–39.

Louwerse MM, Jeuniaux P (in press). How fundamental is embodiment to language comprehension? constraints on embodied cognition.

Louwerse MM, Jeuniaux P, Hoque ME, Wu J, Lewis G (2006). Multimodal communication in computer-mediated map task scenarios. *Proceedings of the 27th Annual Conference of the Cognitive Science Society* (pp. 1717–22). Mahwah, NJ: Erlbaum.

Louwerse MM, McCarthy PM, McNamara DS, Graesser AC (2004). Variation in language and cohesion across written and spoken registers. In K Forbus, D Gentner, T Regier, Eds. *Proceedings of the 26th annual conference of the Cognitive Science Society* (pp. 843–8). Mahwah, NJ: Erlbaum.

Louwerse MM and Zwaan RA (in press). Language encodes geographical information. *Cognitive Science*.

Lund K, Burgess C (1996). Producing high-dimensional semantic spaces from lexical co-occurrence. *Behaviour Research Methods, Instrumentation, and Computers*, 28, 203–8.

Matlock T, Richardson DC (2004). Do eye movements go with fictive motion? In K Forbus, D Gentner, T Regier, Eds. *Proceedings of the 26th Annual Conference of the Cognitive Science Society*. Mahwah, NJ: Erlbaum.

Matlock T, Ramscar M, Boroditsky L (2005). The experiential link between spatial and temporal language. *Cognitive Science*, 29, 655–64.

McNamara DS, Levinstein IB, Boonthum C (2004). iSTART: Interactive strategy trainer for active reading and thinking. *Behavioural Research Methods, Instruments, and Computers*, 36, 222–33.

McNeill D (1992). *Hand and Mind: What Gestures Reveal about Thought*. Chicago, IL: University of Chicago Press.

Meyer DE, Schvaneveldt RW (1971). Facilitation in recognizing pairs of words: Evidence of a dependence between retrieval operations. *Journal of Experimental Psychology*, 90, 227–34.

Ogden CK, Richards IA (1923). *The Meaning of Meaning*. London: Routledge & Kegan Paul.

Pecher D, Zwaan RA, Eds (2005). *Grounding Cognition: The Role of Perception and Action in Memory, Language, and Thinking*. New York, NY: Cambridge University Press.

Pecher D, Zeelenberg R, Barsalou LW (2003). Verifying conceptual properties in different modalities produces switching costs. *Psychological Science*, 14, 119–24.

Peirce CS (1931–58). In C Hartshorne, P Weiss, A Burks, Eds. *The Collected Papers of Charles Sanders Peirce*. Cambridge, MA: Harvard University Press.

Plato (1902). *The Dialogues of Plato. Translated into English with Analyses* and *Introductions* (BMA Jowett, Trans.). New York, NY: Charles Scribner's Sons. (Original work published 360BC).

Pulvermüller F (1999). Words in the brain's language. *Behavioural and Brain Sciences*, 22, 253–70.

Pylyshyn ZW (1984). *Computation and cognition: towards a foundation for cognitive science*. Cambridge, MA: MIT Press.

Quesada JF, Kintsch W, Gomez E (2003). Latent problem solving analysis as an explanation of expertise effects in a complex, dynamic task. In R Alterman, D Kirsh, Eds. *Proceedings of the 25th Annual Conference of the Cognitive Science Society*. Boston, MA.

Richardson DC, Spivey MJ, Cheung J (2001). Motor representations in memory and mental models: embodiment in cognition. *Proceedings of the 23rd Annual Meeting of the Cognitive Science Society* (pp. 867–72). Mawhah, NJ: Erlbaum.

Richardson DC, Spivey MJ, Barsalou LW, McRae K (2003). Spatial representations active during real-time comprehension of verbs. *Cognitive Science*, 27, 767–80.

Rosch E (1973). On the internal structure of perceptual and semantic categories. In TE Moore, Ed. *Cognitive Development* and *the Acquisition of Language* (pp. 111–44). New York, NY: Academic Press.

Roy D, Pentland A (2002). Learning words from sights and sounds: A computational model. *Cognitive Science*, 26, 113–46.

Searle JR (1980). Minds, brains, and programs. *Behavioural and Brain Sciences*, 3, 417–57.

Spence C, Nicholls MER, Driver J (2000). The cost of expecting events in the wrong sensory modality. *Perception and Psychophysics*, 63, 330–6.

Spivey M (2006). *The Continuity of Mind*. New York, NY: Oxford University Press.

Spivey MJ, Geng JJ (2001). Oculomotor mechanisms activated by imagery and memory: wye movements to absent objects. *Psychological Research*, 65, 235–41.

Spivey M, Tanenhaus M, Eberhard K, Sedivy J (2002). Eye movements and spoken language comprehension: effects of visual context on syntactic ambiguity resolution. *Cognitive Psychology*, 45, 447–81.

Spivey M, Tyler M, Richardson D, Young E (2000). Eye movements during comprehension of spoken scene descriptions. In *Proceedings of the 22nd Annual Conference of the Cognitive Science Society* (pp. 487–92). Mahwah, NJ: Erlbaum.

Spivey-Knowlton M, Tanenhaus M, Eberhard K, Sedivy J (1998). Integration of visuospatial and linguistic information in real-time and real-space. In P Olivier, K-P Gapp, Eds. *Representation and Processing of Spatial Expressions* (pp. 201–14). Mahwah, NJ: Erlbaum.

Stanfield RA, Zwaan RA (2001). The effect of implied orientation derived from verbal context on picture recognition. *Psychological Science*, 12, 153–6.

Steyvers M, Shiffrin RM, Nelson DL (2004). Word association spaces for predicting semantic similarity effects in episodic memory. In A Healy, Ed. *Experimental Cognitive Psychology and its Applications* (pp. 237–49). Washington, DC: American Psychological Association.

Tanenhaus MK, Spivey-Knowlton MJ, Eberhard KM, Sedivy JC (1995). Integration of visual and linguistic information in spoken language comprehension. *Science*, 268, 1632–4.

Tseng M, Bergen B (2005). Lexical processing drives motor simulation. In B Bara, L Barsalou, M Bucciarelli, Eds. *Proceedings of the Cognitive Science Society*. Mahwah, NJ: Erlbaum.

van Dijk TA, Kintsch W (1983). *Strategies of Discourse Comprehension*. New York, NY: Academic Press.

Vigliocco G, Vinson DP, Lewis W, Garrett MF (2004). Representing the meanings of object and action words: The featural and unitary semantic space hypothesis. *Cognitive Psychology*, 48, 422-488.

Westermann G, Miranda ER (2004). A new model of sensorimotor coupling in the development of speech. *Brain and Language*, 89, 393–400.

Wiemer-Hastings K, Xu X (2005). Content differences for abstract and concrete concepts. *Cognitive Science*, 29, 719–36.

Wu LL, Barsalou LW (in prep.). Perceptional simulation in property generation.

Wu LL, Barsalou LW (2004). Perceptual simulation in property verification. *Memory and Cognition*, 32, 244–59.

Zwaan RA (2004). The immersed experience: toward an embodied theory of language comprehension. In BH Ross, Ed. *The Psychology of Language and Motivation*, 44. New York, NY: Academic Press.

Zwaan RA, Madden CJ, Yaxley RH, Aveyard ME (2004). Moving words: dynamic mental representations in language comprehension. *Cognitive Science*, 28, 611–19.

Zwaan RA, Stanfield RA, Yaxley RH (2002). Do language comprehenders routinely represent the shapes of objects? *Psychological Science*, 13, 168–71.

Zwaan RA, Yaxley RH (2003). Spatial iconicity affects semantic-relatedness judgements. *Psychonomic Bulletin and Review*, 10, 954–8.

Chapter 16

A well grounded education: The role of perception in science and mathematics

Robert Goldstone, David Landy, and Ji Y Son

16.1 Introduction

One of the most important applications of grounded cognition theories is to science and mathematics education, where the primary goal is to foster knowledge and skills that are widely transportable to new situations. This presents a challenge to those grounded cognition theories that tightly tie knowledge to the specifics of a single situation. In this chapter, we develop a theory of learning that is grounded in perception and interaction, yet also supports transferable knowledge. A first series of studies explores the transfer of complex systems principles across two superficially dissimilar scenarios. The results indicate that students most effectively show transfer by applying previously learned perceptual and interpretational processes to new situations. A second series shows that even when students are solving formal algebra problems, they are greatly influenced by nonsymbolic, perceptual grouping factors. We interpret both results as showing that high-level cognition that might seem to involve purely symbolic reasoning is actually driven by perceptual processes. The educational implication is that instruction in science and mathematics should involve not only teaching abstract rules and equations but also training students to perceive and interact with their world.

Scientific progress has been progressing at a dizzying pace. In contrast, natural human biology changes rather sluggishly and we are using the essentially the same kinds of brains to understand advanced modern science that have been used for millennia. Further exacerbating this tension between the different rates of scientific and neuro-evolutionary progress is that our techniques for teaching mathematics and science are not keeping up with the pace of science (Bialek and Botstein 2004). Politicians, media, and pundits have all expressed frustration with the poor state of mathematics and science education in the United States and worldwide.

There will not be any easy or singular solution to the problem of how to improve mathematics and science education. However, we believe that the consequences of improved science and mathematics education are sufficiently important[1] that it behooves

---

[1] Eric Hunushek, a senior fellow at the Hoover Institution at Stanford University, estimates that improving US students' maths and science grades to the levels of Western Europe within a decade would increase our gross domestic product by 4% in 2025 and by 10% by 2035.

cognitive scientists to apply their state-of-the-science techniques and results to inform-ing the discourse on educational reform. Cognitive scientists are in an uniquely qualified position to provide expert suggestions on knowledge representation, learning, problem solving, and symbolic reasoning. These topics are core to understanding how people utilize mathematical and scientific principles. We also believe that the recent develop-ments in embodied and grounded cognition have direct relevance to mathematics and science education, offering a promising new perspective on what we should be teaching and how students could be learning.

We will describe two separate lines of research on college students' performance on scientific and mathematical reasoning tasks. The first research line studies how students transfer scientific principles governing complex systems across superficially dissimilar domains. The second line studies how people solve algebra problems. Consistent with an embodied perspective on cognition, both lines show strong influences of perception on cognitive acts that are often associated with amodal, symbolic thought, namely cross-domain transfer and mathematical manipulation.

16.2 **Transfer of complex systems principles**

Scientific understanding frequently involves comprehending a system at an abstract rather than superficial level. Biology teachers want their students to understand the genetic mechanisms underlying heredity, not simply how pea plants look. Physics teach-ers want their students to understand fundamental laws of physics such as conservation of energy, not simply how a particular spring uncoils when weighted down (Chi *et al.* 1981). This focus on acquiring abstract principles is well justified. Science often progresses when researchers find deep principles shared by superficially dissimilar phenomena and can describe situations in terms of mathematical or formal abstractions. Finding biological laws that govern the appearance of both snails and humans (Darwin 1859), physical laws that govern both electromagnetic and gravitational acceleration (Einstein 1989), and psychological laws that underlie transfer of learning across species and stimuli (Shepard 1987) are undeniably important enterprises.

Although transcending superficial appearances to extract deep principles has inherent value, it has proven difficult to achieve (Carreher and Schliemann 2002). Considerable research suggests that, in many domains, learners do not spontaneously transfer what they have learnt, at least not to superficially dissimilar domains (Detterman 1993; Gick and Holyoak 1980, 1983). This lack of transfer based on shared deep principles has led to a major theoretical position in the learning sciences called *situated learning*. This community argues that learning takes place in specific contexts, and these contexts are essential to what is learned (Lave 1988; Lave and Wenger 1991). Traditional models of transfer are criticized as treating knowledge as a static property of an individual (Hatano and Greeno 1999), rather than as contextualized or situated, both in a real-world environment and a social community. According to situated learning theorists, one problem of traditional theorizing is that knowledge is viewed as tools for thinking that can be transported from one situation to another because they are independent of the situation in which they are used. In fact, a person's performance on formal tasks is often worse than their performance in more familiar contexts even though, by some analyses, the same abstract tools are required (Nunes 1999; Wason and Shapiro 1971).

Providing a basis for the transfer of principles across domains is a challenge for embodied cognition approaches. Simply put, if cognition is tied to perceiving and inter-acting with particular scenarios, then how can we hope to have transfer from one scenario to another scenario that looks quite different (for an elaboration of this ques-tion, see Sanford, Chapter 10, this volume)? One response, given by several researchers in the situated learning community, is to give up on the prospects of transfer. The very notion of transfer is suspect in that community because of their focus on contextualized knowledge. In one often-reported study, Brazilian children who sell candy may be quite competent at using currency even though they have considerable difficulty solving word problems requiring calculations similar to the ones they use on the street (Nunes 1999). Transfer of mathematical knowledge from candy selling on the streets to formal algebra in the classroom is neither found nor expected by situated learning theorists. So, in order to understand success in the classroom, situated learning theorists place the emphasis on the context of the classroom and to understand success in the streets, the street context is studied.

We share with the situated learning community an emphasis on grounded knowledge. However, we mean something quite different by 'grounding'. For the situated learning community, knowledge is contextualized in the actions, social goals, and physical details of particular concrete scenarios such as selling candy on a Brazilian street. Situated learn-ing theorists equate knowledge with problem solving activities that are cued, in some sense 'grounded', by the features of these concrete scenarios. Generalization is thus a problem-solving behaviour exhibited over a set of scenarios (Greeno 1997). Situated learning theorists typically criticize cognitive theorists for abstracting away from these scenarios to explain generalization. However, for us, knowledge is not simply in the extracted verbal or formal description of a situation, but rather in the perceptual inter-pretations and motoric interactions involving a concrete scenario. While we look to embodied experiences to ground learning, we still believe in the possibility and power of transfer across contexts. It is possible to learn principles in a grounded way that enables the principles to be recognized in the myriad of concrete forms that they can take.

Typically, cognitive theories of transfer are expressed in terms of the acquisition of abstracted formalisms that lead to the direct perception of mathematical structures and patterns. Instead, we believe that learning that transfers to new scenarios and transports across domains, most often proceeding not through acquiring and applying symbolic formalisms but rather through modifying automatically perceived similarities between scenarios by training one's perceptual interpretations. In the following two sections we will: (1) argue for the desirability of cross-domain transfer of scientific principles, and (2) show how this transfer is compatible with a perceptually grounded understanding of science.

16.2.1 Complex systems and transfer

Our claim for the desirability and possibility of transfer of scientific principles across domains is based on the power of *complex systems theories*. A complex systems perspective provides a unifying force, bringing increasingly fragmented scientific communities. Journals, conferences, and academic departmental structures are becoming increasingly specialized and myopic (Csermely, 1999). One possible response to this fragmentation of science is to simply view it as inevitable. Horgan (1996) argues that the age of fundamental scientific theorizing and discoveries has passed, and that all that is left to be done is refining the details of theories already laid down by the likes of Einstein, Darwin, and Newton.

Complex systems researchers offer an alternative to increasing specialization. They have pursued principles that apply to many scientific domains, from physics to biology to social sciences. For example, reaction–diffusion equations that explain how cheetahs develop spots can be used to account for geospatial patterns of Democrats and Republicans in America. The same abstract schema underlies both phenomena – two kinds of elements (e.g. two skin cell colours, or two political parties) both diffuse outwards to neighbouring regions but also inhibit one another. The process of *diffusion-limited aggregation* is another complex adaptive systems explanation that unifies diverse phenomena: individual elements enter a system at different points, moving randomly. If a moving element touches another element, they become attached. The emergent result is fractally connected branching aggregates that have almost identical statistical properties (Ball 1999). This process has been implicated in the growth of human lungs (Garcia-Ruiz *et al.* 1993), snowflakes (Bentley and Humphreys 1962), and cities (Batty, 2005). These examples all describe principles that can be instantiated with highly dissimilar sets of individual elements, but with interactions between the elements that are captured by very similar algorithmic rule sets (Bar-Yam 1997).

Generally speaking, complex systems are systems made up of many units (often called agents), whose simple interactions give rise to higher-order emergent behaviour. Typically, the units all obey the same simple rules, but because they interact the units that start off homogenous and undifferentiated may become specialized and individualized (O'Reilly 2001). Despite the lack of a centralized control, leader, recipe, or instruction set, these systems naturally self-organize (Resnick 1994; Resnick and Wilensky 1993). Many real-world phenomena can be explained by the formalisms of complex adaptive systems, including the foraging behaviour of ants, the development of the human nervous system, the growth of cities, growth in the internet, the perception of apparent motion, mammalian skin patterns, pine cone seed configurations, and the shape of shells (Casti 1994; Flake 1998).

In the following studies, we will focus on one particular complex systems principle, *competitive specialization*, because we have taught this principle in our own undergraduate courses, and because we have conducted controlled laboratory studies on students' appreciation and use of the principle (Goldstone and Sakamoto 2003; Goldstone and Son 2005). A well-worked out example of competitive specialization is the development of neurons in the primary visual cortex that start off homogeneous and become specialized to respond to visually presented lines with specific spatial orientations (von der Malsburg 1973). Another application is the optimal allocation of agents to specialize to different regions in order to cover a territory. In these situations, a good solution is found if every region has an agent reasonably close to it. For example, an oil company may desire to place oil drills such that they are well spaced and cover their territory. If the oil drills are too close, they will redundantly access the same oil deposit. If the oil drills do not cover the entire territory, then some oil reserves will not be used.

Ants and food

The first example of competitive specialization involves ants foraging food resources drawn by a user. The ants follow exactly three rules of competitive specialization: at each time step, (1) a piece (pixel) of food is randomly selected from all of the food drawn by a user, (2) the ant closest to the piece approaches the food at one rate, and (3) all of the other ants move at another rate. In interacting with the simulation, a learner can reset the ants' positions, clear the screen of food, draw new food, place new ants, move ants, start/stop the ants' movements, and set a number of simulation parameters. The two most critical user-controlled parameters determine the movement speed for the ant that is closest to the selected food (called 'closest rate' in Figure 16.1) and the movement speed for all other ants ('Not closest rate').

Starting with the initial configuration of three ants and three food piles shown in Figure 16.1, several important types of final configuration are possible and are shown in Figure 16.2. If only the closest ant moves toward a selected piece of food, then this ant will be the closest ant to *every* patch of food. This ant will continually move to new locations on every time step as different patches are sampled, but will tend to hover around the center-of-mass of the food patches. The other two ants will never move at all because they will never be the closest ant to a food patch. This configuration is suboptimal because the average distance between a food patch and the closest ant (a quantity that is continually graphed) is not as small as it would be if each of the ants specialized for a different food pile. On the other hand, if all of the ants move equally quickly, then they will quickly converge to the same screen location. This also results in a suboptimal solution because the ants do not cover the entire set of resources well – there will be patches that do not have any nearby ant. Finally, if the closest ant moves more quickly than the other ants but the other ants move too, then a nearly optimal configuration is achieved. Although one ant will initially move more quickly toward all selected food patches than the other ants, eventually it will specialize to one patch and the other ants will then be closer to the other patches, allowing for specialization.

An important, subtle aspect of this simulation is that poor patterns of resource covering are self-correcting so the ants will almost always self-organize themselves in a one-to-one relationship to the resources regardless of the lopsidedness of their original arrangement, provided good parameter values are used.

Pattern learning

The second example of competitive specialization involves sensors responding to patterns drawn by the user. Just like the ants/food scenario, the pattern-learning simulation

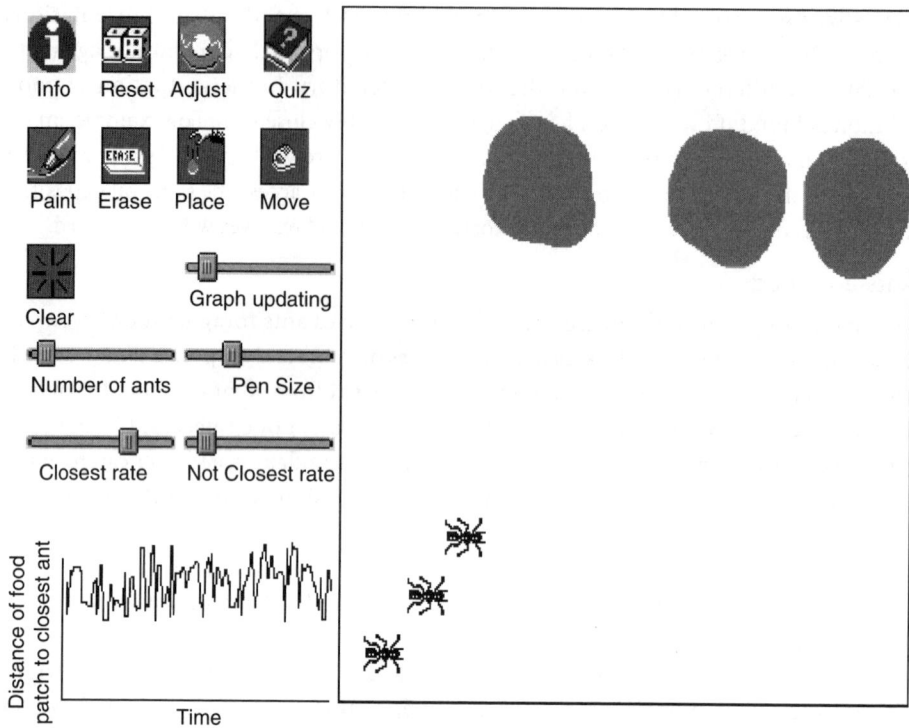

Fig. 16.1 A screen shot of an initial configuration for the 'ants and food' computer simulation. At each time step, a patch of food is randomly selected, and the ant closest to the patch moves toward the patch with one speed (specified by the slider 'Closest rate') and the other ants move toward the patch with another speed ('Not closest rate').

follows the same three rules of competitive specialization. At the beginning of the simulation, the sensors respond to random noise. But at each time step, a pattern is randomly selected from all of the patterns drawn by a user, and the sensor most similar to that pattern adapts to become more similar to that pattern at a particular rate. All of the other sensors adapt towards the selected pattern at another rate. Users can reset sensors to become random again, draw patterns, erase patterns, copy patterns, add noise, start/stop pattern learning, and change a number of parameters. The most important parameters are the rates of adaptation for the most similar and not most similar sensors (Figure 16.3). Note that the final configurations shown in Figure 16.2 also apply to the pattern-learning simulation.

Using these two case studies of the competitive specialization principle, we are now in a good position to state our challenge for grounded educational practices. Our desiderata is a teaching method that will promote transfer from one example of competitive specialization to another. How can students who learn about competitive specialization with the ants and food scenario spontaneously apply what they have learned to the pattern-learning situation? Given the lack of superficial perceptual features shared by

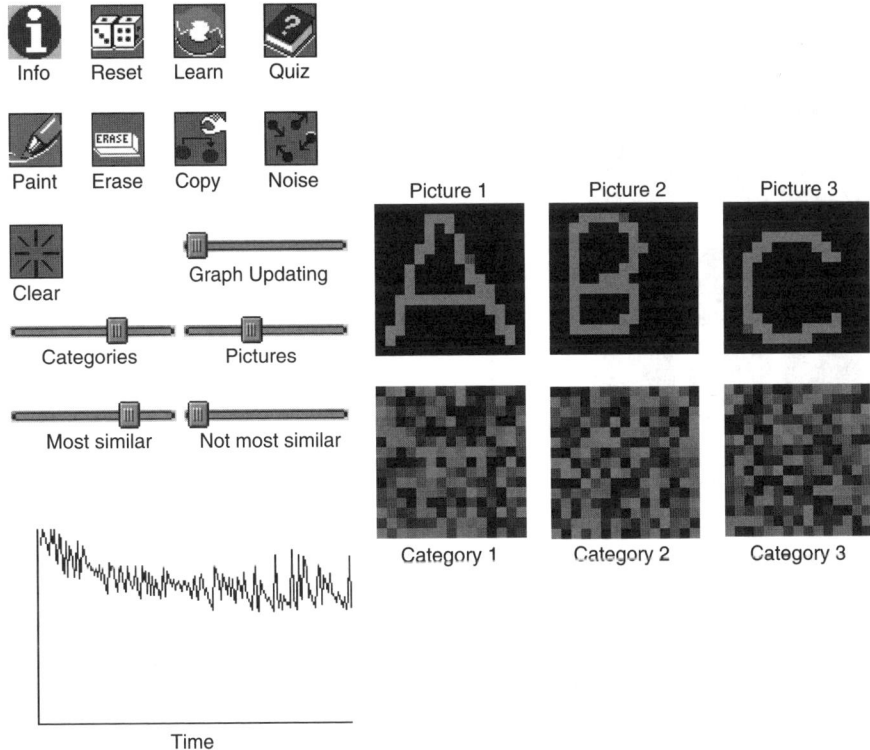

Fig. 16.2 A screen shot of the 'pattern learning' simulation. Users draw pictures, and prior to learning, a set of categories are given random appearances. During learning, a picture is selected at random, and the most similar category to the picture adapts its appearance toward the picture at one rate (specified by the slider 'Most similar') while the other categories adapt toward the picture at another rate ('Not most similar').

Figures 16.1 and 16.2, can the transfer be due to cognitive processes grounded in perception and interaction? The next section sketches out our affirmative answer.

16.2.2 How to teach complex systems principles

Powerful complex systems principles cut across scientific disciplines. We are therefore inherently interested in *transportable* knowledge – knowledge that can be applied to domains significantly beyond those originally presented. In setting the stage for our grounded approach to cognition, we will first describe a traditional alternative approach to transfer.

Transfer via formalisms

It is understandable why so many educators have been drawn to couch their textbook and classroom teaching in terms of formalisms. The formalizations established by algebra, set theory, and logic are powerful because they are domain-general. The same

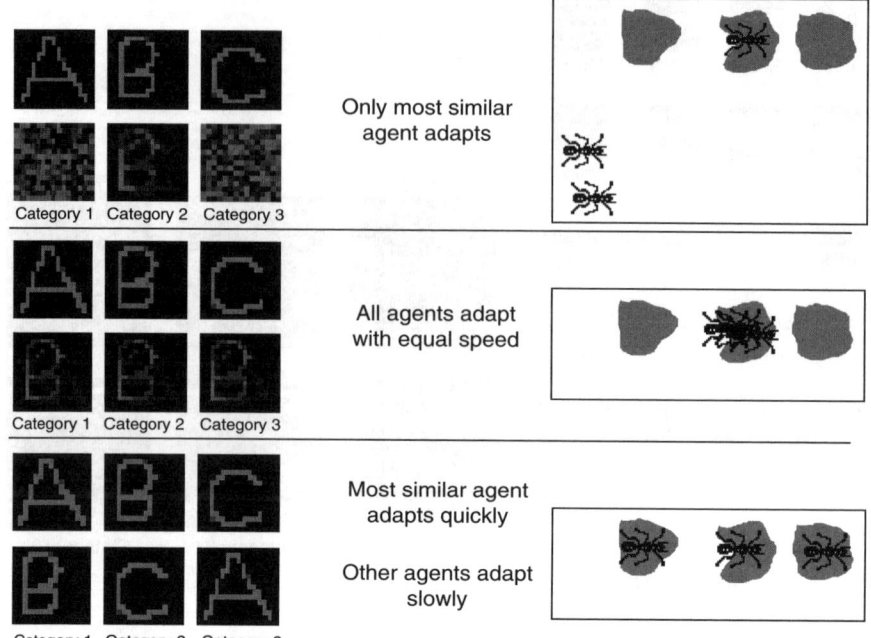

Fig. 16.3 The basis for the isomorphism between the ants-and-food and pattern-learning simulations. If only the most similar agent to a resource adapts, then often a single agent will move toward the average of all of the resources. If all agents adapt equally quickly, they will all move toward the average position. If the agent closest to a resource patch moves much faster than the other agents but all agents move a bit, then each of the agents typically becomes specialized for one resource type.

equation for probability can apply equally well to shoes, ships, sealing wax, cabbages, and kings. Not only is no 'customization' of equations to a content domain required, it is forbidden. The formalism sanctions certain transformations that are provably valid, and once a domain has been translated to the formalism one can be assured that the deductions drawn from the transformations will be valid. Logic and maths are the best examples that we can think of for the kind of symbolic processing that Glenberg *et al.* (Chapter 1, this volume) have in mind when they describe symbols as abstract, amodal, and manipulated by using explicit rules. This is not to imply that we believe that engaging in mathematics is a purely symbolic activity. Crucially, we do not, but it is the best candidate domain for symbolic reasoning because the rules for mathematics are relatively noncontentious compared with, for example, the rules for language. Furthermore, students receive a significant amount of training with the generic, abstract form of mathematical rules and explicitly learn how to transform symbolic representations.

Mathematical formalisms are the epitome of representations that abstract over domains. Given this, there is some reason to believe that they will promote free transfer of knowledge from one domain to another. This notion is endemic in high-school mathematics curricula, which often feature abstract formalisms that, once presented, are only subsequently fleshed out by examples. Systematic analyses of mathematics textbooks have shown that formalisms tend to be presented before worked-out examples and that this tendency increases with grade level (Nathan *et al.* 2002).

Although formalisms may seem plausible candidates for producing transportable knowledge, the literatures from cognitive science and education give three grounds for skepticism. First, the mismatch between how mathematics is presented and how it is conceived by practitioners has often been noted (Lakoff and Nuñez 2000; Thurston, 1994; Wilensky 1991). As teachers' mathematical expertise increases, they increasingly believe that using formalisms is a requirement for solving mathematical problems, even though this kind of strong association is not found in actual students' performance (Nathan and Petrosino 2003). Hadamard has complained that the true heart of a mathematical proof, the intuitive conceptualization, is ignored in the formal description of the proof steps themselves (Hadamard 1949). Scholarly articles contain the step-by-step, formally sanctioned methods, but if one wishes to understand where the idea for these steps comes from, then one must attempt to generate the visuospatial inspiration oneself, without much insight from the published report. In mathematics textbooks, formalisms are either treated as givens, or when they are derived, they tend to be formally derived from other formalisms.

The second reason for skepticism is that formalism-based transfer can only work if people spontaneously recognize when an equation can be applied to a situation. In fact, the connection between equations and scenarios is typically indirect and difficult to see (Chi 2005; Hmelo-Silver and Pfeffer 2004; Nathan *et al.* 1992; Penner 2000). Students often have difficulty finding the right equation to fit a scenario, even when they know both the equation and the major elements of the scenario (Ross 1987, 1989).

The third reason for skepticism is that transfer potential is almost certainly not maximized by creating the most efficient minimalist representation possible, one that leaves out irrelevant content information and captures only relevant structural information. Formalisms are maximally content-independent, but this is the precise property that often leads them to be cognitively inert. They offer little by way of scaffolding for understanding, and may not generalize well because cues to resemblance between situations have been stripped away. Research has shown that the content-dependent semantics of a scenario promotes transfer, and also gives valuable clues about the kinds of formalism that are likely to be useful (Bassok 1996, 2001; Bassok *et al.* 1998).

Grounded transfer

We would like to propose an alternative, intermediate position to the extreme positions that transfer should proceed via formalisms and that remote transfer cannot be achieved at all (Detterman 1993). Our position is that transfer can be achieved, not by teaching students abstract formalisms but through the interpretation of grounded situations. One motivation for this method comes from work with students solving story problems with and without access to physical manipulatives that can be used to act out the story (Glenberg *et al.* 2004). The physical manipulatives helped students to understand and solve the problems, and these benefits transferred to conditions in which students simply imagined using the physical objects (Glenberg *et al.*, 2004). The process of interpreting actual or imagined objects was instrumental in getting students to properly understand the underlying mathematical issues involved in a situation.

Another motivation for this method comes from our observation of students learning principles of complex systems in our classes and laboratory experiments. We have observed that students often interact with our pedagogical simulations by actively interpreting the elements and their interactions. Their interpretations are grounded in the particular simulation with which they are interacting. Furthermore, because the interpretations may be highly selective, perspectival, and idealized, the same interpretation can be given to two apparently dissimilar situations. The process of interpreting physical situations can therefore provide understandings that are grounded yet transportable. Practicing what we preach, we will now ground our notion of interpretive generalization with the competitive specialization principle using an example from our 'complex adaptive systems' undergraduate course. The relevant simulations can be accessed at: http://cognitrn.psych.indiana.edu/rgoldsto/complex/.

Our laboratory and classroom investigations with these two demonstrations of competitive specialization have shown that students can, under some circumstances, transfer what they learn from one simulation to another (Goldstone and Sakamoto 2003; Goldstone and Son 2005). In our experiments, we first gave students a period of focused exploration with the ants-and-food simulation because it embodies competitive specialization in a relatively literal and spatial manner, then we let students explore the pattern-learning simulation. We probe their understanding of the latter simulation both through multiple-choice questions designed to measure their appreciation of the principle of competitive specialization in the pattern learning context, and through a performance-based measure of how quickly students can create parameter settings whereby categories automatically adapt so as to represent the major classes of input patterns.

Using this method, we found that students show better understanding of the pattern-learning simulation when it has been preceded by the ants-and-food simulation than by a simulation governed by a different principle. Student interviews indicate that a major cause of the positive transfer is training perceptual interpretations of grounded situations. After exposure to ants-and-food, several students spontaneously applied the same perceptual representation to pattern learning. The visuospatial dynamics of the ants-and-food simulation are aptly applied. In fact, when originally presenting their general competitive learning algorithm, Rumelhart and Zipser use a visualization along the lines of ants-and-food to give the reader a solid intuition for how their algorithm works. This visualization, shown in Figure 16.4, depicts the adaptation of categories in a high-dimensional space as the movement of those categories in three-dimensional space. Our students' spontaneous visualizations have elements in common with this professional visualization.

Typical elements in our students' visualizations are shown in Figure 16.5. A student was asked to visually describe what would happen when there are two categories and four input pictures that fell into two clusters: variants of As and variants of Bs. The student

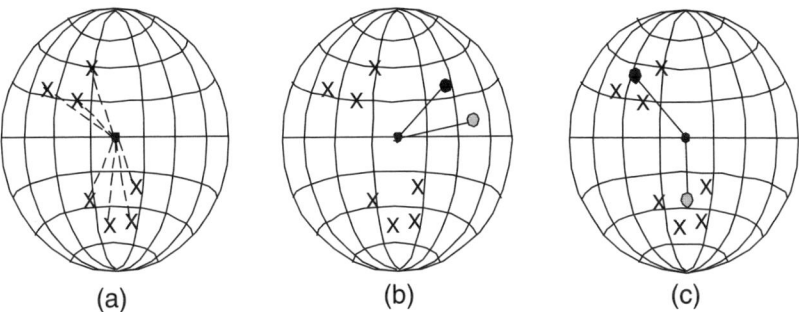

(a)   (b)   (c)

Fig. 16.4 A standard visualization of competitive specialization, adapted from an illustration by Rumelhart and Zipser (1985). Panel A shows the coordinates of seven objects, represented by 'x's, in a three-dimensional space. In panel B, the random starting coordinates for two categories are shown by shaded circles. Panel C shows the resulting coordinates for the categories after several iterations of learning by competitive specialization.

drew the illustration in panel A. In this illustration, adaptation and similarity are both represented in terms of space, much as they are in Figure 16.4. The two categories ('Cat1' and 'Cat2') are depicted as moving spatially toward the spatially-defined clusters of 'A's and 'B's, and the 'A' and 'B' pictures are represented as spatially separated. The context for panel B was a student who was asked what problem might occur if the most similar detector to a selected picture moved quickly to the picture, while the other detectors did not move at all. The student showed a single category (shown as the box with a '1') moving toward, and eventually oscillating between, the two pictures. Again, similarity is represented by proximity and adaptation by movement.

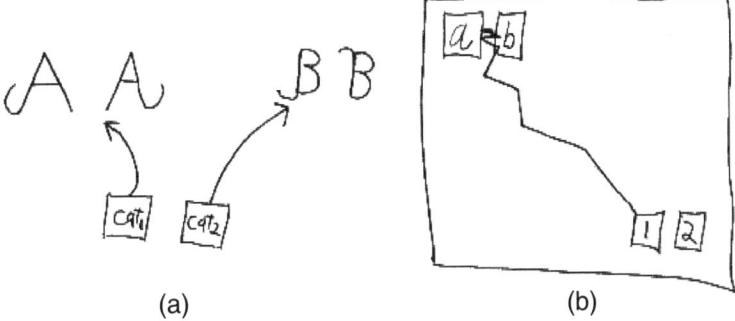

(a)   (b)

Fig. 16.5 Typical visualizations of students expressing their knowledge of pattern learning. In panel A, the student represents the adaptation of categories by moving them through space towards two clusters of stimuli. The similarity of the two variants of 'A' is represented by their spatial proximity. In panel B, a student was asked to illustrate a problem that arises when the closest category to a pattern adapts, but the others do not adapt at all: a single category is shown oscillating between two patterns. Note the similarity to the top panel of Figure 16.3 with the ants-and-food simulation.

Students' verbal descriptions also give evidence of the spatial diagrams that they use to explicate pattern learning. Students frequently talk about a category 'moving over to a clump of similar pictures.' Another student says that 'this category is being pulled in two directions, toward each of these pictures.' A third student also uses spatial language when saying, 'These two squares are close to each other, so they will tend to attract the same category to them.' In this final example, the two square pictures were not physically close to each other on the screen, they were separated by a picture of a circle.

All three of these reports show that students are understanding visual similarity in terms of spatial proximity. The squares are close in terms of their visual appearance and not their spatial distance. Categories do not change their position on the screen, but only in a high-dimensional space that describes pictures. Ants move in visual space, while categories adapt in description space. In doing so, both ants and categories are adapting so that they cover their resources (food or pictures) well. Our students frequently find making the connections between adaptation and motion, and between similarity and proximity to be natural, and much more so after they have had experience with the ants-and-food simulation.

How should we understand these connections? One common approach is to claim that ants-and-food uses space literally, while students' understanding of pattern learning treats space metaphorically or figuratively. Movement in space is viewed as an apt analogy for adaptation. The problem with describing space as figurative in pattern learning is that space is very concrete and grounded for students who draw diagrams like those shown in Figure 16.5. Our alternative proposal is that students are using the literal, spatial models that they learned while exploring ants-and-food to understand and predict behaviour in pattern learning. There is a process where visual similarity is thought of in terms of spatial proximity, but once that mapping has been made, students conduct the same kinds of mental simulations that they perform when predicting what will happen in new ants-and-food situations. So, we do not believe that students abstract a formal structural description that unifies the two simulations. Instead, they simply apply to a new domain the same perceptual routines that they have previously acquired. By one account, the domains are visually dissimilar. However, once appropriately construed, the domains are highly similar visually. In fact, students seem to use superficial spatial properties to construe the ants-and-food simulation.

An interesting aspect of both illustrations in Figure 16.5 is that although space is being used to represent similarity, there are still vestiges of space being used to represent space. In these, and other, student illustrations, the categories are placed below the input pictures, just as they are in the interface shown in Figure 16.2. Even though students have difficulty explicitly describing the general principle that governs both simulations, the perceptual trace left by ants-and-food can prime an effective perceptual construal of pattern learning.

What is most striking about the students' descriptions of pattern learning is the extent of knowledge-driven perceptual interpretation. These interpretations are not simply formalisms grafted onto situations. Rather, the interpretations affect the perception of the simulation elements. Students see single categories trying to 'cover' multiple pictures, pictures being 'close' or 'far' in terms of their appearances, and categories 'moving' toward pictures. To the experienced eye, identical perceptual configurations are visible in the two simulations. Figure 16.3 shows equivalent configurations. The critical point is that these pairs are obviously not perceptually similar under all construals. It is only to the student who has understood the principle of competitive specialization as applied to ants-and-food that the two situations appear perceptually similar. One moral is: just because a reminding is perceptually-based does not make it necessarily superficial. A superficial construal of the left and right columns of Figure 16.3 would not reveal much commonality. However, a dynamic, visual, and spatial construal of pattern learning can be formed that makes the solutions of ants-and-food directly relevant.

In opposition to the traditional schism between perceptual and conceptual processes, our work on transferring complex systems principles suggests that the perceptual interpretation process is key to generating transportable conceptual understandings. Perceptual interpretation requires both a physically present situation and construals of the elements of the situation and their interactions. Simply giving the interpretation is not adequate. Like equations, standalone descriptions are unlikely to foster transfer because of their lack of contact to applicable situations. When we simply give students the rules of pattern learning, they are seldom reminded of ants-and-food. Physically grounding a description is one of the most effective ways of assuring that it is conceptually meaningful. However, giving the grounded situation but no interpretation is inadequate. It is only when the interpretation is added to the presentation of two superficially dissimilar situations that the resemblance becomes apparent. Students are not just seeing events, they are seeing events as instantiating principles. This act of interpretation, an act of 'seeing something as X' rather than simply seeing it (Wittgenstein 1953), is the key to cultivating transportable knowledge.

16.3 Perceptual learning

An important plank of our proposal is that the similarity between situations governed by the same complex systems principle can be used to promote transfer even if the situations are dissimilar to the untutored eye, and even if the similarity is not explicitly noticed. This claim apparently contradicts the empirical evidence for very limited transfer between remote situations (Detterman 1993; Reed *et al.* 1974, but see also Barnett and Ceci 2002 for a balanced evaluation of the evidence). Our claim is that the perceived similarity of situations is malleable, not fixed by objective properties of the situations themselves. It may well be that remotely related situations rarely facilitate each other. However, well designed activities can alter the perceived similarity of situations, and what were once dissimilar situations can become similar to one another with learning (Goldstone 1998; Goldstone and Barsalou 1998). Our hope, then, is not to have students transfer by connecting remotely related situations, but rather to have students warp their psychological space so that formerly remote situations become similar (Harnad *et al.* 1995).

There is already good evidence for this kind of warping of perception due to experience and task requirements (Goldstone 1994; Goldstone *et al*. 2001; Livingston *et al*. 1998; Özgen 2004; Roberson *et al*. 2000).

Graphical, interactive computer simulations (Goldstone and Son 2005; Jacobson 2001; Jacobson and Wilensky 2005; Resnick 1994; Wilensky and Reisman 1999, 2006) offer attractive opportunities for promoting generalization. Principles are not couched in equations, but rather in dynamic interactions among elements (Nathan, Chapter 18, this volume; Nathan *et al*. 1992). Students who interact with the simulations actively interpret the resulting patterns, particularly if guided by goals abetted by knowledge of the principle. Their interpretations are grounded in the particular simulation, but once a student has practiced building an interpretation, it is more likely to be used for future situations. In contrast to explicit equation-based transfer, perceptually-based priming is automatic. For example, an ambiguous man/rat drawing is automatically interpreted as a man when preceded by a man and as a rat when preceded by a rat (Leeper 1935). This phenomenon, replicated in countless subsequent experiments on priming (see Goldstone 2003 for a theoretical integration), is not ordinarily thought of as transfer, but it is an example of a powerful influence on perception due to prior experiences. This kind of automatic shift of perceptual interpretation accompanies engaged interaction with complex system simulations. In these cases, generalization arises, not just from the explicit and effortful application of abstract formalisms, but crtically from the simple act of 'rigging up' a perceptual system to interpret a situation according to a principle, and leaving this rigging in place for subsequently encountered situations.

16.3.1 Perception and idealization

In arguing for an embodied basis for transfer complex systems principles, it is important to clarify what we mean to entail by 'embodied'. To us, embodiment is compatible with idealization. Following Glenberg *et al*. (Chapter 1, this volume), we consider our students' understanding of complex systems principles to be embodied when it depends on activity in systems used for perception and action. Both our computer simulations, and the mental simulations of those simulations are perceptual in that they incorporate spatial and temporal information, and presumably do so by using brain regions that are dedicated to perceptual processing. Arnheim (1970), Barsalou (1999; Barsalou *et al*. 2003; Simmons and Barsalou 2003), Glenberg (Chapter 2, this volume; Glenberg *et al*., 2004), Schwartz (Schwartz and Black 1999), Roy (Chapter 11, this volume) and others have argued that our concepts are not amodal and abstract symbolic representations, but rather are grounded in the external world via our perceptual systems.

Embodied knowledge has a major advantage over amodal representations – they preserve aspects of the external world in a direct manner so that the mental simulations and the simulated world automatically stay coordinated even without explicit machinery to assure correspondence. Dimensions in the model naturally correspond to dimensions of the modelled world. Given this characterization, it is clear that embodied representations neither need to superficially resemble the modelled world nor preserve all of the raw, detailed information of that world (Barsalou 1999; Shepard 1984). Our experience with students' understanding of complex systems computer simulations indicates that simulations lead to the best transfer when they are relatively idealized. Goldstone and Sakamoto (2003) gave students experience with two simulations exemplifying the principle of competitive specialization – ants-and-food, followed by pattern learning. Figure 16.6 shows the design for the experiments. Ants-and-food was either presented using line drawings of ants and food, or simplified geometric forms. Overall, students showed greater transfer to the second scenario when the elements were graphically idealized rather than realistic. Interestingly, the benefit of idealized graphical elements was largest for our students who had relatively poor understanding of the initial simulation. It might be thought that strong contextualization and realism would be of benefit to those students with weak comprehension of the abstract principle. Instead, it seems that these poor comprehenders are particularly at risk for interpreting situations at a superficial level, and using realistic elements only encourages this tendency. Son *et al.* (in press)

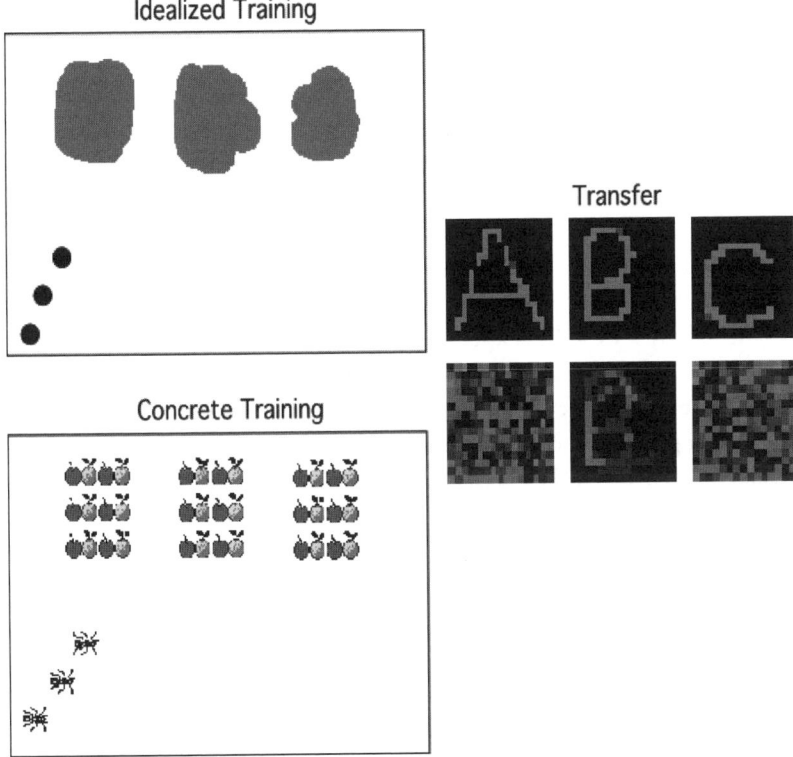

Fig. 16.6 The design of an experiment how concrete visual simulations should be to promote transfer (Goldstone and Sakamoto, 2003). Better transfer performance was found for the idealized, relative to concrete, training condition, particularly for students with poor comprehension of the training simulation.

have found consistent results with rich versus simple geometric objects, and Sloutsky *et al.* (2005) have found similar benefits of idealized over concrete symbols for learning about algebraic groups (see also Wiley 2003 for a review of the benefits and hazards of concretization).

Judy DeLoache's *model room paradigm* (DeLoache 1991, 1995; DeLoache and Burns 1994; DeLoache and Marzolf 1992) has also found that idealization can promote symbolic understanding. A child around the age of 2.5 years is shown a model of a room and watches as a miniature toy is hidden behind or under a miniature item of furniture in the model. The child is told that a larger version of the toy is hidden at the corresponding piece of furniture in the room. Children were better able to use the model to find the toy in the actual room when the model was a two-dimensional picture rather than a three-dimensional scale model (DeLoache 1991; DeLoache and Marzolf 1992). Placing the scale model behind a window also allowed children to more effectively use it as a model (Deloache 2000). DeLoache and colleagues (DeLoache 1995; Uttal *et al.* 1997, 1999) explain these results in terms of the difficulty in understanding an object as both a concrete, physical thing and as a symbol standing for something else.

A natural question to ask is, 'How can we tell whether a particular perceptual detail will be beneficial because it provides grounding or detrimental because it distracts students from appreciating the underlying principle?' The answer depends on the nature of complex systems principles, the learner's cognitive development, and what is easily implemented in the mental simulations. Complex systems models are typically characterized by simple, similarly configured elements that each follow the same rules of interaction. For this reason, idiosyncratic element details can often be eliminated, and information about any element only need be included to the extent that it affects its interactions with other elements. Mental simulations are efficient at representing spatial and temporal information, but are highly capacity-limited (Hegarty 2004a). Under the assumption that a student's mental model will be shaped by the computational model that informs it, the following prescriptions are suggested for building computer simulations of complex systems: (1) eliminate irrelevant variation in elements' appearances, (2) incorporate spatial–temporal properties, (3) do not incorporate realism just because it is technologically possible, (4) strive to make the element interactions visually salient, and (5) be sensitive to peoples' capacity limits in tracking several rich, multifaceted objects.

Another empirically supported suggestion for compromising between grounded and idealized presentations is to begin with relatively rich, detailed representations, and gradually idealize them over time (Goldstone and Son 2005). This regime of 'concreteness fading' was proposed as a promising pedagogical method because it allows simulation elements to be both intuitively connected to their intended interpretations, but also eventually freed from their initial context in a manner that promotes transfer.

16.3.2 Perceptual grounding in mathematics

In the previous section, we contrasted the benefits of presenting scientific principles using perceptually grounded and interactive simulations, rather than traditional algebraic equations. One reason why complex system computer simulations are so pedagogically effective is that they mesh well with human mental models (Gentner and Stevens 1983; Graesser and Jackson, Chapter 3, this volume; Graesser and Olde 2003; Hegarty 2004b). These computer models enact the same kind of step-by-step simulation of elements and interactions that effective mental models do. However, even though we advocate the perceptual grounding provided by simulations for pedagogical use, our position is not that algebraic equations are ungrounded or that they somehow transcend perception. In fact, some of our recent experiments indicate surprisingly strong influences of apparently superficial perceptual grouping factors on people's algebraic reasoning.

To study the influence of perceptual grouping on mathematics, we gave undergraduate participants a task to judge whether an algebraic equality was necessarily true (Landy and Goldstone, 2007). The equalities were designed to test their ability to apply the order of precedence of operations rules. Our participants have learned the rule that multiplication precedes addition. Our instructions and post-experiment interviews confirm that participants know this rule. However, we were interested in whether perceptual and form-based groupings would be able to override their general knowledge of the order of precedence rules. We tested this by having grouping factors either consistent or inconsistent with order of precedence. For example, if shown the stimuli in the top row of Figure 16.7, participants would be asked to judge whether $R * E + L * W$ is necessarily equal to $L * W + R * E$. These terms are necessarily equal, and so the correct response would be 'yes' for this trial. On some trials, the physical spacing between the operators was consistent with the order of precedence – large spacings separated the terms related by addition and small spacings separated terms related by multiplication. A powerful principle of Gestalt perception (Koffka 1935) is that close objects are seen as grouping together. Accordingly, small spaces between terms to be multiplied together would be expected to facilitate grouping them together early, consistent with the order of precedence. On other trials, the physical spacing was inconsistent with the order of precedence, with larger spaces around '*'s than '+'s.

The influence of spacing was profound, as shown in Figure 16.8. When physical spacing was inconsistent with order of precedence rules, six times as many errors were made relative to when the spacing was consistent. However, for problem types where the order of operations did not influence the validity of the equation ('insensitive trials') accuracy was relatively spared, indicating that the deployment of order of operations knowledge was selectively affected by the grouping pressures. Several aspects of the experiment make this influence of perception on algebra striking. First, they demonstrate a genuine cognitive illusion in the domain of mathematics. The criteria for cognitive illusions in reasoning are that people systematically show an influence of a factor in reasoning. The factor should normatively not be used, and people agree, when debriefed, that they were wrong to use the factor (Tversky and Kahneman 1974). The second impressive aspect of the results is that participants continued to show large influences of grouping on equation verification even though they received trial-by-trial feedback.

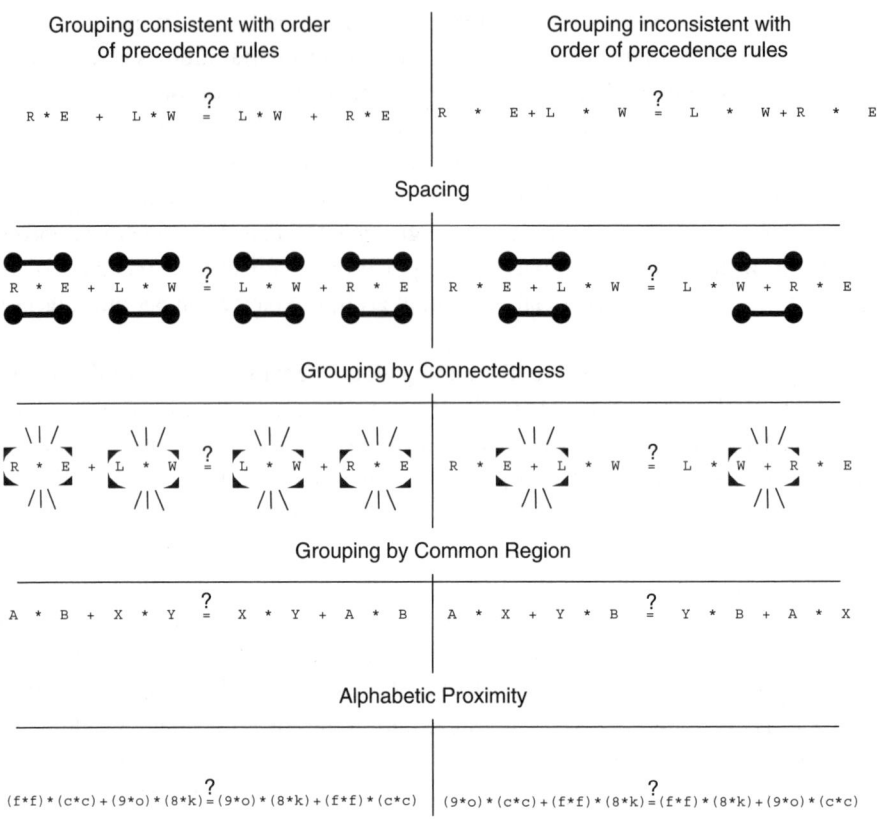

Fig. 16.7 Samples from five experiments reported by Landy and Goldstone (2007). Participants were asked to verify whether an equation is necessarily true. Grouping suggested by factors such as physical spacing, regions suggested by geometric forms, proximity in the alphabet, and functional form similarity could be either consistent or inconsistent with the order of precedence of arithmetical operators (e.g. multiplications are calculated before additions). The above equalities are all true, but participants make far more errors when the perceptual and form-based groupings are inconsistent rather than consistent.

Constant feedback did not eliminate the influence of the perceptual cues. This suggests that sensitivity to grouping is automatic or at least resistant to strategic, feedback-dependent control processes. The third impressive aspect of the results is that an influence of grouping is found in mathematical reasoning. Mathematical reasoning is often taken as a paradigmatic case of purely symbolic reasoning, moreso even than language which, in its spoken form, is produced and comprehended before children even have formal operations (Inhelder and Piaget 1958). Algebra is, according to many people's intuition, the clearest case of widespread symbolic reasoning in all human cognition. Showing that perceptual factors influence even algebraic reasoning provides

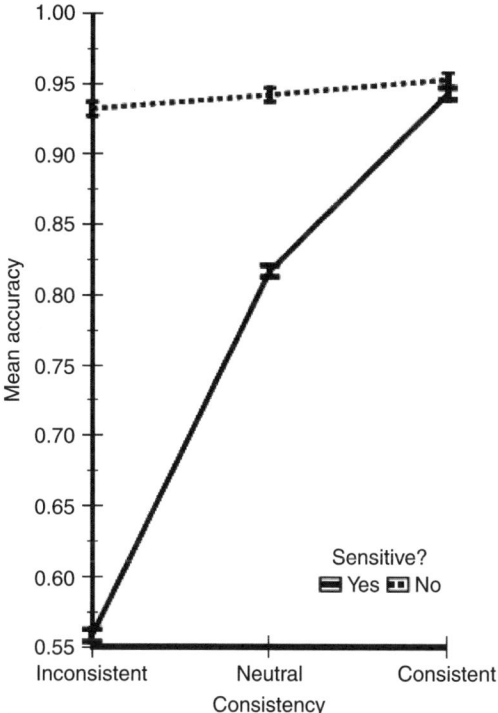

Fig. 16.8 Results from a manipulation of physical spacing in experiment 1 of Landy and Goldstone (2007). When spacing was inconsistent (right side of Figure 16.7) with the order of precedence, the error rate increased more than six-fold compared with when spacing was consistent. This effect was only found when equations were sensitive to physical spacing – trials where different consistent versus inconsistent answers would be given to the question, 'Are the two sides of this equation necessarily equal?' if a participant were influenced by spacing.

prima facie support for the premise that grounding cannot be ignored for *any* cognitive task.

Our further experiments have extended this observed influence of grouping factors on algebraic reasoning. The second and third rows of Figure 16.7 show manipulations of other Gestalt laws of perceptual organization. According to the principle of connectedness (Palmer and Rock 1994), objects that are physically connected to one another have a tendency to be grouped together. Connecting the circles in Row 2 of Figure 16.7 causes them to be grouped together, and experimental results show that when they are grouped together, they tend to group the mathematical symbols with which they are vertically aligned. The third row shows that this grouping pressure does not require physical connection, but rather can also be formed through implied common region.

In both of these examples, even though participants know that the geometric forms are irrelevant to their mathematical task, when the geometric forms are consistent with order of operations, far better algebraic performance results than if the forms are inconsistent. The fourth row of Figure 16.7 shows that alphabetic proximity has an analogous effect to physical proximity. Participants apparently have a tendency to group letters that are close in the alphabet. If close letters, like 'A' and 'B,' are related by an addition operator, this hinders algebraic performance; if the close letters are related by a multiplication operator, this facilitates performance. This influence of alphabetic proximity on mathematical reasoning is surprising because by standard formal accounts all variable names are equivalent and participants, in their explicit description of the task, endorse the statement that the letters used to designate variables are irrelevant. Finally, the last row shows an influence of the functional form of the terms being multiplied and added. The term '$(c * c)$' has a form-based similarity to '$(f * f)$' – they both involve a variable being multiplied by itself. This similarity is sufficient to bias participants to group these terms together. Again, this form-based group helps performance if the terms are related by multiplication, but hinders performance if they are related by addition.

The cumulative weight of these results indicates that even the highly symbolic activity of algebraic calculation is strongly affected by perceptual grouping. Most accounts of the importance of notation in mathematics, and indeed of symbols generally, hinge on their abstract character. Since symbols are largely considered amodal, symbolic reasoning is also easily assumed to be amodal; that is, symbolic reasoning is supposed to depend on internal structural rules which do not relate to explicit external forms. In contrast, the mathematical groupings that our participants create are heavily influenced by groupings based on perception and superficial similarities. The theoretical upshot of this work is to question the assumption that cognition operates generally like systems of algebra or mathematical logic. By traditional symbolic manipulation accounts, to cognize is to apply laws to structured strings where those laws generalize to the shape of the symbols, and that shape is arbitrarily related to the symbols' content (Fodor 1992).

The laws of algebra work in just the same manner. Multiplication is commutative no matter what terms are multiplied. Mathematical cognition could have worked like this too; if it had, it would have provided a convenient explanation of how people perform algebra (Anderson 2005). However, the conclusion from our experiments is that seeing '$X+Y * A$' cannot trivially be translated into the amodal mentalese code $* (+ (G001, G002),$ $G003)$. The physical spacing, superficial similarities between the letters, and background pictorial elements are all part of how people treat '$X * A+Z$.' Although it would be awfully convenient if a computational model of algebraic reasoning could aptly assume the transduction from visual symbols to mental symbols, taking this for granted would leave all of our current experimental results a mystery. We must resist the temptation to posit mental representations with forms that match our intellectualized understanding of mathematics. A more apt input representation to give our future computational models would be a bitmap graphic of the screen that includes the visually presented symbols as well as their absolute positions, spacings, sizes, and accompanying nonmathematical pictorial elements, as well as the associations carried by particular shapes.

16.4 **Conclusions**

We have tried to develop an account of embodied cognition that is consistent with the goal of teaching scientific and mathematical knowledge that is transportable. Toward this end, our chain of argumentation includes several links:

- We argued that understanding complex systems principles is important, both scientifically and pedagogically. Seemingly unrelated systems are often deeply isomorphic, and the mind that is prepared to use this isomorphism can borrow from understanding a concrete scenario.

- If one is committed to fostering the productive understanding of complex systems, then one must be interested in promoting knowledge that can be transported across disciplinary boundaries.

- Observations of students interacting with complex systems simulations indicate that one of the most powerful educational strategies is to have students actively interpret perceptually present situations. A situation's events inform and correct interpretations, while the interpretations give meaning to the events.

- Perspective-dependent interpretations can promote transfer where formalism-centered strategies fail, by educating people's flexible perception of similarity. Transfer, by this approach, occurs not by applying a rule from one domain to a new domain, but rather by allowing two scenarios to be seen as embodying the same principle.

- Even algebraic reasoning is sensitive to perceptual grouping factors, suggesting that perhaps all cognition may intrinsically involve perceptually grounded processes.

Considerable psychological evidence indicates that far transfer of learned principles is often difficult and may only reliably occur when people are explicitly reminded of the relevance of their early experience when confronted with a subsequent related situation (Gick and Holyoak 1980, 1983). This body of evidence stands in stark contrast to other evidence suggesting that people automatically and unconsciously interpret their world in a manner that is consistent with their earlier experiences (Kolers and Roediger 1984; Roediger and Geraci 2005). These statements are, however, reconcilable. A lasting influence of early experiences on later experiences occurs when perceptual systems have been transformed. A later experience will be inevitably processed by the systems that have been transformed by an initial experience, and priming naturally occurs. When a principle is simply grafted onto an early experience but does not change how the experience is processed, there is little chance that the principle will come to mind when it arises again in a new guise. The moral is clear: to reliably make an interpretation come to mind, one needs to affect how a situation is interpreted as it is 'fed forward' through the perceptual system rather than tacking on the interpretation at the end of processing.

An embodied cognition perspective offers promise of scientifically-grounded educational reform in both of our domains of empirical consideration: complex systems simulations and algebraic calculation. In particular, reforms are advocated that involve changes to perception, or the co-option of natural perceptual processes for tasks requiring symbolic reasoning. From this perspective, it is striking that developing expertise in most scientific domains involves perceptual learning. Biology students learn to identify cell structures, geology students learn to identify rock samples, and chemistry students learn to recognize chemical compounds by their molecular structures. In mathematics, perceptually familiar patterns are a major source of solutions for solving novel problems. Progress in the teaching of these fields depends on understanding the mechanisms by which perceptual and conceptual representations mutually inform and interpenetrate one another. Specific hypotheses for future exploration stem from an embodied perspective. How can we modify the perceptual aspects of mathematical notation conventions to cue students as to their formal properties? The minus sign '−' is just as symmetrical as the plus sign '+', but subtraction is order-dependent in a way that addition is not. Should the signs reflect this formal difference? Can physical spacing be used to scaffold learning formal mathematical properties? Can we help students learn the order of precedence by presenting them with 'A + X∗Y' before giving them the more neutral notation of 'A + B ∗ C?'

More generally, our analyses of science and mathematics have led us to reconsider what 'abstract cognition' entails. Abstract cognition should be interpreted akin to abstract art. Abstract art is still based in perception – it would not be art if it couldn't be seen. Similarly, for cognition, abstract is not the opposite of perceptual, nor are 'perceptual' and 'superficial' synonyms. A fertile and novel perspective on abstraction may be that it is the process of a perceptual system interpreting a situation in a useful manner. Interpretation presupposes that there is an external situation to interpret. Oftentimes, the perceptual system will need to be trained in order to avoid superficial or misleading attributes. We believe that viewing abstraction as the deployment of trained and strategic perceptual/motor processes can go a long way toward demystifying symbolic thought, hopefully leading to the development of robust and working models of cognition.

Debate

Arthur Graesser: Part of the problem with transfer from one science domain to another is that the problems may be rather different. If you use one solution you may have an isomorphic problem, but you are really imposing it, trying to make sense of it. If it is good enough, you may say that there is an analogy, and it is that process that helps you perceive things. Given the trajectory of how people make analogies, how would you interpret your data?

Robert Goldstone: You do not want to have rampant analogy making if there are real, important differences between domains. In part, what I am arguing for is not just the cognitive science of transfer, but I am also making claims about what science pedagogy should look like. What are the principles that should be taught in our schools? There I would make the claim for teaching the kinds of complex systems principles that I have described. So, I do agree with you that it is problematic when you have principles that apply in one domain but not another and you try to shoe-horn them into alignment. That clearly happens. But at the same time, exactly the same formalisms can often be successfully applied in different domains. For example, I think some of the most important lessons that we should be teaching in our science classes concern positive and negative feedback systems. Naturally, the nature of these systems will need to be tailored to a scenario, but these skeletal structures arise time after time. Teaching these principles can be a lot more valuable than teaching a student the name for some amino acid because these are principles that at least have the possibility of being relevant for a new situation that a student is likely to come across at a later point. I believe there are a number of such principles: oscillations, reaction–diffusion systems where the presence of one element gives rise to more of a second element that may destroy the first element at the same time as the first element is spreading out, or lateral inhibition between neighbouring elements at the same layer in a network. These principles are general and do arise in different guises. It would be wrong to leave things at that level of description and say that we have completely explained the system. We then need to adapt the characterization to the specifics of the scenario, just as case-based reasoning researchers would have us do.

Max Louwerse: I want to challenge your conclusions that just because something is formal reasoning doesn't make it amodal. I think that's wrong. I think that if it is formal reasoning then it is amodal. What you've pointed to is that there could be performance problems. So, I would want to change the claim to – just because something is formal reasoning does not mean that perception can't affect performance. If I'm teaching predicate calculus in a logic class I'll tell students, 'Here's a rule, *modus ponens*, that is sensitive only to the shapes of symbols.' But if I make the material implication symbol a mile long so that students have to walk several blocks to see what the consequent of the material implication is, they are not going to do as well because they have already forgotten the antecedent. But they engage in formal reasoning when they do reason.

Goldstone: I'm really skeptical of the performance/competence distinction. When you are doing formal reasoning, it is just physical manipulation of chicken scrawls. When you learn logic you're learning that you cross this bit out and replace it with another kind of chicken scrawl. I think, very literally, it is still just concrete and perceptual. You're just learning different types of transformation and translation procedures. In our experiments, we don't just get response-time differences, we get accuracy differences. If you're thinking about it from a formal symbol systems perspective, you would need to say that people are coming up with different algebraic tree structures based upon perceptual properties.

Louwerse: Why is it wrong to think that what I'm teaching my students to do is to behave like a Turing machine? Do you think that a Turing machine would show the same kinds of performance problems that a human being would?

Goldstone: I think it's revealing that Alan Turing described the Turing machine in a particular physical context. He wasn't describing it only in terms of a pure mathematical formalism. He was talking about a specific kind of machine that had specific properties, such as a tape head that is moving back and forth. So, I will take the more radical position that what we are talking about as formal symbol manipulation is what Newell and Simon referred to as physical symbol systems. They are physical. They are based upon the form of the objects. So it matters what those forms are and how they are processed. This relates to something that Friedemann [Pulvermüller] discussed (see Chapter 6). Word representations activate Broca's area because part of their representation is their production. It's not as though words are formal systems. They are being produced and understood as scrawls or sound waves.

Friedemann Pulvermüller: Thanks very much for this important contribution. I want to add a little thing. I'm fully with you in everything you say, especially with regard to education. I'm coming from a neuropsychological domain called aphasia therapy research, and teaching patients language after lesions. It is disastrous to see that the majority of therapists would present words and pictures in a far-removed situation in a 'naming game'. A neurophysiologist had a patient with global aphasia, and in testing her he asked, 'Who is this person sitting to your left?' (it was the patient's daughter). She responded '.. I can't come up with it. I'm sorry, I have tried and tried … My poor Jacqueline, I can't even remember your name.' This is exactly the point you are making. The situation and community of intentions count. The language game that is being played is very important. So it is not just the relation between the picture and word that counts. Noncommunicative language wouldn't even produce the same speech output. With regard to the competence/performance distinction, if we think of the brain in terms of real neuronal networks, we may have difficulty preserving the distinction. It is clear that the neural assemblies and their connections are the implementation of both the rules and the words – the competence of the system. The activity of these same networks is the correlate of performance. So, this distinction does not make too much sense.

Goldstone: The only thing that I would add to that is that, although I am skeptical of a performance/competence distinction, I'm all in favour of having responses based on different types of representations, giving rise to different performances on different occasions. Some of these representations will give rise to output that look more in accord with what we expect by our intellectualized understanding of logic. But I think it would be wrong to say that those that do correspond to our mathematical under-standing are not done by cognitive, physical manipulations.

Deb Roy: I'd like to stay on the topic of performance versus competence and the question of whether we can keep this distinction or not. Of course, probably everyone here is thinking of Chomsky when we think about this distinction. David Marr used slightly different terminology, but his general idea was that when one is trying to understand a system, usually a complex system, it is useful to distinguish what it is doing from how it is doing it. If you don't have a clear idea of what it is doing then it is problematic to ask how it is doing it. So, if we get rid of the labels 'performance' and 'competence' and rephrase the question to ask, 'Is it still worthwhile to keep Marr's distinction?' Or, would you eliminate that distinction as well?

Goldstone: I'm happier preserving a 'what' versus 'how' distinction, and I also still want to be able to talk about errors in processing. We want to be able to say things like, 'What we want to get out is a three-dimensional representation of a scene, but if we use this particular stereopsis algorithm that combines information from the two eyes, then we won't get the desired output.' That is, we do need to be able to say that the operation of an algorithm falls short of its intended computation.

Roy: I'd say that is completely in line with the original performance/competence distinction. You're still on board, but just using different terminology.

Luc Steels: There was more to the original distinction. In a usage-based approach to language, the actual variation and performance is influenced all the time.

Roy: So you're saying that that performance versus competence assumes dynamic versus static systems?

Steels: You can tease out different meanings, but 'competence' ended up being a description of an abstract system without worrying about how using the system constantly impacts the original system.

Roy: Maybe there is a family of interpretations. When I think of Marr's 'what' level, I never imagined that it needs to be fixed. When one thinks of Chomsky, one may think that competence is innate and does not change perhaps.

Goldstone: For me, what is added with the performance/competence distinction is that it suggests that it is useful to ask, 'What would this system be able to do if you took away all of the contextualizing variables that Friedemann just described, or all of the system's performance-based limitations?' However, from my perspective, it's exactly because of those contextual and performance-based factors that the system is able to do anything at all. It doesn't make sense to talk about what the competence of the system would be after extracting out contextual and physical considerations. However, it still might make sense to ask, from a teleological perspective, what the aim of a system is – Marr's 'what is it doing?' question.

Roy: I can live with that.

Author note

Many useful comments and suggestions were provided by Arthur Glenberg, Arthur Graesser, Linda Smith, Manuel de Vega, Uri Wilensky, and in particular Mitchell Nathan. This research was funded by Department of Education, Institute of Education Sciences grant R305H050116, and National Science Foundation REC grant 0527920. Further information about the authors' laboratory can be found at *http://cognitrn.psych. indiana.edu/*.

References

Anderson J (2005). Human symbol manipulation within an integrated cognitive architecture. *Cognitive Science*, 29, 313–41.

Arnheim R (1970). *Visual Thinking*. Faber and Faber: London.

Ball P (1999). *The Self-made Tapestry*. Oxford: Oxford University Press.

Barnett SM, Ceci SJ (2002). When and where do we apply what we learn? A taxonomy for far transfer. *Journal of Experimental Psychology: General*, 128, 612–37.

Barsalou LW (1999). Perceptual symbol systems. *Behavioural and Brain Sciences*, 22, 577–660.

Barsalou LW, Simmons WK, Barbey AK, Wilson CD (2003). Grounding conceptual knowledge in modality-specific systems. *Trends in Cognitive Sciences*, 7, 84–91.

Bar-Yam Y (1997). *Dynamics of Complex Systems*. Reading, MA: Addison-Wesley.

Bassok M (1996). Using content to interpret structure: effects on analogical transfer. *Current Directions in Psychological Science*, 5, 54–8.

Bassok M, Chase VM, Martin SA (1998). Adding apples and oranges: alignment of semantic and formal knowledge. *Cognitive Psychology*, 35, 99–134.

Bassok M (2001). Semantic alignments in mathematical word problems. In D Gentner, KJ Holyoak, BN Kokinov, Eds. *The Analogical Mind* (pp. 401–33).Cambridge, MA: MIT Press.

Batty M (2005). *Cities and Complexity: Understanding Cities with Cellular Automata, Agent-Based Models, and Fractals*.Cambridge, MA: MIT Press.

Bentley WA, Humphreys W J (1962). *Snow Crystals*. New York, NY: Dover.

Bialek W, Botstein D (2004). Introductory science and mathematics education for 21st-century biologists. *Science*, 303, 788–90.

Carraher D, Schliemann AD (2002). The transfer dilemma. *Journal of the Learning Sciences*, 11, 1–24.

Casti JL (1994). *Complexification*. New York, NY: HarperCollins.

Chi MTH (2005). Commonsense conceptions of emergent processes: why some misconceptions are robust. *Journal of Learning Sciences*, 14, 161–99.

Chi MTH, Feltovich P, Glaser R (1981). Categorization and representation of physics problems by experts and novices. *Cognitive Science*, 5, 121–52.

Csermely P (1999). Limits of scientific growth. *Science*, 5420, 1621.

Darwin C (1859). *On the Origin of Species by Means of Natural Selection*. London: Murray.

DeLoache JS (1991). Symbolic functioning in very young children: understanding of pictures and models. *Child Development*, 62, 736–52.

DeLoache JS (1995). Early understanding and use of symbols: the model model. *Current Directions in Psychological Science*, 4, 109–13.

Deloache JS (2000). Dual representation and young children's use of scale models. *Child Development*, 71, 329–38.

DeLoache JS, Burns NM (1994). Symbolic functioning in preschool children. *Journal of Applied Developmental Psychology*, 15, 513–27.

DeLoache JS, Marzolf DP (1992). When a picture is not worth a thousand words: Young children's understanding of pictures and models. *Cognitive Development*, 7, 317–29.

Detterman DR (1993). The case for prosecution: transfer as an epiphenomenon. In DK Detterman, RJ Sternberg, Eds. *Transfer on Trial: Intelligence, Cognition, and Instruction*. Westport, CT: Ablex.

Einstein A (1989). *The Collected Papers of Albert Einstein, Vol. 2: The Swiss Years: Writings, 1900–1909*. Princeton, NJ: Princeton University Press.

Flake GW (1998). *The Computational Beauty of Nature*. Cambridge, MA: MIT Press.

Fodor JA (1992). *A Theory of Content and Other Essays*. Cambridge, MA: MIT Press.

Garcia-Ruiz JM, Louis E, Meakin P, Sander LM (1993). *Growth Patterns in Physical Sciences and Biology*. New York, NY: Plenum Press.

Gentner D, Stevens AL, Eds. (1983) *Mental Models*. Hillsdale, NJ: Erlbaum.

Gick ML, Holyoak KJ (1980). Analogical problem solving. *Cognitive Psychology*, 12, 306–55.

Gick ML, Holyoak KJ (1983). Schema induction and analogical transfer. *Cognitive Psychology*, 15, 1–39.

Glenberg AM, Gutierrez T, Levin JR, Japuntich S, Kaschak MP (2004). Activity and imagined activity can enhance young children's reading comprehension. *Journal of Educational Psychology*, 96, 424–36.

Goldstone RL (1994). Influences of categorization on perceptual discrimination. *Journal of Experimental Psychology: General*, 123, 178–200.

Goldstone RL (1998). Perceptual learning. *Annual Review of Psychology*, 49, 585–612.

Goldstone RL (2003). Learning to perceive while perceiving to learn. In R Kimchi, M Behrmann, C Olson, Eds. *Perceptual Organization in Vision: Behavioural and Neural Perspectives* (pp. 233–78). Mahwah, NJ: Erlbaum.

Goldstone RL and Barsalou L (1998). Reuniting perception and conception. *Cognition*, 65, 231-262.

Goldstone RL Lippa Y and Shiffrin RM (2001). Altering object representations through category learning. *Cognition*, 78, 27-43.

Goldstone RL, Sakamoto Y (2003). The transfer of abstract principles governing complex adaptive systems. *Cognitive Psychology*, 46, 414–66.

Goldstone RL, Son JY (2005). The transfer of scientific principles using concrete and idealized simulations. *The Journal of the Learning Sciences*, 14, 69–110.

Graesser AC, Olde BA (2003). How does one know whether a person understands a device? The quality of the questions the person asks when the device breaks down. *Journal of Educational Psychology*, 95, 524–36.

Hadamard J (1949). *The Psychology of Invention in the Mathematical Field*. Princeton, NJ: Princeton University Press.

Harnad S, Hanson SJ, Lubin J (1995). Learned categorical perception in neural nets: Implications for symbol grounding. In V Honavar, L Uhr, Eds. *Symbolic Processors and Connectionist Network Models in Artificial Intelligence and Cognitive Modelling: Steps Toward Principled Itegration* (pp. 191–206). Boston, MA: Academic Press.

Hatano G, Greeno JG (1999). Alternative perspectives on transfer and transfer studies. *International Journal of Educational Research*, 31, 645–54.

Hegarty M (2004a). Mechanical reasoning by mental simulation. *Trends in Cognitive Science*, 8, 280–5.

Hegarty M (2004b). Dynamic visualizations and learning: getting to the difficult questions. *Learning and Instruction*, 14, 343–51.

Hmelo-Silver CE, Pfeffer MG (2004). Comparing expert and novice understanding of a complex system from the perspective of structures, behaviours, and functions. *Cognitive Science*, 28, 127–38.

Horgan J (1996). *The End of Science: Facing the Limits of Knowledge in Twilight of the Scientific Age*. Reading, MA: Addison-Wesley.

Inhelder B, Piaget J (1958). *The Growth of Logical Thinking from Childhood to Adolescence*. New York, NY: Basic Books.

Jacobson MJ (2001). Problem solving, cognition, and complex systems: differences between experts and novices. *Complexity*, 6, 41–9.

Jacobson MJ, Wilensky U (2006). Complex systems in education: scientific and educational importance and research challenges for the learning sciences. *Journal of Learning Sciences*, 15, 11–34.

Koffka K (1935). *Principles of Gestalt Psychology*. New York, NY: Harcourt Brace Jovanovich.

Kohonen T (1995). *Self-organizing Maps*. Berlin: Springer-Verlag.

Kolers PA, Roediger HL (1984). Procedures of mind. *Journal of Verbal Learning and Verbal Behaviour*, 23, 425–49.

Lakoff G, Nuñez RE (2000). *Where Mathematics Comes From*. New York, NY: Basic Books.

Landy D, Goldstone RL. How abstract is symbolic thought? *Journal 2007 of Experimental Psychology: Learning memory and cognition*, 33, 720–733.

Lave J (1988). *Cognition in Practice: Mind, Mathematics, and Culture in Everyday Life*. New York, NY: Cambridge University Press.

Lave J, Wenger E (1991). *Situated Learning: Legitimate Peripheral Participation*. New York, NY: Cambridge University Press.

Leeper R (1935). A study of a neglected portion of the field of learning: the development of sensory organization. *Journal of Genetic Psychology*, 46, 41–75

Livingston KR, Andrews JK, Harnad S (1998). Categorical perception effects induced by category learning. *Journal of Experimental Psychology: Learning, Memory, and Cognition*, 24, 732–53.

Nathan MJ, Kintsch W, Young E (1992). A theory of algebra word problem comprehension and its implications for the design of computer learning environments. *Cognition and Instruction*, 9, 329–89.

Nathan MJ, Long SD, Alibali MW (2002). The symbol precedence view of mathematical development: a corpus analysis of the rhetorical structure of textbooks. *Discourse Processes*, 33, 1–21.

Nathan MJ, Petrosino A (2003). Expert blind spot among preservice teachers. *American Educational Research Journal*, 40, 905–28.

Nunes T (1999). Mathematics learning as the socialization of the mind. *Mind, Culture, and Activity*, 6, 33–52.

Özgen E (2004). Language, learning, and color perception. *Current Directions in Psychological Science*, 13, 95–8.

O'Reilly RC (2001). Generalization in interactive networks: the benefits of inhibitory competition and Hebbian learning. *Neural Computation*, 13, 1199–242.

Palmer S, Rock I (1994). Rethinking perceptual organization: the role of uniform connectedness. *Psychonomic Bulletin and Review*, 1, 29–55.

Penner DE (2000). Explaining processes: investigating middle school students' understanding of emergent phenomena. *Journal of Research in Science Teaching*, 37, 784–806.

Reed SK, Ernst GW, Banerji R (1974). The role of analogy in transfer between similar problem states. *Cognitive Psychology*, 6, 436–50.

Resnick M (1994). *Turtles, Termites, and Traffic Jams*. Cambridge, MA: MIT Press.

Resnick M, Wilensky U (1993). Beyond the deterministic, centralized mindsets: new thinking for new sciences. Atlanta, GA: American Educational Research Association.

Roberson D, Davies I, Davidoff J (2000). Color categories are not universal: replications and new evidence in favor of linguistic relativity. *Journal of Experimental Psychology: General*, 129, 369–98.

Roediger HL, Geraci L (2005). Implicit memory tasks in cognitive research. In A Wenzel, DC Rubin, Eds. *Cognitive Methods and their Application to Clinical Research* (pp. 129–51). Washington, DC: American Psychological Association.

Ross BH (1987). This is like that: the use of earlier problems and the separation of similarity effects. *Journal of Experimental Psychology: Learning, Memory, and Cognition*, 13, 629–39.

Ross BH (1989). Distinguishing types of superficial similarities: different effects on the access and use of earlier problems. *Journal of Experimental Psychology: Learning, Memory, and Cognition*, 15, 456–68.

Roy D, Pentland A (2002). Learning words from sights and sounds: a computational model. *Cognitive Science*, 26, 13–146.

Rumelhart DE, Zipser D (1985). Feature discovery by competitive learning. *Cognitive Science*, 9, 75–112.

Schwartz DL, Black T (1999). Inferences through imagined actions: knowing by simulated doing. *Journal of Experimental Psychology: Learning, Memory, and Cognition*, 25: 116–36.

Shepard RN (1984). Ecological constraints on internal representations: resonant kinematics of perceiving, imagining, thinking, and dreaming. *Psychological Review*, 91, 417–47.

Shepard RN (1987). Toward a universal law of generalization for psychological science. *Science*, 237, 1317–23.

Simmons K, Barsalou LW (2003). The similarity-in-topography principle: reconciling theories of conceptual deficits. *Cognitive Neuropsychology*, 20, 451–86.

Sloutsky VM, Kaminski JA, Heckler AF (2005). The advantage of simple symbols for learning and transfer. *Psychonomic Bulletin and Review*, 12, 508–13.

Son JY, Smith LB, and Goldstone RL (in press). Simplicity and generalization: short-cutting abstraction in children's object categorization. *Cognition*.

Thurston W (1994). On proof and progress in mathematics. *Bulletin of the American Mathematical Society*, 30, 161–77.

Tversky A, Kahneman D (1974). Judgment under uncertainty: heuristics and biases. *Science*, 185, 1124–31.

Uttal DH, Liu LL, DeLoache JS (1999). Taking a hard look at concreteness: do concrete objects help young children learn symbolic relations? In CS Tamis-LeMonda, Ed. *Child Psychology: A Handbook of Contemporary Issues* (pp. 177–92). Philadelphia, PA: Psychology Press.

Uttal DH, Scudder KV, DeLoache JS (1997). Manipulatives as symbols: a new perspective on the use of concrete objects to teach mathematics. *Journal of Applied Developmental Psychology*, 18, 37–54.

von der Malsburg C (1973). Self-organization or orientation-sensitive cells in the striate cortex. *Kybernetik*, 15, 85–100.

Wason PC, Shapiro D (1971). Natural and contrived experience in a reasoning problem. *Quarterly Journal of Experimental Psychology*, 23, 63–71.

Wilensky U (1991). Abstract meditations on the concrete and concrete implications for mathematics education. In I Harel, S Papert, Eds. *Constructionism* (pp. 193–203). Westport, CT: Ablex.

Wilensky U, Reisman K (1999). ConnectedScience: learning biology through constructing and testing computational theories – an embodied modeling approach. *InterJournal of Complex Systems*, 234, 1–12.

Wilensky U, Reisman K (in press). Thinking like a wolf, a sheep or a firefly: learning biology through constructing and testing computational theories – an embodied modeling approach. *Cognition and Instruction*.

Wiley J (2003) Cognitive and educational implications of visually-rich media: images and imagination. In M Hocks, M Kendrick, Eds. *Eloquent Images: Word and Image in the Age of New Media*. Cambridge, MA: MIT Press.

Wittgenstein L (1953). *Philosophical Investigations* (GEM Anscombe, Trans.). New York, NY: Macmillan.

Chapter 17

Mending or abandoning cognitivism?

Antoni Gomila

17.1 Introduction

The representational–computational theory of the mind, the core of cognitivism, has been the foundation of the cognitive sciences, despite continuing doubts about some, or all, of its postulates, mental symbols being just one of them. The current interest in embodiment makes times ripe, though, for some rethinking of the basics. In this chapter I will explore whether current interest in an embodied, embedded approach to cognition should be carried out as an alternative only to abstract symbols, while keeping most of cognitivism, or rather as a proper alternative to cognitivism. To this purpose, I will highlight the shortcomings and risks of a cognitivist version of the embodied meaning approach, and will finally favour a more radical, interactivist, and dynamical approach to cognition.

17.2 Cognitivism and the problem of understanding

Cognitivism is meant to refer to the set of basic postulates which inform most of the cognitive sciences (Searle 1990; Haugeland 1995, 1998). It is a version of the representational–computational theory of the mind, according to which the mind is an information processor which plays by the formality condition (Fodor 1975; Newell 1980). That is, mental processes are conceived as manipulations of mental symbols in virtue of their form – their *syntax* – while also respecting the semantic relations that hold between these states in virtue of their content. Thus, the brain is viewed as a syntactic engine driving a semantic engine (Block 1990).

The building blocks of cognition, according to this general approach, are these mental symbols, which are conceived as abstract, digital, discrete, amodal, explicit representations. Consequently, cognitivism implies a representational, translational view of language understanding. Understanding is thought to consist in the rule-guided transformation of a set of natural language sentences into a mental representation (a macroproposition, a situation model, etc., according to the particular theory one favours) composed of mental symbols, which supposedly captures the linguistic meaning of the sentences. This tranformation is conceived of as a formal, syntactical, inferential process which ends up with a mental representation expressing the linguistic meaning, or proposition. Hence, the problem arises as to what confers meaning to these internal states and how they gain their semantic contents in the first place, equivalent in expressive power to that of language. This has been called 'the symbol grounding problem' (Harnad 1990).

Since Searle's Chinese room argument, a main part of the discomfort with cognitivism is due to this view of (language) understanding as translation into a mental code; simply substituting one sign (public) for another (mental) through a formal process comes short of providing understanding just by itself. It is not only that algorithmic accounts of this process are still lacking, but rather, as noticed already by Winograd in the 1970s and in the philosophical arena by Wittgenstein in his private language argument, that the mere tokening of a 'mental formula' in one's 'belief box' comes short of amounting to understanding.

However, it should be also kept in mind that many other related problems present insurmountable difficulties for cognitivism. Specifically, in the area of language understanding, the *frame problem*, or as Fodor has called it, the problem of relevance or of nondemostrative inference, is still as challenging as ever. Moreover, it is the reason why no algorithmic account of understanding has been achieved (given the high context-dependency of the process, which is something that formal processes are not good at). The combinatorial explosion of alternatives, plus the lack of algorithmic procedures to reduce the search to just the 'relevant' ones, suggests (for Fodor himself) that cognitive science is limited to modular processes and is not going to be able to account for 'nonmodular' ones (Fodor 1983, 2001). Notice that it is the very allegiance to cognitivism that makes such a problem unsolvable. Therefore, the recent 'massive modularity' program that attempts to avoid the frame problem by claiming that all cognitive processes are modular (Cosmides and Tooby 1992; Sperber 1996; Pinker 1997) may be misguided. If the real problem is cognitivism itself, then the debate on the scope of modularity turns out to be misplaced.

There are many other problems that vex cognitivism, understood as symbolic processing: lack of flexibility (Clark 1989), the problem of common coding of inputs and outputs (Prinz 1990; Turvey 1977), the problem of how to account for double activation of the same symbol in a single sentence (Jackendoff 2002), etc. While we do not need to discuss these other problems in any detail here, it is important to keep in mind that symbol grounding is not the only problem, so that were we to find a solution for it within cognitivism that would not amount to its final success. On the contrary, cognitivism is plagued by these different, related problems – they all have something to do with processing – because of its central commitment to a strict distinction between data structures and operations on them, imported from classical artificial intelligence (AI) teachings. In the last section, I will argue for a different conception of mental representation, which has to be accompanied by a change in the conception of mental processing as symbol crunching to have any chance of overcoming all these problems.

17.3 Peirce on symbol grounding

I was happily surprised by the reference to Peirce in the convenors' position paper (Glenberg *et al.*, Chapter 1, this volume), because I think that Peirce's theory of signs is well suited to clarify the questions involved in symbol grounding and to appreciate the difficulties of cognitivism. To begin with, it seems to me important to realize that while notions of both symbol and grounding can be found in Peirce, they are used in quite a different sense than has become standard usage in cognitive science. For Peirce, symbols are just one kind of sign, the grounding relation of which depends upon public rules of usage. Let's see this in some more detail.

Remember that Peirce was interested in external, public signs in general. His definition of sign states that a sign 'is something that stands for something else in some respect or other for somebody'. His technical term for this something that signifies is 'representamen', and for what is signified, 'object'. The next technical notion, most relevant in this context, is 'ground', which refers to the sort of relation in virtue of which the 'representamen' stands for the 'object'. He distinguishes three kinds of grounding relations: iconic, indexical, and symbolic. As is well known, iconic signs are signs in virtue of resemblance between representamen and objects in some respect for somebody; indexical ones, in virtue of covariation; and symbolic ones in virtue of an arbitrary, conventional agreement (for an introduction to Peirce's theory in terms of its relevance to cognitive science, see Gomila 1996).

It follows from this that, for Peirce, mental states clearly could not be symbols; moreover, it is doubtful whether they could be signs. There is no way in which mental states could be arbitrarily or conventionally viewed as symbols (let's say, that the firing of a cell assembly is taken to mean that the Pope lives in Rome).[1] As a matter of fact, nobody has seriously claimed that the content of mental states is grounded in conventions; however, the practice in AI and in cognitive modelling of hypothesizing states with some determined content by stipulation, which so closely reminds of conventions, is one of the sources of discomfort with symbolic cognitivism that our convenors make explicit (Glenberg et al., Chapter 1). Anyway, what's clear is that cognitivist mental symbols do not correspond to Peircean symbols because they are not grounded conventionally. But do they fare better as Peircean indexes or icons? Mental images would seem to fare better as icons, but, as many will still remember, images were banned from cognitivism in the so-called 'image proposition' debate, in the early 1980s, and were finally accepted only when they were revised as nonanalogical, nonresemblance-based representations, though nonreducible to (cognitivist) symbols (Kosslyn 1994). Current neuroscientific practice speaks openly of brain codes, and the methods used (e.g., event-related potentials [ERP], functional magnetic resonance imaging [fMRI]) look for nomic covariations between external stimuli and neural electrophysiological or hemodynamic patterns, which would look more like indices (DeCharms and Zador 2000).

But remember that Peircean signs require a subject to grasp the grounding relation, and there is nobody inside the head in charge of 'reading' or interpreting the states. This is a second reason why mental states cannot be viewed as Peircean signs: no internal interpreter can be assumed, on pain of falling into the homuncular fallacy. There is a third reason: we have not been able yet to individuate mental states in virtue of their 'form'; in Peircean terms, we cannot yet individuate mental states as 'representamen',

---

[1] As I was preparing the final version of this paper, I came across the recent paper by Held *et al.* (2006), where Vosgerau makes the same claim (Vosgerau 2006, p.270).

so as to apply the type/token distinction characteristic of signs and recognize a mental state token as of the same kind as another. It is not just that such a notion requires a type/token distinction which, *prima facie*, is not clearly applicable to mental signs. As a minimum requirement, a clear sense of what Peirce called 'representamen', the material, iterable form of the sign would require repeatable mental states identifiable as such – as of the same type – quite apart of their meaning relation. However, this is something that we have not yet achieved: our identification of mental state types is through their referent, or their external relata, not their intrinsic form, or internal 'syntax'. The closest we get to such a goal is through work such as Pulvermüller's (Chapter 6, this volume), which in any case proceeds through standard covariational procedures to individuate his neural networks correlated with linguistic signs in standard situations.

Is there any sense in which we could think of mental states as representations? The answer is 'yes', but as natural signs, as informational states, not as signs dependent upon a subject of interpretation, signs that have to be read or understood; hence, not as Peircean signs, but as Gricean natural signs (Grice 1957). And when we turn to natural signs, 'grounding' amounts to carrying information (Dretske 1981), which is spelled out as a nomic correlation, plus some sort of functional relevance of this information for the system. No internal 'interpreter' is needed, and hence there is no lurking regress.

As a matter of fact, the different proposals discussed in this book could be seen from this vantage point as variations on this idea. The disagreement regarding grounding was on the correlated term purported to 'ground' the content of mental or brain states. In *latent semantic analysis* (LSA), covariation with patterns of linguistic use is what 'grounds' the linguistic meaning (Landauer and Dumais 1997; Kintsch 1998); Shapiro's discussion (see Chapter 4, this volume) made reference to Fodor's asymmetric dependency, an attempt to 'ground' meaning in the nomic covariation between mental states and world properties (Fodor 1991); Steels' work (see Chapter 12) could suggest that 'grounding' is achieved by correlation with social practices; and embodied meaning could be viewed as suggesting that linguistic meaning is grounded in the perceptual and motor states coactivated in language processing.

So, at this point, we can get a more precise characterization of what we need to overcome the problems that vex cognitivism, a better view of how the internal states postulated to account for understanding are to be conceived (individuated), how they are supposed to provide understanding, and how they get their representational grip.

17.4 Embodied meaning as a version of cognitivism?

Given these preliminaries, it seems to me that the embodied meaning approach we are considering in this debate offers an alternative only to the classical view of mental symbols as abstract, amodal types, rather than as an alternative to cognitivism overall; in other words, it takes for granted the general idea of cognition as symbol crunching, while proposing a different view of the nature of such symbols – not as abstract, symbolic, etc., but as sensorimotor schemas or bodily patterns of activation, or as built up of such basic elements (a significant ambiguity, see later). In so doing, it comes short of what is needed to overcome the difficulties of cognitivism. Just as the imagery debate – a debate which can be seen as a relevant precedent to the current debate – made clear, an account of mental representation has to be connected to an account of mental processing; a view of mental representation has to go hand in hand with a view of mental processing. In this way, an alternative to abstract symbolic cognitive science has to include an alternative to symbolic processing as syntactic transformation of data structures. Currently, such a processing account is missing, or implicitly cognitivist. In addition, there are even difficulties with the very formulation of the embodied meaning program. In this section, I will try to outline some difficulties that it seems to me affect this approach in this regard, and in the next section I'll suggest what an alternative to the cognitivist processing framework might look like.

The intuition behind the embodied meaning program consists in the idea that the best way to 'ground' the macropropositions that are assumed to express meaning is to exploit their connection to sensory and motor brain states, which 'ground' their content (Barsalou 1999; de Vega 2001; Glenberg and Kaschak 2002; Pulvermüller, Chapter 6, this volume). But the grounding cannot consist exclusively of the simple coactivation in language processing of the same states involved in perceptual or behavioural interaction. There are plenty of such associations in the working brain, most of which do not ground semantic relationships.

To address this stricter requirement, a common, more or less explicit assumption behind this approach is what could be called a *simulationist view* of mental representation. According to simulationism, the very same states involved in perception and action are also involved in 'off-line' cognitive processes, where thinking or memory or imagination involve the same contents as the on-line corresponding processes (i.e., perception, action). In language comprehension, this idea amounts to viewing meaning as reactivation of the brain structures involved in the perceptual or motor interaction with the extralinguistic reality referred to. Therefore, no intermediate state (abstract, amodal) is needed to account for understanding. The 'language of thought' expression is substituted by the activation of those sensorimotor cortices which are supposed to be involved in the perceptual and motor interaction with what the linguistic expression means.

Simulationist approaches have proliferated in different areas: visual imagination (Kosslyn 1994); reasoning (Johnson-Laird 1998, 1983); motor imagination (Jeannerod 2001); emotional decision-making (Damasio 1994) or theory of mind (Gordon 1986, Goldman 2005). Synergistic effects among them contribute to its gathering momentum, and I'm not in a position to discuss simulationism as such. With regard to language comprehension however, it seems unsatisfactory to me. To put it provocatively, the way the embodied meaning program claims to solve the symbol grounding problem seems to me a case of throwing the baby out with the bathwater. In caricature, it could be said that, for symbolism, the grounding problem is to explain how the postulated abstract, amodal, arbitrary representational states get their meaning. But since for embodied meaning there are no such abstract, amodal symbols, there is nothing to ground. End of the problem – of the grounding problem, not of the problem of meaning, of course.

Maybe I'm attacking a straw man for (some version of) embodied cognition. Surely this is not what they want to say; what they want to claim is that the symbolic view of understanding, that of translating linguistic symbols into a 'language of thought' made of abstract symbols, falls short of being a satisfactory account of understanding. Their point, as I understand the drive behind it, is to think of understanding as grounded in the interactions of the organism with its environment, an idea I approve of. However, it is not clear that what they put in place of mental symbols has the proper features to play the role that constitutes understanding. As a matter of fact, they seem to play on an ambiguity, according to which 'grounded' meanings oscillate between symbols supported by sensorimotor schemas and the sensorimotor schemas themselves. In the first case, no different view of understanding is really offered (it is just translation into a language of thought, now grounded – this is, in fact, the view advocated by Harnad in his 1990 paper); in the second interpretation, no account of understanding is really offered because, as in the caricature, the explanandum disappears.

The problem in this case is indeed abstraction. Language expresses abstract meanings and we definitely grasp them. If what we postulate in order to account for how we are able to grasp such abstract meanings is not abstract enough, our account cannot be satisfactory, given the translational view of understanding as background. As a matter of fact, this was the problem of empiricism for years: in trying to ground concepts in images or sensory impressions, abstraction was 'the' problem. To use an example from de Vega *et al.* (2004), in reading a sentence about digging it seems that activity in the motor cortices appears (as if actually digging). However, a straight identification of mental meaning with such a brain pattern would be too simple a proposal: this correlational activity cannot be taken to constitute the content of the word, because contents are abstract but brain activity as such is not. Conversely, if such motor activation amounted to the meaning of the word, then an apraxic person should be unable to understand such a word. Something more sophisticated is needed, even for 'concrete' words, which are, in this sense, also abstract. Thus, there are many different ways to 'dig', or to use the example of my presentation, to 'open' (a door, a box, a theatre, a call for papers,...). Some contributors to the volume (most notably, Pulvermüller, see Chapter 6), try to identify a common brain pattern to all those different cases of 'opening', but even if progress is possible in this direction it is still inappropiate to call such an internal state 'the meaning' – the meaning is the action itself.

For me, the real problem here has to do with the implicit alliance to the translational view of language understanding. Mental symbols are substituted by sensorimotor patterns in the same role as 'internal' translation of the linguistic content, and that move is not big enough to avoid the constitutive problems of cognitivism. The classical objection to such 'imagistic' accounts of meaning is due also to Peirce; it's his comment that there cannot be an iconic sign for negation (while the modern debate started in the 17th Century regarding the nature and necessity of mathematical knowledge). Since there is not a single experience that corresponds to a proposition such as 'not-p' (given that so many may be true in such a situation), negation was thought by Peirce to be symbolic. Barsalou (1999) was also aware of this problem, though he considered it a side problem.[2]

---

[2] Johnson-Laird, on his part, changed his initial view of these 'mental models' as perceptual representations into an abstract view to accommodate negation (Johnson-Laird 2002). I will turn to the notion of 'mental model' in the last section.

But when we consider this tradition, the Peircean notion of 'grounding' doesn't really fit; grounding seems to be understood more like 'constitution' or 'derivation'. Glenberg et al. (Chapter 1, this volume) seem, at some points, to be thinking of this notion in their chapter. The embodied meaning program would consist in the project of deriving or explaining conceptual thought in terms of sensorimotor experience. According to this version of the program, symbols would not be eliminated but 'discharged'. Although they do not mention it, an important resource in this way of understanding the programme would look for multimodal representations as an intermediate step between abstract thought and concrete sensorimotor experience. We can think of very different properties and relations that are clearly multimodal even at the sensory level (examples: rhythm, synchrony, form, and above all, space and time), and it could be claimed that this redundancy gives rise to multi-sensory, or intersensory, representation which 'abstracts' the particulars of any modality (Bahrick *et al.* 2004). Spatial imagery can be amodal in this sense (Schwartz 1999). Again, this is in the empiricist tradition, but its point is not so much to 'ground' the mental representations in the sense of Peirce, Harnad, or Searle, but to construct abstract represen-tations out of less abstract ones, and so on. However, as Fodor has argued insistently (most outstandingly in Fodor 1980), a good account of this process of abstraction is still missing – although a non-Fodorian might believe that statistical approaches have recently been offering new insights in this regard (for an account of the origin of the abstract concept of space out of sensorimotor dependencies, see Philipona *et al.* 2003).

But, while this program is very interesting and important, this is not what we need to ground meaning. For were we to take this seriously as a solution to the symbol grounding problem, we would get the wrong meaning ascriptions: we would have to ascribe sensori-motor experiences or motor programs as meanings, not (truth-valuable) propositions as the content of our mental states. The meaning of 'I drink a beer' is the state of affairs that makes the sentence true, not the movements of my hands and lips, or a representation of those movements either. As another example, would we be willing to say that the laugh a joke provokes is part of its meaning?

There are also a couple of objections having to do with the brain as a grounder of meaning. First, although from the neural point of view it is clearly possible to distin-guish sensorimotor primary areas in virtue of their neural connections to affectors and effectors, respectively, from other associative or executive areas, it seems strange that all meaning must rely on sensorimotor foundations. If everything is sensorimotor, from the meaning point of view, and since linguistic processing involves the whole brain, then it follows that 'sensorimotor' would have different extensions in both cases, or otherwise we could not honour the aforementioned distinction between primary and nonprimary cortical areas.

Second, off-line, simulationist processing seems not to involve the primary sensorimo-tor cortical areas (which would make it possible to distinguish veridical perception from imagination), while the embodied meaning program puts all the weight into the primary brain areas. But I'm not competent enough to press these points.

In conclusion, it seems to me that the embodied cognition program is more concerned with neurological implementation than symbolic cognitivism was, which I think is correct.

As a matter of fact, the implementation of symbolism (in AI) has shown meagre success (as cognitive explanation, not in technological terms), to put it mildly. The development of cognitive robotics as a program to model intelligence has been a consequence of such a stalemate, and has taken the converse approach (starting from the bottom up, where implementation matters a lot, to try to get cognitive abilities out of sensorimotor patterns of interaction). But meaning, I contend, is not neural implementation: it depends upon and is made possible by implementation. Accordingly, I prefer the expression 'embodiment of mental states' rather than 'embodiment of meaning'. It also has the virtue of bringing the problem of mental processing (see Section 17.3) into the open.

17.5 **A more radical, non-cognitivist alternative**

Up until now, I've been arguing that symbol grounding is not the only problem that affects cognitivism; I've suggested that it is in terms of informational states that it has to be addressed (to avoid regress or circularity), and that the embodied meaning program faces the problem of abstraction. Now, I would like to suggest a possible direction language comprehension research might take, which seems to me to offer a way to overcome the problems of cognitivism while suggesting how abstract cognition is possible. It is the option of dynamicism.

For starters, it may be convenient to remind you that representations are introduced as mediator states between meaning and brain: we can understand or think a thought in virtue of a mental state, neurologically implemented, that represents its content. Thus, mental representations are theoretical entities introduced to naturalistically account for certain psychological abilities, those that exhibit intentionality (Markman and Dietrich 2000). It is a fact that we can think abstract, propositional thoughts and understand abstract linguistic expressions, but this does not impose an explanation in terms of corresponding abstract mental symbols (as discrete entities, whose causal powers depend on their 'form', their syntactical properties), or not so abstract sensorimotor patterns. Other options may be available, which may have the added value of a better view of cognitive processing (instead of as symbol crunching). Two other stipulations are relevant, though I cannot discuss them here: language is arguably the canonical way to 'measure', to describe, propositional thought. A practical, implicit, not translational, conception of understanding is in order for many reasons (according to which understanding is not having a mental inscription but practical abilities of how to proceed, in a Wittgensteinian fashion; see Gomila 2002a). Finally, I want to make clear that I agree with the embodied meaning program that questions of implementation and embodiment matter a lot, and this is what makes embodied cognition congenial to dynamicism.

Recall the basics of dynamicism: it proposes a change of metaphors for cognition, from information processing to control; it focuses on sensorimotor and social coordination in real-time, as the outcome of basic processes, from which higher cognitive abilities are thought to emerge; adaptive behaviour is seen as the outcome of the interaction between body and environment, without an internal center of control where the 'instructions' are read and executed; knowledge acquisition is accounted for in terms of self-regulatory and self-organizational processes, not postulation of explicit representations (for an introductory presentation see Beer 2000; Port and van Gelder 1995 was its manifesto). Interesting cognitive properties that dynamic models nicely capture include: dependency upon the details of previous history, upon context, sensitivity to initial conditions, priming, modulation of expectancies, perseverative behaviours in terms of previous motor activation, etc. (Erlangen and Schöner 2002).

Dynamicism has rightly been criticized for its grand antirepresentationalist contentions about cognition on the grounds of modest, meaning-impoverished sensorimotor coordination tasks (Clark and Toribio, 1995), and that, in more complex tasks, some notion of representation will still apply (Clark and Grush 1999). Some of the leading dynamicists agree (van Gelder 1998; Spencer and Schöner 2003); however, it is clearly a noncognitivist kind of representation since it cannot play the formalist role cognitivism attributes to mental symbols (mental processes as syntactic ones, as symbol crunching). This renewed notion of representation is not that of a discrete, stable, static, syntactical entity as in classical cognitivism, but a timed, continuous flux of states in a dynamical system. A way to make clear the appeal of this approach is in the context of motor control, through the notion of *forward models* (Wolpert 1997).

Forward models of control were developed in the early 1960s to overcome the limitations of classical feedback models in cybernetics (Kalman 1960), and were applied in neuroscience in the early 1970s (Ito 1970; Evart 1971). In simple sensorimotor tasks, control is achieved through (sensory, proprioceptive) feedback. Within these models, continuous comparison between the current state (as measured by the proprioceptive feedback) and goal state (specified in a 'vocabulary' compatible with the proprioceptive one), suffices to reduce their difference and eventually reach the goal state.

However, in some cases, where the movement is quick and the context may change fast, the feedback could not be avaible fast enough to help control the movement in time, or effectively. For such cases, forward models offer a better option: in addition to the motor centres that issue a motor sequence or plan, and the sensory or proprioceptive feedback provided from the effectors, an efference copy is sent to a 'model' of the motor system, called the *emulator* because it implements the same input–output function as the body–environmental interaction. While in principle this emulator could consist of a look-up table (a specification of a classical machine-table, of production rules, saying what to do for each input), its interest in control arises when it implements a 'dynamic model' of the subject–world system, an implicit model, the outcome of a previous story of engament and interaction between system and environment. Thus, the emulator offers to the system an anticipation of the expected feedback given such a motor plan, thereby making possible the adjustment of the motor execution before the effective feedback comes. In a forward model, motor control is made efficient through the combination of the information from the sensors, together with the estimation of position offered by the emulator, in order to anticipate the actual state of the system and environment in which the interaction takes place.

An example of such a system may help. Think of long-distance navigation, where every route decision is plotted onto a chart where a scale model of the system (ship's position, target position, accumulated knowledge of the geography – rocks, currents – and information of current state – winds, weather) is represented. In such a way, an expectation of 'where we should be already', given direction and time, can be contrasted with 'where we seem to be', given measures of position – mostly done by global positioning systems nowadays. Of course, here the change of positions in the 'internal' model are executed by the officers in charge, while an internal emulator runs automatically.

The interesting point is that forward models such as this have been useful in accounting for the performance of several motor phenomena: orientation (with the cerebellum as space emulation [Ito 1980; Miall *et al.* 1993]), sensory integration (Wolpert *et al.* 2001), control of saccades (where ocular movements of adjustment to keep visual stability continously take place [Tweed 2003]), response monitoring (Rodríguez-Fornells *et al.* 2002) and neuropsychological syndromes derived from their partial failure (Frith *et al.* 2000; Blakemore *et al.* 2002).

This has led Grush (Grush 2004) to suggest that this cybernetic notion, developed in the area of motor control, could help unify a whole set of areas of research, basically in that it can provide a general model of how both on-line (visual perception, motor control) and off-line processes (motor imagery, visual imagery, and even reasoning, language, and theory of mind phenomena) could be understood in a unified way. His idea is that:

> ... the brain constructs neural circuits that act as models of the body and the environment. During overt sensorimotor engagement, these models are driven by efference copies, in parallel with the body and environment, in order to provide expectations of the sensory feedback, and to enhance and process sensory information. These models can also be run off-line in order to produce imagery, estimate outcomes of different actions, and evaluate and develop motor plans (Grush 2004, p.377).

An important remark is that an emulator system need not 'run' on the same circuits that take charge of the direct interaction with the environment (in motor control at least, the evidence suggest that the cerebellum plays this emulation role), as the simulationist approach contends. This goes against the popular idea of simulation as off-line functioning: that visual imagery takes advantage of the same neural circuits involved in visual perception, except from the primary cortices (Kosslyn 1994), which is echoed with regard to motor imagery and motor control by Jeannerod (1997), and which can also be found in some of the chapters in this book with regard to language comprehension: linguistic understanding of action verbs would exploit the same circuits that the actions themselves to which they refer involve (for example, Pulvermüller, Chapter 6). Barsalou (1999) even introduces the term 'simulators' to express a parallel idea with regard to cognition as imagistic simulation capacities derived from perceptual experience, which would also provide the sense of linguistic expressions.

The difference consists in that emulation theory claims that imagery is produced not just by the activation of the visual or efferent areas, but by the fact that this activation feeds an emulator, which does not need to be restricted to the very same areas. It could be claimed that the difference between simulationism and emulationism is not so important: there might be different brain regions involved in off-line processes, instead of the same, but this does not amount to a different conception of semantic grounding. However, keep in mind that the emulation theory is proposed in the context of dynamic models. In this regard, the interesting question here is how to conceive of the 'forward models' implemented by the emulators. From the technical point of view, they could just consist of a look-up table, i.e., a pairing of classical, symbolic, representations (like 'if … then …' productions). However, since an emulator's function is to mimic the input–output function of some other system, it is clear that an optimal emulator should work on the same state variables as the real system, or at least on an approximation, a scale model.

Grush is not explicit about how to conceive of this articulated model of the body–environmental interaction. Spencer and Schöner's notion of 'dynamic field' (Spencer |and Schöner 2003) is promising to me, but developments in computational neuroscience such as those as presented by Knoblauch (Chapter 7, this volume), which formalize how the cortex could implement these models as complex neural networks, open a new field of possibilities to figure out how the brain might implement such emulator models.

What is crucial, though, is that such emulators, given their time constraints, cannot work as symbol crunchers but as dynamical systems that mimic the relevant variables of the interaction. In other words, from this perspective we are not going to expect specific neural activity correlated with each word; as mentioned before, a strength of dymanicism is that it captures nicely the context-dependency of meaning whereas cognitivism assumes context-independent symbolic types, whose causal interaction was determined precisely by its form.

17.6 Analogue representation of models

The question remains how to conceive of the states of emulators as mental representations, as contentful states, and consequently how language comprehension is to be conceived if not as translation into a mental symbolic code. In this regard, I take it that the representational value of such dynamical models that make emulation possible accords best with what has been called a 'model' analogue versus 'sense' analogue representation. The former is best illustrated by a map, while the latter is typically a mental image (or an off-line simulation of a perceptual or motor experience).

What is meant by analogue representation is a state that carries information in virtue of the resemblance with what it represents. It has generally been understood as perceptual resemblance, and contraposed to propositional, language-like representations. As a matter of fact, it was mental images that drove the general notion, which were thought of as a kind of continuous representation, in opposition to the discreteness and abstractness of propositions (mental symbol composites). It can be claimed, however, that analogue representation (and computation) need not be continuous, but also discrete and abstract. Think, for instance, of a map which marks the size of a population using concentric circles along a scale (less than 10,000 people, one circle; 10,000–100,000 people, two circles; and so on). The map represents the territory by keeping distance relations, and represents population size by 'discretizing' this magnitude along a scale. This suggests the 'map' or 'model' interpretation of analogue representation, even though the term 'model' is already too ambiguous since it's also been used for the perceptual understanding of mental models advocated by Johnson-Laird and his followers (as re-enactment of a perceptual experience, and hence as a kind of 'sense' analogue, and of simulationism). However, he has also lately accepted that symbolic elements may also appear in them, given the outstanding role of negation and quantification in their theory, and its abstractness (Johnson-Laird 2002). It is clear that we would never see such concentric circles in reaching a city, just as the perceptual content of an experience cannot be negative; however, it is still true that the representational relation between city size and number of circles is analogical, in a sense that cuts across the contraposition between continuity and discreteness.

This notion of analogue representation extends naturally into a notion of *analogue computation* (remember that it is useless to solve the problem of representation without solving the problem of computation). Instead of conceiving of computation as rule-based manipulation of explicit symbols, analogue computation involves the instantiation of the relations in focus in another medium, which is designed so as to reproduce their values and their interaction. This is the case of scaled-down models, such as a wind tunnel, where the aerodynamical effects are played out on a small-scale model of the aircraft, which allows one to compute the effects without any explicit algorithm execution. It would also be the case of, say, an electric circuit designed to model the flux of commercial distribution in a country, where wires might be taken to represent routes because of the general, abstract isomorphy between representation and represented. The intrinsic interactions between voltages, resistences, etc., in the model are exploited to compute the interaction of the represented values, while some of the parameters may depend on threshold crossing, for instance (thus introducing discontinuity and discreteness). What makes the representation possible is the (partial) isomorphy between the representing and the represented states and values, not their sensorial or motor resemblance.

To sum up, discrete is not the same thing as digital: the first allows for a unique representation of a set of values; the latter involves disjoint variables (Blachowitz 1997). Embodied representation, I want to suggest, should be conceived under this 'model' analogue representation where it is the relational identity between 'object' and 'representamen', rather than their resemblance, that grounds the representationality and the computational effects rather than its sensible content. In this way, abstraction is allowed and harmless. I take it that Grush's emulation theory of representation can be viewed as an updated version of this notion, in that it grounds representation in the internal 'model' of the interaction between the system and environment that mediates this very interaction, and ultimatelygrounds its representational stance. Just as in an airplane, where control is based both on the values of the relevant variables (wind force, altitude, etc.) and on an on-line emulation of the effects of those variables on the machine, given previous values in a continuous calculation, human cognition is to be seen as relying on these internal models of the interaction. So while these internal models are to be conceived as 'schematic' or 'diagrammatic', they are also temporal and dynamical rather than sensorial, conscious, or imagistic.[3] (Luc Steels commented at some point in the Garachico workshop debate that he uses whatever parameter and magnitude he finds useful, not just continuous variables but discrete as well; it may be alright to represent a magnitude by several intervals rather than a rational number; in any case, relational isomorphy may also play a role).

It is not obvious, however, how this approach could be applied in detail to language comprehension. Some considerations can be advanced. First, this approach makes context matter right from the start: language comprehension will proceed according to practical concerns (a point already underlined by Glenberg (1997) as regards 'memory for action'), which may account for the differences in 'level of processing' (underlined mostly by Sanford, Chapter 10, this volume). Second, understanding is not to be conceived as building an explicit symbolic representation of linguistic content, but rather as the outcome of feeding this linguistic information into an active model of the interactive state (which could account for the difficulties in isolating meaning from other effects of language processing, e.g., emotional, nonverbal, facial, and gestural, as pointed out by Nathan, Chapter 18); in this way, it opens a path to try to deal with the problem of relevance (or frame problem). Third, this continually updated internal 'model' is not to be conceived as either a 'sentence in the language of thought' or the imagistic simulation of a perceptual content; the sensoryimotor cortical activations involved in language processing, as demonstrated by Glenberg, de Vega, Pulvermüller, and others, may be reinterpretable as part of the running of the emulator in the control of one's own activity, given current goals. Fourth, a correlational approach to mental functioning (looking for the neuronal correlates of cognitive functions) should yield to a more interactionist view of neural activity: instead of interpreting neural activation as the coding of meaning, it should be viewed as the activity elicited by the linguistic stimulus in an effort to anticipate what's next and how to behave in such particular situation. The idea here is to carry home at the brain level the well established dictum that 'meanings are not in the head,' but in these patterns of interaction, which allows human experience to make sense in providing continuous anticipation and expectancy of what's next. Finally, it seems to me that computational neuroscience work, such as that discussed by Knoblauch (Chapter 7, this volume), coheres well with the current view of the cortex's main function as anticipation and expectancy generation based on internal models (promising a grand-scale model of the interaction, instead of the egocentric model implemented by the cerebellum in motor control).

There is a further implication as regards the psychological role of linguistic items as material symbols. As Clark (2006) has suggested, an alternative view of understanding (instead of as translation) may focus on its stabilizing and anchoring role of otherwise fluid and context-sensitive modes of thought and reason. I would put it this way: that the

---

[3] In cognitive psychology, it has been the work on mental models by Gentner and Stevens (1983) that has been closer to this interpretation of internal models. Shepard's notion of 'second-order isomorphisms' (Shepard 1975) may be also seen as a precursor to this approach. It was within this program that dynamical aspects of representation were first emphasized (cf. Freyd 1987).

order of explanation should be reversed. Instead of explaining language comprehension in terms of translation into a language of thought (be it abstract or built out of sensori-motor primitives), we should conceive of propositional thought and explicit representation in terms of our linguistic competence, of our practical use of external social symbols (Gomila 2002b). It is the acquisition of this competence that gives our thinking its expressive scope, not the other way around.

Debate

Luc Steels: I just wanted to say, you know, that this kind of emulationism obviously can be very useful. In fact, there's been some people, particularly in Berkeley, looking at embodied language understanding using this simulationist approach. But it is also obvious that it is extremely limited, because we cannot expect to simulate every aspect of our world, you know, to learn to do that. I mean, to do it in a realistic way is just not possible. So, it can play a role, but it cannot substitute for all the rest, I think.

Arthur Glenberg: And what's all the rest?

Steels: Well, for example, if you say there is an airplane flying over here, are you emulating flying and everything? It cannot be; you cannot experience yourself flying. So there are lots of aspects that we could emulate, but it is a small amount of what's actually needed in order to understand, to predict the future, or infer.

Glenberg: The point of contraposing emulation to simulation has to do with this. Simulation, as it has been conceived, relies very much on imagination, as a sort of imagistic construction based on previous perception and active ellicitation. Emulation, on the contrary, was introduced as a sort of nonexplicit, unconscious process that mediates fast and efficient action–perception integration, involving a kind of nonsymbolic model of the body–world interactive system. It is thus conceived as the nonrepresentational background that makes one's experience in the world make sense; or, to put it in another terms, to feel at home in the world. From this point of view, this is what understanding amounts to, not to write an explicit proposition in a sort of belief box, but in that the experience makes sense, that we know how to proceed in such a situation, that we feel at home in the world. It is a distinct idea of understanding, not like the translational one of symbolic cognitivism.

Lawrence Barsalou: I'm interested in asking whether you take a reductionist approach, if the higher-order phenomena different people are talking about can be reduced to these low level dynamic systems. I think the history and philosophy of science show that reductionism doesn't often work; higher-level concepts are needed to explain, or can be ideal for explaining, various phenomena. So I was wondering what your thoughts are on this. Maybe the higher-level concepts will have to be implemented in a different way as was classically conceived, but will still be useful.

Antoni Gomila: I agree. This is not to be understood as a reductionist proposal but as a reinterpretationist one. There is so much interesting work done from the standpoint of cognitivism. I'm not suggesting that it is useless or makes no sense. My reasoning is the following: we have, on the one hand, nice software programs (like AutoTutor) or hardware machines, and are told by cognitivism that the brain works just as these programs and machines. But, on the other hand, neuroscience textbooks do not provide a glimpse as to how this could be so. In the last 50 years, the dominant approach to study cognition has been the top-down one you refer to, hoping that in the end we will be able to find out how this can take place in the brain. However, the flourishing of neuroscience (during the 'brain decade') has not provided such an answer. On the contrary, neuroscience seems to have assumed the cognitivist jargon (of codes, symbols, and so on), without making it coherent at its level. Thus, cognitivism expected neuroscience to make clear how the brain instantiates a Turing machine, but what we've got is a queer situation in which neuroscience doesn't provide such an answer but accepts the idea of the brain as a kind of computational device. This is what seems to me wrong. Dynamicism is appealing to me because it has an offer of bypass such a stange situation. It is more congenial to what we know about how the brain works, and answers the question of how the brain computes: in terms of its dynamic patterns. Dynamic models have been put forward in interactionist terms (by Thelen and Schöner), but also of the brain as a system with no need, for instance, for an executive system.

Author note

I want to thank the convenors for the invitation to participate in the Tenerife meeting, and Manuel de Vega in particular for fruitful, island-to-island conversation on these matters. This work has received financial support from the Spanish Ministry of Education through project BFF2003-129. Thanks also to Paco Calvo, partner in this project, and Larry Shapiro, for helpful comments.

References

Bahrick LE, Lickliter R, Flom R (2004). Intersensory redundancy guides the development of selective attention, perception, and cognition in infancy. *Current Directions in Psychological Science*, 13, 99–102.

Barsalou L (1999). Perceptual symbol systems. *Behavioural and Brain Sciences*, 22, 577–609.

Beer R (2000). Dynamical approaches to cognitive science. *Trends in Cognitive Sciences*, 4, 91–9.

Blakemore SJ, Wolpert D, Frith CD (2002). Abnormalities in the awareness of action. *Trends in Cognitive Sciences*, 6, 237–42.

Blanchowicz J (1997). Analog representation beyond mental imagery. *Journal of Philosophy*, 94, 55–84.

Block N (1990). The computer model of the mind. In D Osherson, E Smith, Eds. *Thinking, Vol. 3: An Invitation to Cognitive Science*. Cambridge, MA: MIT Press.

Clark A (1989). Microcognition. Cambridge, MA: MIT Press.

Clark A and Grush R (1999). Towards a cognitive robotics. Adaptive Behaviour, 7, 5-16.

Clark A, Toribio J (1995). Doing without representing? *Synthèse*, 101, 401–31.

Clark A (2006) Material symbols. *Philosophical Psychology*, 19, 291–308.

Cosmides L, Tooby J (1992). Cognitive adaptations for social exchange. In J Barlow, L Cosmides, J Tooby, Eds. *The Adapted Mind* (pp. 163–227). Oxford: Oxford University Press.

Damasio A (1994). *Descartes' Error: Emotion, Reason and the Human Brain*. New York, NY: Putnam.

DeCharms RC, Zador A (2000). Neural representation and the cortical code. *Annual Review of Neuroscience*, 23, 613–47.

Dretske F (1981). *Knowledge and the Flow of Information*. Cambridge, MA: MIT Press.

de Vega M (2001). From symbolic meaning to embodied meaning [In Spanish]. *Estudios de Psicología*, 23, 153–74.

de Vega M, Robertson D, Glenberg A, Kaschak M, Rinck M. (2004). On doing two things at once: temporal constraints on actions in language comprehension. *Memory and Cognition*, 32: 1033–43.

Erlangen W, Schöner G (2002) Dynamic field theory of movement preparation. *Psychological Review*, 109, 545–72.

Evarts EV (1971). Feedback and corollary discharge. A merging of the concepts. *Neuroscience Research Program Bulletin*, 9, 86–112.

Fodor JA (1975). *The Language of Thought*. New York, NY: Thomas Y Crowell.

Fodor JA (1980). The current status of the innateness controversy. In *Re-Presentations: Essays on the Foundations of Cognitive Science*. Cambridge, MA: MIT Press.

Fodor JA (1983). *The Modularity of the Mind*. Cambridge, MA: MIT Press.

Fodor JA (1991). *A Theory of Content and Other Essays*. Cambridge, MA: MIT Press.

Fodor JA (2001). *The Mind Doesn't Work that Way*. Cambridge, MA; MIT Press.

Freyd J (1987) Dynamic mental representations. *Psychological Review*, 94, 427–38.

Frith CD *et al*. (2000). Abnormalities in the awareness and control of action. *Philosophical Transactions on the Royal Society of London B*, 355, 1771–88.

Gentner D, Stevens A, Eds (1983). *Mental Models*. Hillsdale, NJ: Erlbaum.

Glenberg A (1997). What memory is for. *Behavioural and Brain Sciences*, 20, 1–55.

Glenberg A, Kaschack M (2002). Grounding language in action. *Psychonomic Bulletin and Review*, 9, 558–65.

Goldman A (2005). Simulating Mind: The Philosophy, Psychology, and Neuroscience of Mindreading. Oxford: Oxford University Press.

Gomila A (1996) Peirce y la ciencia cognitiva [In Spanish]. *Anuario Filosófico*, 29, 1345–67.

Gomila A (2002a). Meanings are not in the head. And concepts? [In Spanish]. *Estudios de Psicología*, 23, 131–44.

Gomila A (2002b) The language of conscious thought. Paper presented at the 6th Annual Meeting of the Association for the Scientific Study of Consciousness. Barcelona, Spain, May 31–June 3 2002.

Gordon R (1986). Folk psychology as simulation. *Mind and Language*, 1, 158–71.

Grice HP (1957). Meaning. *Philosophical Review*, 66, 377–88. Reprinted in *Studies in the Way of Words*. Cambridge, MA: Harvard University Press.

Grush R (2004) The emulation theory of representation: motor control, imagery and perception. *Behavioural and Brain Sciences*, 27, 377–96.

Harnad S (1990). The symbol grounding problem. *Physica D*, 42, 335–46.

Haugeland J (1995/1998). Mind embodied and embedded. In *Having Thought: Essays in the Metaphysics of Mind* (pp. 207–37). Cambridge, MA: Harvard University Press.

Ito M (1970). Neurophysiological aspects of the cerebellar motor control system. *International Journal of Neurology*, 7, 162–76.

Ito M (1984). *The Cerebellum and Neural Control*. New York, NY: Raven Press.

Jackendoff R (2002). *Foundations of Language*. Oxford: Oxford University Press.

Jeannerod M (2001). *The Cognitive Neuroscience of Action*. Oxford: Blackwell.

Johnson-Laird PN (1983). *Mental Models*. Cambridge, MA: Harvard University Press.

Johnson-Laird PN (1996). Images, models and propositional representations. In M de Vega, M Intons-Peterson, P Johnson-Laird, M Denis, M Marschark, Eds. *Models of Visuospatial Cognition* (pp. 90–127). New York, NY: Oxford University Press.

Johnson-Laird PN. (2002). Peirce, logic diagrams and the elementary operations of reasoning. *Thinking and Reasoning*, 8, 69–95.

Kalman RE (1960). A new approach to linear filtering and prediction problems. *Journal of Basic Engineering*, 82: 35–45.

Kintsch W (1998). *Comprehension: A Paradigm for Cognition*. Cambridge, MA: Cambridge University Press.

Kosslyn S (1994). *Image and Brain: The Resolution of the Imagery Debate*. Cambridge, MA: MIT Press.

Landauer TK, Dumais ST (1997). A solution to Plato's problem: the latent semantic analysis theory of acquisition, induction and representation of knowledge. *Psychological Review*, 104, 211–40.

Markman A, Dietrich E (2000). In defense of representation. *Cognitive Psychology*, 40, 138–71.

Miall RC, Weir DJ, Wolpert DM, Stein JF (1993) Is the cerebellum a Smith predictor? *Journal of Motor Behaviour*, 25, 203–16.

Newell A (1980). Physical symbol systems. *Cognitive Science*, 4, 135–83.

Philipona D, O'Regan JK, Nadal J-P (2003). Is there something out there? Inferring space from sensorimotor dependencies. *Neural Computation*, 15, 2029–49.

Pinker S (1997). *How the Mind Works*. New York, NY: Norton.

Port R, van Gelder T, Eds (1995). *Mind as Motion: Explorations in the Dynamics of Cognition*. Cambridge, MA: MIT Press.

Prinz W (1990). A common coding approach to perception and action. In O Neumann, W Prinz, Eds. *Relationships Between Perception and Action: Current Approaches* (pp. 167–201). Berlin: Springer.

Rodríguez-Fornells A, Kurzbuch AR, Münte TF (2002). Time course of error detection and correction in humans: neurophysicological evidence. *Journal of Neuroscience*, 15, 9990–6.

Sperber D (1996). *Explaining Culture: A Naturalistic Approach*. Oxford: Blackwell.

Schwartz D (1999). Physical imagery: kinematic versus dynamic models. *Cognitive Psychology*, 38, 433–64.

Searle J (1990). Is the brain a digital computer? Presidential Address to the APA. Proceedings of the American Philosophical Association. Reprinted in P Grimm, G Mar, P Williams, Eds. *The Philosopher's Annual*. Atascadero, CA: Ridgeview Publishing Company.

Shepard R (1975) Form, formation and transformation of internal representations. In R Solso, Ed. *Information Processing and Cognition: The Loyola Simposium* (pp. 87–122). Hillsdale, NJ: Erlbaum.

Spencer JP, Schöner G (2003) Bridging the gap in the dynamic systems approach to development. *Developmental Science*, 6, 392–412.

Turvey M (1977). Preliminaries to a theory of action with reference to vision. In R Shaw, J Bransford, Eds. *Perceiving, Action and Knowing: Toward an Ecological Psychology* (pp. 211–65). Hillsdale, NJ: Erlbaum.

Tweed D (2003). *Microcosms of the Brain: What Sensorimotor Systems Reveal about the Mind*. Oxford: Oxford University Press.

Thelen E, Smith L (1994). *A Dynamic Systems Approach to the Development of Cognition and Action*. Cambridge, MA: MIT Press.

van Gelder T (1998). The dynamical hypothesis in cognitive science. *Behavioural and Brain Sciences*, 21, 615–28.

Vosgerau G (2006). The perceptual nature of mental models. In C Held, M Knauff, G Vosgerau, Eds. *Mental Models and the Mind. Current Developments in Cognitive Psychology, Neuroscience and Philosophy of Mind* (pp. 225–75). Amsterdam: Elsevier.

Wolpert D (1997) Computational approaches to motor control. *Trends in Cognitive Sciences*, 1, 209–16.

Wolpert D, Ghahramani Z, Flanagan JR (2001). Perspectives and problems in motor learning. *Trends in Cognitive Sciences*, 5, 487–94.

Chapter 18

An embodied cognition perspective on symbols, gesture, and grounding instruction

Mitchell J Nathan

18.1 **Introduction**

The need to understand and predict behaviour in complex settings such as the classroom and the workplace elevates the importance of the role of context and communication in building models of cognition. Embodied cognition is an emerging framework for understanding intellectual behaviour in relation to the physical and social environment and to the perception- and action-based systems of the body. By reconsidering cognition with regard to interactions with the world, rather than in terms of the sequestered computational nature of the mind, embodied cognition recasts many of the central issues of the study of thought and behaviour. One of the ways that cognition is seen as embodied is through the close relation of hand gestures with thinking and communication. In this chapter, I investigate how gestures enact symbols and thereby ground the meaning of abstract representations used in instructional settings.

Central to this inquiry are two principles that follow from the theory of embodied cognition (Wilson 2002): that cognition is situated and that cognitive work is off-loaded onto the environment. A third principle – that off-line cognition is body based – is also considered, but reframed to include the influences of social interactions, along with sensorimotor processes, as mediating cognitive behaviour even when others are not present. This last issue is developed more fully in the closing section of this paper.

Instruction is a communicative act. As such, what research reveals about instruction, and the learning that ensues, informs our understanding of the production and comprehension of language. When instruction is examined as discourse, it becomes clear that gesture is ubiquitous (Alibali and Nathan, 2007). A review of the literature (e.g., Church *et al.* 2004; Perry *et al.* 1995; Roth 2003; Singer and Goldin-Meadow 2005; Valenzeno *et al.* 2003) also shows that gesture use is influential. Teachers' uses of gestures have been shown to facilitate learner comprehension of material and performance on later tasks. For example, students are more likely to reiterate ideas from tutors' speech when that speech is accompanied by matching gestures (Goldin-Meadow *et al.* 1999). Preschool students' learning and explanations of bilateral symmetry is enhanced when instruction makes use of appropriate gestures (Valenzeno *et al.* 2003). Gestures also help to elaborate teachers' explanations of complex material (Roth and Lawless 2002) and to direct students' attention to important features of formal mathematical representations (Stevens and Hall 1998). Furthermore, there is evidence that gestures are more often used in instruction when the curricular material is unfamiliar or abstract (Alibali and Nathan, 2007). Taken together, these studies suggest that teachers' gestures can have a substantial, positive impact on students' learning of complex material.

An embodied view of cognition would seem to have a great deal to contribute to our understanding of the nature and influence of gestures. Yet, the literatures on gesture research and on embodied cognition are largely independent. The aim here is to show that embodied cognition can inform research on gesture, and that findings from the gesture literature can contribute to embodied theories of cognition. To foreshadow my position, I argue first that gestures are physically realized forms of communication that have the capability to serve as signs at all of Peirce's (1909) levels of signification, performing iconic and symbolic functions, as well as an indexical role. For example, I show examples of how gesture use in instructional settings physically enacts symbols, and how the reoccurrence of distinct features of instructional gestures provide cohesion over the length of a discourse. This serves to evoke a consistent meaning even when the referent varies or is absent. In this way, gestures exhibit *symbolic off-loading* that helps to convey complex and abstract ideas (Iverson and Goldin-Meadow 1998; Krauss 1998), and to offset the large cognitive demands of high-bandwidth exchanges typical of instructional interactions (Alibali and Nathan 2005).

Secondly, gestures can provide grounding for abstract ideas and representations. Grounding, as an aspect of neural processing, provides an account of how sensorimotor experience gives rise to concepts (Havas, personal communication; also see Pulvermüller, Chapter 6, this volume). Yet, grounding is also used to describe a mapping between an idea and a more concrete referent, such as an object in the world (Glenberg *et al.*, Chapter 1, this volume; Gunderson and Gunderson 2006), in order to facilitate meaning making (e.g., Glenberg and Robertson 1999; Harnard 1990; Lakoff and Johnson 1999; Lakoff and Nunez 2000). One definition of grounding that follows from Peirce (see Glenberg *et al.*, Chapter 1) is to view signs as a means of grounding, in that they connect the interpretant or idea to something tangible, such as an object or event. Similarly, Roy (2005, as presented by Glenberg *et al.*, Chapter 1) suggests that grounding denotes the processes by which an agent (human or machine) relates mental structures to external objects.

In instructional settings, this latter form of grounding is often marked by *linking gestures* that provide conceptual correspondences between familiar and unfamiliar entities. This is particularly evident when the focus of instruction is on learning to understand and produce formal (i.e., conventionalized) representations that have an arbitrary mapping between a representational form and its meaning, as with symbols in Peirce's sense. For example, a teacher may provide a gesture as a link between a mathematical symbol, such as L, and the length of a geometric shape by pointing to L and then tracing the longest side of a rectangle (Alibali and Nathan 2005).

Instructional settings are interesting places to study cognition. They are complex, social environments that involve multiple participants with varied capabilities and goals;

they are relatively uncontrolled environments; subject to real-time constraints on thinking and action; and are focused on intellectually demanding activities such as learning and transfer, teaching, metacognition, and reflective thought that take place over extended periods of time. The evidence presented here is primarily in the form of discourse, using both illustrative excerpts and quantitative content analyses, rather than results from experimental design. While the methodological considerations of studies of classroom instruction and cognition are significant, the considerations of these issues are intended to invigorate the debate and foster progress toward a comprehensive theoretical framework for discussing symbolic and embodied perspectives on cognition as they occur naturally in instructional settings.

After presenting evidence that gestures enact symbols and provide grounding of novel and abstract ideas and representations, some broader themes of embodied cognition are considered. First, I consider these findings with respect to the distinctions made between 'on-line' and 'off-line' aspects of cognition, and their relation to principles of situated cognition, and symbolic off-loading of cognitive work. I then examine these points in relation to computer-based metaphors of cognition. Finally, I revisit the debate within the literature that the primary role of gesture is either for lexical access (chiefly serving the speaker) or communication (chiefly serving the audience; Alibali *et al.* 2000; Krauss 1998). I attempt to use principles of embodied cognition to integrate these two views under a new framework that acknowledges ways that internalized body-based behaviours from social interaction can mediate individual thought processes.

18.2 **Use of signs in instructional gesture**

Let *X* equal *X*.

    – Laurie Anderson (1982), from *Big Science*

In Chapter 1, Glenberg *et al.* orient us to the use of *symbol* as one of the types of signs in Peirce's theory of semiotics.[1] A symbol, in this view, is a sign that takes on meaning through its connection to other entities (other signs, or actual objects or events) via arbitrary rules or conventions. While not identical to its use in Newell's (1990) *physical symbol system hypothesis*,[2] symbols in this sense share many of the properties as they are commonly used in cognitive models. For example, Glenberg and Robertson (2000) note that within cognitive psychology, symbol meaning historically 'arises from the syntactic

---

[1] '... I had observed that the most frequently useful division of signs is by trichotomy into firstly likenesses, or, as I prefer to say, *icons*, which serve to represent their objects only in so far as they resemble them in themselves; secondly, *indices*, which represent their objects independently of any resemblance to them, only by virtue of real connections with them, and thirdly *symbols*, which represent their objects, independently alike of any resemblance or any real connection, because dispositions or factitious habits of their interpreters insure their being so understood.' (Peirce 1909, p. 460–1).

[2] This is in contrast to the embodied view of a perceptual symbol system, where perceptual symbols represent concepts that are contextualized, situated tokens (Barsalou, Chapter 13, this volume; but also see Barsalou 1999; Barsalou *et al.* 1993).

combination of abstract, amodal symbols that are arbitrarily related to what they signify' (p. 379). *Icons* are distinguished from symbols in that they make this connection through resemblance, while an *index* connects to a sign through causal or spatiotemporal relations. Embodied cognition theorists, in contrast, argue that to take on meaning, symbols need to be 'embodied and grounded in sensorimotor experience' (Glenberg *et al.*, Chapter 1, p. 4).

18.2.1 *Language d'action*

In this regard, consider the account of the creation of a sign presented by LeBaron and Streeck (2000). An instructor of a home-repair class teaches a lesson on the use of a scraper tool for adding and removing mortar compound to drywall. He picks up one scraper tool from among several on the counter before him, grasps it as if using it to scrape away excess compound, and performs a vertical up–down motion in midair to demonstrate its use. The actual drywall is not present, and neither is the compound. The instructor then repeats this action with a larger scraper, and then a much smaller one (a putty knife). A bit later, the instructor raises an empty hand shaped as though holding a scraper to show how it can be used to apply mortar compound to (an invisible) drywall surface.

In this example, a manual gesture serves a partly indexical, partly iconic, partly symbolic relation to a tangible event or object. Its indexical role is to the prior episode. Iconically, the instructor is using hand shape and arm motion to evoke the use of a tool. It is also symbolic, not in the sense that it is arbitrary, but because the gestural sign use and meaning transcends the specific event–object that led to its origin. The hand motion comes to be a type, referencing a *class* of tools that all conform to this hand shape and perform a similar function using rhythmic arm motions. Furthermore, for some of the novices in the class, the specifics of drywall and compound are unfamiliar, and the actions do not evoke their own, primary experiences of applying or removing mortar, but rather simply evoke the (ungrounded) conventions of the earlier demonstration. Indeed, the instructor's gesture act is a sign in this discourse.

As Lebaron and Streeck (2000) state, 'The formation of a symbol is a defining moment in the fabrication of shared knowledge because it allows the participants to focus upon and evoke previously shared experiences and to plan and conduct shared activities in their wake' (p. 118). It is a case of *language d'action*, an abstraction of a specific act.

18.2.2 **Catchment**

One of the important attributes of the gesture used to signify the scraping tool is its recurrent shape and motion. McNeill and Duncan (2000) use the term *catchment* to describe those communicative events where distinct features of a gesture recur over the length of a discourse. The recurrence can be signalled by the form of the hand shape, its location, orientation, motion, rhythm, and so on. One interpretation of catchment draws directly on the idea that speech and gesture are mediated by a common set of language production processes (McNeill 1985). Following from this model is the claim that the recurrence of a common idea or image for the speaker is exhibited by the re-enactment of a particular gesture. From this, it follows that the re-enactment marks the recurrence of the original image or thought. Thus, catchments convey a semantic association for the speaker, even when they are directed at different objects or in different contexts. It also follows that catchments can contribute to establishment of coherence for the listener.

Catchment behaviour is evident during instruction and, along with speech, can be regarded as ways that instructors attempt to provide continuity of meaning across representations (Alibali and Nathan 2005). As an example, consider the basketball problem as an entry-level problem in a sixth-grade early algebra unit.

The basketball problem

Consider the following: 'Mr. Robinson and his four daughters want to buy a special, autographed basketball. Mr. Robinson's daughters will each pay the same amount. Mr. Robinson will contribute $18 himself. If the ball costs $42, how much will each daughter pay?'

Over the course of a couple of days, students generated two mathematical accounts of the basketball problem that are recreated in Figure 18.1. The equation on the right side, the *solution equation*, was generated first. It was constructed by the students as a way to summarize their steps of executing an arithmetic solution strategy (Nathan and Koedinger 2000) called *unwinding*. Unwinding is highly effective across a range of tasks and grade levels, from middle school through college (Koedinger *et al.* 2008; Koedinger and Nathan 2004; Nathan *et al.* 2002). During unwinding, the student runs the relations stated in the problem situation backwards and 'unwinds' them by performing the inverse arithmetic operations. But unwinding as a solution method has its drawbacks and can easily be thwarted. It also emphasizes a procedural approach of arithmetic reasoning at the expense of an algebraic approach that highlights the structural relations of the problem. Consequently, mathematics instructors want students to understand a more general and powerful method, *direct modelling*, where the situation of the problem stated above is represented directly by algebraic expressions that explicitly identify the quantitative relations between known and unknown values.

In a later lesson, as a way to introduce the situation equation that directly models the relations in the problem statement, the teacher had students act out a skit of the basketball problem (transcript 1):

Transcript 1: The basketball problem skit

1 **Teacher:** Okay? Let's see it. **2 T:** You got to talk really loudly; you're on stage now. I need everybody, I need the audience to be really quiet.

Situation Equation	Solution Equation
$4 \times \dfrac{\text{amount each}}{\text{daughter paid}} + \$18 = \$42$	$(\$42 - \$18) \div 4 = \text{amount each}$ daughter paid ($6)

Fig. 18.1 Classroom whiteboard depicting the parallel structure of the situation (left-hand side) and solution equations.

3 **T:** This is a performance, please!

4 **Daughters:** Wow look at this basketball!

5 **D:** Cool! It's $42 altogether, wow that's a lot! Can you help pay for it?

6 **T:** (*To actors:*) Nope, now nice and loud now.

7 **Father:** Um, I'll give eighteen dollars and you girls give the rest.

8 **D:** We need more dollars!

9 **T:** (*To actors:*) Okay, freeze!

10 **T:** (*To class:*) Have they described the situation so far if you have the numbers on the board?

11 **T:** Yep, they have, right? They mentioned $42; they asked Dad to help; Dad said he'd kick in $18; and they figured out how much left, left they'd have to pay.

12 **T:** (*To actors:*) OK. I'm going to let you finish off the problem even though we haven't done it on the board yet.

13 **T:** Go ahead and just act out the situation; now what are you going to do?

14 **D:** (*Unintelligible*).

15 **T:** Okay you gotta gotta be really loud.

16 **D:** How much are we each going to have to pay?

17 **D:** (*The four daughters in unison:*) Well 24 divided by 4 is 6.

18 **D:** (*To other actors:*) So we … (*unintelligible*). So all of us have to… (*giggling*)

19 **T:** Talk it over with Dad, maybe finish it off here …

20 **D:** (*All four in unison:*) We're each gonna pay six dollars.

21 **F:** (*Unintelligible*).

22 **T:** Okay can you say that *really* loud again Matt, cos that was really good. (*Unintelligible*) listen over there.

23 **F:** Altogether plus the $18 dollars that'd be $42, so we can buy the... basketball

24 **T:** Give 'em a hand.

25 **Class:** (*Applause*)

The teacher relied on students' understanding of this relatively simple situation – made apparent through roleplay – to show that the situation equation is really a mathematical summary of the known and unknown quantities and their relations described in the story problem. It was not merely the retelling of the story, but the acting itself that helped establish the proper situation model and gave meaning to the situation equation (lines 4–7). For example, known and unknown information was differentiated through the use of numbers or letters, respectively (lines 8–10). Even operations were enacted. The students playing the four daughters grounded the multiplicative relation (4x) by announcing in unison how much each would pay (line 20). The physical separation of the daughters and the father is also consistent with the separation of the quantities that they each contribute to additive portion of the algebraic model. This enactment of the basketball problem shows the value of body-based structures and spatial relations for thinking about analytical expressions that model the situation.

There is empirical evidence that this approach is effective for improving learning.[3] This work on simulation relates to recent work by Glenberg and colleagues (Glenberg and Mehta, Chapter 2, this volume; Glenberg *et al.*, in press) using real and imagined manipulation to enhance arithmetic story problem solving with elementary school children. That work frames the challenge for story-problem solving as performing proper indexing, and scaffolding the indexing so it actually happens. It shares the focus on establishing proper mapping. As Glenberg *et al.* hypothesize, 'when the structure of the mental model can be easily mapped onto the mathematical operations, performance is dramatically improved.' Learning gains with the ANIMATE system for solving algebra story-problems provides further support for this instructional approach. In that study (Nathan et al. 1992), those who learned to generate and manipulate algebraic expressions with reference to a running animation of the referent situation described by the story-problem showed greater test performance, more story inference making, and more self-correction than users in comparison groups with and without the simulations.

After establishing the two equations shown in Figure 18.1, the teacher invited the students to consider the similarities between the two symbolic representations in Figure 18.2. She set out to foster links between these two different symbolic representations, and to convey a mapping between the solution steps (solution equation) and the structural relations (situation equation) that she believes will help students to understand the meaning of each mathematical representation in terms of the other. To establish the links, the teacher uses gesture and speech within instructional communication. One thing to note is how the teacher tries to avoid the problem of symbol-to-symbol mapping that leaves the meaning of new symbols ungrounded (Glenberg and Robertson 2000; Harnad 1990).

Initially, students focused on the common elements: as originally written, they each had $42, $18, and a 4. The teacher used these observations to highlight how elements of the two representations map from one side of the board to the other even though their mathematical relations were different (see transcript 2).

Transcript 2: Relating the situation and solution equations (with indices to gestures in [i]).

1 **Teacher:** How are they different? I'd like to keep going on what G3 started to say; how they're written. Tell me something more about how they're written that's different. Boy-1?

2 **B1:** They're backwards.

3 **T:** What do you mean they're backwards?

4 **B1:** Um, like 18 and 42 are in the end of that [0] equation.

---

[3] Grounding abstract representations may be most helpful during the beginning stages of learning to work with formalisms because they also tend to model relatively simple mathematical relations. However, as the complexity of the relations increases, there may be greater advantages incurred when working with amodal representations (Koedinger *et al.*, 2008).

5 *[Gesture-0: Boy-1 points to the situation equation]*

6 **T:** Okay so the 18 [1] and the 42 are sort of at the end over here

7 *[1: Teacher gestures with flat right hand, palm down by sliding her hand under the situation equation, moving from 42 to equal sign, to 18 and back to 42. Her motion freezes under 42 as she says 'end.']*

8 **B1:** And the four [2] and the six are at [3] the beginning of that one.

9 *[2: Teacher moves over to the solution equation and gestures with flat right hand; then, with palm still down, slides hand, in order: under 42, minus sign, 18.]*

10 *[3: Teacher makes slight break in the underline motion when reaches the division symbol, then, 4, equal, and 6.]*

11 **T:** [4] Okay

Having established that the two equations were made up of some of the same elements, the teacher asked students to consider what was different between the two equations (see transcript 2, line 1). This prompt elicited a far deeper awareness from students of how the representations were related. The teacher's use of linking gestures (discussed in more detail below) surfaced when a student (line 2) described the relationship between the two equations as 'backwards'[4]. In response, the teacher probed further (line 3). The first boy (B1) referred to 'that equation' (line 4) and used a pointing gesture to provide a reference (i.e., the situation equation) for his deictic statement. However, the teacher provided a more specific reference (lines 6–7) as she walked over to the situation equation and slid her whole hand (see Figure 18.2a) to underscore the specific elements (the plus-18 and the equals-42) that B1 referred to. While B1 was still talking, the teacher then walked over to the other side of the board and performed the same hand and arm motion – a catchment (see Figure 18.2b) – under the difference expression (\$42 – \$18) of the solution equation (line 9). The act served more than an indexical role for the speaker's words, it also established through the recurrent hand shape that a correspondence existed for the two expressions across the different representations. This conveyed a structural mapping across the two equations, much like that discussed in the analogical reasoning literature (e.g., Gentner 1983).

18.3 **Grounding and linking**

It seems implicit, but worth stating overtly, that in order to serve a grounding role worldly objects and events need to already have some familiarity or meaning assigned to them. In line with this, Lakoff and Nunez (2001, p. 49) argue that for a symbol to be understood it must be associated with 'something meaningful in human cognition that is ultimately grounded in experience and created via neural mechanisms.' With this in

---

[4] These students have not previously discussed mathematical terms like 'inverse relations', and it is understandable that they would need to use colloquial terms like 'backwards' to try to describe what they see.

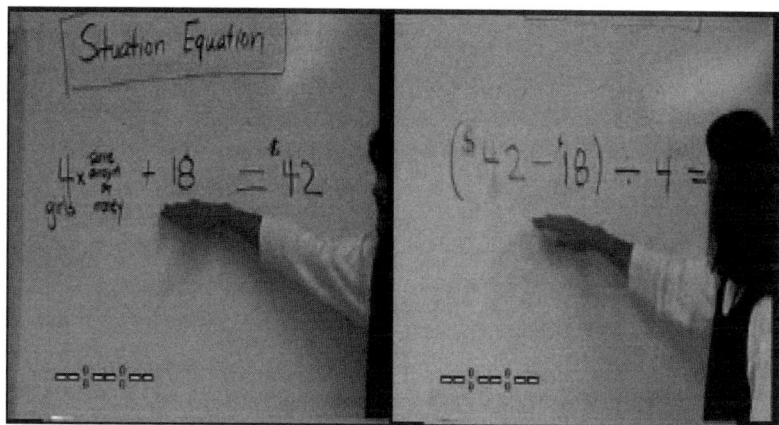

Fig. 18.2 The teacher used an arbitrary gesture to index elements in the situation equation in panel A with a student's speech. She exhibited a catchment by repeating this same gesture for corresponding elements of the solution equation in panel B to establish their relationship across the representations.

mind, it is worth surveying some instructional interactions to see how educators introduce symbols and help students assign meaning to them.

18.3.1 Grounding instruction

It is difficult to imagine the study of classroom learning and instruction – especially in secondary and college-level courses on mathematics and science – without the presence of external symbols (systems of notation) and representations. External representations – particularly *formal representations* such as variables, equations, graphs, and schematics – figure prominently in the classroom and in curricular materials. Much of the focus of mathematics and science education centres on manipulating, interpreting, and producing external representations (Laurillard 2001).

External representations also figure prominently in Peirce's triadic theory of signs. 'A sign, or *representamen*, is something which stands to somebody for something in some respect or capacity.' (Peirce, c. 1897). This description of external representation clarifies its distinction from internal, mental representations, while providing its connection. As Peirce states, 'It [the sign] addresses somebody, that is, creates in the mind of that person an equivalent sign, or perhaps a more developed sign. That sign which it creates I call the *interpretant* of the first sign.' (Peirce, c. 1897).

Grounding as a connection or mapping is commonly inferred when studying communicative situations. Grounding is generally attributed through elaborated speech, the display and manipulations of objects and external representations, and through various forms of gesture, including pointing and representational gesticulation. Instructional settings are particularly rich with attempts to provide meaning for signs, since these are sensorially demanding contexts where new concepts and formal representations are often being introduced and applied.

My colleague, Martha Alibali, and I have used the term *grounding instruction* to highlight ways in which teachers exhibit grounding in the classroom (Alibali and Nathan 2005). One such instance of grounding instruction draws again from the basketball problem episode.

Transcript 3: Teacher grounds a student's speech (with indexes to gestures in [i]).

52 Teacher: [10] Adding, subtracting, multiplying, dividing. How are some of those the same and some of those different?

53 *[Gesture-10: counts each operation off with her fingers]*

54 T: B3? You had your hand up. No? B1?

55 B1: Um, the timesing and dividing are... timesing was there and dividing's there... [11]

56 *[11: Teacher walks over from the right (solution equation) side of the board to the situation equation side and points to the multiplication symbol.]*

A student notes (without explicitly saying it) that the inverse operations of multiplication and division appear in corresponding locations across the two representations. Here, the teacher links B1's reference to 'timesing' (transcript 3, lines 55–56) by walking over to the situation equation and pointing beneath the multiplication operator.

From a conversational standpoint, gesturing to elaborate another speaker's speech is unusual, though my colleagues and I have seen this in other datasets of classroom discourse (Alibali and Nathan, 2007; Nathan *et al.*, 2007). One reason the teacher might do this is to clarify what B1's utterance actually meant – 'timesing' is not a standard mathematical term –while avoiding interrupting him. But the move is pedagogical and communicative, as well as indexical. It linked the word with the sign for the entire class, and was not merely a clarification request between interlocutors as is typically found within conversation analysis (e.g., Sacks *et al.* 1974). This move suggests the teacher's desire to elicit students' ideas to convey mathematical content, thereby engaging other students while still maintaining control over the mathematical meaning of that speech. In this way, linking can bring together the best that student and teacher have to offer.

The linking gesture has the potential to be a powerful pedagogical technique for grounding the meaning of unfamiliar symbols and representations. Linking gestures were evident in a detailed study of a middle school algebra classroom (Alibali and Nathan, 2007). The focal lesson involved illustrations of pan balance scales, like that seen in Figure 18.3, with various combinations of objects of unknown weight. As the teacher described it, 'We're going to translate some of these... pans into equations.' Three pan balances were presented in order: two that balanced (and led to equations; see Figure 18.3), and one that did not (and led, therefore, to an inequality).

After a class discussion of balance and weight, the teacher set out to show the parallel structure between the pan balance and algebraic sentences. In particular, she wished to highlight how balance related to equality (and to the equal sign), how lack of balance related to inequality (and to the greater-than sign), how collections of the same object related to scalar multiplication, how collections of different objects related to addition (and to the '+' operator), and how the unknown weight of any object related

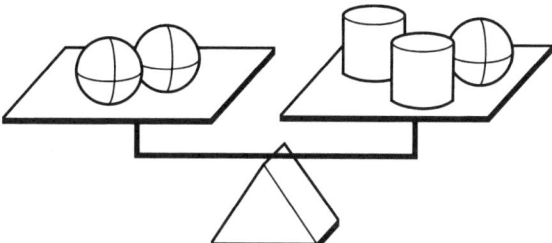

Fig. 18.3 Illustration of a pan balance used in the lesson.

to the use of variables. Second, the teacher supported students' constructions of algebraic models of the pan balances. Third, she demonstrated how equations could be simplified through meaningful operations on the pan balance, such as cancelling out identical variables (objects) from opposite sides of the algebraic sentence (removing identical objects from both sides of the pan balance), and how variables (objects of unknown weight) could be substituted for one another.

In that study, the unit of analysis was the verbal utterance. Each utterance was coded as speech alone, or speech with writing, pointing, or representational gestures (a combination of iconic and metaphoric gestures as they are described by McNeill 1992). An utterance could be accompanied by any number of gestures. In addition, each gesture referent was coded as referencing either the pans (considered the most concrete referent), the algebraic symbols and equations written during the lesson, or the links the teacher made between the pans and symbols (the most abstract referent).

During the 23-minute lesson, 158 utterances were collected that addressed the instructional topic and occurred when the teacher was on camera. Some form of gesture accompaniment, mostly pointing, occurred in 74% of the teacher utterances. The teacher also exhibited linking behaviour. Through careful placement and timing of her gestures, the teacher established a connection between the physical point of balance and the equal sign. Normally, we would not expect the equal sign to be a novel symbol for middle school students, since this is generally introduced in early elementary grades during arithmetic lessons. However, the role of the equal sign in algebra is broader then when used in arithmetic. Specifically, algebraic equations use the equal sign to convey an *equality relation* between two expressions (i.e., the two sides of the equation), and this relation must hold even when the value of these expressions may be unknown. In contrast, the arithmetic meaning of the equal sign indicates that one side of an equation produces the value given in the other side when the rules of computation are properly followed.

How did the teacher set out to convey this alternative interpretation of the equal sign symbol? First, she used pointing, indexing the fulcrums as 'wedges'. In speech she made the link that the wedges will operate 'as the equals sign.' She then wrote an equal sign directly under, and in close proximity to a wedge, setting up the relation by location and establishing the line of symmetry defined by the pan balance as also serving as the line of symmetry for the forthcoming equation.

I want to provisionally claim that the teacher intended these accompanying actions to serve as a form of grounding for her students. That is, I assume that the link she established by drawing the equal sign in that way was meant to be interpreted by students as connecting the mathematical idea of the equality relation to the physical concept of balance. The scale was chosen because it was believed to be familiar and meaningful to the students (an assumption that was checked earlier in the lesson), and it physically embodied the notion of an equality relation (as opposed to producing an answer) through physical balance of equal weight (actually torque) (cf. Lakoff and Nunez 2001).

Grounding instruction of this sort varied with the instructional referent during the lesson. Verbal utterances by the teacher that made links between the equations and the pans was more likely to be accompanied by some form of grounding, through gesture and writing, than were utterances that referred exclusively to algebraic equations or pans: $\chi^2(1, 158) = 4.13, p < 0.05$. The number of grounding acts per utterance also differed significantly across referents, $F(2, 155) = 3.60, p < 0.03$. The teacher used significantly more grounding acts per utterance when talking about pan–equation links than when talking about pans or algebraic equations by themselves.

The teacher's use of grounding instruction also changed over the course of the lesson. The mean number of grounding acts per utterance was high for the initial pan balance equation (mean = 0.78, standard error = 0.08). It then decreased during the second problem of the same type (0.40, 0.07), but increased again for the third, novel problem, which extended the principle to an inequality (0.67, 0.08): $F(2, 260) = 5.35, p < 0.01$. This pattern suggests that the teacher used gestures most often when she introduced materials that she expected to be new and unfamiliar (i.e., on problems 1 and 3) and this practice faded when she believed that the material had become familiar to students. Thus, there is support for the claim that the teacher's use of gestures was to serve a grounding role during instruction.

Lastly, Alibali and Nathan (2007) found that the frequency of grounding instruction was influenced by student demands. The teacher was more likely to gesture after a student utterance, such as a question, than before (56% vs. 31%). Further, the mean number of gestures per utterance for the teacher was significantly greater immediately following student utterances than immediately preceding them (0.69 vs. 0.33, t(35) = 3.17, p < 0.01).

It appears that student needs, in the form of questions or exposure to new and abstract material, trigger more gesturing on the part of the teacher. This is noteworthy, since there is empirical evidence that gesture use by teachers can improve student learning and subsequent gesture production in children (Cook and Goldin-Meadow 2006). We now consider the findings reported above in terms of the theory of embodied cognition, and explore their implications for understanding cognitive behaviour in complex settings.

18.4 Discussion

My aim for this chapter was to show that embodied cognition can inform research on gesture, and that findings from gesture research can contribute to embodied theories of cognition. The findings on instructional gesture presented here included the investigation of cognitive behaviour *in situ* and considered the influences of social and environmental interactions on that behaviour. We saw that instructors can use the physical and dynamic forms that gestures take to enact symbols and to reinstate them (using catchments) at different points in the discourse. Furthermore, those enactments can be transported across contexts and they can denote a class of entities (e.g., scraper tools). In addition to observing examples, classroom data were used to show when gesture served a grounding function. For example, we saw that the referents of a lesson that were more abstract and less familiar were more likely to be the subject of gestures, particularly gestures we expect to serve a grounding role that, presumably, help students understand the lesson and foster meaning making. We also saw that the teacher's use of gesture was responsive to student needs and occurred more frequently after students made demands on the teacher. Now we re-examine these findings of the nature and role of gesture in terms of contemporary theory of embodied cognition.

In reviewing the central tenets and claims of embodied cognition, Wilson (2002) makes a critical distinction between 'on-line' and 'off-line' aspects of cognition. On-line behaviour includes those forms of activity that are rooted in the environment within which it occurs, so that cognitive behaviour is affected by situational and contextual factors such as real-time demands, spatial constraints, and feature-based affordances (e.g., Goldstone et al., Chapter 16, this volume; Kirsch and Maglio 1994). During on-line cognitive behaviour, the mind helps navigate through the world, both metaphorically and literally. Because of this, Wilson (2002) notes, 'the mind can be seen as operating to serve the needs of a body' (p. 635). On-line aspects of cognition are very important for learning, as well as situated performance, and constitute one road to the development of artificially intelligent agents that learn to engage in purposeful interactions with their environment and other agents (Roy, Chapter 11; Steels, Chapter 12).

Still, Wilson suggests exercising caution so as not to overstate the situated nature of cognition. While much of what people do is on-line, she argues that a signature aspect of cognition is the ability to operate off-line, drawing on mental representations of the world that are removed from the events and things that are the object of attention, such as planning, recall, and running mental simulations. From an embodiment perspective, off-line processes may utilize purely mental processes and structures to reason about the world, but to do so they engage the same sensorimotor processes involved in situated cognition and action, when the objects of cognition are distal or even imagined. In this case, as Wilson (2002) notes, the body now serves the mind.

By Wilson's account, off-line behaviour 'is by definition not situated.' (p. 626). However, Wilson applies a narrow view of situated behaviour, one that is not universally accepted by cognitive scientists (e.g., Greeno *et al.* 1996). Obviously, all cognitive processing occurs in some situational context. An alternative to Wilson's view that off-line processing is not situated is to distinguish between the referent situation that is being modelled (but not indeed present) by the off-line processes, from the new setting within which off-line reasoning is taking place. The climate and available associations of the immediate context, even during so-called 'off-line processing', can be shown to influence one's reasoning. One unfortunate but apparently quite robust example of this is when perceptions of stereotype threat impede the intellectual performance of under-represented minorities when they are in mixed-race or mixed-gender testing situations (Inzlicht and Ben-Zeev 2000; Steele 1997; Steele and Aronson 1995). In these cases, members of stereotyped groups may, when their minority status is primed by the current situation, invoke other, threatening settings that serve to hamper their performance. Furthermore, the referent (but distal) situation can also influence off-line cognition. Consider the case of planning: it is rare indeed for a plan to be completely specified. Contingency planning (Suchman 1987), which is highly responsive to the immediate context, is commonplace enough – often planned for, explicitly, in marking out the specified and unspecified portions of a plan – as to be considered a fundamental part of planning. One consideration is that there may, in fact, be no off-line cognition at all. Through enactment, we are really on-line, even when removed from the situation.

The select research on gesture reviewed here also showed support for the claim that some of the cognitive work involved in instruction is off-loaded onto the environment and makes direct use of the environment to support cognitive activity (see Steels, Chapter 12, this volume; Schwartz and Black 1999). Symbolic off-loading, in particular, can help mitigate the inherent limitations of memory and attention, and also allow agents in the environment to ground the meaning of novel and abstract ideas and representations. In this vein, we saw how teacher use of linking gestures made explicit connections between novel and familiar entities. For example, the novel relational interpretation of the equal sign was connected to the fulcrum of a balance scale. We also observed how the introduction of an algebraic use of letters was made by connecting the letters to the weights of the objects they were intended to represent. As a final example of off-loading, we saw how gestures provided explicit links between representations. One particularly elaborate way this was shown was in the analogical mapping constructed by a teacher to establish an element-by-element correspondence between two equations representing different aspects of an algebraic characterization of a problem situation. Computer animations and simulations also provide this kind support and appear to be a natural way for students to learn the meaning of formal representations (Goldstone *et al.*, Chapter 16, this volume; also see Nathan *et al.* 1992). In the language of embodied cognition, one can say that these communicative forms are attempts to ground these new ideas and symbols, in that they establish connections that potentially service comprehension and inference-making processes.

The prevalence of gesture use during instruction, and communication more generally, raises an issue fundamental to the validity of computer-based metaphors of intellectual behaviour, as depicted in classical *information processing theory* (Newell 1990). In its canonical form, an information processor sits at the core of the cognitive architecture. Outside of this 'central processing unit' (CPU) are 'peripherals' – sensory inputs that provide information from the perceptual systems, and actuators that are output systems for acting upon the world (Eisenberg 2002; Wilson 2002). But what about the case when a hand gesture is the primary goal of a speech act, as when a teacher wants to make a linking gesture between a familiar object and a new representation? Data on teaching shows that the use of gesture is tightly woven to teachers' language production. Its role in this complex cognitive activity is not peripheral, but may in fact be central to the peda-gogical and communicative goals of the instructor. By drawing on the metaphor of mind-as-CPU (and body-as-peripheral), the cognitive science community marginalizes the influences of body-based behaviours, and unnecessarily demarcates what is and is not legitimate phenomena within the study of cognition.

The important role of gesture during instruction, and communicative settings more broadly (McNeill 1992), highlights its importance as a visual component of speech production in socially mediated settings. For example, speakers gesture often in commu-nicative settings, and do so more when their listeners can see them (Alibali and Don 2001; Alibali *et al.* 2001). Yet gesture also occurs in settings where speakers are not in visual contact, such as when people are speaking on the phone or otherwise out of visual range, and when the speakers are congenitally blind and have never seen gesturing modelled (Iverson and Goldin-Meadow 1997). For these reasons, some have suggested that gesture is not primarily intended to serve the audience, but rather to aid the speaker with lexical access, particularly when the utterances deal with spatial phenomena (Krauss 1998). Both competing claims about the primary role of gesture can be seen as compati-ble with embodied cognition, particularly that cognition is off-loaded onto the environ-ment, and that even off-line cognition co-opts sensorimotor processes and structures that operate during situated action. However, it is difficult to reconcile them with one another (Alibali *et al.* 2000). In these closing remarks, I consider how a broader treatment of embodiment that includes the social as well as the physical environment can point the way to an overarching framework for gesture production that addresses both commu-nicative and non-communicative functions.

There is growing awareness of the social nature of human thought and learning (e.g., Greeno *et al.* 1996). One avenue recently explored is how gesture serves to create a common frame of reference or common ground among speakers (Nathan and Alibali 2007). It has long been acknowledged that the establishment of common ground – what many scholars refer to as *intersubjectivity* – is an essential aspect of effective communica-tion, and that speakers devote a great deal of effort and time ensuring this at a discourse level, even though they may fail to agree on the specific matter at hand (for a review, see Nathan *et al.*, 2007).

I propose that communicative and noncommunicative functions of gesture can be brought under one theoretical frame that addresses intersubjectivity in both individual and socially situated settings. Contemporary cognitive neuroscience is exploring basic interpersonal processes such as imitation, empathy, and the ability to impute the inten-tions of others – all behaviours that hinge on intersubjectivity – in an effort to under-stand fundamental mechanisms of both individual and social behaviour. During discourse, speakers learn that establishment of intersubjectivity is an effective way to further their social aims during conversations. Researchers studying the behaviour of mirror neurons (e.g., Rizzolati *et al.*, 2001) hypothesize that these neural circuits permit people to directly understand the actions of others because these actions evoke in us the bodily states that we would normally occupy if we initiated those same actions. Mirror neurons are specially evolved and selected areas of brain circuitry 'that allow us to appreciate, experience, and understand the actions we observe, the emotions and the sensations we take others to experience' (Gallese 2003a, p. 525), by constituting them in intersubjective relation to our own actions and feelings. From this empathetic response, speakers constitute *a shared manifold of intersubjectivity* with those with whom they interact (Gallese 2003b). The use of gesture and other body-based behaviours add to the establishment of this shared manifold to the degree that they invoke similar sensations in the listener/observer. This allows interlocutors to provisionally enter into a shared social space, even though their ideas and interpretations may differ from one another.

Through social interaction, and by observing the social interactions of others, I hypothesize that interlocutors internalize the social and cognitive behaviours that lead to effective discourse. This includes, in no small measure, gestures and other actions that facilitate the appropriate selection of words so that the shared manifold is maintained. As proposed within sociocultural theory (e.g., Vygotsky 1978; Wertsch 1979) these interpersonal interactions help to determine the development of intrapersonal (i.e., individual) processes. The body-based behaviours that were so effective in socially mediated thinking are co-opted to fulfil the goals of individual thinking, such as planning and lexical access. Those gestures that foster intersubjectivity among speakers also assist us in articulating our own thoughts to ourselves (and others) in a clear manner. In this way, we are drawing on socially mediated processes such as gesture even when we are seemingly 'alone with our thoughts.'

Debate

Friedemann Pulvermüller: Thank you very much for your interesting talk. It seems to be clear that gestures somehow help in understanding an explanation. However, it could be at different levels. There is no doubt that gestures help, for example, in structuring the discourse by making emphasis – this is the word I want to attend to. Or I could make an index using a hand gesture that says that something needs to be kept in mind, in the case of understanding a formula. So certain processes can be symbolized or indexed, but on the other hand explaining or symbolizing addition and multiplication might be extremely difficult or impossible. So can you maybe comment on that?

Mitchell Nathan: It's true, I don't think a teacher would only want to use gesture as their only source of instruction or gesture even in speech. I didn't show a video of this skit but there's a really nice part here where the four daughters all say in unison, 'We need to contribute our money to buy this basketball.' And there's sort of this element of oh wow, that's like multiplication, you know. Here's these four girls at the same time, conveying this idea. As a teacher you wouldn't by any means want to narrow down your repertoire of how to convey the range of ideas, even just in mathematics, and gesture isn't going to be suitable for everything. I agree. What's remarkable is how pervasive it is, and it's also a little bit of a concern how kind of under the radar it's gone for so long. For instance, we do very little in teacher education around what teachers should gesture. We don't talk about it at all – I don't really know of any programs that do talk about it. However, it's clear that it could be used to one's advantage tremendously.

Walter Kintsch: Should we spank the students when they invent something as intelligent as the 'unwinding' strategy? Because you really don't want them to do this. I mean, it works, but to scale up you want them to use the equation.

Nathan: A big debate that I think that happens in education, and certainly relevant to algebra instruction, is where you want people to end up versus how you want them to get there. So I think we have a lot of agreement about where we want everyone to end up. But I think it's pretty clear that if we try to do it directly by teaching these formalisms, they don't get there. So, for instance, think of the teacher who believes that symbolic reasoning precedes story-problem solving, and reserves story problems until students reach a certain level of equation manipulation. Well, some of those students are never gonna get across that threshold. They're never going to get to exercise the reasoning they may have for solving a story problem, because the teacher feels like they're sparing this student from this onerous task that they'll never be able to accomplish because they have a developmental model that puts story-problem solving as farther out than equation solving. But in fact, we have some developmental data that compares the developmental models, symbol precedence (i.e., symbol first) versus verbal precedence. And it shows that the verbal precedence is a far better fit for our data across two datasets. So it would be a mistake for a teacher to withhold story-problem solving. They're not doing the student a favour. In fact, what we show in the intervention studies is that you can leverage students' reasoning on these more accessible verbal problems and get them to do the symbolic representation, and even get them to solve nonlinear problems in 7th and 8th grade, which often they don't get to do anyway. So I think there's a lot of promise for this method. I understand your intuition, and I think we should always be careful about it, but I think that as you look at the developmental trajectory issue, you start to realize that we need to be looking at how to move people. And that's why this progressive formulization is such an appealing idea.

Antoni Gomila: My first comment is that it looks like what is more important for algebra teaching is probably reading comprehension. Language understanding through texts, maybe even more than gestures, I don't know. The other thing is that your talk nicely illustrates the point that Wertheimer already made in his classical productive thinking theory, explaining that to understand is to know how to proceed, in a given context. The context can be very different, and not just to repeat the same situation again and again to apply the same equation again and again, but to be able to codify a representation of a problem in more abstract terms.

Nathan: Yeah, I think that's a great connection, thank you. That's nice.

Francis Quek: Ah, just a couple of things. Number one, you showed that when teachers do the equation the first time there are more gestures, and then the second time the number of gestures drop. And you started by talking about catchments that tend to hold certain features to show that there's continuity of ideas. Did you see that?

Did you see that although the number of gestures declined, the form of the gesture was the same so the student could be cued in that it's the same problem?

Nathan: I can't say that I've systematically looked at it across the entire dataset. But I can tell you we see lots of examples of this catchment phenomenon. And what's interesting to me about catchments – I don't know if Duncan and McNeill talk about it quite this way – but there's a way in which a repeated hand form that's meant to convey the same idea across different settings is like a symbol, Right? And one of my favourite examples is Streeck and LeBaron's work with a vocational education teacher showing students how to use a scraper for a drywall, like putting the compound on the wall. He picks up the tool and introduces this tool idea to a room that's basically novices, but then later just the hand form, so there's no tool in his hand anymore; the hand form is used to reinvoke the idea of the tool. And then it gets used to reinvoke other tools that have the same function but are not exactly the same tool. So, like, a spackling device, which is not so different than a drywall compound tool, also gets referenced by this hand form. I think it's an incredibly powerful idea. So we do see it, but we haven't had a cause to systematically pull it out of that dataset. Unless, you know, you have some hypotheses we could talk about.

Quek: One last question: I just wanted to introduce this so we might discuss this later. We have talked a lot about pointing and deixis, but pointing at a reference involves three things, right? There is the thing pointed to, the field of reference, and the point of view of reference, which is actually very odd for instruction, whether I'm looking from here or from there. And good teachers would actually shift their shoulders – Randi Engle from Stanford has been working on this – they would actually give a different viewpoint just by subtly pointing at the same thing but from a different position. So if you're looking at something from a viewpoint of inequality as opposed to an equality, the teacher might do this change of position, and that cues a student in to know that something new is happening. And, of course, the third aspect of pointing action is *how* you point. So the question I have, given that there's so much pointing in the work that you see, have you looked at not so much the field of reference but source of reference. That might actually be more telling.

Nathan: We've looked at how pointing is involved in linking. I'm not sure if that's exactly the same thing you're asking for. We have looked at how it's used to link across two representations, say an equation and, like, an equation on one side and an inverse of that equation on the other side. Martha Alibali has some nice data on the formula for area and the parts of a shape of the object, essentially to dereference the variables in the area formula. Is that what you're asking?

Quek: Randi Engle did this. She showed teachers explaining a lock mechanism to a student. And, of course, when explaining a lock mechanism you actually look at the lock from the top, from a viewpoint of the tumblers of the key. And she showed that teachers would consistently shift positions. The teacher would consistently use different viewpoints to cue the students in, and that's one of the hardest things to do in teaching, to explain the changes in abstraction and changes in viewpoint. And this seems to be much better cued in gesture than in speech.

Nathan: I'm not sure how much cause that would come up for inscriptions like equations where the viewpoint probably doesn't change that much from teacher to student. So, because she's looking at three-dimensional objects and kind of moveable systems, it seems that it's more relevant.

Quek: Thank you very much.

Author note

I would like to thank my colleague Martha Wagner Alibali for her extensive contributions to this program of research. I also wish to thank members of the STAAR research collaborative. In addition, David Havas was very generous with his time and feedback during our frequent conversations on these issues and reviews of an earlier draft of this chapter. I thank Rob Goldstone for his thoughtful review of this work. Finally, my warmest regards to Amy French for her openness to the scrutiny that comes with participating in classroom research. Preparation of this paper was funded in part by funds from the University of Wisconsin School of Education. This research was funded through a grant entitled 'Understanding and Cultivating the Transition from Arithmetic to Algebraic Reasoning,' by the Interagency Educational Research Initiative, an alliance of the National Science Foundation, the US Department of Education's Institute of Educational Sciences, and the National Institute of Child Health and Human Development within the National Institutes of Health.

References

Alibali MW, Don LS (2001). Children's gestures are meant to be seen. *Gesture*, 1, 113–27.

Alibali MW, Heath DC, Myers HJ (2001). Effects of visibility between speaker and listener on gesture production: some gestures are meant to be seen. *Journal of Memory and Language*, 44, 169–88.

Alibali MW, Nathan MJ (2007). Teachers' gestures as a means of scaffolding students' understanding: evidence from an early algebra lesson. In R Goldman, R Pea, BJ Barron, S Derry, Eds. *Video Research in the Learning Sciences* (pp. 349–365). Mathwah, N.J. Erlbaum

Alibali MW, Nathan MJ (2005). Teachers use gestures to link multiple representations of mathematical information. In 'Gestures and Embodied Action in Teaching, Learning, and Professional Practice.' Paper presented at the Annual Meeting of the American Educational Research Association, Montreal, Quebec, Canada.

Anderson L (1982). Let X = X. On *Big Science* [LP record]. New York, NY: Warner Brothers.

Barsalou LW (1999). Perceptual symbol systems. *Behavioural and Brain Sciences*, 22, 577–609.

Barsalou LW, Yeh W, Luka BJ, Olseth KL, Mix KS, Wu L (1993). Concepts and meaning. In K Beals, G Cooke, D Kathman, KE McCullough, S Kita, D Testen, Eds. *Chicago Linguistics Society 29: Papers from the Parasession on Conceptual Representations* (pp. 23–61). Chicago, IL: University of Chicago, Chicago Linguistics Society.

Church RB, Ayman-Nolley S, Mahootian S (2004). The role of gesture in bilingual education: does gesture enhance learning? *International Journal of Bilingual Education and Bilingualism*, 7, 303–19.

Cook SW, Goldin-Meadow S (2006). The role of gesture in learning: Do children use their hands to change their minds? *Journal of Cognition and Development*, 7, 211–32.

Eisenberg M (2002). Output devices, computation, and the future of mathematical crafts. *International Journal of Computers for Mathematical Learning*, 7, 1–44.

Gallese V (2003a). The manifold nature of interpersonal relations: the quest for a common mechanism. *Philosophical Transactions of the Royal Society B: Biological Sciences*, 358, 517–28.

Gallese V (2003b) The roots of empathy: the shared manifold hypothesis and the neural basis of intersubjectivity. *Psychopatology*, 36, 171–80.

Gentner D, Stevens A (1983). *Mental Models*. Hillsdale, NJ: Erlbaum.

Glenberg AM, Jaworski B, Rischal M (submitted). Improving reading improves math.

Glenberg AM, Jaworski B, Rischal M, Levin JR (in press). What brains are for: action, meaning, and reading comprehension. In D McNamara, Ed. *Reading Comprehension Strategies: Theories, Interventions, and Technologies*. Mahwah, NJ: Erlbaum.

Glenberg AM, Robertson DA (2000). Symbol grounding and meaning: a comparison of high-dimensional and embodied theories of meaning. *Journal of Memory and Language*, 43, 379–401.

Goldin-Meadow S (1997). When gestures and words speak differently. *Current Directions in Psychological Science*, 6, 138–43.

Goldin-Meadow S, Kim S, and Singer M (1999). What the adult's hands tell the student's mind about math. *Journal of Educational Psychology*, 91, 720-730.

Greeno JG, Collins AM, Resnick LB (1996). Cognition and learning. In D Berliner, R Calfee, Eds. *Handbook of Educational Psychology* (pp. 15–46). New York, NY: Macmillan.

Gunderson JP, Gunderson LF (2006). Reification: what is it, and why should I care? In *Performance Metrics For Intelligent Systems (PERMIS) Workshop* (pp. 39–46), Gaithersburg, MD: National Institute of Standards and Technology.

Harnad S (1990). The symbol grounding problem. *Physica D*, 42, 335–46.

Hostetter AB, Bieda K, Alibali AW, Nathan MJ, Knuth EJ (2006). Don't just tell them, show them! Teachers can intentionally alter their instructional gestures. In R Sun, N Miyake, Eds. *Proceedings of The 28th Annual Conference of the Cognitive Science Society* (pp. 1523–8). Mahwah, NJ: Erlbaum.

Inzlicht M, Ben-Zeev T (2000). A threatening intellectual environment: why females are susceptible to experiencing problem-solving deficits in the presence of males. *Psychological Science*, 11, 365–71.

Iverson J, Goldin-Meadow S (1997). What's communication got to do with it? Gesture in blind children. *Developmental Psychology*, 33, 453–67.

Iverson JM, Goldin-Meadow S (1998). Why people gesture when they speak. *Nature*, 396, 228.

Kirsh D, Maglio P (1994). On distinguishing epistemic from pragmatic action. *Cognitive Science*, 18, 513–49.

Koedinger KR, Alibali M, Nathan MJ (2008). Trade-offs between grounded and abstract representations: evidence from algebra problem solving. *Cognitive Science* 32, 366–397.

Koedinger KR, Nathan MJ (2004). The real story behind story problems: effects of representations on quantitative reasoning. *Journal of the Learning Sciences*, 13, 129–64.

Krauss RM (1998). Why do we gesture when we speak? *Current Directions in Psychological Science*, 7, 54–60.

Lakoff G, Johnson M (1980/2003). *Metaphors We Live By, Second Edition*. Chicago, IL: University of Chicago Press.

Lakoff G, Nunez RE (2000). *Where Mathematics Comes From: How the Embodied Mind Brings Mathematics into Being*. New York, NY: Basic Books.

Laurillard D (2002). *Rethinking University Teaching: A Conversational Framework for the Effective use of Learning Technologies*, Second edition. New York, NY: Routledge.

LeBaron C, Streeck J (2000). Gestures, knowledge, and the world. In D McNeill, Ed. *Language and Gesture: Window into Thought and Action* (pp. 118–38). Cambridge: Cambridge University Press.

McNeill D (1985). So you think gestures are nonverbal? *Psychological Review*, 92, 350–71.

McNeill D (1992). *Hand and Mind: What gestures Reveal about Thought*. Chicago, IL: University of Chicago Press.

McNeill D, Duncan SD (2000). Growth points in thinking-for-speaking. In D McNeill, Ed. *Language and Gesture: Window into Thought and Action* (pp. 141–61). Cambridge: Cambridge University Press.

Nathan MJ, Alibali MW (2007). Giving a hand to support the mind: how gesture use enables intersubjectivity in the classroom. In the symposium "Mechanisms by which gestures contribute to Establishing common ground: Evidence from Teaching and Learning, presented to the International Society for Gesture Studies (ISGS). Third International Conference. Evanston, IL.

Nathan MJ, Eilam B, Kim S (2007). To disagree, we must also agree: how intersubjectivity structures and perpetuates discourse in a mathematics classroom. *Journal of the Learning Sciences* 16, 525–565.

Nathan MJ, Kintsch W, Young E (1992). A theory of algebra word problem comprehension and its implications for the design of computer learning environments. *Cognition and Instruction*, 9, 329–89.

Nathan MJ, Koedinger KR (2000). Teachers' and researchers' beliefs about the development of algebraic reasoning. *Journal for Research in Mathematics Education*, 31, 168–90.

Nathan MJ, Stephens AC, Masarik DK, Alibali MW, Koedinger KR (2002). Representational fluency in middle school: a classroom based study. In Mewborn D, Sztajn P, White D, Wiegel H, Bryant R, Nooney K, Eds. *Proceedings of the 24th Annual Meeting of the North American chapter of the International Group for the Psychology of Mathematics Education*. Columbus, OH: ERIC Clearinghouse for Science, Mathematics, and Environmental Education.

Newell A (1990). *Unified Theories of Cognition*. Cambridge, MA: Harvard University Press.

Perry M, Berch D, Singleton J (1995). Constructing shared understanding: the role of nonverbal input in learning contexts. *Journal of Contemporary Legal Issues*, 6, 213–35.

Peirce CS (c. 1897). A Fragment, CP 2.228, c. 1897. Retrieved November 14, 2006 from http://www.helsinki.fi/science/commens/terms/fallibilism.html

Peirce CS (1909). 'A Sketch of Logical Critics', EP 2:460-461. Retrieved November 14, 2006 from http://www.helsinki.fi/science/commens/terms/fallibilism.html

Roth W-M (2003). Gesture–speech phenomena, learning and development. *Educational Psychologist*, 38, 249–63.

Roth W-M, Lawless D (2002a). Scientific investigations, metaphorical gestures, and the emergence of abstract scientific concepts. *Learning and Instruction*, 12, 285–304.

Sacks H, Schegloff E, Jefferson G (1974). A simplest systematics for the organization of turn-taking in conversation. *Language*, 50, 696–735.

Schwartz DL, Black T (1999). Inferences through imagined actions: knowing by simulated doing. *Journal of Experimental Psychology: Learning, Memory, and Cognition*, 25, 116–36.

Singer MA, Goldin-Meadow S (2005). Children learn when their teacher's gestures and speech differ. *Psychological Science*, 16, 85–9.

Suchman L (1987). *Plans and Situated Actions*. Cambridge: Cambridge University Press.

Steele CM (1997). A threat in the air: how stereotypes shape the intellectual identities and performance of women and African–Americans. *American Psychologist*, 52, 613–29.

Steele CM, Aronson J (1995). Stereotype threat and the intellectual test performance of African–Americans. *Journal of Personality and Social Psychology*, 69, 797–811.

Stevens R, Hall R (1998). Disciplined perception: learning to see in technoscience. In M Lampert, M Blunk, Eds. *Talking Mathematics in School: Studies of Teaching and Learning* (pp. 107–49). Cambridge: Cambridge University Press.

Valenzeno L, Alibali MW, Klatzky RL (2003). Teachers' gestures facilitate students' learning: a lesson in symmetry. *Contemporary Educational Psychology*, 28, 187–204.

Vygotsky LS (1978). *Mind and Society: The Development of Higher Mental Processes*. Cambridge, MA: Harvard University Press.

Wertsch JV (1979). From social interaction to higher psychological processes: a clarification and application of Vygotsky's theory. *Human Development*, 22, 1–22.

Wilson M (2002). Six views of embodied cognition. *Psychonomic Bulletin and Review*, 9, 625–36.

Chapter 19

Reflecting on the debate

Manuel de Vega, Arthur C Graesser, and
Arthur M Glenberg

19.1 Introduction

This chapter presents a comprehensive view of the main topics of debate in the book, and includes some positive reflections and conclusions. It is organized into the following sections:

- Section 19.2 presents the main positive argument in favour of a symbolic framework, namely the claim that many forms of meaning consist of symbolic computations, including the statistical models that are based on word covariation, without need of perceptual grounding.

- Section 19.3 discusses experimental evidence that supports the embodiment approach, with special attention to the action-sentence compatibility effect (ACE) experiments that demonstrate motoric activation while processing action verbs.

- Section 19.4 focuses on abstract meaning, which is usually considered a challenge for embodied theories because of the claim that abstract words and sentences do not activate sensorimotor experience with enough systemacticity to be useful.

- Section 19.5 addresses the problem of reference, which is at the heart of the grounding problem, and how symbolic and embodied frameworks can account for referential processing.

- Section 19.6 discusses the neuroscience data that show how perceptual and motor areas in the brain are recruited in the processing of meaning, the role of the traditional linguistic areas (Broca and Wernicke), and other higher-order processing regions of the brain.

- Section 19.7 explores how meaning could be reduced to computations, and how the fields of artificial intelligence (AI) and robotics might contribute to clarifying some issues in the current debate.

- Section 19.8 focuses on how the embodiment and the symbolic frameworks can be incorporated into theories of complex learning and education, such as the building of complex explanations or the extent to which the acquisition of abstract concepts is facilitated or interfered by perceptual–motor experience.

◆ Section 19.9 proposes a minimal account of meaning that may be acceptable for all participants in the debate, identifies the disagreements, and stimulates productive research venues.

Most sections are organized in terms of arguments, counterarguments, and conclusions, and use the discussants' opinions as raw material. The discussants generally kept a coherent point of view along the different topics of the debate. Many of the discussants made an effort to understand the opponents' data and points of view. That is, defenders of embodiment qualified their analysis and approached some aspects of symbolism, whereas discussants who were sympathetic to a symbolic stance attempted to explain how grounding on perceptual experience takes place. Nonetheless, the positions of many of the participants remain sufficiently distinctive that their research programs and hypotheses evolved in different trajectories.

19.2 Meaning as symbolic computation and symbol covariation

Cognitive science has a rich history of theoretical frameworks in the symbolic camp. Interestingly, many of the computational symbolic models that were developed in the 1970s and 1980s (Anderson 1983; Card et al. 1983; Laird et al. 1987; Miller and Johnson-Laird 1976; Norman and Rumelhart 1975; Ortony *et al.* 1988) explicitly attempted to explain the representations and processing of perception, action, emotions, and other aspects of embodied experience in addition to language, categorization, decision making, problem solving, and other cognitive facilities that appear to be inherently symbolic. Much has been learned about psychological foundations of symbolic models over the last 50 years of cognitive science. As a consequence, none of the participants in the present debate would argue for the psychological plausibility of classical symbolic models that are detached from referential content, such as propositional calculus and predicate calculus. None of the participants would claim that meaning is always abstract and is uncorrelated with perception and action. There are more debatable issues that separate the symbolic and embodied frameworks.

Much of this section addresses one of the controversies that raised some stimulating discussions in the debate. The controversy addressed the role of statistical models that capture covariations among symbols. In particular, to what extent can statistical symbolic models that rely on symbol covariation, such as *latent semantic analysis* (LSA; Landauer et al. 2007), play an explanatory role in the comprehension of meaning? Are such models sufficient, or is it necessary to ground the representations in embodied perceptual–motor experience?

19.2.1 In favour of covariation

Among those discussants with a symbolist approach, some advanced the position that meaning is based on word co-occurrence. One salient example is LSA (Landauer, and Dumais 1997; Landauer *et al.* 2007), a statistical theory of knowledge representation that considers the co-occurrence of words in documents, based on large corpora with millions of words. LSA also has mathematical techniques to reduce dimensionality (called singular value decomposition) and thereby reflects a type of contextual co-occurrence in which word A is similar to word B because they both occur in context C. Let's call this approach *statistical symbolism* to distinguish it from the traditional forms of symbolism that dominated the more philosophically oriented cognitive scientists in the 1970s and 1980s (Fodor 1983; Pylyshyn 1986; see Shapiro, Chapter 4, this volume, for a review) and from those building computational symbolic models during the same period (Anderson 1983; Card *et al.* 1983; Kosslyn and Shwartz 1977; Laird *et al.* 1987; Miller and Johnson-Laird 1976; Norman and Rumelhart 1975).

The statistical and traditional symbolic approaches have several differences. First, statistical symbolism uses words (which are unquestionably symbolic but also observable) as units of analysis. This focus on words as a unit of analysis is addressed in this section, although it should be acknowledged that researchers could conceivably integrate word units with units of perception and action (see Graesser and Jackson, Chapter 3, this volume). In contrast, traditional symbolist approaches were based on the doctrine of the language of thought (e.g. Fodor 1983; Pylyshyn 1986; see Shapiro, Chapter 4 for a review). Theoretically, language of thought operates with mental symbols which, unlike words, are conceptual, abstract, and sometimes amodal, although they share with natural language some formal properties, such as the potential arbitrariness of symbols and syntactic rules.

Second, the statistical symbolism approach is more empirical than the traditional symbolic approaches. For instance, the impressive ability of LSA to predict psychological phenomena emerges directly from the words themselves or, more accurately, from the statistical properties of words taken from natural language corpora. In contrast, the formal features of meaning (symbols and composition rules) in the language of thought approach were designed by the researchers. For example, in language of thought implementations with propositional representations, propositions are constructed according to rules and intuitions by the researchers (and applied by hand by trained collaborators). The resulting propositional languages run the risk of showing differences among researchers. For instance, Kintsch and van Dijk's (1978) propositional analysis labels predicates (main verbs, adjectives, adverbs) and arguments (nouns, pronouns) with English words, whereas the syntax is underspecified and does not capture such formal distinctions as subject versus predicate phrases. In Anderson's (1983) ACT theory, the propositions have a different syntax that does explicitly capture subject–predicate distinction, whereas nouns and main verbs have corresponding explicit words as in the case of Kintsch and van Dijk. In Norman and Rumelhart's (1975) ELINOR model, words are decomposed into theoretical primitives and the syntactic analyses are more recursive and complex. As a consequence of these different assumptions about representations, the structural properties of sentences differed quite a bit among propositional analyses.

The turn of symbolism from *a priori* formal analyses to a mathematical analysis of natural language corpora was good news for the current debate because covariation models provide an empirical foundation that is broadly acceptable. For instance, some discussants in the symbolic band demonstrated that LSA models can make empirical predictions of spatial knowledge (e.g. geographic distances), metaphor understanding, and even some typical embodiment effects reported in the literature (see Louwerse and Jeuniaux, Chapter 15, this volume). On the other hand, some embodiment researchers examined some of the implications of data produced by LSA and ultimately demonstrated that the semantics emerging from word covariation, although an important psychological phenomenon, is not deeply identifiable with meaning (see Glenberg and Mehta, Chapter 2).

Although statistical symbolists accept in principle that grounding on perception and action can happen, they do not capitalize on the functional power of embodied processes in on-line comprehension of language. Instead, computations on symbols, words, or other representations that are disembodied or not fully embodied can explain much of meaning in most language comprehension tasks. Let's summarize their arguments.

Symbols reflect world knowledge

Statistical symbolism shows that the mathematical parameters in a space of multiple dimensions obtained from LSA mirror world knowledge in a sort of second-order isomorphism and, consequently, explicit activation of embodied features is not always necessary. For instance, Louwerse and Jeuniaux (Chapter 15, this volume) claim: 'language is structured in such a way that many relationships that can also be found in the embodied world are structured in language,' or Kintsch (Chapter 8) argues that Symbols are not defined by the real world, but symbols mirror the real world. An isomorphism exists between symbolic representations and the world of action and perception.

Symbolist proponents in this book provide multiple examples of how world knowledge is reflected by covariation data obtained by LSA. Thus, the cosine of distance among pairs of cities correlates with real distances in the world, or the cosine between wasp wings and butterfly wings is lower than between wasp wings and bee wings. These isomorphic capacities of word covariation methods are a strong argument, as we will see, for sometimes defending a purely symbolic computation of meaning.

On-line comprehension is too fast and/or complex to be embodied

The symbolic approach argues that grounding on embodied cognition, although sometimes occurs, is not feasible in many comprehension tasks. For instance, comprehension of a complex text requires an on-line process of multiple features of the linguistic code (graphemes, phonemes, morphemes, words, parsing), the construction of a referential situation model, and the global gist of the text and sections of text. The deeper meaning requires integration of ideas, the formation of local and global coherence, inferences, pragmatic aspects of the communication, and so on. Under such processing burden, symbolists suggest that on-line activation of embodied features often does not fit the temporal constraints of reading (Louwerse and Jeuniaux, Chapter 15; Kintsch, Chapter 8; Graesser and Jackson, Chapter 3; and Sanford, Chapter 10, this volume).

For instance, Louwerse and Jeuniaux (Chapter 15) say 'That some particular words trigger visual/motor imagery or other embodied representations is common in certain circumstances but it is argued that efficiency in language processing simply does not yield a continuous systematic grounding.' A similar argument is carefully developed by Graesser and Jackson (Chapter 3) who quantify the time needed for on-line updating of situation models: 'It takes approximately 300–1000 milliseconds to construct a new referent in a discourse space, several hundred milliseconds to move an entity from one location to another, a few hundred more milliseconds to have the mental camera zoom in on an entity within a crowded mental space, and so on' and they conclude: 'These considerations on timing and complexity make it doubtful that all referring expressions and clauses in the text have fully embodied representations during reading. It would take an order of magnitude increase in reading time to achieve this.' Barsalou, Santos, Simmons, and Wilson (Chapter 13), who are in many ways defenders of embodiment theories, provide data that confirm that words associations (covariations) are computed faster than situational or embodied features of meaning.

Finally, Sanford (Chapter 10) also expresses doubts about on-line simulation derived from a quite restrictive notion of simulation. He accepts that motor simulation or resonance observed in ACE experiments involves a sort of shallow simulation of the generalized direction of movement. However, this gross simulation differs considerably from fine-grained simulation of actions involving specific motoric programs (e.g. body-part motions) that would be involved in a real simulation. According to Sanford, fine-grained simulations do not occur on-line, a conclusion substantially similar to Graesser and Jackson's position.

Notice that these arguments do not reject the possibility that high-resolution simulations could take place eventually or under special conditions. For example, high-resolution simulations are prone to occur when ideas are in the discourse focus (as opposed to presupposed), when the experimental goals dictate such simulations, when there is not a burden on working memory, and when the stimuli are so simple that there is the luxury of thorough processing (Graesser and Jackson, Chapter 3). What they claim is that, in most cases, such simulations are not compatible with the temporal constraints of on-line comprehension of a large number of text constituents in normal reading of normal texts.

Symbolic computation is a shortcut to meaning

The shortcut argument is a conclusion from the two previous premises. If word covariations reflect world knowledge, and grounding on embodied representations is not feasible in on-line comprehension, then statistical symbolism seems an appropriate alternative because computing semantics of covariation is allegedly fast and sufficiently reliable. Louwerse and Jeuniaux (Chapter 15) make explicit this conclusion: 'symbols can, but do not always have to, be grounded. That is, language is structured in such a way that many relationships that can also be found in the embodied world are structured in language. Language thereby provides a shortcut to the embodied relations in the world.'

A different argument of how symbols are shortcuts for meaning is Kintsch's claim that computation using arbitrary symbols (e.g., words) is faster than computation using nonarbitrary symbols (see Chapter 8). Thus, some studies with young children demonstrate that they tend to confuse the iconic representation of objects (e.g., pictures) with the real thing, whereas they do not have such confusions when the task is mediated by words. The moral is that in some occasions an arbitrary representation is a better cue for meaning than an iconic or embodied representation.

Meaning could derive from word relationships alone

Some words do not seem to derive their meaning from grounding on embodied experience. Instead, their meaning seems purely conceptual and resulting from their relations to other words or concepts. Whereas the meanings of most concrete words are learnt in perceptual and manipulative contexts, it seems that the meaning of words like justice, truth, confidence, brotherhood, etc., are not necessarily derived from such concrete scenarios. Concrete concepts can also sometimes be learned from verbal definitions rather than from grounding on experience. Thus, Louwerse and Jeuniaux (Chapter 15) claim that '… there is also compelling evidence that symbols can derive meaning from other symbols. For instance, we know that a rose is a flower because "rose" appears in linguistic contexts similar to the contexts in which "flower" appears. The same is true for associations between "perfume" and "smell" and for "thunder" and "loud".'

Sanford (Chapter 10) proposes that antonymous relationships are not grounded but rely on what he calls interdefinition: 'The need for interdefinition is ubiquitous: for adequate communication, we need to know that 'left' is the opposite of 'right', and that 'north' is the opposite of 'south'. But we do not normally need to engage in on-line embodied cognition to use these words or these facts … the claim that (lexical) meaning is exclusively embodied must be false, since relationships between terms need to be known, and these are essentially abstractions.' Finally, Kintsch (Chapter 8) demonstrated that a LSA-based model with dependency parsing algorithms is able to make sophisticated semantic operations working exclusively with linguistic contexts. Included in these demonstrations are metaphor comprehension ('my lawyer is a shark'), sentence completion ('if you try, you can _____ any problem'), and homograph resolution ('bark' in the context of dog, or in the context of tree).

The existence of abstract words and sentences is a favourable argument for symbolist theories and posits a real challenge for the embodiment approach. Given the crucial importance of this subject we will return to it in a subsequent section.

19.2.2 **Against covariation**

A positive consequence of the current debate is that none of the discussants deny the empirical facts provided by the adversaries. For instance, the observation that humans are sensitive to subtle covariations among words or among semantic features was accepted implicitly or explicitly by those supporting an embodied approach. However, some of them presented arguments that minimize or qualify the role of such covariations in meaning (Glenberg and Mehta, Chapter 2; Barsalou et al., Chapter 13; Shapiro, Chapter 4; and Zwaan, Chapter 9, this volume). Let's see their claims.

Ungrounded covariations do not provide meaning

Covariation among ungrounded symbols is not always the basis of meaning (Barsalou *et al.*, Chapter 13; Glenberg and Mehta, Chapter 2; Shapiro, Chapter 4). Thus, Barsalou *et al.*

agree with statistical symbolism that computing covariations in the lexicon is a robust psychological phenomenon, which occurs by default as part of meaning computations. But covariations only provide a form of shallow meaning. Real meaning is necessarily grounded on perception and action (simulations), although certainly these simulations, which are deeper processes, could be activated with a certain delay with respect to lexical covariations.

Glenberg and Mehta (Chapter 2) provided the most direct challenge to the claim that covariation alone can account for meaning. Unlike Searle's mental experiment of the Chinese room, they report three actual experiments in which covariation of ungrounded symbols was used as an independent variable to see whether participants were able to derive their meaning. For instance, in one experiment, Glenberg and Mehta selected a semantic domain familiar to the participants, the domain of two-wheeled vehicles, and made empirical observations of their frequency on campus. Later on, they created an interface in which the participant could explore how frequent two-wheeled vehicles categories were, and how they related to several descriptors such as ISA relations, properties, parts, and so on. However, in this learning task, neither the semantic labels for the categories nor the descriptors were provided. Thus, the symbol for a particular category and the symbol for a particular descriptor were ungrounded, but these symbols did co-occur in frequencies that matched their real-world co-occurrence.

The students' task was to learn the co-occurrence relations and to use those relations to discover the semantic domain. The results showed, in general, that participants were skilful in using co-occurrence to learn which symbols went with which others. However, they failed completely in the domain identification task. That is, after learning at least part of the covariation structure among the symbols, they could not then use that structure to identify that the symbols referred to two-wheeled vehicles. Therefore, meaning did not emerge from covariation, even though covariation was computed accurately.

Covariation of words is a by-product of experiential covariation

Words (or features) covariations are not meaning but correlates with meaningful experience. Several contributors (Barsalou et al., Chapter 13; Zwaan, Chapter 9; and Glenberg and Mehta, Chapter 2) emphasize that the primary phenomenon of covariation occurs among experiential or referential traces. Thus, perceptions, actions, and internal states (emotions, intentions, etc.) co-occur in our experience. Among these experience traces are words themselves which have a rich perceptual quality (phonetic, prosodic, motoric, orthographic, and the like). Covariation is thus a global phenomenon that includes words (as perceptual and motoric events), their grounded referents, and associated internal states. This multimodal network of covariations allows obtaining statistical regularities from just one aspect of it (e.g., words), but this does not mean that the co-occurring experiential traces are then discarded when computing meaning.

It is remarkable here how the same phenomenon (lexical covariations) is evaluated differently by statistical symbolists and defenders of embodiment. For instance Barsalou *et al.* (Chapter 13) agree with statistical symbolists that covariations in the lexical system and in the simulation system mirror each other, but they consider that 'One reason is that people constantly hear language that corresponds to perceived situations'.

Zwaan (Chapter 9) elaborates on this: 'Just as linguistic traces can co-occur in time and space, so referential traces can co-occur in time and space (Zwaan and Madden 2005). For example, tomatoes and lettuce can be found together in salads; staplers and notepads on desktops; and guitars and drums in recording studios. Given that language most often describes situations in the real world, or in realistic environments, L–L co-occurrences tend to reflect R–R co-occurrences. Techniques such as LSA can capitalize on such second-order correlations between co-occurrences.'

Meanings derived from words are also embodied

A counterargument to the statement that new meanings derive from word co-occurrence alone is that even new meanings are indirectly grounded in embodied representations (Zwaan, Chapter 9; Glenberg and Mehta, Chapter 2; and Pulvermüller, Chapter 6). Using the combinatorial properties of perceptual symbols, Barsalou (1999) explains how embodied knowledge can be used to simulate an unknown entity. For instance, cues such as 'What has horns like a giraffe, a deer head, a horse neck, and legs like a zebra?' can easily lead to a schematic image of the animal (an okapi) that is sufficient to recognize it in the flesh or in a picture (Zwaan, Chapter 9).

An example of the role played by grounding on learning new meanings is the case of innovative denominal verbs reported by Kaschak and Glenberg (2000). A sentence like 'Lyn crutched Tom her apple to prove her point' could be difficult to understand in isolation. However, it was easily understood in a context in which the protagonists are baseball players bragging about batting skill, and one of them (Lyn) is on crutches because of a twisted ankle. Nonetheless, she claims that she can hit anything using anything, and 'Lyn crutched Tom her apple to prove her point.' Of course, this is a genuine case of meaning derived from word combinations, except that the new meaning is more easily explained by deriving and meshing the affordances of crutch and apple and the like, than by LSA covariation data.

Pulvermüller (Chapter 6) speculates on an embodied neural mechanism to explain how new meanings could be obtained 'parasitically' from linguistic context: 'A new word would activate its form-related perisylvian cell assembly, while neurons outside perisylvian space processing aspects of the semantic of context words are still active. The correlated activation of the semantic neurons of context words and the form-related perisylvian neurons of the new word may lead to linkage of semantic features to the new word form.'

Embodied meaning can be fast and unconscious

Statistical symbolism suggests that computing word covariances is much faster than activating simulations. This can be true as *Barsalou et al.* (Chapter 13) demonstrated in their own research: activation of linguistic relations is faster than activating simulations. However, the argument can be made that some kinds of simulations are activated very fast (in the range of a few hundred of milliseconds) as we will see in a latter section. Discussants in the embodiment side suggest that fast embodied simulations can be unconscious as many other cognitive processes which occur fast and automatically. The lack of awareness could produce the subjective impression that embodied grounding does not occur (Glenberg and Mehta, Chapter 2; and Barsalou et al., Chapter 13).

19.2.3 Conclusions

Table 19.1 summarizes the arguments in favour of statistical symbolism and the counter-arguments provided by the embodiment defenders. Statistical symbolism theories go beyond old-fashioned symbolic theories in paying attention to the statistical properties of language and conceptualization. They also differ from computational symbolic theories that attempt to explain the representation and processing of perceptual, motor, and other modality-specific modules of cognition. The statistical symbolic theories were originally inspired by mathematical analysis of texts that could account for developmental data, inductive learning, and such psycholinguistic phenomena as word frequency effects, similarity, and semantic priming. Statistical symbolism also converges with connectionist models in proposing that psychological properties emerge from the computation of co-occurrences (Glenberg, de Vega, and Graesser, Chapter 1), although many connectionists prefer to deal with neuron-like subsymbolic units rather than words or concepts. The covariance discussants in this volume do not provide direct neurological arguments, but there are covariance theorists who make the case that covariance is compatible with the powerful associative properties of the brain emphasized by some current neuroscience theories. For example, Damasio's (1989) time-locked circuits, neo-Hebbian cell assemblies (Pulvermüller, Chapter 6; and Knoblauch, Chapter 7), and functional neuronal connectivity (Just, Chapter 5) all exploit the correlational properties of neuronal activation. Therefore, it is possible that statistical symbolist theories may end up providing an explanatory account of neurological data with human participants.

Contributors within the embodiment approach accept some of the ideas advanced by the symbolic approach. They agree with the trivial observation that natural language is literally symbolic (e.g. composed of arbitrary symbols syntactically organized) and, most importantly, that word covariation plays a role in language processing. However, the interpretation of these facts differs substantially between symbolists and embodiment theorists. First, statistical symbolists claim that exploiting covariations is a shortcut for meaning, and in many circumstances grounding is not necessary. By contrast, some embodiment people believe that covariations do not provide meaning and grounding is

Table 19.1 Statistical symbolism: pros and cons

Statistical symbolism arguments	Embodiment counterarguments
Symbols reflect world knowledge	Yes, but only because words correlate with experience
Embodied cognition at deeper levels is too slow for most on-line comprehension	Some kinds of embodied simulations are fast and unconscious
Symbols are short-cuts for meaning. Meaning does not need to have embodied grounding	Meaning frequently or always requires embodied grounding
Some meaning is derived from symbol relationships	Combinatorial properties of perceptual and motoric symbols are capable of producing new meanings

always necessary for real comprehension. Second, the very notion of 'word' differs in both approaches. Statistical symbolists consider that words are just pointers to amodal symbols which may or may not mirror embodied experience; in this sense, statistical computations are made on mental symbols and/or meaning is an abstract property emerging from words but not identifiable with words. By contrast, the embodiment theorists emphasize that words are perceptual events with rich sensory features (visual, auditory, and motor) and that some of these features migrate to the encoded meaning representations (Barsalou *et al.*, Chapter 13; Pulvermüller, Chapter 6; and Zwaan, Chapter 9). Thus, Pulvermüller demonstrates that listening to words elicits activity in the superior temporal cortex (related to the phonetic analysis of the linguistic signal) but also the somatotopic precentral gyrus (related to articulating speech) (see Chapter 6). If words are action–perception multimodal events rather than abstract symbols, then covariations could be computed on the perceptual traces of those events, combined with the perceptual traces of other concurrent events (objects, actions, agents, etc.), so the boundaries between symbols and embodied traces end up becoming imprecise (Zwaan, Chapter 9).

19.3 Meaning as embodied cognition

19.3.1 In favour of embodied meaning

The embodiment approach is supported by empirical data consistent with the idea that meaning recruits sensorimotor knowledge. Dozens of carefully designed experiments, both with behavioural and neurological methods, have provided impressive evidence for embodied meaning. For instance, some classical studies in the 1980s and 1990s have shown that comprehenders can build situation models or 'proxy-situated' scenarios that implement spatial, temporal, causal, or emotional features of situations (Sanford, Chapter 10). Other experiments clearly demonstrated the activation of visual features (shape, size, colour, or orientation), or other sensory features of objects (sounds, taste, smell, or touch) when people understand concrete nouns and sentences. Finally, recent experiments show that motor programs are mobilized when we understand verbs or sentences describing actions.

The embodiment contributors also have gone beyond simple descriptions of embodiment phenomena. They have provided theoretical interpretations of the empirical facts in a number of paradigms, developed the quasi-neurological notions of resonance (Zwaan, Chapter 9), incorporated Hebbian assemblies as a neuronal mechanism for embodiment in the brain (Pulvermüller, Chapter 6), reused the Gibsonian notion of affordance as the core of word meaning (Glenberg and Robertson 2000), developed the hybrid notion of perceptual symbols and simulators (Barsalou 1999), and the like.[1] Let's see their arguments.

---

[1] Although some symbolic models offer a disembodied approach to cognition, there are dozens of computational symbolic models that attempt to model the modality-specific mechanisms. However, it is the embodiment framework that is taking the lead in unveiling how many of the constraints of perception and action end up shaping our meaning representations when we comprehend language.

Embodied meaning is activated rapidly

We already mentioned how symbolists and embodiment defenders strongly disagree on the speed of processing of embodied meaning. Some experiments suggest that at least some embodied parameters of meaning are processed very fast. For instance, the ACE studies first performed by Glenberg and Kaschak (2002) indicated that motoric simulations are activated on-line while understanding sentences that describe actions. De Vega (Chapter 14) presents a set of experiments with a variant of the ACE procedure that showed meaning–action interference as early as 100 milliseconds after the action verb onset. Other studies also show a fast activation of visual motion simulations when people understand sentences describing motions, as the interaction between sentence meaning and concurrent moving stimulus indicates (Kaschak et al. 2005). In the same vein, Pulvermüller (Chapter 6) describes studies with magnetoencephalography (MEG) showing that action verbs trigger activation over specific premotor areas in the brain after only 100–200 milliseconds.

The fast set up of embodied simulation is a convincing phenomenon. However, some of the same ACE studies have also shown that motoric activation associated with action sentences is short-lived and fades in a few milliseconds (de Vega, Chapter 14). In addition, automatic activation of sensorimotor features associated with words or phrases is just an aspect of embodied meaning. More complex simulations, such as updatable scenarios with multiple situation parameters, may require resource demanding processes and probably are much slower (Graesser and Jackson, Chapter 3; and Sanford, Chapter 10).

Embodied meaning as neuronal resonance

A very plausible neuronal mechanism for embodied meaning is resonance or the reactivation of sensorimotor areas when words are processed. For instance, the ACE effects could be explained by the interconnection between the brain areas responsible for processing word forms (e.g., Broca's and Wernicke's areas) and the motoric areas in the brain. The connectivity among language areas and motoric areas is a consequence of the history of co-activation of action verbs and particular motoric events in the individual's experience, which creates Hebbian cell assemblies (Pulvermüller, Chapter 6; and Zwaan, Chapter 9). The same resonant circuitry can be established between concrete nouns and sensory areas in the brain which usually process colour, shape, sound, smell, and the like. Of course, resonance alone is a very basic phenomenon corresponding to the 'vocabulary' of the embodied code or, in Barsalou's (1999) terms, the perceptual symbols.

To account for the high-order meaning at the level of sentence comprehension or discourse integration, some combinatorial operations on perceptual symbols or 'resonances' should take place in the brain. The functional description of these operations is related to the notion of 'simulators' (Barsalou 1999) or 'meshing of affordances' (Glenberg and Kaschak 2002). The brain structures responsible for these combinatorial properties have not yet been specified.

As discussed in the previous section, however, statistical symbolic theorists would argue that covariation and associationism provide the most relevant aspect of brain mechanism. Indeed, associationism and covariation are at the essence of Hebbian learning theory.

Embodied meaning as proxy-situated cognition

A rather different view of embodied meaning, and one that is closer to the symbolic camp, is the notion of *proxy-situated representations* proposed by Sanford (Chapter 10; see also Barsalou et al., Chapter 13). Sanford assumes that during comprehension the incoming information is related to situation-specific background knowledge that markedly differs from propositional or other arbitrary representations, and that often involves sensorimotor parameters. Proxy-situated representations are close to the traditional symbolic notions of scenario or schema. Sanford is, however, quite skeptical about the functional role of resonance on the on-line construction and updating of fine-grained simulations.

Embodied meaning prepares for action and specifies reference

One argument is that embodied meaning has functional advantages over symbolist accounts. The basic idea that we will develop in a subsequent section is that embodied meaning provides a better interface with the world than symbolic meaning. First, embodied meaning prepares one for action (Barsalou et al., Chapter 13; and Zwaan, Chapter 9). In Barsalou et al.'s terms: 'The presence of situational information prepares agents for situated action. Rather than only representing the focal knowledge of interest, as in a dictionary or encyclopaedia entry, a category representation prepares agents for interacting effectively with its members.' In some cases, such as negative sentences, it might also induce interlocutors to refrain from action, such as Zwaan (Chapter 9) suggests when we listen to the sentence: 'Don't eat those red berries or you'll get very sick.' The idea of 'preparation for action' is also germane to the notion of meaning as affordance mobilization postulated by Glenberg and Robertson (2000). Thus, nouns corresponding to objects would activate automatically motor affordances. For instance, the word 'paper' activates motoric affordances such as 'getting', 'tearing', 'throwing', or 'writing', depending on the particular context.

A second functional advantage of embodied meaning is that it facilitates the process of specifying reference. A pragmatic function of human communication is providing enough cues for the addressee to specify the particular object, event, or state being referred to (de Vega, Chapter 14). Given that many word referents are perceptual, a perceptual-like meaning might be more efficient interfacing system than an abstract amodal code. Embodied meaning has the advantage of using a common code with perceptual experience, and mapping from one to the other is thus potentially straightforward. In contrast, specifying reference in a purely symbolic system would require a 'translation' function (e.g., correspondence rules) to map amodal symbols into perceptual experience, and vice versa. These translations may be computationally possible but, in principle, less parsimonious than using embodied experiences directly.

19.3.2 Against embodied meaning

Embodiment experiments are not conclusive

The symbolists accept the positive experimental evidence supporting embodiment. However, they also consider embodied representations to be constructed only under specific conditions, such as when the information to be comprehended is directly in the discourse focus, when there are adequate working memory resources and sufficient time to build the embodied representations, when the experimental goals favour the construction of such representations, and/or when the experimental stimulus is sufficiently limited (few distractions) that the only alternative is to process the verbal material at such a level (Graesser and Jackson, Chapter 3; and Louwerse and Jeuniaux, Chapter 15). Many of the experiments are set up to meet these conditions and are therefore not discriminating in testing the alternative theoretical predictions of the embodiment and symbolic frameworks. The strongest confirmation of embodiment would occur under conditions in which the text is not in the discourse focus (i.e., presupposed information), there is a load on working memory, the task requires quick processing, the task goals are not pitched at embodied processing, and the stimulus is complex enough to accommodate alternative processing goals. Experimental support for embodiment rarely meets these conditions so the status of available evidence is at best debatable. Thus, symbolists see embodiment data as occurring under predictable and specific conditions rather than being ubiquitous in comprehension. In some cases, symbolists have been more radical in their criticisms of embodiment experiments by proposing alternative explanations for their results (Sanford, Chapter 10; Kintsch, Chapter 8; Louwerse and Jeuniaux, Chapter 15).

A corollary of the statement that symbols mimic world knowledge is that they could explain/predict the typical results of embodiment experiments. This is exactly what Kintsch proposes in Chapter 8. He claims, for instance, that Glenberg and Robertson's (2000) results could be predicted by a combination of LSA and dependency tree algorithms, without need of grounding words on conceptual affordances. For instance, Kintsch's symbolist model predicts that the sentence 'She gave him a red beanbag to play with' is more sensible than 'She gave him a large refrigerator to play with' (as demonstrated by Glenberg and Robertson), as the cosines among sentence fragments are higher in the former sentence than in the latter sentence.

Louwerse and Jeuniaux (Chapter 15) consider that some experiments supporting embodied meaning are biased by experimental task demands. For instance, in some experiments there are nonlinguistic cues (e.g. simultaneous visual animations or motoric responses) that could be responsible of activating the intended sensorimotor features. In other cases, a 'deep' processing of meaning (including sensorimotor grounding) is prompted by the task instructions. For instance, Zwaan and Yaxley (2003) observed that people judged the relationship between pairs of partitive words faster when they were presented in iconic positions ('attic' above 'basement') that when they were not ('basement' above 'attic'). However, as Louwerse and Jeuniaux demonstrated, such embodiment effects only emerge in tasks demanding 'deep' processing, whereas in shallow processing task such as lexical decision there was no effect of spatial iconicity. Louwerse and Jeuniaux's demonstration that shallow processing is not embodied is acceptable to some researchers who are sympathetic to embodiment (Barsalou et al., Chapter 13).

Resonance is sometimes an epiphenomenon

The symbolists accept the overwhelming evidence of sensorimotor resonance from words during comprehension, but they consider the fast-acting resonance a temporary phase in the total balance of comprehension processes. Thus, the fast resonance observed in the ACE experiments may correspond to the 'promiscuous' construction phase in Kintsch's *construction–integration theory* (see Chapter 8), rather than the integration phase which encodes the meaning. In Graesser and Jackson's words: 'much of this automatic activation of representations end up dying away and never make it to the integration phase that establishes a more coherent representation of the meaning of the text.' It is therefore important to differentiate meanings that are quickly activated from words from meanings that reflect the deeper interpretations of sentences and discourse. It is the latter embodied meanings that symbolists doubt are constructed quickly, whereas there is no reason to doubt the embodiment of initial activations.

Kintsch's own position on this subject is an exercise of scientific pragmatism. He also assumes that a multimodal activation of images, somatic markers, words, schemas, and the like may occur at some point in the comprehension process. But he concludes that a comprehensive theory on how all these complex systems operate together is beyond the scope of current cognitive science. So, for practical reasons, he opts for exploiting the regularities emerging from the linguistic code, forgetting for the moment the other 'complexities' of meaning, namely, the embodiment processes. Even Sanford, who assumes proxy-situated process of meaning in some comprehension tasks, is quite skeptical about the functional role of resonance or activation of sensorimotor processes (see Chapter 10). He suggests that such processes are not real-time components of comprehension but side effects of processing. In other words, his position is close to the epiphenomenon view of resonance.

Finally, Gomila (Chapter 17) has a nihilist opinion regarding neurological resonance. He considers that brain correlates cannot be identified with the content of meaning. For instance, the activation of premotor areas in the brain when we understand the verb 'digging' does not imply that the meaning of the word is there. If that were the case, an apraxic patient wouldn't be able to understand action verbs. Of course, this is an empirical question, but apraxic patients would not provide a definite answer. The embodiment approach would predict a selective comprehension difficulty for action descriptions in apraxics, but still these patients would have some comprehension of these sentences because action traces are multimodal, involving visual motions, and sounds in addition to motoric features. A recent study confirms this idea: none of us have motoric resonance when seeing a dog barking but still we may have an embodied representation of barking in the visual and auditory modalities (Buccino *et al.* 2004).

Preparing for action and perceptual reference can be misleading for some tasks

Researchers in the symbolic camp could potentially counter the claim that actions are the most functional level of representation. Actions may work for a large class of everyday experiences, but a more functional representation may consist of causal explanations and deep mental models that transcend everyday actions. Graesser and Jackson (Chapter 3) discuss the importance of abstract notions such as 'force' and the role of causal explanations when solving everyday problems requiring conceptual physics. Indeed, the everyday actions run the risk of creating persistent misconceptions (i.e., the impetus fallacy) that block adaptive problem-solving activities. So there is some debate over the levels of representation (action, perception, explanation, planning) that are most functional for the survival of organisms.

In the same vein, the symbolic camp could counter the claims about the functional advantages of perceptual references. Sometimes perceptual similarity between the properties of the current problem, and a previous solution runs the risk of blocking important analogies and causal schema in problem solving (Forbus et al. 1995; Goldstone and Sakamoto 2003).

19.3.3 **Conclusions**

Table 19.2 summarizes the arguments in favour of embodiment and the counterarguments against embodiment. There may be a feeling of disorientation when putting together all the information reviewed up to now. However, we have hopefully provided some clarity in the claims, arguments, counterarguments, and disagreements. Both symbolists and embodiment defenders present compelling data and arguments for their cause. Although these approaches have a long tradition in cognitive science, both have made recent progress. The new statistical symbolists have substituted for the intuition-based formal rules (e.g., propositions, production systems, and the like) a more inductive symbolism which takes advantage of the undeniable associative power of human cognition. On the other side, embodied meaning is becoming a mature idea in cognitive science. Empirical demonstrations of sensorimotor grounding are well established, and the new dual-task paradigms have shown automatic activation of embodied features. In addition, embodiment has found favourable terrain in some neuroimaging studies that provide neurological evidence of perceptual and motoric resonance for words.

In this book, the two approaches not only have presented arguments and data to support their own views, but they also have attempted to explain the data supporting the opposite view. Thus, embodiment theorists accept that the brain computes word co-occurrence and these computations play a role in comprehension. But they propose that

Table 19.2 Embodied meaning: pros and cons

Embodiment arguments	Symbolist counterarguments
Embodied meaning is activated rapidly.	Not so for deeply encoded representations and extended combinatorial simulations.
Embodied meaning involves a resonance process.	Resonance is epiphenomenal in the sense of being a temporary activation that may or may not be incorporated in encoded meanings.
Embodied meaning involves, in some cases, a proxy-situated representation.	Proxy-situated representations are basically symbolic.
Embodied meaning prepares for action and facilitates reference specification.	Action and perceptual cues for reference have some utility for function, but also some liabilities.

the statistical processing of meaning is a 'shallow' process rather than real 'deep' comprehension. Embodiment theorists also recognize some exciting challenges in explaining abstract representations. On the other hand, symbolists recognize that grounding meaning on embodied perception and action is important, but this grounding is either a temporary epiphenomenal activation (e.g. motoric resonance) or a costly and slow process that occurs only under specific conditions on-line. In addition, some symbolists try to explain the embodiment results by means of purely symbolic computations or specific task demands.

Some convergence between the symbolism and embodied meaning also exists. The two approaches emphasize how associative processes contribute to meaning. Thus the mechanisms postulated by embodiment theories to explain resonance in the brain rely on the same principle of covariation applied by symbolists to explain meaning. Of course, the parallelisms are limited. Statistical symbolism exploits only symbol-to-symbol covariations, with limited attention to the brain mechanisms, whereas resonance is a crossmodal process of covariation among words and sensorimotor structures broadly distributed in the brain.

Which is more economical with respect to processing time? Unfortunately, applying parsimony principles for processing does not make a choice easier because both theories tell a story of processing parsimony adapted to their own theoretical biases. For instance, statistical symbolists assume that computing covariances on symbols is faster than, and functionally close to, creating embodied simulations (Kintsch, Chapter 8; Louwerse and Jeuniaux, Chapter 15). Therefore, it might be more parsimonious to compute meaning by relying exclusively on the fast and efficient symbols, and suppressing a superfluous and costly embodied grounding. On the other side, embodiment discussants (Zwaan, Chapter 9; and Pulvermüller, Chapter 6) have also a parsimonious proposal. Everything in language comprehension is embodied, not only the meaning of words, but also the representations of words themselves. With this latter perspective, we have a broad multimodal network in the brain for quickly processing the motoric, visual, and auditory features of words, as well as their meaning correlates in sensorimotor areas. But there is no place for amodal symbols.

19.4 Abstract meaning

Many experiments that support the embodiment approach have used concrete words or sentences referring to objects, actions, or emotional reactions as experimental materials. These are important contents in human cognition and language, but people also use language to describe experiences beyond those provided by concrete words and sentences. Symbolist feel quite comfortable with abstractions; indeed, radical symbolists consider that all symbols are ultimately abstract. By contrast, abstraction seems a challenge to the embodiment approach because abstract words often do not appear to be grounded in embodied experiences. The best hope for the embodiment approach is to demonstrate that abstract words and sentences are also grounded on perception and action, which is rather counterintuitive.

Several discussants set up arguments on the difficulty of explaining abstract meaning in terms of perceptual grounding. Unfortunately, 'abstract' or 'abstraction' are themselves abstract and polysemic terms. For instance, the two first definitions of 'abstract' in the Encarta English Dictionary are: (1) not relating to concrete objects but expressing something that can only be appreciated intellectually, and (2) based on general principles or theories rather than on specific instances. These definitions contain in turn many abstract terms ('relating', 'concrete', 'intellectually', 'general', 'appreciated') and are certainly difficult to ground on any specific perceptual experience. To better organize this important topic we will distinguish different aspects of meaning which are related to abstraction. These aspects include: (a) granularity, (b) abstract content words, and (c) abstract linguistic markers. This divide-and-conquer strategy delimits the problems for embodiment theories, as well as suggesting some difficulties for symbolic theories.

19.4.1 Granularity

Granularity specifies the level of detail or scope of conceptual or linguistic descriptions. Speakers and listeners can focus on coarse- or fine-grained conceptualizations of events and situations. Compare, for instance, the difference in granularity between 'There was somebody in the street' and 'There was a man with a hat at a corner of the street'. The issue of granularity has been discussed by several participants with a focus on action words. Sanford (Chapter 10) was concerned with the degree of granularity acceptable for considering that an embodied simulation occurs. One might consider the ACE results, for example. An interaction was reported between a directional hand motion and the concurrent understanding of a transfer sentence. The critical point to be made is that this interaction is not specific enough because the transfer sentences used in the experiments described different kinds of hand/arm motions ('I showed you the book' versus 'He threw me a stone') or information transfer ('I told you the story') that do not share the same effectors. The motoric simulation is coarse grained in the sense that it involves merely a directional parameter of motion rather than recruiting specific motor programs (Sanford, Chapter 10; and Gomila, Chapter 17). However, Sanford assumes that an 'embodied' simulation must be fine-grained, involving specific effectors or motoric programs and preserving analogically the dynamics and kinematics of motions. Under such strict criteria of embodiment, he is tempted to discard that ACE data show a real simulation. Rather they illustrate 'some sort of an abstraction, even if that abstraction is originally grounded in a specific action or situation' (see Chapter 10).

Some participants within the embodiment approach also noticed that motor simulations can be coarse grained (de Vega, Chapter 14; Pulvermüller, Chapter 6; and Knoblauch, Chapter 7). But their interpretation of this fact differs from that of the symbolist. For instance, Pulvermüller reports that the brain architecture that controls actions includes different levels of 'abstraction', from the motor and premotor areas to the 'abstract' disjunction neurons placed in the prefrontal cortex. When a coarse-grained description of an action (e.g. 'open') is given, the disjunction neurons communicate to a range of different motor programs compatible with its meaning. Similar circuitry has been also reported to process abstract visual information, including both visual feature neurons and high-order disjunction neurons along the inferior temporal and the parahippocampal cortex (Pulvermüller, Chapter 6; and Knoblauch, Chapter 7). Therefore, an embodied framework can readily accommodate a wide range of grain sizes and need not be committed to the most detailed level.

Whether we accept that *embodied meaning* recruits abstract processes or that *abstract meaning* could be modal or partially embodied seems a matter of where to set the cut-off point for defining categories. We all agree that there could be coarse-grained perceptual and motoric meanings. If we get a very strict criterion of what an embodied simulation should be, then we will be inclined to claim that coarse-grained descriptions of perceptual and motoric events are essentially abstract. By contrast, if we stress that coarse-grained descriptions always recruit sensorimotor processes, even though 'abstract neurons' might be involved, then we may conclude that coarse-grained meaning is still embodied. In practical terms it doesn't matter much, because Sanford's strict criterion still recognizes that coarse-grained abstractions are modality specific (Chapter 10), and Pulvermüller's abstract neurons involved in simulations are a sort of logical operator with a symbolic flavour (Chapter 6). In other words, we could choose between the half-empty and the half-full bottle, according to our theoretical preferences.

19.4.2 **Abstract words**

The study of abstract and concrete words (mainly nouns) has a longstanding tradition in cognitive psychology (Glaser 1992; Paivio 1971). How do we process abstract words such as 'truth' or 'idea' that apparently do not refer to any physical entity? This question seems to be a fundamental challenge for embodied theories. An answer was given by Paivio's *dual code theory*: concrete words are processed both by the symbolic (verbal) and the embodiment (imagery) systems, whereas abstract words are processed just by the symbolic system. Many experiments in the last decades supported this dual approach (see Barsalou *et al.*, Chapter 13, for a critical review). Concrete words are better recalled, recognized faster, and produce a stronger electrophysiological signature (N400 component) than abstract words (Kounios and Holcomb 1994).

The hybrid account of meaning postulated by Paivio has been recently challenged by embodiment defenders. Thus, some contributors (notably Barsalou et al., Chapter 13; and Zwaan, Chapter 9; see also Barsalou and Wiemer-Hastings 2005) propose that abstract and concrete words, although differing in the content and complexity of their meanings, are both grounded on situated knowledge, and some of the situated knowledge is embodied. The critical difference is that concrete words activate a specific set of sensorimotor traces, whereas abstract words activate a heterogeneous set of traces because they are compatible with a much broader range of situations. For this reason, when both kinds of words are presented without context (e.g., list of words or pairs of words) for a memory task, abstract words fail to retrieve a particular context and are processed superficially, resulting in worse recall. The abstract word disadvantage disappears, however, when they are incorporated into rich linguistic or nonlinguistic contexts. For instance, the classical experiments of Bransford and Johnson (1973) showed that a text involving a coarse-grained description composed of many abstract words was better remembered and understood when it was preceded by a title that made it possible to ground the abstract words in a specific kind of situation (e.g., washing clothes, a parade).

Even the neurological differences between concrete and abstract words fade when words are included in contexts, or when the participants actively engage in generating contexts. In such cases, abstract words produce as strong a N400 as concrete words (Holcomb *et al.* 1999), and neuroimaging shows an activation of content-related brain areas ('simulators') in addition to the most typically linguistic area of Broca (Barsalou *et al*, Chapter 13). Levels of processing strongly modulate how abstract words are processed. When the processing is shallow, such as in a word recognition task, concrete words provide a better grounding on situation knowledge and abstract words are processed almost exclusively by the linguistic system, as Paivio suggested. However, lists of isolated words are not representative of ordinary language comprehension tasks. In natural settings, abstract words appear in rich contexts that promote deep comprehension and, apparently, as much grounding as concrete words.

The above data and arguments could be considered inconclusive by some symbolists. First, the idea that abstract words are sometimes grounded on situated knowledge does not mean necessarily that situated knowledge is embodied. By contrast, situated knowledge could rely on symbolic or abstract schemas. Secondly, the data that suggest that context provides a concrete instantiation (embodied) for abstract words is also compatible with a statistical symbolic approach: the effect of context could be explained by degree of constraints. That is, highly constrained material is processed differently from minimally constrained material.

What is clear is that concrete and abstract words are not entirely equivalent. First, they differ in content: abstract words more frequently convey mental states or events as referents, such as beliefs, wishes, roles, etc. (Barsalou et al., Chapter 13). Second, abstract words are more cross-situational in comparison with the situation-specificity meaning of concrete words (Zwaan, Chapter 9), although when used in particular contexts they could retrieve a particular context as much as concrete words (Barsalou et al., Chapter 13). Finally, abstract concepts are more complex in terms of their contingent relations.

19.4.3 Abstract grammar

The formal accounts of grammar in the Chomsky tradition assume that linguistic constructions obey rules that may be arbitrary and independent of semantics. Cognitive linguists assume, however, that grammar is motivated by semantic principles and these principles determine the fundamental concepts in human cognition (Tomasello 2003). For instance, temporality is incorporated into the temporal adverbs and verb morphology, whereas spatiality is expressed by means of locative prepositions. Although adverbs, prepositions and morphology are quite abstract linguistic markers, they could refer to perceptual dimensions which are compatible with an embodied approach to grammar. For example, English speakers using locative prepositions of direction ('in front of', 'behind', etc.) exploit the egocentric coordinates of their body and choose figure and ground entities with particular perceptual and functional properties (de Vega, Chapter 14).

In the same vein, when using temporal adverbs ('while', 'after', etc.) speakers contrast two events (a figure and a ground) and apply a sort of temporal metric to choose the longer duration event as the ground (de Vega, Chapter 14). Finally the double-object construction is a syntactic frame to express the transfer of a physical object (or abstract information) between an agent and a receiver (Glenberg and Kaschak 2002). In summary, some linguistic markers are semantic devices that promote construction of situated meaning.

However, there are grammatical constructions which appear much more abstract, and posit a challenge to embodiment theories. For instance, there are linguistic markers that could be characterized as 'logical operators' rather than referring to perceptual entities or relations. This is the case of the negation 'not', the disjunctive 'or', the adversative 'but', or the conditional sentences expressing counterfactual meanings. Some discussants in the embodiment approach consider that these linguistic markers are indeed processed as embodied simulations. For instance, Zwaan and Madden (2005) propose that negation in a sentence like 'the eagle was not in the nest' prompts the reader to initially represent the negated situation (the eagle in the nest) and later on the actual situation (the nest without the eagle). In the same vein, counterfactual sentences such as 'If I had been generous, I would have lent you the *Harry Potter* book' determine a momentary simulation of the transfer action in spite of the fact that such action is not described as actual (de Vega, Chapter 14). Negation and counterfactuals seem to demand the management in time of a dual meaning: the simulation of a 'false' scenario immediately followed by (or simultaneous to) an alternative simulation of the 'real' scenario.

Here, embodiment discussants do not have a unitary view. De Vega (Chapter 14) proposes that abstract linguistic markers like negations or counterfactuals instruct addressees to manage alternative meanings (in principle, embodied) derived from linguistic constructions, and this 'management function' could be considered genuinely symbolic or disembodied. However, this is a different sort of symbolism than the traditional proposals such as propositional theories. The latter are general theories that explain all features of meaning as a symbolic computation, whereas de Vega's management function is a high-order operator to deal with multiple representations which themselves are embodied. Pulvermüller's (Chapter 6) neurological account is close to this view when he proposes that disjunctive neurons in the frontal cortex perform abstract operations such as switching on and off alternative representations. In contrast, Zwaan (Chapter 9) and Glenberg and Mehta (Chapter 2) do not consider abstract or symbolic operations to be necessary in any case. For instance, Zwaan explains that grammatical markers are a by-product of resonance processes. For example, the word 'but' will activate all instances of having experienced the input 'but'. From the symbolist perspective, there would be mental models, situation models, or mental spaces created from the hypothetical content.

19.4.4 Conclusions

The issue of abstract meaning is a hot potato for the embodiment account, although some aspects of abstract meaning can be equally annoying for the symbolist approach. For instance, word covariation computed by LSA and other similar algorithms does not generate the subtle meanings of words like 'but', 'not', or 'because'. These closed-class words occur so frequently and in such varied linguistic contexts that covariations are not very informative. However, there are many other symbolic models that would not fall prey to this problem, particularly the vast array of models in computational linguistics that have deep semantic processing (Jurafsky and Martin 2000).

Much more research is needed to get a clear view of how we understand abstract meaning, but we can draw some tentative conclusions. First, abstract meaning is not a unitary phenomenon. Participants in this book discussed several facets of abstract meaning that are related to the symbolic/embodied debate: abstract nouns, abstract transfer verbs, a few closed-class words, degrees of granularity, negation, and counterfactual meaning. Each of these kinds of abstractions should be analysed separately both by symbolists and embodiment people because they have different functional and representational characteristics. Second, abstract words are not empty tokens, but tackle some meaningful dimension of experience, such as space, time, or introspective reactions. For instance, many abstract words have an emotional valence (e.g., 'truth' is positive and 'problem' is negative), or the double-object constructions express a directional transfer. Third, the context (linguistic and nonlinguistic) that usually accompanies abstract words reduces their indeterminacy. That is, abstract meaning in context becomes situated or otherwise more constrained.

19.5 Referential meaning

Meaning has multiple facets that can be studied independently (Roy, Chapter 11; and de Vega, Chapter 14). One facet is the 'intensional' character of meaning, namely how words or symbols relate to each other. Symbolist theories have been very successful with this intensional perspective, developing associative networks of concepts and semantic features, explaining conceptual similarity, or, more recently, building covariation models that obtain semantic regularities inductively. Another facet of meaning is its referential or 'extensional' character, namely how words (and symbols) refer to external entities in the world. The latter extensional meaning is emphasized by embodiment theories and too often neglected by symbolic theories. To organize this section we will distinguish two kinds of referential problems: referring to the current situation and referring to absent situations.

19.5.1 Referring to the current situation

Language is undoubtedly grounded in perception/action at least when we communicate on-line about the current environment. We know, for example, that grounding words in the immediate perception and actions is an important mechanism for learning word meaning at early stages of language acquisition. Young children and adults communicate in contexts in which words directly refer to objects and actions, while adults provide multiple embodied cues, such as gestures and gazes, to specify referents in the scenario. Adults frequently communicate in settings in which they need to refer to the immediate environment and establish grounding in that environment before deep understanding can be achieved (de Vega, Chapter 14; Zwaan, Chapter 9; and Roy, Chapter 11). For instance, Roy claims that 'to understand the meaning of the assertion that "there is a cup on the table" includes the ability to translate this utterance into expectations of how the physical enviroment will look, and the sorts of things that can be done to and with the environment if the utterance is true.' The role of gestures, gazes and body alignment with referred objects is evident not only to refer to immediate perceivable objects, but also to abstract entities (Nathan, Chapter 18).

It is difficult to conceive how an ungrounded symbolic system could perform the truth verification operations involved in reference to the current situation. Computational symbolic models routinely attempt to establish such grounding, although difficulties emerge when the world becomes progressively more complex. Defenders of statistical symbolic models also accept the need for perceptual grounding of on-line reference. Thus, Kintsch (Chapter 8) claims that 'The meaning of a word would be its referent in the real world, as in the perceptual symbol system of Barsalou (1999) and similar proposals,' and Graesser and Jackson (Chapter 3) accept that embodied processes are required under certain circumstances: 'when tasks and tests encourage embodied activities,' or 'when there is visual-spatial grounding (e.g., an established spatial layout).'

The robots presented and discussed in this book are systems with basic referential capabilities with respect to the immediate environment. Shapiro (Chapter 4) mentions Brooks' robots which are programmed with very simple routines to react to perceptual features of the environment without need to provide them with internal maps, schemas or any other sort of representation. Brooks' robots exhibit sophisticated behaviours and learn to exploit the structure of their environment without a well structured spatial representation. Most spectacularly, Luc Steels' Talking Heads (Chapter 12), Deb Roy's Ripley (Chapter 11), or Andreas Knoblauch's Bot (Chapter 7) go beyond this insect-like behaviour and communicate about the immediate environment as we will see later.

19.5.2 Referring to absent situations

Grounding language in the immediate perceptual experience is just one case of reference. In many conversational situations, and in all text comprehension tasks, people understand displaced referents that are unrelated to the current perceptible situation (de Vega, Chapter 14). Thus, we can refer to and understand descriptions of events beyond the here-and-now with minimal or no help of perceptual cues (except words). We can also refer to introspective or mental states such as beliefs, wishes, and emotions which have fewer perceptual correlates.

The embodiment/symbolism debate is particularly active at this level of reference. The embodiment theories posit that displaced reference is still grounded on sensorimotor processes. Thus, understanding the sentence 'Katy drove her car to New York' would activate successively fleeting visual images of a woman into a car, a motoric activation of driving, a brief vision of New York skyscrapers, and perhaps an attenuated emotion associated with the experience of driving in heavy traffic. A neural description of these embodied processes is that the sensorimotor systems in the brain, which are used in real-world situations, are partially reactivated to simulate the described situation. Symbolists consider the fleeting embodied processes to be sometimes superfluous or optional for comprehension, whereas there are other conditions when deeper encoded meanings require embodiment. Statistical symbolists also argue that fast computation of ungrounded symbols is sufficient for understanding in many circumstances.

Searle's Chinese room is a metaphor on the logical insufficiency of ungrounded symbols to deal with reference (see Shapiro, Chapter 4, for a complete discussion of the topic). The inhabitant of the Chinese room (let's say Shapiro himself) cannot establish any correspondence between the Chinese characters and their referents (either in the current or displaced situation) because he has no information about the external world. Under these circumstances it is clear that grounding is necessary. As Shapiro notes about the Chinese room: the 'view that a symbol can become meaningful to an interpreter solely in virtue of knowing its relations to other symbols whose meanings are not known, [can be called] the symbolist's folly.'

How does statistical symbolism work in the Chinese room? The answer is not any better than traditional symbolism. Shapiro could be provided with statistical facilities that compute covariations among Chinese symbols and organize the data into a 'semantic' multidimensional space. Shapiro would be informed that the cosines between Chinese individual characters or string of characters represent semantic similarity, coherence and so forth. No matter, statistical symbolism never would open the door to leave the room, and connect symbols with the real world. Shapiro's situation in the Chinese room would be exactly the same than that of the participants in Glenberg and Mehta's (Chapter 2) experiments when they were exposed to symbolic covariations that mimicked real-world events. The results showed that participants learned the covariations, but never discovered the referents of these covariations.

19.5.3 Symbolism and reference

Are the symbolism drawbacks to explaining referential meaning insurmountable? Some discussants with a background in philosophy of mind (Gomila, Chapter 17; and Shapiro, Chapter 4) and AI/robotics (Graesser and Jackson, Chapter 3; Knoblauch, Chapter 7; Roy, Chapter 11; and Steels, Chapter 12) don't think so. Some of them propose an intriguing possibility: how meaning is represented (e.g., in amodal or perceptual symbols) is an entirely independent question of whether meaning is grounded. In other words, the grounding problem illustrated by the Chinese room is not an intrinsic problem for symbolism. It might be possible to build a symbolic system grounded on perception and action and therefore the current debate would be basically misleading (Graesser and Jackson, Chapter 3; Knoblauch, Chapter 7; Roy, Chapter 11; and Steels, Chapter 12).

It could be argued that all computations rely on amodal symbols or subsymbols at some level of description (Shapiro, Chapter 4; and Knoblauch, Chapter 7). For instance, if we try to describe the computations of meaning in the human brain, then 'perceptual symbols are not similar to those things they represent' (Shapiro). Indeed, nobody would seriously expect that object shapes and orientations were projected analogically on the neuronal tissue, or that processing colour red makes neurones redder. Theories about neural computation like neo-Hebbian approaches or connectionist models could be more easily described as symbolic or subsymbolic systems even if they are used to support an embodiment account of meaning, as Pulvermüller does in Chapter 6. In the same vein, implementing robots, which are embodied systems, can only be done by running programs with computational symbols (Roy, Chapter 11; Knoblauch, Chapter 7; and Steels, Chapter 12).

At this point it seems that symbolism has won the match in the sense that computations are symbolic whether in the brain or in embodied robots. Then what are we talking about when some of us defend the idea of perceptual symbols? The answer is that theories of embodiment (perceptual symbols, simulations or resonances) are only embodied at a high-order level of description (e.g., functional architecture) rather than at the level of microcomputations (Shapiro, Chapter 4; and Knoblauch, Chapter 7). This idea was implicit in the initial definition of embodiment by Glenberg, de Vega and Graesser (Chapter 1): 'Linguistic symbols are embodied to the extent that the meaning of the symbol (the interpretant) to the agent depends on activity in systems also used for perception, action, and emotion.' To put it in neurological terms, words are embodied because the same brain circuitry is used to process their meaning and to perceive, and act; or, in AI terms, words are embodied because they are referred directly to perceptual and motoric events performed by the robot or the interlocutor. A fine-grained description of how the brain (or a conversational robot) computes perception-action and meaning, however, would essentially be symbolic or subsymbolic.

19.5.4 Conclusions

There are a few substantial conclusions derived in this section. First, statistical symbolist theories cannot account for reference. Second, this failure of statistical symbolism is particularly remarkable for the most basic level of reference to the immediate situation. Thus, a perceptual reference test ('is the cup on the table?') cannot be solved within a purely statistical symbolic system. Third, the symbolism drawbacks only emerge when symbolism tries to extend beyond its proper level of description. At a fine-grained level of description processes are 'symbolic' or 'subsymbolic', but the high order architecture (the algorithmic level) of the system is embodied and grounded in perception, action, and emotions.

If statistical symbolism does not have much to say about reference, why has it been so successful? The answer is that other interesting semantic tasks can be performed by means of ungrounded symbolic mechanisms. For instance, LSA models are useful to compute gross cues of text coherence, to judge similarity among linguistic units and the like. Kintsch (Chapter 8) describes how LSA implemented with parsing algorithms predicts sophisticate semantic phenomena like the resolution of homographs meaning in context, or the interpretation of metaphors. Graesser and Jackson (Chapter 3), on their side, describe how the symbolic AutoTutor holds a mixed initiative dialogue with the student 'by asking and answering questions, giving hints, filling in missing pieces of information, and correcting misconceptions.' These are impressive achievements, no doubt, but the referential processing is extremely limited and sometimes nonexistent. The referential processing of AutoTutor is most salient in the version in which learners manipulate controls in a microworld with interactive simulation and view the outcome of the simulation. The words do refer to objects and events in the microworld and are thereby referentially grounded, but AutoTutor is limited by the fact that it cannot ground referents that are not anticipated ahead of time by AutoTutor (or the designer of the system).

Ungrounded symbolic computations cannot provide truth verification operations for concrete statements such as 'there is a cup on the table' in a particular situation. Instead, the best that ungrounded symbolic computations can do is judge the truth of intensional semantic statements, like 'a table is furniture', or judge how correct a student's answer is when compared with an academic database. Ungrounded symbolic computations might be able to evaluate the predicative goodness or plausibility of concrete statements like 'there is a cup on the table', but this is not enough to decide how well it describes a particular situation.

19.6 **Meaning in the brain**

In a trivial sense, meaning is embodied because it is processed in the brain and the brain is an organ or the body. However, this general statement is not informative or relevant to the current debate (Glenberg, de Vega, and Graesser, Chapter 1). Instead, the embodiment/symbolic debate arises in the neurosciences arena when discussing the specific brain circuitry involved in meaning. The neurological ideal for symbolists would be finding a modular system in the brain specialized in processing amodal symbols functionally dissociated from motoric, perceptual, or emotional structures. By contrast, the embodiment's expectation is finding a partial reactivation of motoric, perceptual and emotional cortical areas as part of linguistic meaning. It should be acknowledged that the neuroimaging methods have been much more employed by the embodiment than the symbolic approach in the present volume. Thus, two of the discussants who presented neuroimaging data in this book (Barsalou et al., Chapter 13; Pulvermüller, Chapter 6) defended many components of an embodied view of cognition. The third discussant (Just, Chapter 5) has a hybrid proposal. Let's see their arguments and data, as well as the symbolists' point of view.

19.6.1 **The embodied brain**

The neuroimaging experiments discussed in this book considerably differ in linguistic materials, task demands, and the theoretical framework employed. Thus, Pulvermüller (Chapter 6) used action words in passive listening tasks, and he developed a neo-Hebbian cell assemblies' account of his results. Barsalou et al. (Chapter 13) also used lexical items in their experiments but in the context of associative tasks, and their theoretical goal was to investigate how language, at certain stage, produces situated simulations. Finally, Just (Chapter 5) used sentence comprehension tasks of different kinds and with different kind of populations (mainly autistic and controls) to show that meaning is broadly distributed in the brain both in sensorimotor and high-order regions. Despite these differences, all three chapters provide strong evidence that meaning is to some extent embodied in the brain.

The most radical embodiment view was defended by Pulvermüller (Chapter 6). According to him, there is no amodal brain area to process the symbolic meaning of words. He rules out that Wernicke's or Broca's areas perform this function, as we will see later. Instead, words trigger the activation of specific sensorimotor areas in the cortex related to their particular meaning. For instance, passively listening to action verbs describing mouth ('lick'), hand ('pick'), or leg actions ('kick') activate somatotopic areas in the motor and premotor cortex related to mouth, hand, and leg motion, respectively. In addition, Pulvermüller traced the temporal pattern of activation for these words by means of magnetoencephalography, observing an initial activation of Wernicke and Broca, followed a few milliseconds later by activation in specific somatotopic areas. Finally, he used transcranial magnetic stimulation to stimulate weakly right hand motoric area in the brain and observed that participants responded faster to hand action words than leg action words, whereas stimulation in the leg motoric area produced the opposite results.

Barsalou *et al.* (Chapter 13) clearly dissociated the linguistic process of words from their perceptual and situated representation of meaning. For instance, in one experiment they gave participants target words to mentally generate word associates, imagining related situations, or generating properties. As expected, the word association task determined an activation of perisylvian areas in the left hemisphere. The situation association task produced activation of imagery related areas in the brain (precuneous, right middle temporal gyrus, the right middle frontal gyrus, and the left fusiform gyrus), and the properties generation task activated both kind of regions. A temporal analysis of the BOLD signal indicated that the 'linguistic' regions (e.g., Broca) were activated earlier than the imagery-related areas.

How are the classical perisylvian areas characterized by embodiment approaches? Pulvermüller (Chapter 6) and Barsalou *et al.* (Chapter 13) assume a crucial role of the superior temporal lobe (Wernicke) and the inferior frontal gyrus (Broca) of the left hemisphere in language processing. However, they do not think that these structures are in charge of processing words meaning, but just of linguistic properties of words. Thus, Pulvermüller considers that Wernicke's area is responsible for the phonetic analysis and recognition of spoken words, and the Broca's area controls the motoric programs of speech. In addition, he demonstrated that both areas are interconnected cell assemblies because the input to one of them (e.g., auditory words, processed in the Wernicke's area) determines coactivation of the other (Broca's area). On their side, Barsalou *et al.* attribute to these areas the capability not only to processing linguistic forms but also to generating linguistic associations, based on sound similarity, root similarity, antonym, synonym, etc. These all are important functions that contribute to speeded language processing (e.g., by creating lexical expectations during comprehension). However, according to Barsalou *et al.*, they involve a relatively shallow processing that cannot be identified with meaning.

19.6.2 **The symbolic brain**

The classical neurological model of language with two relatively well defined 'modular' structures in the left hemisphere seems close to the neurological ideal of symbolism (e.g., Fodor 1983). However, as we have seen in Section 19.6.1, there is no evidence that these structures process language beyond the physical and formal features of words, which imply to some extent sensorimotor processes. Nonetheless, there are other brain structures that could be candidates for processing symbolic amodal meaning. One possibility would be the existence of a central 'concept area' or 'convergence zone' at which concepts are related to each other and abstract information is extracted from them. However, as Pulvermüller (Chapter 6) discusses, there is no evidence of a single convergence zone for meaning.

Another possibility, defended by Just (Chapter 5), is that some abstract or symbolic meanings are processed in structures of the frontal cortex. Just provides a set of empirical arguments, based on his neuroimaging studies. A summary of his results follows: (1) readers or listeners of sentences with high-imagery value activate imagery-related regions (e.g., intraparietal sulcus[2]) whereas low-imagery sentences only activate 'symbolic' regions (perisylvian structures in the left hemisphere, and areas in the prefrontal cortex); (2) autistic individuals, unlike nonautistics, activate imagery regions in both high-imagery and low-imagery sentences; (3) the comprehension of novel metaphors mobilize imagery-related areas (parietal regions) whereas the comprehension of frozen or familiar metaphors only activates ordinary language structures (e.g., Wernicke's and Broca's areas, dorsolateral prefrontal cortex bilaterally, etc.); and (4) the categorization of tools and dwellings can be identified in motor and visual areas in the brain by means of machine-learning algorithms. In summary, Just accepts that embodied imagery processes are involved in some, but not all, comprehension tasks. When the linguistic materials are concrete and the task demands a visuospatial problem solving (e.g., Is it true that 'The number eight when rotated 90 degrees looks like a pair of spectacles'), then it is most likely that embodied processes take place. But for more 'abstract' tasks and materials sensorimotor areas do not fire and the 'symbolic' perisylvian and prefrontal areas make most of the job. An intermediate case is that of metaphors: frozen metaphors are processed linguistically, whereas novel ones activate sensorimotor processes.

But the question can be raised whether the prefrontal lobe is a symbolic processing system. The question requires much more study, because different prefrontal areas have been associated with disparate cognitive processes, such as theory of mind, working memory, discourse integration, conflict resolution, emotional evaluation, etc. Generally speaking, all these operations seem quite abstract and amodal, but we should not forget that the prefrontal cortex has rich connectivity with the sensorimotor cortex as well as subcortical structures in the brain, and thus prefrontal processing could consist of a deep elaboration of embodied information.

---

[2] The 'imagery' brain areas reported by Just (parietal regions) differ from the 'simulation' areas described by Barsalou et al. (precuneous, right middle temporal gyrus, etc). Such differences could result from different task demands, as well as the specific regions of interest selected by the researchers in their analysis.

19.6.3 **Conclusions**

Neuroimaging data and other neuroscience techniques provide an extensive body of information, including three-dimensional maps of brain functions and temporal cues of these functions. These techniques are potentially useful to decide between alternative psychological models, like embodiment and symbolic accounts of meaning. The results of neuroscience data discussed in the book clearly support two kinds of brain structures related to language: language-specific and meaning-specific structures. The former correspond to the classical Broca's and Wernicke's areas, whereas the latter are much more broadly distributed in the brain, including perceptual and motoric areas, which are selectively activated depending on the particular content of the words. We might be tempted to consider that both symbolism and embodiment are supported by neurological data. Nevertheless it would be naïve to expect that the neurosciences could solve all problems at once. Not only do the same set of data admit different interpretations, but the neuroscience data are currently fragmentary and partially contradictory.

The reinterpretation of data occurs on both sides of the debate. First, we have seen that the classical language areas of Broca and Wernicke, perfect candidates for symbolic amodal processing, are ruled out as such by embodiment discussants (Pulvermüller, Chapter 6; and Barsalou et al., Chapter 13). According to embodiment researchers the role of these areas is just to processing the physical features of words or, at most, the statistical properties of their traces. Second, there are similar theory-driven reinterpretations for the embodiment data (Gomila, Chapter 17; Graesser and Jackson, Chapter 3; and Louwerse and Jeuniaux, Chapter 15). Even accepting that some sensorimotor activation occurs in word processing, symbolists consider it of minor functional relevance with respect to the activations that end up being part of the encoded meaning representations. Thus, Gomila (Chapter 17) suggests that such activation is epiphenomenal because the core of meaning is necessarily abstract. Louwerse and Jeuniaux (Chapter 15) propose that sensorimotor activation is so weak and so far away from the 'symbolic' regions in the brain that it soon fades without entering working memory.

Another recurring reinterpretation of the embodiment neurological data relies on Kintsch's construction–integration theory of discourse comprehension (Graesser and Jackson, Chapter 3). The 'promiscuous' construction phase could activate a variety of representations including those generated by sensorimotor areas in the brain. However, in the integration phase most of these representations are suppressed except those in the symbolic systems, whatever they are. Beeman's neurological theory of discourse comprehension fits well with Kintsch's proposals (Beeman 1998). He established a division of labour between the two brain hemispheres. The left hemisphere involves fine-grained semantic fields for word, whereas the right hemisphere has coarser but broader semantic fields for content words. In the construction phase, there is a broad activation of both hemispheres. Specifically, the right hemisphere provides loose associations that could be useful, whereas in the integration phase, the left hemisphere 'selects' the appropriate information and the remaining fades. Although Beeman does not apply his theory to decide the relative role of embodiment and symbolic processes, one could make the case that the right-hemisphere coarse processing includes the activation of perceptual properties, whereas the left hemisphere has fine-grained processing of words and a selective capability that is more typical of symbolism.

Localizing functions in the brain is an important achievement, but the three neuroscientists who participated in the debate shared a concern for the temporal dimension of brain activity. Localization and temporality together will provide the most complete functional picture of how meaning is processed in the brain. Given that many structures of the brain seem to cooperate in language comprehension, it is important to know the degree of synchrony or sequence among them, and the persistence of their activation. Neuroimaging techniques, especially fMRI, provide an excellent spatial detail of brain processes by recording the haemodynamic activity in the brain. However, haemodynamic responses have a poor temporal resolution, as they are delayed a few seconds after the neuronal events and it takes them more than 20 seconds to return to baseline. Alternative techniques such as ERP, MEG, and TMS are examined convergently to grasp the temporal dimension of brain processes. Thus, Pulvermüller (Chapter 6) observed with MEG the strong temporal correlation between Wernicke's, Broca's, and somatotopic areas in the brain, which indicates the interconnectivity among long-distance structures. In addition, the development of fMRI event-related designs and the careful analysis of time series in the BOLD signal allow a coarse interpretation of temporal changes in activation. In this way, Just (Chapter 5) found a functional connectivity or synchrony between Wernicke's area and imagery-related structures when people understand high-imagery sentences, and Barsalou *et al.* (Chapter 13), using the same procedure, obtained a faster activation of linguistic areas than of simulation areas in a properties generation task.

The neuroscience of language is a fast growing and exciting field that is providing strong evidence of embodied meaning. New data relevant to the debate are coming from laboratories everyday. For example, it has recently been reported that there is an activation of the medial temporal cortex when people perceive moving visual stimulus, and also when they read sentences describing motions (Rüschemeyer et al., submitted). On the other hand, the discovery of mirror neurons could also provide new light to the role of embodied meaning and even to the evolutionary origins of language (e.g., Gallese and Lakoff 2005; Rizzolatti and Craighero 2004, 2007). Mirror neurons are multimodal processors that combine visual, auditory and other modalities information with motoric processes in the brain. But they are also relatively 'abstract' processors related to the high-order conceptualizations of intentions, empathy, and the like.

19.7 Meaning as computation

In the AI field, the embodiment/symbolic debate may require a reconceptualization (Knoblauch, Chapter 7; Roy, Chapter 11; Shapiro, Chapter 4; and Steels, Chapter 12). If we accept that meaning is the result of computations, and computations are necessarily symbolic (e.g., they use a programming language that operates on symbols), then we might conclude that meaning is symbolic. This argument seems appropriate when a researcher tries to build an artificial system, such as a robot that is able to process linguistic meaning. But such a symbolic system would not necessarily be grounded in perception and action. It depends. As we have mentioned in Section 19.5.3, several contributors in this book argued convincingly that computational symbolism and grounding can be perfectly compatible (Shapiro, Chapter 4; Steels, Chapter 12; Roy, Chapter 11; Knoblauch, Chapter 7; and Graesser and Jackson, Chapter 3), but it depends on details about the computational architecture and goals of the system. On one hand, systems such as LSA, many versions of AutoTutor, and some connectionist models are not grounded in perception and action. On the other hand, robots that are equipped with appropriate sensors and effectors may not suffer the semiotic solitude in the Chinese room that was discussed by Searle and Harnad. Let's see why.

19.7.1 Disembodied artificial intelligence

Some AI models are ungrounded or disembodied. Two prototypical examples were described in Chapter 1: LSA (Landauer and Dumais, 1997) and the connectionist model of Rogers *et al.* (2004). The two models differ in theoretical background an implementations (symbolist versus connectionist), but they are equally ungrounded systems that utilize statistical covariation as a method to compute meaning. Ungrounded computations create a sufficiently rich worry on their own to deserve full attention of many researchers. As Roy (Chapter 11) suggests: 'I am not sure whether there is really a debate to be had between the two camps [symbolism and embodiment], but rather a difference of opinion in choosing which parts of an immensely complicated overall problem to focus on.' Exploiting symbolic relations, or, for that matter, computing on massive subsymbolic networks, allows one to explore the organization of semantic memory, semantic similarity processes, words associations, and other psychological phenomena, but these do not typically involve reference. The choice of ungrounded models might be correct if it is accompanied by an appropriate definition of their application domain. The discussion rises when ungrounded symbolic models pretend that they are theories that accommodate grounded meaning.

An interesting case of AI is AutoTutor, as presented by Graesser and Jackson in Chapter 3. AutoTutor is a computer tutor that simulates human tutors and that helps individuals learn about abstract difficult topics by holding a conversation in natural language. We will comment on the instructional implications of AutoTutor in the current embodied/symbolic debate in the next section. The immediate focus is on its implementation. AutoTutor incorporates a number of symbolic technologies, including LSA, syntactic parsers, and the like. Some versions of AutoTutor also have advanced interface components that could be considered as typically embodied. For instance, on the perception end, there is speech recognition hardware and software and sensors to detect and interpret (as emotional cues!) the user's facial expressions and changes in body position. On the action end, there is a realistic dynamic tutor face that moves muscles in synchrony with speech and changes facial expressions according to the anatomical and physiological constraints of human face. AutoTutor is, according to Graesser and Jackson, a symbolic–embodied hybrid. However, AutoTutor does not go the full distance in having a sufficiently embodied semantics. AutoTutor does adaptively interpret, guide, and correct the student's responses by using its symbolic background knowledge, semantic pattern matching and completion algorithms, and some referential processing capabilities. The symbolic components include LSA, syntactic parsers, and semantic processors that are entirely symbolic. AutoTutor's dialogue planner does attempt to ground student contributions by aligning them with expectations and misconceptions in its curriculum scripts (with mechanisms that are analogous to people attempting to align referring expressions to objects, people, and events in the immediate situation). However, AutoTutor does not ground each noun, verb, preposition, and other important words in the perceptions and actions that correspond to either: (a) the referential domain knowledge (such as physics or computer literacy) or (b) the speech participants of the conversation (namely the learner and AutoTutor). AutoTutor is therefore lacking in an important dimension of embodied semantics. AutoTutor is an ongoing project, so perhaps it can be fortified with perceptual/motor grounding at these levels. The version of AutoTutor with interactive simulation is a first step in achieving these important forms of grounding.

19.7.2 Robots as embodied creatures

Robots are by definition embodied machines in the sense that they interact autonomously and directly with their environment. Robots that have been successfully programmed to communicate about the environment are embodied meaning systems because they must be able to map words into perceptual and motoric processes, and vice versa. We have in this book three examples of conversational robots with outstanding capabilities. This subsection will describe briefly their performance, whereas the next subsection will comment their contribution to the debate.

First, Roy (Chapter 11) built a robot called Ripley that emulates the descriptive and directive speech acts of toddlers, consisting of compositions of symbols that relate to the current environment and goals. Ripley's world is composed of a few small three-dimensional colourful objects, such as cups or apples. A description of Ripley in Roy's own words is: 'Ripley is an interactive robot that integrates visual and haptic perception, manipulation skills, and spoken language understanding. The robot uses a situation model, a kind of working memory, to hold beliefs about its here-and-now physical environment. The mental model supports interpretation of primitive speech acts. The robot's behaviours are governed by a motivation system that balances three top-level drives: curiosity about the environment, keeping its motors cool, and doing as told (which includes responding to directives and descriptives).'

Ripley's language skills are based on embodied semantics. Thus, verbs like 'look' and 'give' are grounded in sensorimotor programs (e.g., moving and focusing the camera, grasping and approaching an object, etc.). Even adjectives are grounded on action; thus, 'the red one' triggers a searching action for a red object, and 'the heaviest' (when implemented) determines that Ripley systematically weighs the available objects to decide which one is the heaviest. This embodied semantics differs considerably from other forms of symbolic semantics, such as word covariation patterns or implementations of semantic networks.

Second, Knoblauch (Chapter 7) describes another conversational robot (called Bot) that also executes simple orders in a restricted environment such as 'Bot put plum to green apple.' Its most distinctive feature is that the programming mimics some anatomical and computational properties of the brain. Namely, it uses a six-layer architecture mimicking cortical layers, and is implemented as Hebbian cell assemblies which are likely mechanisms of learning in the brain. Word meaning is also embodied, because a prompt like 'Bot put plum to green apple' triggers not only grammatical routines but also action goals, and specific motor programs.

Finally, Steels' 'Talking Heads' is a rather different robotic experiment. The main goal was to show that two robots with software agents are able to dynamically establish a common semiotic map of the environment. In other words, in the course of interaction they agree how to segment the set of stimuli into categories (e.g., based on shape or colour), and how to call these categories without human intervention, just as a consequence of aligning or coordinating their behaviour. Here embodied semantics is not carefully designed and programmed by computer scientists, but it is 'socially' constructed by the robots in the course of their aligned interaction with the physical layout. The robots are two cameras that can reorient themselves to 'look' at particular stimulus in front of them. They play a guessing game in which the speaker uses a label to name an object (e.g., a red square in the upper left side), and the hearer moves the head trying to find the target. The process is reminiscent to language acquisition in children in the context of mother–child interaction.

These three systems combine important computational requirements necessary to produce embodied meaning:

- They have the appropriate perceptual and motoric systems to connect with the world. For instance, Ripley has auditory, visual, and haptic systems.

- They have software facilities for language recognition, and speech synthesis to produce the linguistic signals.

- Some of them (Ripley and Bot) have basic syntactic parsers to understand and produce sentences.

- All of them establish functional connections between words, actions, and perception, an important requirement of embodied meaning.

- They create and update representations or beliefs of world states. This is a great accomplishment because even in their reduced environment, dynamic changes occur after the robot or the human's actions.

- Ripley and Bot also represent internal states such as goals and expectations, which are necessary to respond to human prompts.

- Talking Heads have alignment capabilities which are similar to those shown by human interlocutors in face to face communication.

- Talking Heads learn beyond pre-programmed knowledge, and Bot shows pattern completion capabilities.

19.7.3 **Robots and reference**

Conversational robots definitely are implementations of embodied meaning that over-come the grounding problem. Their performance is human-like, although constrained to very simplified environments and linguistic repertoires. Their implementations can be described at an algorithmic level with typical cognitive psychology vocabulary: mental models, expectations, goals, beliefs, working memory, parsing, and the like. However, these robots are far from science fiction robots with complete cognitive and linguistic capabilities. In Section 19.5 we described two kinds of referential meanings in human language: on-line reference to the here-and-now situation and displaced reference to absent situations. All three examples of robots in this book are quite skilful at referring to ongoing events in the here-and-now, but none was programmed to refer to previous events or to future events that are outside of the scope of its immediate goals and surroundings. In other words, they do not use language for displaced reference.

We must notice that even talking about current objects and events is a sophisticated cognitive task that goes beyond a simple transduction of physical signals into motor programs. For instance, to respond to human prompts, Ripley and Bot build up and update mental models of the current environment and create internal states like goals and expectations. These operations differ from those of a Gibsonian system like Brooks' robots discussed by Shapiro (Chapter 4). Gibsonian robots are programmed with very simple routines to react to features of a very constrained environment. After some prac-tice the robots develop quite complex and adaptive behaviours, such as navigating through a busy environment, avoiding objects, and changing courses. And they do all that without schemas, mental maps, or any other kind of representation. The Brooks robots, however, do not scale up by intelligently generalizing what they learn in their constrained environments (carefully designed or selected by the programmer) to new environments in the wild. Just as important, Gibsonian robots do not talk or respond to human orders. The moral is that conversation, even when referring to the here-and-now, requires more than a nonrepresentational Gibsonian creature navigating in the environ-ment. Conversation requires three-way interactions among the robot, the interlocutor and the environment. Under these computational demands, goals, representations, and meaning emerge as necessary algorithmic descriptors of robots' functional architecture.

But could a robot solve more complex conversational tasks like referring to past events? Let's suppose that Roy wants to program Ripley for a minimal displaced communication. After executing successively the orders 'give me the red cup,' and 'put the blue cup at your left,' Ripley should be able to answer questions like: 'What did you do before putting the blue cup at your left?,' or 'where was the red cup before you gave it to me?' Maybe Roy and his colleagues could modify Ripley to extend its communication skills beyond the here-and-now by providing the robot with a long-term episodic memory. Such an ability would probably involve a computational explosion because the robot would not only have to update the situation model to accomplish on-line reference, but it would also need to keep a record of past situation models organized in time, and it would need efficient routines for retrieving the memories. The current version of Ripley includes updating mechanisms that delete outdated object schema tokens from situation memory when sensorimotor evidence does not support the existence of an object. For instance, if an apple has been moved from position A to position B, the token of the apple in A is deleted. In this way computational explosions are avoided, but at the cost of inability for displaced reference.

Current conversational robots have reached extraordinary goals. Perceiving, acting, and talking about a current (although simplified) environment are not trivial performances. They are the outcome of great conceptual integration and even greater implementation efforts by outstanding researchers. Nonetheless, additional efforts are necessary to build a robot capable of displaced reference which would approach human discursive capabilities.

19.7.4 Conclusions

The contribution of AI and robotics to the symbolic/embodiment debate has been illuminating. Let's summarize some conclusions.

Some symbolic artificial intelligence models are ungrounded

This is not a problem itself because ungrounded systems can perform symbolic operations which are psychological and technologically relevant. Problems only arise when these models exceed their application domain and try to be a comprehensive theory of linguistic meaning.

Conversational robots are both symbolic and grounded

The symbolic aspect of conversational robots is the technical fact that they are implemented in computational symbols. Conversational robots employ embodied semantics, as words refer directly to perceptual and motoric events in the layout, the robot or the interlocutor. In other words, robotics has solved the grounding problem using internal symbols to refer to the external environment.

Conversational robots are 'representational' systems

Talking about a situation requires modeling objects and events, and updating the models as the situation changes. It also requires motivational states such as expectations and goals. Complementary to the latter, nonrepresentational or Gibsonian robots interact adaptively with the environment (e.g., navigating), but they are unlikely to ever be conversational.

Current conversational robots are constrained to on-line reference to the here and now. As we have seen, displaced reference is a challenge for talking robots because keeping retrievable memory traces of past events probably involves a computational explosion.

19.8 Meaning and instruction

Researchers taking an embodied approach often study the on-line comprehension of narrative materials, whereas researchers with a symbolic approach often study the process of learning expository and academic texts. The biases in the choice of materials and research goals could explain part of the theoretical divergences. Narratives are understood with relatively low cognitive cost, and they mainly describe episodic and experiential information. Therefore, it seems intuitively appropriate to propose embodied meaning for narrative materials, with contents close to our everyday experience. For instance, reading a good novel might produce in the reader a feeling of immersion in the narrative world, including quite vivid visual images and emotional involvement in the protagonists' affairs. Our visual images, empathic emotions, and even pragmatic involvement can be even more intense in everyday conversations. By contrast, academic texts generally do not describe episodic events about everyday experiences but rather abstract knowledge that is generic, decontextualized, and unrelated to ordinary experience. Academic texts are not read for fun but effortfully studied for learning. In this section we will discuss some arguments and data from the symbolic and the embodiment accounts in the instructional arena (Goldstone, Landy, and Son, Chapter 16; Graesser and Jackson, Chapter 3; Kintsch, Chapter 8; Nathan, Chapter 18).

19.8.1 Symbolic teaching

Symbolic theories have been more frequently used than embodiment theories to create instructional technologies that improve learning or that directly tutor students during the comprehension and learning of academic content. Returning to Graesser and Jackson's (Chapter 3) discussion of AutoTutor – a demonstration of how a predominantly symbolist technology could be successfully applied to create instructional programs – it is timely to recall that most of the modules in AutoTutor are symbolic, but it is technically most accurate to classify it as a hybrid system. For example, to interpret students' responses, AutoTutor uses LSA and 'computational linguistics to evaluate the extent to which the student assertions cover the content of the correct principles and misconceptions. If a student assertion (or combination of assertions) has a high conceptual overlap with a particular correct principle, that principle is counted as covered. If there is a high overlap with a misconception, then AutoTutor immediately corrects the misconception.'

AutoTutor incorporates symbolic systems and some forms of grounding (as discussed earlier), but it does not go the distance in having a fully embodied simulation of meaning. Two embodied aspects of AutoTutor are that: (1) provides the student with an interface that incorporates some physical parameters of human communication, and (2) provides visual simulations of the target concepts that can be manipulated, perceptually observed and discussed. Thus, the tutor interface is an animated human face that changes its expression and its verbal messages in a fashion that is sensitive to the interpreted verbal and nonverbal responses of the student (including facial expressions and body positions). The dynamic visual simulations of physical processes, such as the effect of inertia forces on a car, are presented on the computer screen to 'ground' the abstract knowledge; the parameters of the embodied situation can be manipulated by the learner, such as the mass and speed of vehicles.

The most crucial question is how people learn abstract concepts like those taught in academic contexts. The answer perhaps depends on the particular content that is learned.

History, for instance, has plenty of narrative content that the student can understand using the general machinery involved in narrative comprehension. Other content areas, such as physics and mathematics, are very abstract so learning may end up being more disembodied (although see section 19.8.2 for different opinions). Finally, content areas such as biology, geometry, and chemistry may involve abstractions but also sensorimotor knowledge like recognizing and transforming visual patterns, or manipulating substances.

Graesser and Jackson (Chapter 3) discuss the case of learning physics by means of AutoTutor and they conclude that students do not understand very deeply the theoretical concepts of physics, at least in the first stages of learning: 'They are satisfied if they can recognize the meaning of content words (nouns, adjectives, verbs) and can interpret sentence fragments.' Graesser and Jackson do not think that this shallow learning is the appropriate standard for comprehending and learning physics. Instead, they believe that learners need to be challenged with difficult problems that encourage them to construct explanations and deep mental models, an activity that is also advocated for the comprehension of both narrative and academic information according to the constructivist theory (Graesser *et al.* 1994).

Graesser and Jackson agree with the embodiment approach that mature comprehension needs a deeper representation that is often embodied. Indeed, this motivated them to fortify AutoTutor with grounded information in conversation and interactive simulation. However, they also reported two puzzling empirical findings that should give moments of pause for embodiment theorists. First, some of the persistent misconceptions in Newtonian physics (such as the impetus fallacy) can be attributed to our everyday interactions with the embodied world. Thus, embodiment runs the risk of being a breeding ground for misconceptions (Ploezner and VanLehn 1997). Perhaps more abstract representations and facilities (such as formulae) are needed to transcend the persistent misconceptions that are shaped by our embodied experiences. Second, there were small gains in deep learning when students had the addition of the interactive simulations and microworlds, the facility that was designed to encourage embodiment. Perhaps more scaffolding is needed to show students how to use the embodied learning environments. In fact, those students who spent more time manipulating parameters and observing the simulated outcomes did learn more, which is good news for the embodiment camp. Nevertheless, the result does suggest that is not sufficient to plant the learner in an environment that is rich in perceptual motor activities and expect deep learning to occur.

Kintsch (Chapter 8) also suggests that, at least in some circumstances, comprehension is facilitated by abstract symbols and disrupted by embodied experience. He mentions some developmental data that suggest that toddlers understand better a task when it is presented with symbols than with visual icons. The rational is that icons interfere with conceptualization because they are confused with the objects they depict and even can trigger motoric acts which are irrelevant for the task (for instance, children try to sit on a picture of a chair). In the same vein, Kintsch claims that 'Teaching math with manipulatives can be counterproductive, because the children learn to manipulate objects, without ever inferring abstract mathematical principles'.

19.8.2 **Embodied teaching**

Two chapters in this book present evidence that even teaching mathematical or abstract concepts could benefit from embodied cognition. Goldstone et al. (Chapter 16) explore two examples of how grounding influences mathematical conceptualization: positive transfer and cognitive illusions. Nathan's (Chapter 18) analyses demonstrate how gestures are actively used by teachers to ground symbols (even mathematical ones) on perception, and how these gestures improves students' comprehension and learning.

Goldstone *et al.* (Chapter 16) start with a critical analysis of teachers' implicit beliefs about mathematical thinking. For instance, teachers believe that solving math problems is a formal process of applying equations and deriving procedures from them in a step-by-step fashion. However, these formal procedures are just the way mathematical ideas are expressed in books and journals, whereas 'the true heart of a mathematical proof, the intuitive conceptualization, is ignored in the formal description of the proof steps themselves.' And, according to Goldstone et al., the intuitive conceptualization of mathematics is grounded on sensorimotor information that can be associated with positive transfer of knowledge, but also associated with errors and cognitive illusions.

In one experiment, Goldstone et al. showed students a computer simulation of ants and food in which three ants follow a few simple rules to approach the food at different speed, depending on distance. The student could control the speed of the ants and their closeness to the food and see the consequences on the ants' behaviour. In this way, the student learned how to optimize these parameters for the ants' food-eating behaviour. In another task, students received a pattern-learning program designed to recognize three patterns drawn by the user, and the student's task was to optimize the program's learning. Goldstone *et al.* found that students better understood the pattern-learning simulation when it was preceded by the ants-and-food simulation, indicating a positive transfer, because the same principle of competitive specialization underlies the two simulation games. What is important here is that transfer was not based on learning abstract principles explicitly, but it was based on playing with sensorimotor simulations. Goldstone *et al.* accept that learning abstract principles could be an instructional goal. However, they consider that such principles could be induced from sensorimotor instantiations (e.g., simulation games) and transferred to new domains rather than being derived from formal descriptions.

One important kind of perceptual cue consists of gestures. Nathan (Chapter 18) emphasizes that teaching is communication setting in which teachers gesticulate spontaneously. These gestures are used to physically enact different referents including indexical, iconic, and symbolic meanings. An interesting kind of gesture is *catchment*, which reoccurs over the length of the teaching discourse, to mark the recurrence of a common idea or image for the speaker. This recurrence of the referential grounding provides coherence both for the listener and the speaker. An example of a catchment use is a teacher grounding a pupil's explanation by using a gesture to index elements in an equation depicted in the blackboard. Later on, the teacher can use this same gesture for corresponding elements of the solution equation in another part of the blackboard, thus establishing a relationship across the representations.

But grounding can also contribute to cognitive illusions in mathematics. For instance, Goldstone et al. showed that students who are familiar with the rule that multiplication precedes addition commit more errors judging an equation when the perceptual grouping of the elements is inconsistent (a) than when it is consistent with the precedence rule (b):

(a) $R * E+L * W = L * W+R * E$

(b) $R * E + L*W = L*W + R*E$

This cognitive illusion is automatic, persistent, and resistant to instruction. The phenomenon apparently reinforces Kintsch's argument, mentioned in Section 19.8.1, that perceptual cues in the instructional setting can interfere with comprehension. Another interpretation is, however, that perceptual cues are always there and they are exploited automatically by students. Teachers should be aware of this and provide perceptual cues which are consistent with the abstract rules of mathematics.

19.8.3 Conclusions

A number of conclusions emerge for the embodiment/symbolic debate when we look at the instructional arena, but collectively they give no clear advantage to either the symbolic or embodied frameworks. Obviously mathematics and other formal disciplines involve abstract concepts that are sometimes devoid of any simple physical instantiation or, alternatively, they are compatible with many physical instantiations. But teaching and learning mathematics are not necessarily disembodied processes, as knows every teacher who has struggled to help learners by giving them concrete examples, showing visual depictions, and engaging in conversations with substantial embodied grounding. There are other ways to encourage embodiment. First, there is some agreement that including teachers' facial expressions (Graesser and Jackson, Chapter 3) and gestures (Nathan, Chapter 18) in teaching mathematical and physical concepts allows learners to ground the concepts in embodied cues of human communication which are favourable for learning. Second, concepts in mathematics and physics are expressed by means of notational systems which are physically grounded on perceptual, procedural, and motoric experience (Goldstone et al, Chapter 16.). Third, grounding concepts on embodied activities (e.g., simulation games) facilitates transfer of abstract principles to new situation problems that markedly differ (Goldstone et al., Chapter 16; see also Glenberg *et al.*, 2007), particularly for those learners who take the time and effort to manipulate these simulations and observe what happens (Graesser and Jackson, Chapter 3).

The situation is paradoxical because abstract concepts are often grounded on perception and action in instructional settings, but this grounding appears to sometimes facilitate and sometimes interfere with learning. Grounding on irrelevant or misleading perceptual and procedural information proves detrimental to performance (Kintsch, Chapter 8; and Goldstone *et al.*, Chapter 16) and explains some of the persistent misconceptions about science (Graesser and Jackson, Chapter 3). Perhaps the most relevant practical question is how to design the instructional setting to get the best match between the abstractions to be learned and the sensorimotor procedures used to ground them. For instance, perceptual cues like visual grouping should be consistent with the conceptual structure or the sequence of procedures to apply to avoid interfering cognitive illusions. Another important issue is to improve teachers' metacognition on mathematical and scientific thinking. They should become aware that learning abstract principles or solving problems sometimes needs to be grounded on well designed perceptual procedures, and overcome the formalist prejudice that mathematical thinking is a disembodied rule-based process. However, they should also be aware of conditions when abstract symbols, procedures, and formulae are needed to transcend persistent misconceptions that are created from everyday embodied experiences.

19.9 **Conclusions from the debate**

In this book, members of several cognitive science communities that sometimes don't communicate very much debated their points of view on a fundamental issue: is meaning embodied or symbolic? Some leaned heavily towards the embodied framework, others were sympathetic to defending or salvaging the symbolic position, but most were somewhat agnostic with mixed views. They made efforts to honestly discuss each other's arguments and data, trying to convince their skeptics but also appreciating dissenting views. Sometimes the symbolists and embodiment people belonged to different traditions in cognitive science, so their disagreement on an issue reflected different standards of science.

At the risk of oversimplifying the differences, we can offer a couple of contrasts. Symbolists consider that a priority of cognitive science is the validation of their theories by building computational systems and formally specifying the theoretical components. By contrast, embodiment researchers are more inclined to consider the ecologically naturalistic interactions of people in their worlds and neuroscience evidence, whereas they rarely engage in computational models and simulations. These disciplinary differences between the two bands make more valuable their effort to establish a rational debate on the subjects. In spite of their divergent perspectives, or maybe because of them, some interesting insights arose.

This final section organizes a set of conclusions trying to establish a minimal account of meaning that could be acceptable for all, to isolate the main disagreements, and to propose research venues to deal with them. We will start with some relatively trivial conclusions that have the advantage of being more likely acceptable to everybody. Then we will offer other less evident conclusions which are more informative, but also more polemic.

Language is symbolic and embodied

Linguistic symbols (words) are arbitrary in the sense that they do not resemble their referents except for the extreme case of onomatopoeia. Words are also abstract labels for categories or types. Finally, words are combined into high-order units by means of syntactic principles and patterns. All these are typical features of symbolic systems. Linguistic symbols prompt initial activations that are strongly embodied. That is, the encoded symbols have perceptual features (phonetic or orthographic) and motoric features (articulatory). The activation of linguistic symbols are not amodal but multimodal.

Meaning is not language

Meaning is conveyed by language, but it is something other than words or sentences. Meaning is a construction that often involves a mapping or alignment function between linguistic symbols and referents. The current debate was not on the symbolic or embodied nature of words (they are both), but on the symbolic or embodied nature of meaning.

Meaning in conversational settings is often embodied

On-line reference to the current situation is strongly embodied in many situations because the objects, people, and spatial setting are immediately present. Words are directly grounded to perceptual and motoric objects and events. In addition to words, speakers employ multiple bodily cues to help the addressees to establish referents. They use iconic and pointing gestures, change their facial expressions, move their bodies into a particular stance, and use other paralinguistic cues complementing words.

On-line reference in conversational settings typically recruits a communication interface facility to ground words into perception and action. In other words, a statistical symbolist model is not enough to process on-line meaning because it lacks this grounding. On the other hand, sometimes communication breaks down or is incomplete by virtue of the lack of grounding or misalignments between the meanings of speech participants.

Displaced reference is detached from current perception/action

When we use language to refer to memory-based or fictional events, meaning is detached from the here and now, and the embodiment of meaning is less obvious. Whereas in on-line reference the gestures and gazes are aligned to referents, in displaced reference they are not, even when we describe perceptual and motoric events. Displaced reference is a fruitful terrain for the debate. The embodiment doctrine proposes that even in displaced reference people build virtual simulations of the referred situation whereas some of the symbolist approaches, namely the statistical symbolists, propose that computing relationships among ungrounded symbols is sufficient.

Covariation is a general principle of meaning

Nobody doubts that the human brain is, among other things, a powerful associative mechanism. But the application of this associative principle to linguistic meaning underlies different theoretical approaches. Statistical symbolism emphasizes the computation of word covariations, and extracts meaning from those covariations. Embodiment theories extend this capability to the crossmodal covariations among different experiential traces, or among experiential traces and words. Finally, some neural models mention the temporal correlation or synchrony among close and distant brain areas, which indicates functional connectivity among them.

Paradoxically, despite the emphasis of statistical symbolism on the covariation principle, this principle could be more widely used in a neurological-embodiment framework. Theories based on symbol-to-symbol covariations, although useful for certain purposes, are over-simplifications. Covariation is a much more general property of human experience and neural processes beyond word co-occurrence.

There are language-specific and meaning-specific areas in the brain

The previous reflections are strongly supported by the new data of neuroscience. The linguistic functions of the classical perisylvian areas in the left hemisphere have been confirmed by neuroimaging results. These language-specific areas are always activated to process the sounds of words (left superior temporal cortex and angular gyrus) and the articulatory programs of speech, including gestures (left inferior frontal cortex). However, there is no evidence that perisylvian areas process symbolic meaning. Instead, meaning specific areas are much more broadly distributed in the brain, and consist of perceptual, motoric, and emotional areas which are selectively activated depending on the particular content of words or sentences.

Meaning is representational

Some antisymbolic doctrines are also anti-representational views (see reviews by Shapiro, Chapter 4; and Gomila, Chapter 18). They propose that all that is needed are a few dynamic system routines and everything else emerges from interacting with the environment. Unfortunately, this radically Gibsonian perspective seems only appropriate to explaining adaptive behaviour in a very constrained physical environment (hand-picked by the researcher) rather than the complexities of meaning or memory. It is significant that Brooks' Gibsonian robots are not programmed to speak, only to navigate. Instead, conversational robots agent like the ones described in this book require internal states such as goals and expectations. We could speculate that a robot capable of displaced reference would need even more sophisticated representations of events and situations stored in and retrieved from their memories.

The embodiment–abstraction continuum

Abstraction is a general feature of human language and cognition. As Paivio (1971) once suggested, there is a concrete–abstraction continuum. At one extreme of the continuum we have concrete words like 'table', which are embodied because they can be grounded on a perceptual experiences, but which are also 'abstract' because they apply to an infinite class of objects (an abstract category). At the other extreme there are words like 'no', 'justice' and the like which offer a serious challenge to embodiment theories and, to some extent, also to symbolic theories. Thus embodied theories find it difficult to discover sensorimotor or situated processes in these words, at least when they are presented in isolation. Statistical symbolism finds that these words co-occur with many others in very different contexts, and therefore their symbolic definition might be fuzzy (this is the reason why LSA does not pay much attention to syntactic markers).

We do not claim to have a solution on how abstraction is understood, but we can propose a tentative pragmatic principle for some forms of abstraction: 'You can use abstract words and sentences to the extent that other contextual and situational cues specify sufficiently the referents'. Demonstratives like 'this' or 'that' are very abstract terms, for example. The only meaning they convey in isolation is closeness (in space or time) of a referred entity to the speaker. But we can use demonstratives in face-to-face conversational settings to refer to very concrete and specific objects (e.g., a jar), because other bodily cues such as pointing gestures and gazes accompany their use. In contrast, when we describe the same object in a telephone conversation, the demonstrative alone is not sufficient and more 'concrete' words need to be added to the message (e.g., 'this jar,' or 'the jar near me').

Resonating, modelling, and timing

A recurrent point of discussion was whether embodied meaning is possible given the time constraints and limited cognitive resources necessary to keep discourse comprehension on-line. Symbolists consider that embodied simulations are too slow and resource consuming to be routinely generated in deep comprehension. To clarify this point we will distinguish between two sorts of embodied representations: (a) resonance of sensorimotor processes, and (b) situation models. These embodied representations differ in automaticity, persistence of traits, awareness, and cognitive effort.

Resonance of sensorimotor features has a relatively fast onset after word presentations (in 100 milliseconds or so), is automatic and unconscious, and fades quite soon (in 200 or 300 milliseconds). From a neural perspective, resonance between language-specific and sensorimotor areas is established as Hebbian cell assemblies by recurrent co-activation of these assemblies in our experience. Instead, situation models are higher-order simulations that combine perceptual features (some times resonances) and proxy-situated, updatable representations. Situation models extend across clause and sentence limits, and are not automatically generated. We do not have much information about the neural bases of situation models (see Schmalhöfer and Perfetti 2007), although they could rely on brain subsystems associated with parsing and updating situational representations (e.g., hippocampus, posterior cingular areas, etc.).

Future discussions and research programs could gain clarity by distinguishing between the two levels of embodied representations. Resonance is fast, low-cost, and automatic, and therefore is compatible with the temporal constraints of on-line comprehension. But modelling is typically resource consuming, slow, and more optional. However, to get real comprehension, there needs to be some modelling process, even of gross granularity; otherwise, the linguistic processes and resultant learning would be shallow.

Instructional designs require careful analyses of embodiment parameters

Several facets of teaching need to be considered. First, in many teaching situations there is a face-to-face communication between teacher and students. Second, as a consequence of face-to-face communication, embodiment cues are inevitably present in the teaching situations (gestures, face expressions, gazes, visual notations, motoric procedures). Third, embodied cues could favour or be detrimental to comprehension and learning depending on their degree of consistency with the abstract concepts to be taught. Thus, appropriate embodied cues like catchment gestures or visual diagrams could enhance learning, whereas inconsistent cues such as misleading visual grouping or excessively iconic information proves detrimental to learning. Also, everyday experiences often plant misconceptions that can be readily transcended by more abstract representations and tools. To conclude, whatever your theory of meaning is, instructional programs should design the teaching environment in a manner that provides embodied cues that are consistent and well synchronized with the target contents, or that provide abstract artifacts that correct or transcend the inevitable inconsistencies.

References

Anderson JR (1983). *The Architecture of cognition*. Cambridge, MA: Harvard University Press.

Barsalou L, Wiemer-Hastings K (2005). Situating abstract concepts. In D Pecher, R. Zwaan, Eds. *Grounding Cognition: The Role of Perception and Action in Memory, Language, and Thought* (pp. 129–63). New York, NY: Cambridge University Press.

Barsalou L (1999). Perceptual symbol systems. *Behavioural and Brain Sciences*, 22, 577–660.

Beeman M (1998). Coarse semantic coding and discourse comprehension. In M Beeman, C Chiarello, Eds. *Right Hemisphere Language Comprehension: Perspectives from Cognitive Neuroscience* (pp. 255–84). Mahwah, NJ: Erlbaum.

Bransford JD, Johnson MK (1973). Considerations on some problems in comprehension. In WG Chase, Ed. *Visual Information Processing*. New York, NY: Academic Press.

Buccino G, Lui F, Canessa N *et al.* (2004). Neural circuits involved in the recognition of actions performed by nonconspecifics: an fMRI study. *Journal of Cognitive Neuroscience*, 16, 114–26.

Card S, Moran T, Newell A (1983). *The Psychology of Human–Computer Interaction*. Hillsdale, NJ: Erlbaum.

Damasio AR (1989). Time-locked multiregional retroactivation: a systems-level proposal for the neural substrates of recall and recognition. *Cognition*, 33, 25–62.

Fodor J (1983). *The Modularity of Mind*. Cambridge, MA: MIT Press.

Forbus K, Gentner D, Law K (1995). MAC/FAC: a model of similarity-based retrieval. *Cognitive Science*, 19, 141–205.

Gallese V, Lakoff G (2005). The brain's concepts: the role of the sensorimotor system in reason and language. *Cognitive Neuropsychology*, 22, 455–79.

Glaser WR (1992). Picture naming. *Cognition*, 42, 61–105.

Glenberg AM, Jaworski B, Rischal M, Levin JR (2007). What brains are for: Action, meaning, and reading comprehension. In D. McNamara (Ed). Reading comprehension strategies: Theories, Interventions, and Technologies (pp. 221–240). Mahwah NJ: Lawrence Erlbaum Publishers.

Glenberg AM, Kaschak MP (2002). Grounding language in action. *Psychonomic Bulletin and Review*, 9, 558–65.

Glenberg AM, Robertson DA (2000). Symbol grounding and meaning: a comparison of high-dimensional and embodied theories of meaning. *Journal of Memory and Language*, 43, 379–401.

Goldstone RL, Sakamoto Y (2003). The transfer of abstract principles governing complex adaptive systems. *Cognitive Psychology*, 46, 414–66.

Graesser AC, Singer M, Trabasso T (1994). Constructing inferences during narrative text comprehension. *Psychological Review*, 101, 371–95.

Holcomb P, Kounios J, Anderson J, West W (1999). Dual-coding, context–availability, and concreteness effects in sentence comprehension: an electrophysiological investigation. *Journal of Experimental Psychology: Learning, Memory, and Cognition*, 25, 721–42.

Jurafsky D, Martin JH (2000). *Speech and Language Processing: An Introduction to Natural Language Processing, Computational Linguistics, and Speech Recognition*. Upper Saddle River, NJ: Prentice-Hall.

Kaschak M, Glenberg A (2000). Constructing meaning: the role of affordances and grammatical constructions in sentence comprehension. *Journal of Memory and Language*, 43, 508–29.

Kaschak MP, Madden CJ, Therriault DJ *et al.* (2005). Perception of motion affects language processing. *Cognition*, 94, B79–89.

Kintsch W, van Dijk TA (1978). Towards a model of text comprehension and production. *Psychological Review*, 85, 363–94.

Kosslyn SM, Shwartz SP (1977). A simulation of visual imagery. *Cognitive Science*, 1, 265–95

Kounios J, Holcomb P (1994). Concreteness effects in semantic processing: ERP evidence supporting dual-coding theory. *Journal of Experimental Psychology: Learning, Memory, and Cognition*, 20, 804–23.

Laird J, Newell A, Rosenbloom P (1987). SOAR: an architecture for general intelligence. *Artificial Intelligence*, 33, 1–64.

Landauer TK, Dumais ST (1997). A solution to Plato's problem: the latent semantic Analysis theory of acquisition, induction and representation of knowledge. *Psychological Review*, 104, 211–40.

Landauer T, McNamara D, Dennis S, Kintsch W, Eds (2007). *Handbook of Latent Semantic Analysis*. Mahwah, NJ: Erlbaum.

Miller GA, Johnson-Laird PN (1976). *Language and Perception*. Cambridge, MA: Harvard University Press.

Norman DA, Rumelhart DE; LNR Research Group (1975). *Explorations in Cognition*. San Francisco, CA: WH Freeman.

Ortony A, Clore G, Collins A (1988). *Cognitive Structure of Emotions*. Cambridge: Cambridge University Press.

Paivio A (1971). *Imagery and Verbal Processes*. New York, NY: Holt, Rinehart, & Winston.

Ploetzner R, VanLehn K (1997). The acquisition of informal physics knowledge during formal physics training. *Cognition and Instruction*, 15, 169–206.

Pylyshyn Z (1986). *Computation and Cognition*. Cambridge, MA: MIT Press.

Rizzolatti G, Craighero L (2007) Mirror neurons and language. In G Gaskell, Ed. *Oxford Handbook of Psycholinguistics*. Oxford: Oxford University Press.

Rizzolatti J, Craighero L (2004). The mirror neuron system. *Annual Review of Neuroscience*, 27, 169–92.

Rogers TT, Lambon RMA, Garrard P, *et al.* (2004). The structure and deterioration of semantic memory. A neuropsychological and computational investigation. *Psychological Review*, 111, 205–35.

Rüschemeyer SA, Glenberg AM, Kaschak MP, Friederici AD (submitted). Listening to sentences describing visual motion activates MT/V5.

Schmalhöfer F, Perfetti C, Eds. *Higher Level Language Processes in the Brain: Inferences and Comprehension Processes*. Mahwah, NJ: Erlbaum.

Tomasello M. (2003). *Constructing a language: A usage-based theory of language acquisition*. Harvard: Harvard University Press.

Zwaan R, Madden C (2005). Embodied sentences comprehension. In B Pecher, RA Zwaan, Eds. *Grounding Cognition. The Role of Perception and Action in Memory, Language and Thinking*. Cambridge: Cambridge University Press.

Zwaan R, Yaxley, R. (2003). Spatial iconicity affects semantic–relatedness judgments. *Psychonomic Bulletin and Review*, 10, 954–8.

Author Index

Subject Index

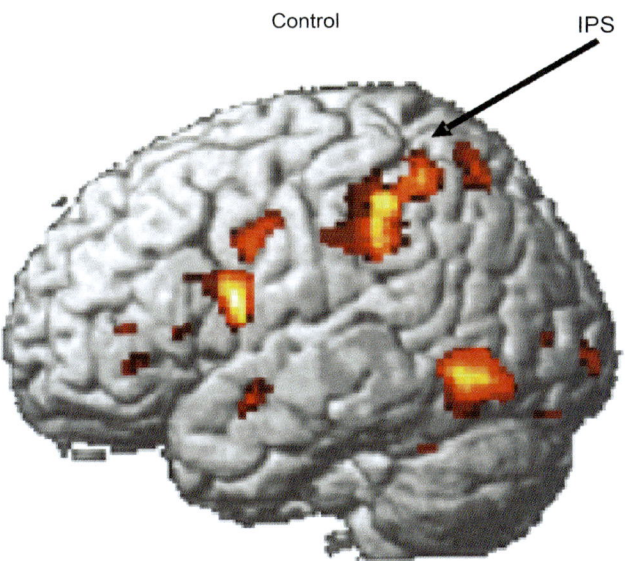

Plate 5.1 High- minus low-imagery effect of visual sentence comprehension in control participants, displaying a large imagery effect in parietal, prefrontal, and inferior temporal areas. IPS = intraparietal sulcus.

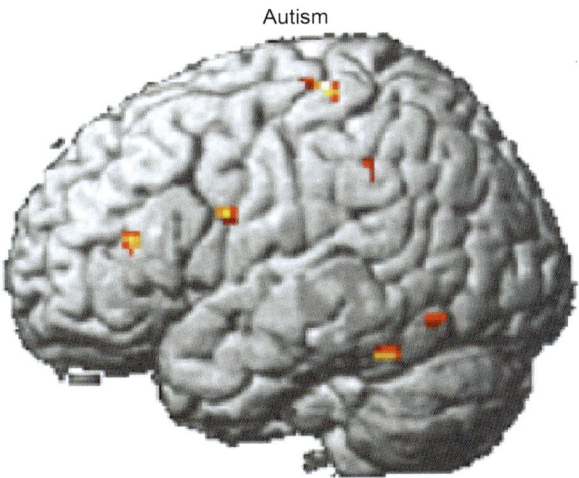

Plate 5.2 High- minus low-imagery effect of visual sentence comprehension in people with autism, displaying a negligible imagery effect.

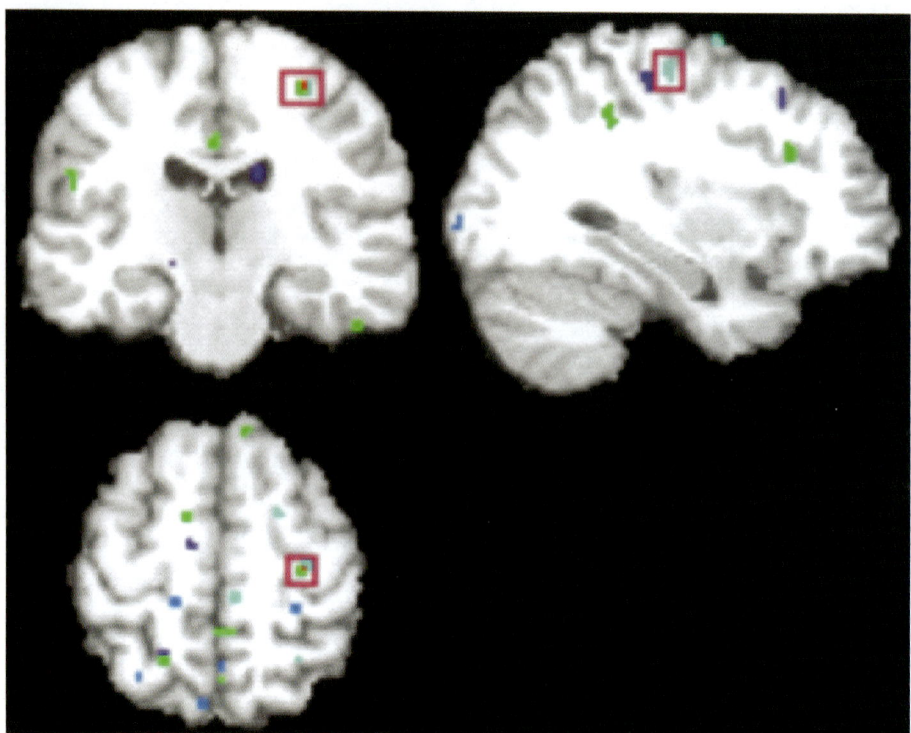

Plate 5.3 Voxels in the motor region used by the classifier for word identification in several participants.

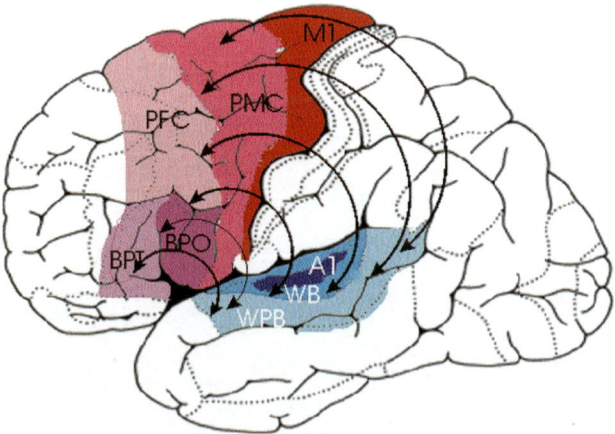

Plate 6.1 Schematic illustration of the cortical systems for language and action. Long-distance connections between the language and action systems are indicated. Inferences on cortico–cortical links in humans are based on neuroanatomical studies in monkeys (after Pulvermüller 2005). A1 = core region of the primary auditory cortex; BPO = Broca area, pars opercularis; BPT = Broca area, pars triangularis; M1 = primary motor cortex; PFC = prefrontal cortex; PMC = premotor cortex; WB = auditory belt region in Wernicke area; WPB = auditory parabelt region in Wernicke area.

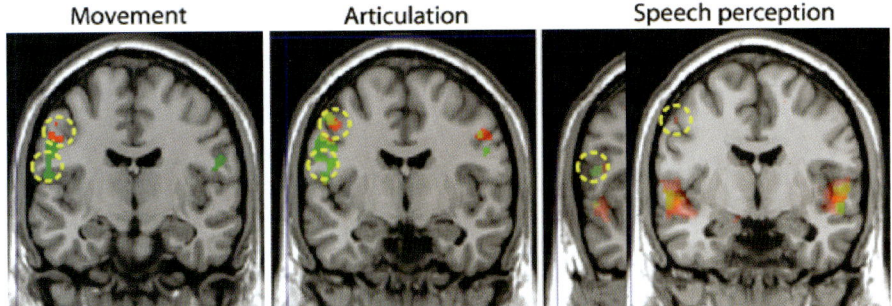

| Movement | Articulation | Speech perception |

Plate 6.2 Somatotopic activation of precentral cortex during motor movements, silent articulation and speech perception. (A) Movement of the lips (red) contrasted with movements of the tongue tip (green). (B) Silent articulation of CV syllables starting with a *[p]* (red) versus a *[t]* sound (green). (C) Listening to meaningless syllables starting with *[p]* (red) or *[t]* (green). Similar somatotopic motor/premotor activation patterns were found for lip- and tongue-related actions and perceptions, supporting the embodiment of phoneme processing in action–perception networks (modified from Pulvermüller *et al.* 2006).

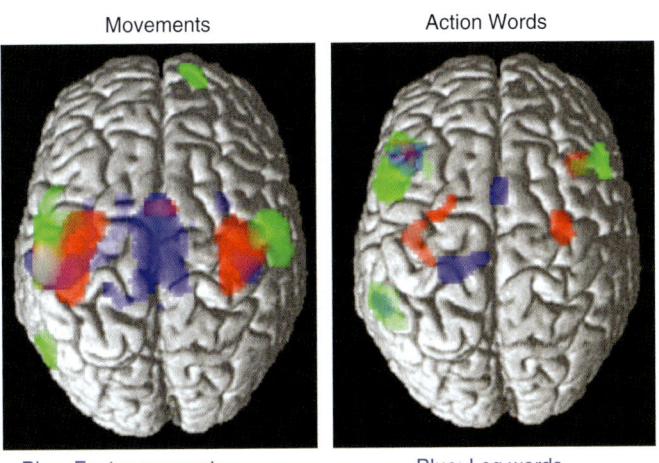

| Movements | Action Words |

Blue: Foot movements
Red: Finger movements
Green: Tongue movements

Blue: Leg words
Red: Arm words
Green: Face words

Plate 6.3 Cortical activation (functional magnetic resonance imaging) during motor movements and during passive reading of action words. Overlapping activation is elicited by both leg- (blue), arm- (red), and face-related (green) movements and words, indicating a common neural substrate for the processing of actions and the meaning of action words (after Hauk *et al.* 2004).

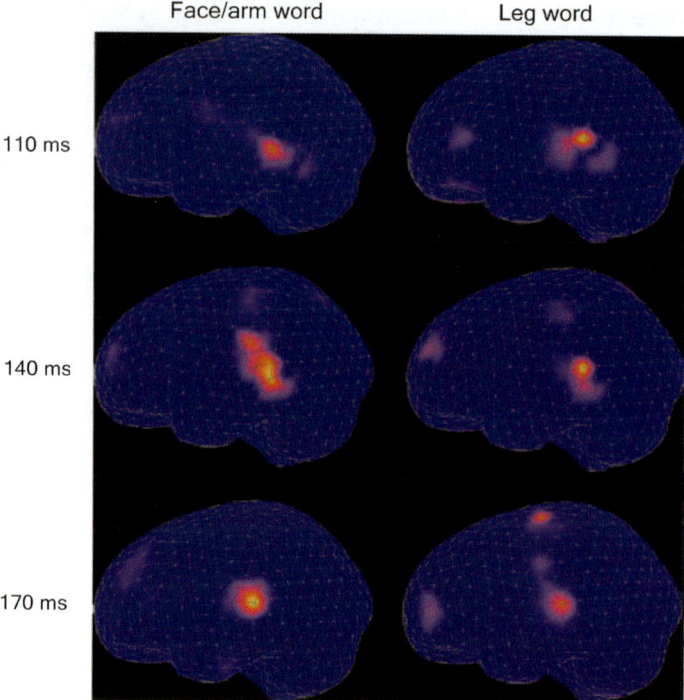

Plate 6.4 Cortical activation (magnetoencephalogram) elicited by face/arm (left) and leg-related words (right) at different times after spoken action words could be uniquely recognized. Note the slight upward movement of the inferior central source for the face/arm word and the delayed appearance of the superior central source for the leg word. These activation time courses may reflect the travelling of neuronal activity in distributed neuronal assemblies that represent and process words with different action-related meanings (after Pulvermüller, Shtyrov et al. 2005).